AAAS
Science
Book List
Supplement

Other AAAS reference sources

AAAS Science Book List, Third Edition. Compiled by Hilary J. Deason
AAAS Science Book List for Children, Third Edition. Compiled by Hilary J. Deason
AAAS Science Film Catalog. Compiled and edited by Ann Seltz-Petrash and
    Kathryn Wolff
Science for Society: A Bibliography, Sixth Edition. Prepared by Joseph M. Dasbach
Science Books & Films (a quarterly review journal). Shari Finch, Editor

AAAS
Science
Book List
Supplement

A selected and annotated list of science and mathematics books which supplements the AAAS Science Book List (3rd ed.; 1970) for secondary school students, college undergraduates, teachers and nonspecialist readers.

Compiled and edited by

Kathryn Wolff
and
Jill Storey

American Association for the Advancement of Science
Washington, D.C. 1978

Library of Congress Cataloging in Publication Data

Wolff, Kathryn
  AAAS Science book list supplement.

  Extends the coverage of the AAAS Science book list, by H. J.
Deason, from 1969 through 1977.
  Includes indexes.
  1. Science—Bibliography. I. Storey, Jill Deborah, joint au-
thor. II. Deason, Hilary J. The AAAS Science book list. III. Ameri-
can Association for the Advancement of Science. IV. Title.
Z7401.W64   1978   [Q158.5]   016.5   78-6540
ISBN 0-87168-218-4

AAAS Publication 78–5

Full reviews of the books listed in this volume were published in the magazine *Science Books &
Films* and its predecessor *Science Books: A Quarterly Review*.

© Copyright 1978
American Association for the Advancement of Science
1515 Massachusetts Avenue, NW, Washington, D.C. 20005

Printed in the United States of America
by
Corporate Press, Washington, D.C.

Contents

(Subjects are listed according to Dewey Decimal classification)

The Social Sciences

Language

Pure Sciences

Technology

Indexes

A personal note to our readers

Any compiler of the "best" (or even the "better") of any large category of things begins with enthusiasm and ends in frustration. The enthusiasm stems from a belief in the need, an awareness of available resources, a desire to be helpful. The frustration develops as the task is carried toward an end which cannot also be a completion. Especially in a major compilation such as this, covering books from many different scientific disciplines, choices must be made and limits reluctantly accepted. The pressure for closure *now* and the need to produce, in a reasonable time, a book of manageable size which can be sold at less than an astronomical price take their toll. Of the many thousands of books which could have been included here, we finally selected only 2842.

It may be that the value of the compilation we provide here is obscured in our own eyes because we see how much more thorough the coverage might have been in the many fields we sampled. But as we look over the final selection, we are pleased to see how many outstanding books and notable authors are included in these pages. So despite all, we believe that what we have done is of value to the teachers and librarians whom we have constantly addressed in imagination as we sorted, annotated, selected, and compiled what we sincerely hope is a useful book.

KATHRYN WOLFF
JILL STOREY

December 1977
Washington, D.C.

Introduction

This *AAAS Science Book List Supplement* is a guide to collateral and reference texts and to recreational reading in all of the sciences and in mathematics for junior and senior high school students, college undergraduates, teachers, and nonspecialist adults. It is a supplement to the *AAAS Science Book List, Third Edition* which was published in 1970 and which is still in print. There were two earlier *SBL* editions (1959 and 1964), both now out of print, and all three *Lists* have been widely used throughout the English-speaking world as acquisition guides for secondary school, college undergraduate, and public libraries to develop collections in the sciences that will serve their varied readers most effectively.

These *AAAS Science Book Lists* have been published over a period of time during which a substantial revolution occurred in public interest in science, in the teaching of science and mathematics at all levels, and in the training and certification of science and mathematics teachers. During this time, good teachers came to realize that science instruction should not be limited to a single textbook supposedly containing immutable fact or to a single laboratory manual covering only known results. Instead, such teachers have used extensive supplementary readings and have sought texts and references which show the sciences as dynamic, open-ended processes, continuously developing from experiment and observation and continually changing as new knowledge expands our understanding of that which is already known.

In these earlier *Book Lists,* as in this *Supplement,* we have sought to list those books which show correct information about all the multiple fields of science. We have especially sought to include the books which show the excitement of experimentation, testing, discovery, evaluation and reevaluation—the processes that most clearly characterize the science developments which have so strongly influenced recent human history.

Preparation of the *AAAS Science Book Lists* evolved as a natural activity of the American Association for the Advancement of Science. As the oldest and largest (120,000 individual members, nearly 300 affiliated societies) interdisciplinary scientific organization in the Western World, AAAS has long been interested in science education. It is one of the Association's charter obligations ("to improve the effectiveness of science in the promotion of human welfare, and to increase public understanding and appreciation of the importance and promise of the methods of science in human progress"), and it finds one means of expression in the production and dissemination of these *Science Book Lists.*

The *Lists* grew out of a joint AAAS–National Science Foundation "Traveling High School Library" (1955–1962), in which AAAS circulated multiple sets of 200 "best" science and mathematics books among a large number of secondary schools in the United States which at that time had inadequate or nonexistent science collections. This project, as well as a similar program for elementary schools, included annotated catalogs, and these catalogs came into widespread use as buying guides for both schools and libraries. The many requests AAAS received for a more comprehensive list led to the publication of the first *AAAS Science Book List* of 900 titles in 1959. The second edition, published in 1964, contained annotations of 1376 titles and a print run of 17,000 was exhausted in less than 2 years.

These two *Lists* were prepared by requesting from all major publishers copies of appropriate books in all of the sciences, then sending the books out to scientists for individual review and acceptance or rejection as suitable for the *Lists.* Additional titles were suggested by some reviewers, and copies of these books were also secured for review. By this tedious, cumbersome, and time-consuming process, the final *Lists* were compiled. The difficulties inherent in this kind of massive, one-time review process made obvious the need for a continuing, current, and comprehensive science book review process. The need was especially acute since many, if not most, published reviews of science books (other than those at the professional level) were concerned chiefly with readability and style, and often gave no reliable indication of the quality of the scientific and technical content. As a result, in the spring of 1966 AAAS began the publication of a new book review periodical, *Science Books: A Quarterly Review (SBQ)*.

SBQ began reviewing some 200 new science books in each issue, with all reviews prepared by scientists proficient in the particular discipline of the book under review. In *SBQ*, books for readers of all ages, from kindergarten through college (and their teachers), were reviewed, and some of the books were "recommended" or "highly

Table 1. Brief comparison of *AAAS Science Book List, Third Edition*, and *AAAS Science Book List Supplement*. Columns show numbers of books in each category.

Subjects	Third Edition	Supplement
Information and Communication Sciences	11	63
Philosophy and Ethics	—	29
Psychology	70	158
Sociology	—	173
Cultural Anthropology	27	22
Natural Resources	17	89
Social Pathology and Services	—	46
Education	—	37
Linguistics	8	10
Science in General	157	112
Mathematics	229	204
Astronomy	142	105
Physics	147	87
Chemistry	90	105
Mineralogy	17	12
Geological Sciences	151	149
Paleontology	22	16
Physical Anthropology, Evolution, Genetics	44	77
Biology	260	218
Botany	98	77
Zoology	314	319
Medical Sciences	145	206
Engineering Sciences	185	174
Agricultural Sciences	105	91
Domestic Sciences	—	24
Industrial Technologies	48	10
Urban Planning and Architecture	42	25
Photography	14	25
Geography and Exploration	44	74
Archeology	54	80
Other categories		25

recommended'' for particular age groups. In the preparation of the *Third Edition* of the *SBL*, these titles were included, augmenting the content and replacing the out-of-date items of the *Second Edition*. The final compilation for the *Third Edition*, which contains 2441 titles, lists many of the classic works in science which are not too technical for most school and public library collections. It also annotates texts, popular science trade books, and special books for the hobbyist, the citizen concerned with science and public policy matters, the interested general reader in any of at least 35 fields, and of course, for students and their teachers.

Since 1970, the explosion in scientific knowledge, and in scientific publishing, has continued apace. *Science Books: A Quarterly Review* grew fatter and became *Science Books & Films,* which now carries more than 1200 reviews per year. As we embarked on the production of a fourth edition of the *AAAS Science Book List*, we realized that a supplement would be far more useful than a revision of the *Third Edition,* especially since many more recent books could be listed in a supplement of affordable size than could be included in a completely new edition. Thus this supplement to the *Third Edition* came into being, containing some 2850 books published since 1969 and selected from among those reviewed in *SBQ* or *SB&F* and recommended for a particular age or ability level. As in the *Third Edition,* the arrangement is by the most recent Dewey Decimal classification system. (Categories from the 18th edition, as amended, were used in the *Supplement.*) Again, biographies have been included in the various subject categories instead of being collected together in a special section by themselves. Where it seemed that some confusion could arise or some books be overlooked, we have attempted to provide cross-references in the text and additional references in the index. Nevertheless, the interdisciplinary nature of so many science-related topics and of so many science classes has made it very difficult to provide the extensive cross-referencing necessary in cases such as the environmental sciences, which can involve (and be classified under) categories such as conservation, sociology, oceanography, ecology (in three different places), natural history, public health, engineering, economics, political science, and even others.

In scope, the *Third Edition* emphasizes the physical and biological sciences and their applications, as can be seen in Table 1. The *Supplement* includes not only these categories but reflects the increased emphasis on the social and behavioral sciences in school curricula and, we believe, the interests of the general reader as well. Most books listed in both the *Third Edition* and the *Supplement* are the casebound editions, but paperback editions are noted if they are also available. In some instances, softcover titles have been listed because they were considered important acquisitions and were available in no other edition. A standard citation form which includes publishers' names has been employed throughout, but the publishers' mailing addresses, included in the *Third Edition,* have been omitted in the *Supplement.* (Readers will find current addresses for all publishers cited in the *Literary Market Place: The Directory of American Book Publishing* which is updated and published yearly by R. R. Bowker Company, 1180 Avenue of the Americas, New York, N.Y. 10036.)

The preparation of this *Supplement* was a substantial undertaking, and its successful completion was due to the efforts of many people. In particular, we wish to acknowledge the assistance of the many reviewers and the staff of *SB&F.* To name the reviewers individually here would require many pages, but they are listed in the various issues of the magazine, and we thank them all. Among the staff members of *SB&F,* special thanks are due to Ann Seltz-Petrash and Susan Gordon. Other members of the AAAS staff who made substantial contributions to the preparation of this volume are Joan Wrather and Arleen Rogan, both of whom contributed many hours of work in the early phases of annotation preparation. We are also grateful to Faith Little, who undertook the colossal task of compiling the indexes. And finally, we wish to thank Anne Holdsworth, who guided the book through production and without whose assistance the road to publication would have been much, much rockier.

How To Use This Book

Books are arranged alphabetically within their Dewey Decimal categories. The Table of Contents lists, in order, every Dewey category used in the book. The Title and Subject Index also lists the Dewey categories alphabetically by first word, by all key words in a category, and by general subject area when appropriate (for example, *Fuels and Energy* is indexed under *Fuels, Energy,* and *Resources*).

Books should be ordered directly from the publisher (or distributor if one is given). Current addresses for all publishers and distributors cited can be found in the *Literary Market Place: The Directory of American Book Publishing* which is updated and published yearly by R.R. Bowker Company, 1180 Avenue of the Americas, New York, N.Y. 10036.

The citations for individual books contain the information about the book supplied by the publisher at the date of publication. The various parts of the citation form and the abbreviations used are as follows:

(1) **Black, Allison E.,** (2) **(Ed.).** (3) *Basic Biology, 3rd rev. ed.* (4) (Illus.; trans.) (5) University Press (6) (distr.: Marvin), (7) 1977 (8) (c.1972). (9) xii+381pp. (10) $8.95; $4.95(p). (11) 77-3746. (12) ISBN 0-99999-338-8; 0-99999-339-X. (13) Index; gloss.; bib.; CIP. (14) **JH-SH-GA-C-P** (15) (Annotation).

1. Author's or editor's name.
2. Editor.
3. Title, 3rd revised edition.
4. Illustrated; translated.
5. Publisher's name.
6. Distributor's name.
7. Date of publication.
8. Copyright date (given only if the book was copyrighted more than one year prior to the publication date).
9. Number of pages in the book—front matter and main text.
10. Price. The first price is for the casebound edition; the second is for the paperback edition if one exists; prices are those listed at the date of publication.
11. Library of Congress number.
12. International Standard Book Number. The first number is for the casebound edition, the second for the paperback.
13. Indicates that the book contains an index, a glossary, a bibliography, and Library of Congress Cataloging in Publication data.
14. Reading level(s) as assigned by the subject matter specialist who reviewed the book:
    - **JH** Junior High School (grades 7-9)
    - **SH** Senior High School (grades 10-12)
    - **GA** General Audience
    - **C** First two years of college
    - **P** Professional level
15. Annotations are based on much longer reviews published in *AAAS Science Books & Films,* back issues of which are available either from the American Association for the Advancement of Science or from University Microfilms, Ann Arbor, MI 48106.

Generalities

001 Knowledge and Its Extension

001.5 Information and Communication

De Fleur, Melvin L., and Sandra Ball-Rokeach. *Theories of Mass Communication, 3rd ed.* (Illus.) McKay, 1975. xv + 288pp. $3.95. 74-112656. ISBN 0-679-30293-X. Index.

    SH-C After a brief history, interactions between the press, motion pictures and the American "mass" society are considered, followed by a projection of the effects of emerging electronic communication media which, with feedback, are considered able to produce a communications revolution of the order of mass printing and broadcasting.

Fuchs, Walter R. *Cybernetics for the Modern Mind.* (Illus.) Macmillan, 1971. 357pp. $6.95. 76-119159.

    SH Describes the development of programming languages and the artificial, unambiguous languages of modern science. Explores the idea that the developments of modern science are shaped by the Indo-European dialects under which these developments occurred. The later chapters deal with the binary number system, Boolean logic, stored programs and elementary computer architecture.

Kochen, Manfred, (Ed.). *Information for Action: From Knowledge to Wisdom.* (Illus.) Academic, 1975. xv + 248pp. $12.50. 75-3968. ISBN 0-12-417950-9. Indexes;CIP.

    C-P This work addresses itself to WISE (World Information Synthesis Encyclopedia) or the idea of a "world brain." In the first part, the basic issues and priorities of information science and technology and their relation to social problems are discussed. The second section contains papers on the utilization of knowledge and information.

Michie, Donald. *On Machine Intelligence.* Halsted/Wiley, 1974. xi + 199pp. $6.95. 74-8057. ISBN 0-470-60150-7. CIP.

    C-P This brief, readable anthology provides the reader with information on artificial intelligence and quiets any emotional fears about the potentiality of machine intelligence. Recommended for students and professionals in computer science, information management, and electronic design and engineering.

Parsegian, V.L. *This Cybernetic World of Men, Machines, and Earth Systems.* (Illus.) Doubleday, 1973. xx + 254pp. $1.95(p). ISBN 0-385-00612-8.

    C-P Parsegian develops cybernetic concepts through the use of numerous concrete examples, including a home heating system, a pendulum, the neurological functions of the body, a computer and society. The examples are carefully linked together to form a whole picture. The subject is beautifully presented. Most useful as collateral reading or reference.

Silverstein, Alvin, and Virginia Silverstein. *Bionics: Man Copies Nature's Machines.* (Illus.) McCall, 1970. 74pp. $4.50. 74-117017. ISBN 0-8415-2013-5.

    SH-C Bionics is the study of sensory and other systems in living creatures as a model for the design of similar artificial systems. Many examples from nature show how the operation of the biological clocks and compasses of various organisms all contribute to the construction of effective, nonliving devices. The style is highly simplified, but it is informative and interesting.

Wickelgren, Wayne A. *How to Solve Problems: Elements of a Theory of Problems and Problem Solving.* (Illus.) Freeman, 1974. 262pp. $10.00; $4.95(p). 73-15787. ISBN 0-7167-0846-9; 0-7167-0845-0. Index;CIP.

C-P   Wickelgren presents theoretical and practical analyses of problems and problem solving reflecting recent advances in artificial intelligence and computer simulation of thought. Intended both as a text and for the general reader; useful to students and workers in the fields of mathematics, science, engineering and business.

001.5436 CRYPTOGRAPHY

Gardner, Martin. *Codes, Ciphers and Secret Writing.* (Illus.) Simon&Schuster, 1972. 96pp. $4.95. 72-82218. SBN 671-65201-X. Bib.

JH   Simply written, the examples are fully explained and practice puzzles are provided. Covers such topics as transposition ciphers and communication with extraterrestrial life. "Code" is used interchangeably with "cipher" and no examples of codes are given.

Pallas, Norvin. *Code Games.* (Illus.) Sterling, 1972. 112pp. $3.50. 78-167656. ISBN 0-8059-4518-4.

JH-SH-C   As a source book for party entertainment, as enrichment in mathematics or language arts classes, and as good fun for friends who like a little mental exercise, the codes illustrated and the instructions on ways of varying them are fascinating.

001.56 NONVERBAL COMMUNICATION

See also 410-419 Linguistics.

Fast, Julius. *Body Language.* M. Evans (distr.: Lippincott), 1970. 192pp. $4.95. 72-106592.

SH-C   The author analyzes and develops a very fascinating explanation of nonverbal communication (by gestures, winking, smiling, frowning, standing and sitting posture, etc.) based on Hall's four distance zones (intimate, personal, social and public). Not an indepth study.

Harrison, Randall P. *Beyond Words: An Introduction to Nonverbal Communication.* (Illus.) Prentice-Hall, 1974. xi+210pp. $7.95. 73-17202. ISBN 0-13-075141-9; 0-13-076133-8(p). Gloss.;index;CIP.

C   Harrison discusses the theory of nonverbal communication, which he defines as "the exchange of information through nonlinguistic signs." Topics covered here include nonverbal clues, nonverbal communication, communication systems, code systems, systems for prediction, codes (sounds, faces, hands and body), time, space and object, and media messages. Would make a fine text.

Krames, Lester, et al. (Eds.). *Advances in the Study of Communication and Affect: Vol. 1: Nonverbal Communication.* Plenum, 1974. x+212pp. $14.95. 74-6325. ISBN 0-306-35901-4. Index;CIP.

C-P   This book brings together state-of-the-art papers by investigators from various disciplines who research both human and nonhuman communication. Especially valuable to biological and psychological investigators in communication.

McGough, Elizabeth. *Your Silent Language.* (Illus.) Morrow, 1974. 128pp. $4.95. 74-4253. ISBN 0-688-21820-2; 0-688-31820. Index;CIP.

JH-SH   This introduction to nonverbal communication ("kinesics") covers eyes, facial expression, the extremities, posture, courting, touching, space and cross-cultural aspects of body language. Drawings enhance the text.

Spiegel, John P., and Pavel Machotka. *Messages of the Body.* (Illus.) Free Press/ Macmillan, 1974. xiv + 440pp. $17.95. 73-10572. ISBN 0-02-930400-8. Index;CIP.

  C A commentary on observational and conceptual problems associated with investigations of the movements, postures, and gestures of the human body. Also offers an interdisciplinary survey of strategies for researching the nature of nonverbal communication.

001.6 DATA PROCESSING

See also 301.5 Social Institutions.

Cagan, Carl. *Data Management Systems.* (Illus.) Melville/Wiley, 1973. xii + 141pp. $12.95. 73-11036. ISBN 0-471-12915-1.

  C-P This book provides an excellent foundation for understanding data base and introduces the reader to the principles of generalized data management programs upon which the industry is basing its solutions to data base problems. Still highly recommended, although some of the information is already out of date.

Cashman, Thomas J., and William J. Keys. *Data Processing & Computer Programming: A Modular Approach.* (Illus.) Canfield, 1971. xiii + 498pp. $10.95. 79-160832. ISBN 0-06-382360-8.

  SH A comprehensive introduction to data processing from unit record equipment through punched card computer systems to magnetic tape and direct access systems. A basic text for an introductory course for nonscience students, especially in business and accounting.

DeRossi, Claude J. *Exploring the World of Data Processing.* (Illus.) Reston, 1975. xiv + 256pp. $9.95. 74-28054. ISBN 0-87909-259-9. Gloss.;index;CIP.

  SH-C The scope of the book is tremendous: history, computer center operations, EAM equipment, programming from machine to high level languages, and much more. Conversational style, many illustrations.

Doyle, Lauren B. *Information Retrieval and Processing.* (Illus.) Melville/Wiley, 1975. xv + 410pp. $17.95. 75-1179. ISBN 0-471-22151-1. Index;CIP.

  C-P An excellent revision of Becker and Hayes' *Information Storage and Retrieval* (1963), reflecting the needs of more sophisticated audiences. Incorporates substantial new material; covers problems of growth, publishing, reproduction, and storage; explores use of telecommunication and on-line systems, new ways to meet information crises, and evaluation and user-oriented topics. An ideal text for students and professionals in information and library science.

Flores, Ivan. *Data Structure and Management.* (Illus.) Prentice-Hall, 1970. x + 390pp. $13.95. 73-125290.

  SH-C A very thorough introduction to the organization and processing of data for computer manipulation. Through the use of an operating system (OS/360), the author presents methods by which data can be organized, related, and used to their fullest. For the layman and the student with little background.

Harmon, Margaret. *Stretching Man's Mind: A History of Data Processing.* (Illus.) Mason/Charter, 1975. xv + 239pp. $8.95. 74-3403. ISBN 0-88405-282-6. Gloss.;index;CIP.

  SH-C An informative and well-written semitechnical book on the history of data processing. Starting with the abacus and continuing through to modern computers, the author covers many lesser known inventions. Useful as a brief, comprehensive reference.

Health, Education and Welfare. *Records, Computers, and the Rights of Citizens: Report of the Secretary's Advisory Committee on Automated Personal Data Systems.* MIT Press, 1973. xxxv + 344pp. $2.95(p). 73-13449. ISBN 0-262-58-025-X; 0-262-08-070-2(p). Index.

   C   Report covers threats to individual privacy precipitated by development of computerized information systems and recommends safeguards to protect personal information, including restrictions on the use of social security numbers as identifiers. Recommended for students, citizens, and anyone worried about the invasion of privacy problems posed by computers.

Lott, Richard W. *Basic Systems Analysis.* Canfield, 1971. x + 260pp. $10.00. 72-143694. ISBN 0-06-385320-5.

   SH-C   A highly generalized approach to the analysis of business systems. Offers the student a brief introduction to many of the elements of designing business-oriented systems, such as inventory control, payroll or management-information systems. Often-ignored aspects of systems analysis, such as effect upon personnel, selling of the system, and training of personnel are presented.

Martin, James. *Principles of Data-Base Management.* (Illus.) Prentice-Hall, 1976. xvi + 352pp. $18.50. 75-29054. ISBN 0-13-708917-1. Gloss.;index;CIP.

   C-P   An excellent discussion of the reasons for developing an information system around data bases, and of data organization and structure. Emphasis is on different management considerations, information quality, and management interaction with the data base. Includes an overview of the CODASYL and DL/1 IBM data description languages. Assumes some background in computer technology.

Murdick, Robert G., and Joel E. Ross. *Information Systems for Modern Management, 2nd ed.* (Illus.) Prentice-Hall, 1975. xv + 671pp. $14.95. 74-8541. ISBN 0-13-464602-9. Index;CIP.

   C-P   This is a comprehensive introductory text on management information systems designed to teach the practicing manager and the pragmatic student of management to appreciate and "think" systems. Useful as an undergraduate text.

Stern, Robert A., and Nancy B. Stern. *Principles of Data Processing.* (Illus.) Wiley, 1973. x + 630pp. $11.95. 72-6849. ISBN 0-471-82324-4.

   SH-C   The emphasis here is on applications rather than concepts. Included are an introduction to the appearance and function of the various kinds of forms and machinery used in automated data processing (ADP); data processing software; an introduction to COBOL, RPG, FORTRAN, PL/1 and BASIC; systems analysis and design; and two case studies (an accounts receivable system and an inventory system).

Westin, Alan F., and Michael A. Baker (Eds.). *Databanks in a Free Society: Computers, Record-Keeping and Privacy.* Quadrangle, 1972. xxi + 522pp. $12.50. 75-183193. ISBN 0-8129-0292-0.

   SH-C   This National Academy of Sciences study deals with the problems of due process and privacy posed by recordkeeping in an increasingly computerized society. Includes an introduction to computer technology; profiles and trend sketches of 55 computerizing organizations; forecasts of future developments in recordkeeping and the law; and policy implications. Excellent appendixes.

Wooldridge, Susan. *Computer Output Design.* (Illus.) Petrocelli/Charter, 1975. xiii + 262pp. $11.95. 75-17672. ISBN 0-88405-308-3. Index;CIP.

   SH-C   The book is an excellent description of the operational and systems problems of medium-to-large volume output. Included are types of output devices, techniques for error checking and correction, audit trails, turnaround documents, types of reproduction, security and protection and choosing software and documentation standards. Fine for beginning students.

001.64 ELECTRONIC DIGITAL COMPUTERS

See also 621.38195 Computers.

Dickey, Larry W. *Introduction to Computer Concepts: Hardware and Software.* (Illus.) Prentice-Hall, 1974. x + 342pp. $15.95. 73-16452. ISBN 0-13-480004-4.

    C Successfully introduces both hardware and software concepts, with no prerequisite beyond high school algebra assumed. Evolved from a Bell Telephone Laboratory course, the book refers to the IBM 360/20 and UNIVAC 1108 computers and COBOL and FORTRAN languages. Suitable for a descriptive first course in computers.

Greenberger, Martin, (Ed.). *Computers, Communications, and the Public Interest.* Johns Hopkins, 1971. xix + 315pp. $12.50. 74-140671. ISBN 0-8018-1135-X.

    C Papers from a lecture and discussion series on computers and their impact on society. Topics included the various uses to which computers are put and the future of these uses. Discussants included Nobel laureats, administrators, legislators, scientist-advisers, and consumer advocates. The papers are well edited, with all technical terms explained. The book can be recommended as a collateral source for any course treating the impact of science on society.

Hawkes, Nigel. *The Computer Revolution.* (Illus.) Dutton, 1972. 216pp. $7.95. 70-166165. ISBN 0-525-08405-3.

    SH-C An informed nontechnical review of the contemporary status of computers and their applications in Great Britain and the United States. The topics are "From Abacus to ENIAC," computers in business, science and the arts, computer construction, learning and intelligence, simulation, and the future. The material is well presented and excellently illustrated.

Katzan, Harry, Jr. *Information Technology: The Human Use of Computers.* (Illus.) Petrocelli, 1974. xii + 350pp. $11.95. 74-1453. ISBN 0-88405-059-9. Index;CIP.

    C Katzan presents a lucid, well-written overview of programming languages; discusses the philosophy of computation and systems design; and gives new insights into computer hardware. Final chapters are devoted to philosophical applications and concepts which broaden the reader's awareness of the technology of automated information systems. Useful for both coursework and as a reference.

Katzan, Harry, Jr. *Introduction to Computer Science.* (Illus.) Petrocelli/Charter, 1975. xii + 500pp. $14.95. 75-5751. ISBN 0-88405-309-1. Index;CIP.

    SH-C This book, intended for an introductory course in computer science, covers nearly everything that beginning students could understand, including fundamental concepts, computer systems, software, and other topics.

Martin, James. *Introduction to Teleprocessing.* (Illus.) Prentice-Hall, 1972. xv + 267pp. $10.50. 74-38242. ISBN 0-13-499814-6.

    SH-C-P Martin introduces the reader to the basic requirements, methods, hardware and software necessary to accomplish data transmission and includes an index of basic concepts. The book may be used as a text and is highly recommended as a reference to anyone involved in data transmission.

Ouellette, R.P., et al. *Computer Techniques in Environmental Science.* (Illus.)

Petrocelli/Charter, 1975. viii + 248pp. $14.95. 74-23805. ISBN 0-88405-281-8. Index;CIP.

C-P An excellent survey of the use of computers in environmental applications. Many case studies detail computer uses in monitoring and scientific and administrative functions. Perhaps more useful to a computer scientist than to an environmentalist but a good addition to university libraries.

Scientific American. *Computers and Computation: Readings from* Scientific American. (Illus.) Freeman, 1971. vii + 283pp. $10.00; $4.95(p). 78-170396. ISBN 0-7167-0937-6; 0-7167-0936-8.

SH-C Twenty-six articles are reprinted here, grouped into five coherent sets, with an "overview" introduction to each set. Seven articles are on fundamentals, five on games, music, and artificial intelligence, four on mathematics of, by, and for computers; six on computer models of the real world; and four are on uses. The wealth of illustrative material characteristic of *Scientific American* is particularly valuable here.

Tanenbaum, Andrew S. *Structured Computer Organization.* (Illus.) Prentice-Hall, 1976. xix + 443pp. $18.50. 74-30322. ISBN 0-13-854505-7. Index;CIP.

C This is an excellent, basic, sophomore-level text on assembly language, computer organization, systems and general hardware/software topics.

Weizenbaum, Joseph. *Computer Power and Human Reason: From Judgment to Calculation.* (Illus.) Freeman, 1976. xii + 300pp. $9.95. 75-19305. ISBN 0-7167-0464-1; 0-7167-0463-3(p). Index;CIP.

C-P Weizenbaum contends that computers gain their power because they can accept only certain types of data and allow only certain types of questions. Thus computers should not be programmed to do such things as making judicial or psychiatric judgments because abstract concepts cannot be adequately programmed. Must reading for scientists, philosophers and humanists.

001.642 COMPUTER PROGRAMMING

See also 519.7 –.8 Programming and Special Topics.

Albrecht, Robert L., et al. *Basic.* Wiley, 1973. ix + 325pp. $3.95(p). 72-11700. ISBN 0-471-02048-6.

JH-SH-C A programmed text which should be useful for self-study of BASIC at any educational level. Requires no previous knowledge of computer operation, design or programming, and the examples (from statistics to string processing) require only elementary mathematics.

Bates, Frank, and Mary L. Douglas. *Programming Language/One: With Structured Programming, 3rd ed.* (Illus.) Prentice-Hall, 1975. xiv + 336pp. $12.50; $9.95(p). 75-2113. ISBN 0-13-730457-9; 0-13-730473-0. Index;CIP.

SH-C This edition contains major changes which conform to the ANSI BASIC-II standard while incorporating the concepts of structured programming. Requires no previous exposure to computers and only high school algebra. Arranged by concepts rather than formal language constructs. Can be used without an instructor. A superb text.

Brier, Alan, and Ian Robinson. *Computers and the Social Sciences.* (Illus.) Columbia Univ. Press, 1974. 285pp. $12.50; $6.50(p). 74-12052. ISBN 0-231-0391-4-X; 0-231-03915-8. Index;CIP.

SH-C The book is intended to help social scientists and their students use computers. Sections on development and design describe differences in analog and digital computers and include a nontechnical description of hybrid computers. The material is presented clearly and is up to date.

Conway, Richard, and David Gries. *A Primer on Structured Programming: Using PL/I, PL/C, and PL/CT.* (Illus.) Winthrop, 1976. xii + 397pp. $8.95(p). 75-40276. ISBN 0-87626-688-X. Index;CIP.

SH-C   This technical primer contains material of interest to both beginner and expert. Discusses fundamental concepts, including program testing, program structure, development and independent subprograms. Many sample programs are in the PL/C language.

Corlett, P.N., and J.D. Tinsley. *Practical Programming, 2nd ed.* (Illus.) Cambridge Univ. Press, 1972. xii + 264pp. $4.95(p). 75-161295. ISBN 0-521-08198.

SH-C   This introductory text also includes basic principles of numerical analysis, and simple concepts in number theory, statistics, and data processing. Two new chapters on FORTRAN have been added. Includes exercises and problems, and illustrative examples which can be readily programmed for firsthand computer experience.

DeRossi, Claude J. *Learning COBOL Fast: A Structured Approach.* (Illus.) Reston, 1976. xi + 212pp. $8.95(p). 75-41485. ISBN 0-87909-447-8. Index;CIP.

C-P   This small book gives enough of the basics of COBOL programming to do reasonable problems, but without too many details. For more complex problems, the reader will be able to describe the problem in COBOL terms to a professional programmer. Most suitable for self-study.

Friedmann, Jehosua, et al. *FORTRAN IV.* (Illus.) Wiley, 1975. xii + 452pp. $5.95(p). 74-34044. ISBN 0-471-28082-8. Index;CIP.

C   A remarkably detailed book which begins at the very beginning and covers input and output, control, and variables especially well. An excellent text, either for FORTRAN alone or in a general data processing course.

Gildersleeve, Thomas R. *Computer Data Processing and Programming.* (Illus.) Prentice-Hall, 1970. xii + 170pp. $8.50. 76-99959. ISBN 0-13-16528-6.

SH   A lucid, elementary introduction to computer data processing and programming in the RPG (Report Program Generator) language. Good discussion of the motivation for data processing by varied means, both for batch and real-time applications.

Harkins, Peter B., et al. *Introduction to Computer Programming for the Social Sciences.* (Illus.) Allyn&Bacon, 1973. x + 258pp. $5.95(p). 72-89238.

C-P   Flow-charting is briefly described, as are FORTRAN conventions. The bulk of the text consists of programmed examples, and appendices provide details, including remarks on debugging programs. There is no information on how to formulate one's own problems, but the collection of sample programs may be adapted to the student's needs. (See also *Science*, Vol. 182, 28 Dec. 1973, concerning the care to be exercised in computerizing social problems.)

Hirsch, Seymour. *BASIC: A Programmed Text.* Melville/Wiley, 1975. ix + 502pp. $8.95(p). 75-6806. ISBN 0-471-40045-9. Gloss.;index;CIP.

SH-C   The author describes key elements of BASIC with the help of easy-to-understand examples in short frames, each of which gives a small unit of information. Quizzes with answers for each section are particularly valuable as a study aid.

Hume, J.N.P., and R.C. Holt. *Structured Programming Using PL/1 and SP/k.* (Illus.)

Reston, 1975. xii + 340pp. $9.95; $7.95(p). 75-23350. ISBN 0-87909-793-0; 0-87909-792-2. Index;CIP.

SH-C   The authors adeptly weave structured programming techniques together with the fundamentals of Pl/1 by employing a series of structured programming subsets of PL/1 and SP/k (k = 1 to 8), which systematically enlarge a student's programming ability. An excellent teaching tool for introductory courses.

Kennedy, Michael, and Martin B. Solomon. *Ten Statement Fortran plus Fortran IV: Sensible, Modular, and Structured Programming with WATFOR and WATFIV, 2nd ed.* (Illus.) Prentice-Hall, 1975. xxvii + 579pp. $9.95(p). 74-18468. ISBN 0-13-903385-8. Index;CIP.

C   A very good introduction to FORTRAN, this text is particularly useful for those students using IBM systems with WATFOR and WATFIV compilers. The concept of structured programming is introduced, and a chronology of computer developments is supplied.

Lane, Ron. *An Introduction to Utilities.* (Illus.) Petrocelli/Charter, 1975. xiv + 162pp. $9.95. 75-19284. ISBN 0-88405-285-0. Index;CIP.

C-P   Lane has organized a selected set of functions of IBM System/360 and System/370 Utilities which use OS, VS or VM operating systems. He presents a small subset of the available functions that provide for most of the IBM user's needs. Rules for all utility functions are summarized in an appendix.

Ledgard, Henry F. *Programming Proverbs.* (Illus.) Hayden, 1975. 134pp. $5.65(p). 74-22058. ISBN 0-8104-5522-6. Index;CIP. *Programming Proverbs for FORTRAN Programmers.* 130pp. 74-22074. ISBN 0-8104-5820-9.

C-P   Concise and useful handbooks for students and professional programmers wishing to write correct, readable, well-structured and easily maintained programs. The author includes over 20 older maxims which will lead to high quality programs. Included are use procedures, proper use of intermediate variables, warnings, use of good mnemonic names and examples in the five most common langauges or in FORTRAN only.

Maurer, Ward Douglas. *Programming: An Introduction to Computer Techniques.* (Illus.) Holden-Day, 1972, xiii + 335pp. $12.95. 70-188126. ISBN 0-8162-5453-2.

SH-C   Directed toward a second course in programming, this book provides comprehensive coverage of both assembly and algebraic language programming concepts. The book is well organized and contains many good examples and illustrations. Students should be familiar with the binary number system, flow-charting and one algebraic language, such as FORTRAN, PL/1, BASIC, ALGOL, etc. Access to a computer for programming exercises is recommended.

Mott, T.H., Jr., S. Artandi and L. Struminger. *Introduction to Programming for Library and Information Science.* Academic, 1972. xvii + 231pp. $8.95. 78-182675. ISBN 0-12-508750-0.

C-P   An elementary introduction to the Boolean logic which is necessary for selective retrieval of text materials stored in a computer system. The authors have chosen PL/1 as the programming language which best suits the needs of a text processor and cover programming fundamentals, character manipulation, arrays and sets, and the basic concepts of Boolean algebra. There is an analysis of the document retrieval problem as well as an introduction to subroutines, functions and data file management. Highly recommended provided students have access to a PL/1 compiler.

Peterson, W. Wesley. *Introduction to Programming Languages.* (Illus.) Prentice-Hall, 1974. ix + 352pp. $12.95. 74-2468. ISBN 0-13-493486-5. CIP.

C-P Peterson discusses ALGOL, BASIC, FORTRAN, PL/1, APL, COBOL, SNOBOL and LISP, giving each one a different degree of emphasis depending on the variety of applications to which it is suited. The book is concerned with the day-to-day use of programming languages. A fine text for any serious computer science student or professional programmer.

Rudd, Walter G. *Assembly Language Programming and the IBM 360 and 370 Computers.* (Illus.) Prentice-Hall, 1976. xviii + 553pp. $16.95. 75-17826. ISBN 0-13-049536-0. Gloss.;index;CIP.

C Rather complete coverage of assembly language programming for the mathematics-oriented beginning programmer. Covers IBM 360/370, OS and DOS.

Stark, Peter A. *Computer Programming Handbook.* (Illus.) Tab, 1975. vii + 506pp. $12.95; $8.95(p). 75-24688. ISBN 0-8306-5752-5; 0-8306-4752-X. Index.

SH-C The author describes computing machines and related equipment, what they do and what the programmer does with them, and then develops three approaches to programming using successively higher levels of language. Useful as an introduction to any of the three language levels, or as collateral reading for high school or introductory vocational courses.

001.94 MYSTERIES

See also 133 Parapsychology.

Byrne, Peter. *The Search for Big Foot: Monster, Myth or Man?* (Illus.) Acropolis, 1975. 263pp. $8.95. 75-13943. ISBN 0-87491-159-1. Index;CIP.

SH-C Byrne traces the historical background of the search for "Sasquatch," reviews present discoveries, matches anecdotes with "reliable" observations, discusses the three most "reliable" episodes, assesses the likelihood that the coastal U.S.-Canada area could support and "hide" Bigfoot, provides a discussion on the validity of a film of Bigfoot obtained in 1967, and offers some conclusions.

Cohen, Daniel. *The Magic Art of Foreseeing the Future.* (Illus.) Dodd, Mead, 1973. 192pp. $4.95. 72-6882. ISBN 0-396-06718-2. Bib.

JH-SH-C A clever and very readable introduction to ways of attempting to foresee the future. Oracular pronouncements, astrology, numerology, the physiognomonic and morphologic methods of character diagnosis, necromancy, dream interpretations, prophets, and studies of ESP and other psychic phenomena are covered.

Corliss, William R. *Strange Phenomena: A Sourcebook of Unusual Natural Phenomena: Vol. G-1.* (Illus.) Corliss, 1974. vi + 277pp. $6.95(looseleaf). 73-91248. ISBN 0-9600712-1-0. Indexes.

JH-SH-GA-C *Strange Phenomena* focuses on those observations which do not fit into the normal web of scientific theory. Includes UFO observations as well as other unexplained occurrences, all of which are integrated into the author's useful categorization of strange sightings.

Kusche, Lawrence David. *The Bermuda Triangle Mystery—Solved.* (Illus.) Harper &Row, 1975. xvi + 302pp. $10.00. 74-1828. ISBN 0-06-012475-X. Index;CIP.

SH The author recounts the legend of the triangle and examines the individual incidents in chronological order. A 15-page bibliography is included, as are maps and photographs.

Napier, John. *Bigfoot: The Yeti and Sasquatch in Myth and Reality.* (Illus.) Dutton, 1973. 240pp. $8.95. 71-179857. SBN 0-525-006658-6.

SH-C-P   Napier evaluates the voluminous and confused "evidence" of human-
like creatures living in the Himalayas and the forests of northwestern America, as well as the mythology concerning these creatures. Entertaining and valuable.

Place, Marian T. *On the Track of Bigfoot.* (Illus.) Dodd, Mead, 1974. 156pp. $4.25. 73-15357. ISBN 0-396-06883-9. Index;bib.

JH-SH   Bigfoot, Sasquatch, Oh-Mah—the known facts are presented, as are the
stories of individuals who claim to have encountered, directly or indi-
rectly, the creatures. The question of their existence is left for the reader to decide. Excellent illustrations and maps.

Ryan, Peter. *UFOs and Other Worlds.* (Illus.) Penguin, 1975. 48pp. $1.75(p). ISBN 0-14-061017-0. Index.

JH   The authors give straightforward, scientifically sound accounts of some UFO
sightings and explain most by natural events. Hypothetical (and reasonable) explanations are offered for those sightings not so explainable.

Sagan, Carl, and Thornton Page (Eds.). *UFO's—A Scientific Debate.* (Illus.) Cornell Univ. Press, 1973. xxxi + 310pp. $12.50. 72-457. ISBN 0-8014-0740-0.

SH-C   This collection of papers from a symposium on unidentified flying objects
(UFO's) sponsored by AAAS includes Philip Morrison's paper on the nature of scientific evidence, Carl Sagan's crystal-clear views, sociological perspec-
tives, the psychology and epistemology of UFO sightings, the abilities and limitations of witnesses, and much more. A tremendously valuable book.

Salisbury, Frank B. *The Utah UFO Display: A Biologist's Report.* (Illus.) Devin-Adair, 1974. xiv + 286pp. $7.95. 74-75389.

JH-SH-C   Salisbury subjects UFO sightings to rigorous scientific analysis and
clearly develops reasons for rejecting the views of both ardent UFO-
ologists and ultraconservative scientists. Recommended for the general reader and as collateral reading in science or philosophy.

Tomas, Andrew. *On the Shores of Endless Worlds.* (Illus.) Putnam's, 1974. xii + 222pp. $7.95. 73-93747. SBN 399-11343-6. Index.

GA-SH-C   Incorporates the author's own experiences and technical data derived
from astrochemical research to underscore the possibility of visitors to
earth from outer space millions of years ago. Author concludes that more research is needed to locate indisputable evidence of prehistoric landings on earth. Recommended reading for courses in astronomy, physics, and biology—the book will also intrigue nonspecialized readers.

020 Library and Information Sciences

Donohue, Joseph C. *Understanding Scientific Literatures: A Bibliometric Approach.* (Illus.) MIT Press, 1973. xiii + 101pp. $10.00. 72-10334. ISBN 0-262-04039-5.

C-P   The thesis is advanced that through the measurement of regularities in the
citing of sources in a body of thought it is possible to identify an essential
nucleus of authors and journals and to promote the effective management of a litera-
ture in libraries. The field of information science is analyzed as an example, and various techniques and formulas are applied in order to identify graphically the literature nucleus of the field. Detailed in its presentation, complicated in its techniques, this book is not for the casual reader.

Fussler, Herman H.   *Research Libraries and Technology: A Report to the Sloan Foundation.* Univ. of Chicago Press, 1974. xi + 91pp. $5.95. 73-81481. ISBN 0-226-27558-2.

C-P   Fussler examines the problem of information handling in large university libraries and reports on library studies already done. The possibilities of new technology from the copying machine to the holograph are discussed. An excellent review, suitable for any library with research needs.

Morse, Grant W.   *Concise Guide to Library Research, revised 2nd ed.* Fleet Academic, 1975. 262pp. $12.50; $5.50(p). 74-21358. ISBN 0-8303-0143-7; 0-8303-0148-8. Indexes.

C   This is a successful attempt at teaching students how to use the library. It should be available to every freshman college student.

Strauss, Lucille J., Irene M. Shreve and Alberta L. Brown.   *Scientific and Technical Libraries: Their Organization and Administration, 2nd ed.* (Illus.) Wiley, 1972. x + 450pp. $14.95. 71-173679. ISBN 0-471-83312-6.

C   The earlier editions were considered handbooks for working librarians; this edition is a class text. The experienced technical librarian will also find much of interest here. This book is a must for all special libraries, technical or not, or for anyone wishing to set up a special library in his or her organization.

030 General Encyclopedic Works

For specific subjects, see individual disciplines; for major general works, see AAAS Science Book List, 3rd ed. [*Washington, D.C.: AAAS, 1970*] *and* Encyclopedia Buying Guide 1975–1976: A Consumer Guide to General Encyclopedias in Print [*NY: Bowker, 1976.*].

Crowley, Ellen T., Christopher Crocker, Donna Wood, et al.   *Acronyms, Initialisms, & Abbreviations Dictionary, 5th ed.* Gale Research Co., 1976. xiv + 757pp. $38.50. 76-10036. ISBN 0-8103-0502-X.

SH-C   Most acronyms are not understood outside the specialized field or activity in which they developed, and their proliferation is a source of great confusion. AIAD covers most fields, from aerospace to sports to associations to science and technology to government, and many more, both domestic and international. There is an annual supplement as well as a dictionary of reverse acronyms, initialisms, and abbreviations. Useful to a clientele of very diversified interests and occupations.

Field Enterprises.   *The World Book Encyclopedia, 22 Vols.* (Illus.) Field Enterprises Educational, 1976. 12,815pp. (total). $299.00. 75-1722. ISBN 0-7166-0076-5. Index.

JH-SH-C   Major articles in science give a solid, standard foundation; offer visual aids for clarity and interest; encourage related reading in other articles and sources; present an outline for overview; and ask specific questions for review. The last volume is a research guide and index, containing a subject index, a brief introduction to literature research, and reading and study guides. Vocabulary ranges from simple to complex, depending on the expected grade level of the reader. The facts are accurate, but most of the science experiments and projects will probably be ineffective if undertaken without help from a knowledgeable adult. Good organization; ample cross-references.

069 MUSEUM SCIENCE

Allen, Jon L.   *Aviation and Space Museums of America.* (Illus.) Arco, 1975. 287pp. $12.00. 73-91258. ISBN 0-668-03426-2; 0-668-03631-1(p).

SH   Locations and concise descriptions are given for 57 "museums" in the U.S. and Canada, ranging from large collections such as the Smithsonian's National Air and Space Museum and the U.S. Air Force Museum at Dayton, Ohio, to small aerodromes where flying circuses of vintage aircraft are maintained and flown. A valuable guidebook to a first-hand study of the development of aviation and space technology.

Martin, Lynne. *Museum Menagerie: Behind the Scenes at the Nature Museum.* (Illus.) Criterion, 1971. 90pp. $3.95. 76-136215. ISBN 0-200-71790-1.

JH This excellent introduction to the traditional natural history museum covers the collection of specimens, their preservation and storage, as well as exhibit preparation, with careful attention to techniques employed. Numerous anecdotes told in a conversational manner, a good bibliography of books, films, and periodicals, and a list of museums are included.

Mills, John Fitzmaurice. *Treasure Keepers.* (Illus.) Doubleday, 1974. 160pp. $7.95. 73-82573. ISBN 0-385-04491-7; 0-385-04644-8. Gloss.;index.

SH-C Deals with eight aspects of the contemporary museum: (1) the problems caused by effective public display versus protection of the objects; (2) the scientific skills of conservationists and where such skills may be learned; (3) methods of dating; (4) "restoration"; (5) detection of forgeries; (6) defense against and recovery from extreme disaster; (7) the evolution of the museum; and (8) present-day innovations. A useful and consistently instructive volume.

Williams, Patricia M. *Museums of Natural History and the People Who Work in Them.* (Illus.) St. Martin's, 1974. 120pp. $5.95. 73-93926.

JH-SH-C Williams attempts to explain the scope, activities, methods and techniques of anthropology, botany, geology, and zoology in their roles as parts of the natural history museum. A major portion of the book is devoted to the work of a museum staff in conducting research programs, public education activities and preparing publications. A list of some of the natural history museums in 48 states is appended. An overview for the general reader and for students in a beginning museology course.

Philosophy and Related Disciplines

133 PARAPSYCHOLOGY

Aylesworth, Thomas G. *ESP.* (Illus.) Watts, 1975. 63pp. $4.33. 74-26797. ISBN 0-531-00826-6. Gloss.;index;CIP.

SH Many case studies of ESP experiences and scientific investigations into the phenomena are reported. Precognition, telepathy and clairvoyance are considered in some detail.

Christopher, Milbourne. *ESP, Seers, & Psychics.* (Illus.) Crowell, 1970. x + 268pp. $6.95. 78-127607. ISBN 0-690-26815-7.

SH Written by the head of the Occult Investigation Committee of the Society of American Magicians, this book is highly recommended to those who are curious about the techniques and tactics used by fortune-tellers and spiritualists of all sorts. The book is illustrated with handbills or photographs of famous seers and psychics of the past. It provides a means of bringing those who seek magic back to evidential science.

Rogo, D. Scott. *Parapsychology: A Century of Inquiry.* Taplinger, 1975. 319pp. $12.50. 74-22890. ISBN 0-8008-6236-8. Index.

SH-C   While not a definitive text, this book is a carefully compiled survey of the entire field of parapsychology. Using an historical approach, the author presents material in a balanced, factual way. In conclusion, various theories explaining phenomena are presented. A reasonably complete overview for beginning students.

Shumaker, Wayne. *The Occult Sciences in the Renaissance: A Study in Intellectual Patterns.* (Illus.) Univ. of Calif. Press, 1972. xxi + 284pp. $15.00. 70-153552. ISBN 0-520-02021-9.

C-P   This interesting, comprehensive and scholarly account of astrology, witchcraft, white magic, alchemy and Hermetic theology provides insight into the thinking of the people today who have turned away from demonstrable facts, reasoned conclusions and the essential skepticism of science. Not for the casual reader.

140 Philosophy—Specific Viewpoints

Arieti, Silvano. *The Will to Be Human.* Quadrangle, 1972. vi + 279pp. $8.95. 72-78504.

C-P   Integrates hundreds of references from psychology, psychiatry, sociology, ethology, philosophy, literature, politics, history and religion into a searching inquiry into the complex problem of human will. Arieti traces the evolution and origin of will from early animal life and spontaneous movement through its development into "the most specifically human characteristic" which has both a psychological and a sociological dimension.

Becker, Ernest. *Escape from Evil.* Free Press/Macmillan, 1975. xix + 188pp. $9.95. 75-12059. Index.

SH-C-P   Becker considers evil, guilt, and inequality, but does so with such interesting cultural and anthropological analogies that the nonphilosopher will be fascinated.

Brabazon, James. *Albert Schweitzer: A Biography.* (Illus.) Putnam's, 1975. 509pp. $12.95. 74-30545. SBN 399-11421-1. Index.

C   The influence of current thought on Schweitzer and his influence on his contemporaries is discussed. Brabazon describes the man and the period in which he lived, and offers a valuable insight into Schweitzer's theologic and philosophic ideas. Recommended to the general reader.

Disch, Robert, (Ed.). *The Ecological Conscience: Values for Survival.* Prentice-Hall, 1970. xv + 206pp. $2.45(p). 71-130009.

C   This well-chosen anthology considers the scientific, sociological and political aspects of ecological values. The general theme is that man's values, based on social and religious standards, must change to alter the consequences of "progress." Valuable for both biology and sociology curricula.

Eiseley, Loren. *The Night Country.* (Illus.) Scribner's, 1971. xi + 240pp. $7.95. 78-162747. ISBN 0-684-12568-4.

C   The author has recreated episodes that influenced his early thinking, and in his mature reflections as anthropologist, naturalist, and philosopher, he searches for deeper meanings in human experiences. In superb prose and unconventional form, the volume comprises a most revealing autobiographical sketch of an eminent scholar.

Hamilton, Michael P., (Ed.) *This Little Planet.* Scribner's, 1970. 241pp. $6.95. 76-120363.

SH-C   The book is a collection of articles by scientists and theologians on pollution, scarcity and conservation, with the background theme the lecture by Lynn White (UCLA) who argued that the Judaeo-Christian tradition encouraged the desecration of the world because God had created man "in his own image" to "subdue" the earth. The need for revising and extending theological concepts to proceed hand-in-hand with developments in the natural and social sciences is recognized. The book deserves a wide audience.

Hudson, Liam. *Human Beings: The Psychology of Human Experience.* (Illus.) Anchor/Doubleday, 1975. 232pp. $2.95(p). 73-9031. ISBN 0-385-01403-1. Index;CIP.

C-P   This book is recommended for a college course in philosophy of science and as an adjunct to an introductory psychology text. Hudson argues for a systematic epistemology of human experience, termed "hermeneutic," in which people are not just objects for abstract investigation. Case examples from psychology, literature and the arts showing the complex and hierarchical approach necessary to predict and understand the "psychology of human experience" are included.

Livingston, John A. *One Cosmic Instant: Man's Fleeting Supremacy.* Houghton Mifflin, 1973. 243pp. $5.95. 73-4686. ISBN 0-395-14012-9.

SH-C   An examination of ethics and values, especially biological ones, in the light of the current human condition by a noted ornithologist and naturalist. Well-written and thoughtful; a good resource for transdisciplinary courses.

Needleman, Jacob. *A Sense of the Cosmos: The Encounter of Modern Science and Ancient Truth.* Doubleday, 1975. 178pp. $6.95. 73-83660. ISBN 0-385-00010-3. Index;CIP.

C-P   Needleman attempts to show modern science as an important extension of one's self and suggests that we can learn from our increasing knowledge of the complete human being. He discusses the universe, the science of medicine, the fear of death, the science of living things, physics, psychotherapy, the sacred, and magic. Useful for collateral reading for college students studying physics, philosophy, religion and the humanities.

Nisbet, Robert. *The Social Philosophers: Community and Conflict in Western Thought.* (Illus.) Crowell, 1973. xii + 466pp. $10.00; $5.95(p). 72-83132. ISBN 0-690-74406-4; 0-690-74405-6.

C-P   Nisbet traces in the writings of the philosophers of earlier times the origins of some of the queer things that have happened over the past four millenia. His interesting and well-defended thesis is that much of philosophy is directed toward a way of thinking that will allow people to live as social animals and in some sort of close proximity to each other.

Roth, John K. (Ed.). *The Philosophy of Josiah Royce.* Crowell, 1971. viii + 421pp. $8.95. 76-146287. ISBN 0-690-61839-5.

SH-C   At a time when urbanized Americans must take a more comprehensive environmental viewpoint if they are to survive, it is well to have available an easily managed selection of the writings of Josiah Royce. In his later writings Royce moved away from the Absolute to an unlimited community of finite individuals—a community in which all people participate for the good of all. Royce's challenging social and ethical insights are capably assembled and assessed.

Von Ditfurth, Hoimar. *Children of the Universe: The Tale of Our Existence.*

(Illus.;trans.) Atheneum, 1974. viii + 301pp. $10.95. 73-91629. ISBN 0-689-10588-6. Index.

SH-C Translation of a German science journalist's treatment of astronomical discoveries, their relationship to phenomena occurring in the extraterrestrial universe, and their impact on man's existence and evolution on earth. This wide-ranging popularization of astrophysical theory relates transfers of intergalactic matter to medieval allegory and modern-day religion. Provocative metaphysics and excellent science.

150 Psychology

Cohen, Ira S., (Ed.). *Perspectives on Psychology: Introductory Readings.* (Illus.) Praeger, 1975. xiii + 464pp. $6.95(p). 73-8396. ISBN 0-275-51080-8; 0-275-88710-3 (p). CIP.

SH-C A selection of 35 interesting readings on topics covered in the usual beginning psychology course. Each article is put in context by the editor's introduction, and subjects range from sanity to yogi control of psychic heat. Should be in high school libraries as a stimulus for better students; also useful as a supplementary text for beginning college psychology courses.

Dallett, Kent. *It's All in Your Mind: Understanding Psychology.* (Illus.) National Press, 1973. ix + 244pp. $6.95; $4.95(p). 72-97839. ISBN 0-87484-270-0; 0-87484-269-7.

C A fresh and intriguing look at the basic subject matter of psychology (behaving humans, observing the behavior of other humans) in three side-by-side, interrelated tracks of material. Much material is covered, primarily in reference to the reader's own experience. Requires considerable ability to abstract and conceptualize the issues of psychology, but is an excellent resource for the thinking, general reader.

Diamond, Solomon, (Ed.). *The Roots of Psychology: A Sourcebook in the History of Ideas.* Basic, 1974. xvii + 781pp. $24.95. 72-76919. SBN 465-06740-9. Indexes.

C-P More than 280 topics are arranged chronologically in 28 chapters: Topics include dreams, mental illness, social psychology, child development, and the concept of man through levels of intelligence. A fine reference work for the general reader.

Fantino, Edmund, and George S. Reynolds. *Introduction to Contemporary Psychology.* (Illus.) Freeman, 1975. 610pp. $12.95. 74-2301. ISBN 0-7167-0761-6. Gloss.;index;CIP.

SH-C This unique introductory text mixes both traditional and newer psychologies with mentalistic and behavioristic approaches. Basic psychological processes such as memory and perception, and major areas of psychology are covered separately. Recommended for all kinds of introductory psychology classes and for pleasant reading by advanced students.

Gibbons, Don E., and John F. Connelly. *Selected Readings in Psychology.* (Illus.) Mosby, 1970. xi + 273pp. $4.95. 71-124054.

SH-C Articles and book excerpts intended for adjunctive class assignments are classified according to major introductory psychology headings. The authors exclude the commonly cited works and include lesser known material. Interesting and useful in everyday relationships and in self-understanding.

Harrison, Albert A. *Psychology as a Social Science.* (Illus.) Wadsworth, 1972. xviii + 523pp. $10.50. 75-187500. ISBN 0-8185-0028-X. Gloss.

C   This introductory text emphasizes personality and social psychology. The author presents the material in a personal, conversational form without loss of intellectual content.

Lindzey, Gardner, Calvin Hall and Richard F. Thompson. *Psychology.* (Illus.) Worth, 1975. xiii + 802pp. $13.95. 74-21032. ISBN 0-87901-036-3. Gloss.;indexes. *Study Guide; Teacher's Manual.*

C   This is a readable, comprehensive introduction to the study of psychology, with particular emphasis on human behavior. Such varied topics as alcoholism, color vision, psychosurgery, love and meditation are discussed. The usual chapters on learning, perception, thinking, development, intelligence, personality, attitudes, behavior disorders and their treatment are also included.

London, Perry. *Beginning Psychology.* (Illus.) Dorsey, 1975. xvi + 655pp. $12.50. 74-12929. ISBN 0-256-01677-1. Gloss.;indexes;bib.

C   Biological bases of behavior, perception, learning, motivation, emotion, personality, adjustment, behavior disorders and social behavior are written about in a stimulating way. Sex is discussed frankly and openly. A useful text for an introductory psychology course.

McNeil, Elton B. *Being Human: The Psychological Experience.* (Illus.) Harper&Row, 1973. viii + 371pp. $8.95. 72-6266. ISBN 0-06-012902-6; 0-063854406. Bib.

SH-C   Treats human psychology in the context of contemporary American culture. Chapters cover the range of societal fads and problems, from tarot cards to swingers, from the generation gap and the quest for identity to marriage and death, and from criminal violence to modern day utopias.

Whittaker, James O. *Psychology of Modern Life.* (Illus.) Human Sciences/ Behavioral, 1976. 436pp. $9.95; $6.95(p). 74-12622. ISBN 0-87705-234-4. Gloss.;index;CIP.

SH   This general treatment covers most major concepts developed this century, and Whittaker interprets the material in terms of the adolescent's situation. Discussions on emotions, social relations, adjustment, and study habits are particularly well adapted to mid-teen students. A good introduction to quantitative psychology and one of the best available texts for this age group.

150.19 SYSTEMS, SCHOOLS, AND VIEWPOINTS

Carpenter, Finley. *The Skinner Primer: Behind Freedom and Dignity.* (Illus.) Free Press/Macmillan, 1974. xvi + 224pp. $7.95; $2.45(p). 73-16603. ISBN 0-02-905310-2. Index;CIP.

C-P   This unbiased examination of Skinnerian psychology examines the meaning, value, and paradoxical nature of the concept of freedom. The thorough discussion of cognitive freedom makes this book especially valuable for high school teachers as well as college students.

Droz, R., and M. Rahmy. *Understanding Piaget.* (Trans. from the French, c. 1972) International Univ. Press, 1976. xv + 227pp. $10.00. 75-18509. ISBN 0-8236-6690-5. Index;CIP.

C-P   This intensive study guides the reader to the appropriate writings of Piaget and summarizes his main ideas by topic. The overview of Piaget's work is a useful reference. Includes an annotated bibliography and abstracts of his major works in the original French and in translation.

Evans, Richard I. *Carl Rogers: The Man and His Ideas.* Dutton, 1975. lxxxviii + 195pp. $10.95; $3.95(p). 74-23270. ISBN 0-525-07645-X. Index;CIP.

C   A personal and extemporaneous view of Rogers, several essays by him, a joint symposium with Skinner and a chronological bibliography are provided. Useful supplement for psychology of personality courses.

Evans, Richard I. *Jean Piaget: The Man and His Ideas.* (Trans.) Dutton, 1973. lxi + 189pp. $8.95. 73-79550. SBN 0-525-13660-6; 0-525-47360-2.

C-P   Reviews major points of view in psychology, integrates Piaget's concepts of cognitive development with a more general ontology, and relates Piaget the man to the historical foundations of psychology. Also deals with interdisciplinary research, cybernetics, Piagetian concepts and education.

Fancher, Raymond E. *Psychoanalytic Psychology: The Development of Freud's Thought.* Norton, 1973. xi + 241pp. $8.95. 73-1273. ISBN 0-393-01101-1; 0-393-09356-5. Index;bib.

SH-C-P   Traces the development of Freud's thought up to the time of his death. In a lively and authentic manner, Fancher covers the climate of culture in which Freud developed; the clinical problems he solved; the metaclinical, scientific problems he confronted; and the extension of his thinking from 19th century neurology to questions of creativity, sociology, biology, personality development, etc.

Hall, Elizabeth. *From Pigeons to People: A Look at Behavior Shaping.* (Illus.) Houghton Mifflin, 1975. 130pp. $6.95. 75-17030. ISBN 0-395-21894-2. Gloss.;index;CIP.

SH-C   This is an excellent and lucid introduction to the growth and application of principles of learning applied to individuals in a variety of settings. It specifically introduces basic behavioral nomenclature through specific case material. Recommended as a supplementary text for a high school or junior college course in general psychology.

Nevin, John A., (Ed.). *The Study of Behavior: Learning, Motivation, Emotion, and Instinct.* (Illus.) Scott, Foresman, 1973. 418pp. $10.95. 72-92940. ISBN 0-673-05430-6.

C-P   An extensive treatment of conditioning (classical and operant), reinforcement, punishment, stimulus control (discrimination learning) and Skinnerian experimental methods. The information load is high, but the writing style of each author is crisp and understandable. A very useful reference for teachers of introductory psychology or the first learning theory course.

Perls, Fritz. *The Gestalt Approach and Eye Witness to Therapy.* Science&Behavior, 1973. xv + 206pp. $6.95. 73-76971. ISBN 0-8314-0034-X.

SH-C-P   "The Gestalt Approach" is Perls' attempt to restate in simple language his basic theory of gestalt therapy. "Eye Witness to Therapy" is on films of gestalt therapy. Gestalt principles come to life as Dr. Perls works with various people's dreams. A must for all psychology students and gestalt theory practitioners.

Rachlin, Howard. *Introduction to Modern Behaviorism.* (Illus.) Freeman, 1970. ix + 208pp. $6.00. 70-117974; ISBN 0-7167-0928-7.

C   This historical introduction to modern behaviorist psychology presents the basic concepts and summarizes some of the directions in which this mode of thought appeared to be moving. Especially useful as an adjunct to a first course in learning or motivation.

Roazen, Paul. *Freud and His Followers.* (Illus.) Knopf, 1975. xxxiii + 599pp. $15.00. 73-20782. ISBN 0-394-48896-2. Index;CIP.

C   A fascinating chronology of Freud's life and the psychoanalytic movement, largely based on interviews with his contemporaries. The human context of Freud's ideas and his relationships with women and disciples is amply presented. This splendid book will provide a rare learning experience for any reader.

Sexton, Virginia S., and Henryk Misiak (Eds.).   *Historical Perspectives in Psychology: Readings.* (Illus.) Brooks/Cole, 1971. ix + 452pp. $4.95. 77-155898. ISBN 0-8185-0012-3.

C-P   A sampling of the literature of psychology covering history; interdisciplinary relationships and the growth and development of the profession; theories of psychology; characteristics of American psychology; psychology in various other countries, particularly in Europe; and a summary of possible future trends.

Storr, Anthony.   *C.G. Jung.* Viking, 1973. xii + 116pp. $5.95. 72-81254. SBN 670-1094-3; 670-01962-3(p). Index.

C-P   This easily understandable book is more about Jung's ideas than his life, though the two are inextricably interwoven. Gives a continuing contrast between Jung and Freud.

Voss, James F.   *Psychology as a Behavioral Science.* (Illus.) Goodyear, 1974. xvi + 298pp. $8.95. 73-84223. ISBN 0-87620-735-2. Gloss.;index.

C   A well-written introductory survey which stresses the behavioral aspects of psychology, including perception, learning, cognition, and motivation.

Wheeler, Harvey, (Ed.).   *Beyond the Punitive Society: Operant Conditioning: Social and Political Aspects.* Freeman, 1973. viii + 274pp. $8.95. 73-1269. ISBN 0-7167-0785-3.

C   Contains papers from a symposium sponsored by the Center for the Study of Democratic Institutions. Gives an overview of operant conditioning and behaviorism, an evaluation of B.F. Skinner's ideas and a response by Skinner. The dilemma—who controls the controller—should have received greater emphasis, but otherwise it is a good source for discussions, seminars and the general reader. Impressive list of contributors.

Zusne, Leonard.   *Names in the History of Psychology: A Biographical Sourcebook.* (Illus.) Hemisphere (distr.: Halsted/Wiley), 1975. xvii + 489pp. $17.95. 74-26643. ISBN 0-470-98676-X. Index;CIP.

C   This is a sourcebook of the eminent psychologists and others who contributed significantly to the development of the science of psychology from 1600-1967. Zusne focuses on the philosophy, methods, discoveries and innovations of each individual. A landmark reference book.

152 PHYSIOLOGICAL AND EXPERIMENTAL PSYCHOLOGY

Andreas, Burton G.   *Experimental Psychology, 2nd ed.* (Illus.) Wiley, 1972. xiv + 608pp. $12.50. 78-171910. ISBN 0-471-02905-X

SH-C   A good popular introduction to experimental psychology. The treatment of complex issues is clear and straight-forward without oversimplification. Good examples from both classic and current research are given.

Denenberg, Victor H. (Ed.).   *Readings in the Development of Behavior.* (Illus.) Sinauer, 1972. ix + 483pp. $6.95(p). 78-181-990. ISBN 0-87893-151-1.

C-P   Samples a wide literature, with thoughtful editorial comments. Social processes and learning, which would ordinarily be given prominence, receive far less weight than do biological and chemical variables. For the advanced student of physiological psychology and animal behavior; includes classic studies.

Gardner, Howard.   *The Shattered Mind: The Person After Brain Damage.* (Illus.) Knopf, 1975. xiv + 458pp. $10.00. 74-7740. ISBN 0-394-49315-X. Index;CIP.

C-P   This book is an interesting account of a neuropsychologist's travail in understanding the functions of the normal brain by the study of ideational and emotional behavior of people with brain injuries. The book is richly documented with case histories. Written for the advanced student and the specialist.

Hergenhahn, B.R.   *A Self-Directing Introduction to Psychological Experimentation, 2nd ed.* (Illus.) Brooks/Cole, 1974. vi + 441pp. $9.50(p). 74-77344. ISBN 0-8185-0130-8.

   C   One of the few good laboratory manuals for beginning experimental psychology courses. It is completely self-contained, with numerous references and a series of provocative questions following each experiment. Will aid students in becoming more sensitive and insightful researchers.

Jeeves, Malcolm.   *Experimental Psychology: An Introduction for Biologists.* (Illus.) Arnold (distr.: Crane, Russak), 1974. 59pp. $2.50(p). ISBN 0-7131-2447-4; 0-7131-2448-2(p).

   SH-C   Should interest a wide range of students and professionals. The major theme is the application of scientific methods to behavioral phenomena: perception, memory, etc. Classical as well as contemporary experimental methods and issues are included. Useful text for psychology courses.

Kimble, Daniel P.   *Psychology as a Biological Science.* (Illus.) Goodyear, 1973. xviii + 227pp. $7.95. 73-78901. ISBN 0-87620-665-8. Bib.

   C   This book clearly demonstrates that the biological perspective on psychology is an important topic, but the dry cellular neurophysiology and anatomy may fatigue interest before one reaches the juicier chapters on sexual behavior, regulation of food and water consumption, learning, movement and sleep. These sections coherently present experiments and results directly applicable to human behavior.

Lee, Philip R., Robert E. Ornstein, et al.   *Symposium on Consciousness.* (Illus.) Viking, 1976. 182pp. $10.00. 75-30642. ISBN 0-670-68903-3. CIP.

   C   Split-brain experiments, electroencephalogram (EEG) activity of the cerebral hemispheres, mystic experience and a theoretical approach to discrete states of consciousness are covered, as well as implications of an understanding of consciousness for medical practice and the science of human behavior.

Leukel, Francis.   *Introduction to Physiological Psychology, 3rd ed.* (Illus.) Mosby, 1976. xii + 514pp. $14.75. 75-33031. ISBN 0-8016-2974-8. Gloss.;index;CIP. *Study Guide, 2nd ed.* x + 175pp. $6.50(p). ISBN 0-8016-2968-3.

   C   This introduction to the physiological bases of behavior for students with no presumed biological background is simplified yet accurate. Emphasis is on basic physiology and the endocrine system, with a more extensive discussion than usual of the psychological dimensions of physiological sensation, and a discussion of practical implications.

152.1 SENSORY PERCEPTION

Beck, Jacob.   *Surface Color Perception.* (Illus.) Cornell Univ. Press, 1972. xiv + 206pp. $11.50. 76-38118. ISBN 0-8014-0704-4. Gloss.;index.

   C-P   Concerned with the description, theory and analysis of the perception of color as an attribute of a surface or surfaces. The classic concepts of the roles of color constancy, modes of color appearance, contrast, assimilation, adaptation, illumination and surface texture are methodically, objectively and lucidly reviewed.

Furst, Peter T. (Ed.).   *Flesh of the Gods: The Ritual Use of Hallucinogens.* (Illus.) Praeger, 1972. xvi + 304pp. $10.00; $4.95(p). 78-143970.

   SH-C   The ten papers deal with a wide variety of naturally occurring substances used by primitive peoples for religious experiences and healing. More botanical than pharmacological information is given. A very thought-provoking collection, recommended as collateral reading or for reference in religion, botany, social science or the psychology of mind-altering drugs.

Geldard, Frank A. *The Human Senses, 2nd ed.* (Illus.) Wiley, 1972. xi + 584pp. $12.50. 72-37432. ISBN 0-471-29570-1.

   SH-C-P   Structure, function, and physiological reactions of the major sensory modalities are covered, including the physical nature of relevant stimuli, anatomical structure, and neural function for the eye, ear, skin, kinesthetic senses, smell, and taste from the receptor to the cortex; and complex excitatory-inhibitive relationships in each receptive field and its higher neuronal pathways. Comprehensive, well-organized and thoroughly referenced.

Gregg, James R. *The Sportsman's Eye: How to Make Better Use of Your Eyes in the Outdoors.* (Illus.) Winchester, 1971. vii + 210pp. $6.95. 73-150385. ISBN 0-87691-035-5.

   SH-C   An unusually comprehensive and reliable coverage of visual problems, their explanations, and numerous helpful practical solutions. The identification of each technical topic with familiar recreational activities makes for pleasant reading and easy grasp of related concepts. Appropriate for students in recreation, physical education, and health courses.

Gregory, R.L., and E.H. Gombrich (Eds.). *Illusion in Nature and Art.* (Illus.) Scribner's, 1974. 288pp. $19.95. 73-21146. ISBN 0-684-13800-X. Index.

   SH-C   The authors deal with such questions as "What is an illusion?" "How is it produced?" "Is it in the eye of the beholder or is it the real thing?" Useful as a reference and for interesting collateral reading.

Jonas, Doris, and David Jonas. *Other Senses, Other Worlds.* Stein&Day, 1976. 240pp. $8.95. 75-11816. ISBN 0-8128-1841-5. Index;CIP.

   SH-C   This distinguished mix of fact and fantasy, science and philosophy, examines the sensory systems of present species and then imagines placing such systems in more highly-evolved creatures, typically in other solar systems. Sensory evolution is thoroughly integrated with the evolution of morphology and with the physical conditions of some particular environment and galaxy. Supports the idea of evolution of intelligence. Useful as a supplementary text in animal behavior courses but most applicable in the area of sensation and perception.

Jonas, Gerald. *Visceral Learning: Toward a Science of Self-Control.* Viking, 1973. 154pp. $6.95. 72-12064. SBN 670-74703-3.

   SH-C-P   Jonas traces the development of the visceral learning research program of Professor Neal E. Miller of New York's Rockefeller University.

Rock, Irvin. *An Introduction to Perception.* (Illus.) Macmillan, 1975. xii + 580pp. $13.95. 74-3806. ISBN 0-02-402490-2. Indexes;CIP.

   C-P   Rock seeks answers to the problem of why we see what we think we see. Despite a phenomenological approach, it is still a challenge to find methods for testing perception and cognition. The text is very well organized and clearly written. A good library and teacher reference.

153 INTELLIGENCE AND CONSCIOUS MENTAL PROCESSES

Cole, Michael, John Gay, Joseph A. Glick, and Donald W. Sharp. *The Cultural Context of Learning and Thinking: An Exploration in Experimental Anthropology.* (Illus.) Basic, 1971. xx + 336pp. $10.00. 73-158446. ISBN 0-465-01498-4.

   C-P   Several aspects of cognition among U.S. school children and those of the Kpelle tribe in Liberia are compared and significant differences are established and related to the specific cultural context in which learning takes place. In brief, it was found that people are good at things that are important to them and which they have occasion to do often.

Dobb, Leonard W.   *Patterning of Time*. (Illus.) Yale Univ. Press, 1971. xiv + 472pp. $15.00. 72-151572. ISBN 0-300-01454-6.

   C   A monumental work dealing with the effects of time on the lives of people. Theorems lead to theses on how time is patterned by humans, and studies show how time concepts affect actions, how different cultures and people perceive time, how the biological clock is related to behavior, and how time concepts develop.

Gibson, Eleanor J.   *Principles of Perceptual Learning and Development*. (Illus.) Appleton-Century-Crofts, 1969. viii + 537pp. $8.50. 72-77536. ISBN 0-390-36145-3.

   C-P   The diversity of content is staggering and impressive. Gibson summarizes the principal evidence and theories of perceptual learning and development from an evolutionary standpoint and in humans specifically. The account is comprehensive, succinct, and full of insight. A classic.

Halacy, D.S., Jr.   *Man and Memory*. Harper&Row, 1970. 259pp. $6.95. 74-95962.

   SH-GA-C   A charming, intelligent, informative book on the psychology and biology of learning and memory. Not a "how-to-remember" book, but rather a clear and entertaining presentation of important scientific research.

Lindsay, Peter H., and Donald A. Norman.   *Human Information Processing: An Introduction to Psychology*. (Illus.) Academic, 1972. xv + 737pp. $11.95. 70-182657.

   C   Emphasis is on the way things are learned, and specific models of particular phenomena are developed. Chapters deal with problems in perception; language; learning and cognitive development; motivation; and problem solving. There are two appendixes, on measurements of physiological mechanisms and on operating characteristics. Provides an exciting approach to experimental psychology.

Ornstein, Robert E. (Ed.).   *The Nature of Human Consciousness: A Book of Readings*. (Illus.) Freeman, 1973. xiii + 514pp. $5.95(p). 73-4431. ISBN 0-670-50480-7; 0-7167-0790-X(p).

   SH-C   Extends the boundaries of academic psychology from observation and experimental and logical manipulation to include idiosyncratic, receptively experienced psychological states. The selection of articles is judicious and includes classic and current authors. Traditional perception, sensation, social psychology, brain physiology and biofeedback articles introduce the student to sound approaches to the problem of consciousness and its exotic (altered) states. Stimulating selections on Eastern religious methods.

Ornstein, Robert E.   *On the Experience of Time*. (Illus.) Penguin, 1970. 126pp. $1.45.

   SH-C   A careful analytical review of the literature and nine experiments on perception of time. Four time modes (rhythm and brief intervals, duration, temporal perspective, and simultaneity and succession) are examined, as are various theories of temporal perception, including "biological clocks" and "memory storage clocks") are examined.

Ornstein, Robert E.   *The Psychology of Consciousness*. (Illus.) Freeman, 1972. xii + 247pp. $3.50(p). 72-4432. ISBN 0-670-58198-4; 0-7167-0797-7(p).

   SH-C-P   Explores the unconscious and conscious mind by bringing together material from scientific and philosophical thought. The dual nature of the mind is stressed: one the verbal, rational and analytic; the other the intuitive, artistic and creative. There are illuminating experiments which readers can perform upon themselves.

Tart, Charles T.   *States of Consciousness*. (Illus.) Dutton, 1975. xi + 305pp. $12.50; $4.95(p). 75-5940. ISBN 0-525-20970-0; 0-525-47406-4. Index;CIP.

   C-P   An interesting, speculative book about changes in reported experience, including ordinary waking, sleep, hypnosis, alcohol and marijuana intoxication, and meditative states. Tart's theory that altered states are induced by two basic operations—disrupting forces and patterning forces—is integrated with other psychological knowledge. Much of the material, although plausible, is highly speculative.

153.9 INTELLIGENCE AND APTITUDES

Block, N.J., and Gerald Dworkin (Eds.). *The IQ Controversy: Critical Readings.* (Illus.) Pantheon, 1976. xiii + 557pp. $15.95; $6.95(p). 75-38113. ISBN 0-394-49056-8; 0-394-73087-9. Index;CIP.

    C-P The articles collected here focus on the Lippman-Terman debate of the 1920s, genetic components of IQ differences, social and political consequences of IQ differences, logical analysis of the concepts of IQ and heritability, and the social-political consequences of the controversy. This is a valuable—but unbalanced—collection.

Cohen, Daniel. *Intelligence: What is It?* (Illus.) Evans (distr.: Lippincott), 1974. 159pp. $4.95. 73-80178. ISBN 0-87131-127-5.

    JH-SH-C-P Cohen discusses what is meant by intelligence; what is the evidence; and how good is the support for any statement made in this field. He emphasizes how important it is to study learning, perception, and cognitive styles to understand how both humans and animals adapt to their worlds.

Fine, Benjamin. *The Stranglehold of the I.Q.* Doubleday, 1975. vi + 278pp. $7.95. 75-5260. ISBN 0-385-01576-3. Index;CIP.

    GA-C Although considerable research is reported, this book is more for the general reader. Fine argues that we should depend less on performance measures and more on creating true learning situations. Could be used as supplementary reading for a course in testing.

Herrnstein, R.J. *I.Q. in the Meritocracy.* Atlantic-Little, Brown, 1973. x + 235pp. $7.95. 73-304. ISBN 0-316-34864-9.

    C A closely reasoned, nontechnical statement concerning IQ, beginning with a bit of historical background, continuing with an examination of the concept and uses of intelligence, and concluding with an analysis of the nature-nurture problem and its social significance. The thrust is that the very procedures employed to achieve a more equalitarian society may bring about a society even more sharply differentiated by IQ.

Jensen, Arthur R. *Genetics and Education.* (Illus.). Harper&Row, 1972. vi + 378pp. $10.00. 72-866-36. ISBN 06-012192-0.

    C This collection of Jensen's articles (which take the position that there is a significant heritability for the potential for education) contains a long preface which details chronologically the events pursuant to the initial submission of his very controversial 1969 *Harvard Educational Review* article entitled "How much can we boost IQ and scholastic achievement?"

Liungman, Carl G. *What is IQ? Intelligence, Heredity and Environment.* (Illus.;trans. c. 1970.) Cremonesi (distr.: Atheneum), 1975. vi + 234pp. $15.95. ISBN 0-86033-003-6. Index;bib.

    C This exceptionally well arranged and written book argues that intelligence tests apply only to a special population group; other tests apply to other groups; and intelligence is not genetically related.

Loehlin, John C., et al. *Race Differences in Intelligence.* Freeman, 1975. xii + 380pp. $12.00; $5.95(p). 75-1081. ISBN 0-7167-0754-3; 0-7167-0753-5. Indexes;CIP.

C   A compendium of fact and commentary which will stand for years as a principal reference on the topic. Appendixes cover explanations of method, details of calculation, summaries of principal investigations, references, etc. Recommended as a source to biology, psychology and sociology teachers, as collateral reading, and for the mature general reader.

Seagoe, May V.   *Terman and the Gifted*. (Illus.) Kaufman, 1975. xiii + 258pp. $10.00. 75-19063. ISBN 0-913232-27-0. Index;CIP.

SH-C   An interesting biography of Lewis M. Terman, a pioneer in the study of mental testing, gifted people, school hygiene, sex differences and marriage. This is an excellent introduction to the history of important and characteristically American facets of the growth of psychology.

154 SUBCONSCIOUS STATES AND PROCESSES

See also 612.82 The Brain and Sleep Phenomena.

Barber, Theodore X., et al.   *Hypnosis, Imagination and Human Potentialities*. Pergamon, 1974. ix + 189pp. $9.50; $5.00(p). 73-19539. ISBN 0-08-017932-0; 0-08-017931-2. Indexes;CIP.

SH-GA-C-P   Phenomena which are reported to have occurred in the "hypnotic state" are examined and tested, and the results of these extensive studies are translated into everyday psychological terms. Of interest to the general reader as well as the professional.

Buranelli, Vincent.   *The Wizard from Vienna: Franz Anton Mesmer*. (Illus.) Coward, McCann&Geoghegan, 1975. 256pp. $8.95. 75-24072. SBN 698-10697-0. Index;CIP.

C   The author puts Mesmer and mesmerism into historical context and traces some of the ramifications in psychiatry, politics and literature. This is a college-level book, recommended for the reader with some background in psychology and philosophy. Very lively and generally clear.

Deming, Richard.   *Sleep, Our Unknown Life*. Nelson, 1972. 142pp. $4.95. 72-5872. ISBN 0-8407-6230-5.

JH-SH-C   The marvels of sleep, with material on dreams, sleep habits, insomnia, drugs and sleep, and such topics as enuresis, snoring and sleep-learning. Well referenced and provocatively written. Suitable for supplementary reading in a high school psychology or junior high science course.

O'Nell, Carl W.   *Dreams, Culture, and the Individual*. Chandler&Sharp, 1976. xi + 88pp. $2.50(p). 76-513. ISBN 0-88316-523-6. Index;CIP.

GA   O'Nell discusses the topic from many angles, from sex, age and cultural differences in dream content to physiological, psychological and social functions of dreams. Well-documented and nontechnical, but with ample references.

Singer, Jerome L.   *The Inner World of Daydreaming*. Harper&Row, 1975. xi + 273pp. $10.00. 74-1856. ISBN 0-06-013907-2. Index;CIP.

C   Singer surveys a quarter of a century's research about daydreaming: frequency, content and structure of daydreaming; experiments on the complex interplay of attention to internal and external events; a review of models of the function of daydreaming; evaluation of Freud's catharsis theory; the origins of daydreaming in the make-believe of children; and the role of daydreaming in considering future courses of action. Useful for nonspecialists with a psychology background.

155 DIFFERENTIAL AND GENETIC PSYCHOLOGY

See also 301.43 –.45 Social Science categories.

155.2 INDIVIDUAL PSYCHOLOGY

Baughman, E. Earl. *Black Americans: A Psychological Analysis.* (Illus.) Academic, 1971. xxi + 113pp. $6.95. 70-152748.

   C   A succinct, scholarly, lucid, and objective essay on some of the major psychological processes and traits characteristic of black Americans. Presents superior discussion on intelligence (including Jensenism), scholastic performance, self-esteem, rage and aggression, psychopathology, and socialization and family structure.

Chown, Sheila M., (Ed.). *Human Ageing: Selected Readings.* Penguin, 1972. 397pp. $3.95(p).

   C-P   Some 30 papers on cognition and personality in aging. All contributions are based on original research. Highly technical, but nonspecialists can learn much from it.

Hauser, Stuart T. *Black and White Identity Formation: Studies in the Psychological Development of Lower Socioeconomic Class Adolescent Boys.* Wiley-Interscience, 1971. xv + 160pp. $9.95. 77-138910. ISBN 0-471-36150-X.

   C-P   An exceptionally well-thought-out 4-year effort to apply objective measures to a complex and highly theoretical psychoanalytic concept. Discussion of the role of negative identification, of diffusion of identity in latency, and of the effects of absent or poorly esteemed fathers is especially pertinent.

Lanyon, Richard I., and Leonard D. Goodstein. *Personality Assessment.* Wiley, 1971. xii + 267pp. $8.95. 75-140552. ISBN 0-471-51740-2.

   C   A guide to the field for college students in testing courses, and especially for educators, ministers and social workers. Difficult concepts are treated both accurately and clearly. Relevant to all social sciences.

Macaulay, J., and L. Berkowitz. *Altruism and Helping Behavior.* Academic Press, 1970. x + 290pp. $11.50. 74-86370.

   C   An interesting collection of research summaries in the area. It varies from the very readable to the extremely dull, but in all cases the information is presented in a relatively nontechnical and thorough manner. For introductory psychology or sociology students.

Osmond, Humphry, et al. *Understanding Understanding.* Harper&Row, 1974. 223pp. $7.95. 72-11876. ISBN 0-06-013239-6.

   C   A new typology of personality is described based on thinking, feeling, intuition, and sensation, and on the distinction introvert/extrovert. While convincingly applied to a number of well-known people, the typology is inadequate for atypical patterns. The book is stimulating, insightful and readable, but sometimes superficial and dogmatic.

Sarason, Irwin G. *Personality: An Objective Approach, 2nd ed.* (Illus.) Wiley, 1972. xvi + 601pp. $10.95. 79-175797. ISBN 0-471-75406-4.

   C   A good presentation sequencing the study of personality into five sections: 1) various theoretical positions, 2) methods of assessment, 3) experimental procedures and findings, 4) developmental processes influencing personality and 5) maladaptive behavior. A balanced, eclectic approach most desirable for undergraduate students in psychology.

Selg, Herbert, (Ed.). *The Making of Human Aggression: A Psychological Approach.* (Illus.;trans.) St. Martin's, 1976 (c.1971). 202pp. $7.95. 75-9499. CIP.

C   A carefully written critical review of current theories of and research into
aggression. Presents psychobiological, psychoanalytical, learning and ethological evidence to show that human aggression is a learned behavior. Includes references to early and recent literature. A useful reference beyond the usual introductory
textbook.

Wiedeman, George H., and Sumner Matison (Eds.). *Personality Development and
Deviation: A Textbook for Social Work.* International Univ. Press, 1975. xvi + 253pp.
73-89439. ISBN 0-8236-4070-1. Indexes.

C   This introductory resource book for social work students is based on traditional Freudian personality theory. There is an interesting chapter on symptom
formation as well as diagnostic categories using clinical case examples. Useful reference provided other points of view are considered.

Wiggins, Jerry S., K. Edward Renner, Gerald L. Clore, and Richard J. Rose. *The
Psychology of Personality.* (Illus.) Addison-Wesley, 1971. xii + 705pp. $10.50. 70-
136129.

C   Four points of view typify theoretical approaches to personality—biological,
experimental, social, and psychometric-trait. The authors elaborate the central
assertions about personality developed with each point of view; adequate and apt
examples from research or clinical literature illustrate assertions. Then the authors
examine four crucial issues from each point of view. Overall, an excellent introductory text that is readable, fair,and covers enormous territory.

155.3 SEX PSYCHOLOGY

See also 301.41 The Sexes and Their Relations.

Gadpaille, Warren, J. *The Cycles of Sex.* Scribner's, 1975. xiv + 496pp. $17.50;
$6.95(p). 75-5981. ISBN 0-684-14216-3; 0-684-14224-4. Index;CIP.

C   This essentially psychoanalytic explanation of sexual development pertains to
the life cycles of the individual from fetal development until death. The author
attempts to integrate relevant aspects of sociology, cultural anthropology, embryology, endocrinology, neurophysiology, etc.

Green, Richard. *Sexual Identity Conflict in Children and Adults.* (Illus.) Basic, 1974.
xxii + 327pp. $15.00. 73-76589. SBN 465-07726-9.

C-P   Detailed studies of adults who want to change their sex are presented, and
there are also studies of children who before puberty show signs characteristic of the opposite sex. Ethics of intervention are considered. An important book in a
specialized field; not easy reading.

Maccoby, Eleanor Emmons, and Carol Nagy Jacklin. *The Psychology of Sex Differences.* Stanford Univ. Press, 1974. xiii + 634pp. $18.95. 73-94488. ISBN 0-8047-
0859-2. Index.

C-P   Researchers in the area and anyone interested in sexual equality should read
this excellent analysis of what is and is not known about sex differences. The
authors present several theoretical frameworks for analyzing the data on intellect,
achievement, social behavior and origins of psychological sex differences. While no
definitive conclusions are reached, this is an extremely important book.

Money, John, and Anke A. Ehrhardt. *Man and Woman, Boy and Girl: The Differentiation and Dimorphism of Gender Identity from Conception to Maturity.* (Illus.)
Johns Hopkins Univ. Press, 1973. xiv + 311pp. $12.50; $3.50(p). 72-4012. ISBN
0-8018-1405-7; 0-8018-1406-5.

C-P   This multidisciplinary approach includes clinical details of almost a thousand cases of hermaphroditism and sexual disorders. The interplay of heredity and environment in the development of gender identity is emphasized. Teachers and parents will find it profitable.

Schaeffer, Dirk L., (Ed.).   *Sex Differences in Personality: Readings.* (Illus.) Brooks/Cole, 1971. vi + 186pp. $4.65(p). 76-139686.

C   Articles on research and theory in the area of sex differences in people, covering perception of self and others, intrapersonal and interpersonal behavior, and sex-related behavior, all classified in terms of their relevance to either Freudian or social learning theories of development.

Wagner, Nathaniel N. (Ed.).   *Perspectives on Human Sexuality: Psychological, Social and Cultural Research Findings.* Behavioral, 1974. x + 517pp. 74-2422. ISBN 0-87705-147-X; 0-87705-148-8. Index;CIP.

C   These 23 research articles provide the reader with a thorough basis to study human sexuality. Excellent social science material.

155.4 DEVELOPMENTAL PSYCHOLOGY—CHILDREN

Despert, J. Louise.   *The Inner Voices of Children.* (Illus.) Brunner/Mazel, 1975. 162pp. $4.95(p). 75-16253. ISBN 0-87630-106-5. CIP.

C   The text and photographs permit the reader to see the world of infants to teens through the developing eyes of healthy, "normal" children. Topics include the birth experience and the newborn's needs, mobility and exploration and toddlerhood with its temper tantrums, siblings and the inevitable potty training. Recommended for all parents and anyone whose work relates to children.

Elkins, David.   *Children and Adolescents: Interpretive Essays on Jean Piaget.* Oxford Univ. Press, 1970. xi + 160pp. $5.95. 70-109931.

C   A collection of graceful essays by one of the foremost interpreters of Jean Piaget's work. Captures the essence of Piaget on this topic.

Feshbach, Seymour, and Robert D. Singer.   *Television and Aggression.* Jossey-Bass, 1971. xviii + 186pp. $8.50. 70-138457. ISBN 0-87589-083-0.

C   An important book about an important question: What effect do aggressive TV programs have on the behavior of children? The authors have observed actual behavior in normal settings with a controlled TV diet. The findings are complex and show the danger of glib generalizations in this area. May be heavy going in places for nonprofessionals, but worth the effort.

Gilbert, Sara D.   *Three Years to Grow: Guidance for Your Child's First Three Years.* (Illus.) Parents', 1972. x + 256pp. $5.95. 72-2314. ISBN 0-8193-0616-9.

SH-C   This descriptive account of the first three years of life will help parents cope with those difficult and important developmental crises in a more relaxing and understanding manner.

Lindgren, Henry Clay.   *Children's Behavior: An Introduction to Research Studies.* (Illus.) Mayfield, 1975. 437pp. $6.95(p). 74-33860. ISBN 0-87484-317-0. Gloss.;index.

C   This collection of readings is intended for students who have no familiarity with psychological research or its interpretation. Topics focus heavily on cognitive development and its modification by social and cultural influences. For anyone interested in child development or educational psychology courses.

Nash, John. *Developmental Psychology: A Psychobiological Approach.* (Illus.) Prentice-Hall, 1970. xii + 583pp. $9.50. 74-86520. ISBN 0-13-208363-9.

C Nash's view of child development is that biological systems shape development as much as environment does. In his view, virtually all behavior originates from innate, but not inflexible, programs of development. Useful to serious students with a background in psychology.

Neisser, Edith G. *Primer for Parents of Preschoolers.* (Illus.) Parents', 1972. 320pp. $5.95. 72-2315. ISBN 0-8193-0618-5.

SH-C Almost all areas of a preschooler's life are discussed, as are the enormous variations encountered in the behavior of a 3- to 5-year old child. Family interactions, sibship coping mechanisms, special problems in only or multiple birth situations, permissiveness, controls and limit-setting are examined.

Scientific American. *The Nature and Nurture of Behavior: Developmental Psychobiology: Readings from Scientific American.* (Illus.) Freeman, 1973. 143pp. $7.00; $2.95(p). 72-11800. ISBN 0-7167-0868-X; 0-7167-0867-1.

SH-C A collection on prenatal development and the capacity of newborns. Includes research on visual development related to birth, and effects of critical events in shaping basic systems of the organism, such as sex, stimulation, arrested vision, and perceptive and sensory feedback. Research on environmental determinants of behavior, such as imprinting, sensory and social stimulation and deprivation, environmental experience, and innate intelligence, is also described.

Singer, Robert D., and Anne Singer. *Psychological Development in Children.* (Illus.) Philadelphia: Saunders, 1969. xv + 437pp. $8.00. 73-81828.

C Child psychology is covered from a developmental point of view, with emphasis on the importance of early and continuous stimulation in a social framework which functions to internalize norms.

Watson, Robert I., and Henry Clay Lindgren. *Psychology of the Child, 3rd ed.* (Illus.) Wiley, 1973. xiii + 491pp. $10.95. 72-4538. ISBN 0-471-92240-4. Indexes.

C This book contains theory and methodology as well as references to everyday experience. Classic and current research is well integrated in coverage of principles of child development; socialization, social learning and personality development; infancy; the preschool years, and parental influences, peer influences, and school relationships.

Weiner, Irving B., and David Elkind. *Child Development: A Core Approach.* (Illus.) Wiley, 1972. vii + 247pp. $6.00. 78-37368. ISBN 0-471-92571-3.

C An unusual, somewhat controversial approach to presenting the psychology of child development. Each of four stages (infancy, preschool, middle childhood, adolescence) is discussed from the point of view of physical and mental growth, personality and social development, individual and group differences, and abnormal development.

Wender, Paul H. *The Hyperactive Child: A Handbook for Parents.* Crown, 1973. viii + 120pp. $3.95. 72-96670. ISBN 0-517-50322.

C Contains useful descriptions of the characteristics and development of hyperactive children, recommended treatment and excellent advice on finding competent professional help. The advice to parents about handling hyperactive children is sensible and clear and relies extensively on rewarding desirable and ignoring undesirable behavior.

White, Burton L. *The First Three Years of Life.* (Illus.) Prentice-Hall, 1975. xiii + 285pp. $10.00. 75-22159. ISBN 0-13-319178-8. Index;CIP.

SH-C   White guides parents through aspects of recent research in child behavioral development. He considers seven phases of a child's educational development and makes explicit suggestions and behavioral recommendations.

155.45 EXCEPTIONAL CHILDREN

See also 371.9 Special Education and 362.3 Mental Retardation.

Baroff, George S. *Mental Retardation: Nature, Cause, and Management.* (Illus.) Hemisphere (distr.: Halsted/Wiley), 1974. xvi + 504pp. $17.95. 74-14877. ISBN 0-470-05404-2. Indexes;CIP.

C-P   The title of this outstanding book accurately describes its contents. It will be useful for several audiences. As a comprehensive text for courses on mental retardation, it must be counted as one of the best. Concerned laypersons will find it a lucid and balanced source of information.

Blodgett, Harriet E. *Mentally Retarded Children: What Parents and Others Should Know.* Univ. of Minnesota Press, 1971. 165pp. $5.95. 72-152301. ISBN 0-8166-0612-9.

SH-C-P   For those parents of the retarded who have worked through their feelings sufficiently, but also for medical students, physicians in primary care, college and possibly senior high school students, this is a very useful and accurate, clearly written volume on the nature, education, and sociology of persons with mental retardation.

Burlingham, Dorothy. *Psychoanalytic Studies of the Sighted and the Blind.* International Univ. Press, 1972. vi + 396pp. $15.00. 76-184213. ISBN 0-8236-4510-X

C-P   Examines the complexities of the mother-child relationship, the role of the mother in the analyses of children, and the advantages and disadvantages of including the mother in the child's treatment. The difficulties of raising a blind child are examined.

Cantwell, Dennis P. (Ed.). *The Hyperactive Child: Diagnosis, Management, Current Research.* Spectrum (distr.: Halsted), 1975. xii + 209pp. $12.95. 74-34352. ISBN 0-470-13441-0. Index;CIP.

C   This is a well-written, balanced attempt to explain the world of the hyperactive child from an interdisciplinary viewpoint. Recommended for parents and others wishing a general introduction to the subject.

Graziano, Anthony M. *Child Without Tomorrow.* (Illus.) Pergamon, 1974. xv + 290pp. $12.00; $6.95(p). 73-3394. ISBN 0-08-017085-4; 0-08-0177220-0. Index;CIP.

C   The author chronicles his experiences developing and implementing a program based on behavior modification as applied to severely disturbed children. First-hand experience is used to underscore the potentially destructive role of mental health agencies' interruption of innovative, but successful, treatment programs. Useful to the mental health professions as well as to parents of disturbed children.

Lane, Harlan. *The Wild Boy of Aveyron.* (Illus.) Harvard Univ. Press, 1976. 351pp. $15.00. 75-34080. ISBN 0-674-95282-0. Index;CIP.

C   This highly readable, exciting book begins with the first publication of newly discovered and translated contemporary source material about Victor, a "wild" boy found in France in 1800. Lane discusses significant issues in philosophy of education, education of the deaf and of the retarded from a behavior modification point of view.

155.5 PSYCHOLOGY OF ADOLESCENTS

Albrecht, Margaret. *Parents and Teen-Agers: Getting Through to Each Other.* (Illus.) Parents', 1972. x + 288pp. $5.95. 72-2318. ISBN 0-8193-0624-X.

JH-SH   Albrecht speaks to both parents and adolescents about the maze of complex behaviors and differential growth processes that both groups must mutually negotiate. A relaxed overview of relevant issues, ranging from the "identity" problem, sex and drugs, to the formation of adolescent groups and the escape of some adolescents into marriage with its multi-faceted difficulties.

Comer, James P., and Alvin F. Poussaint. *Black Child Care: How to Bring up a Healthy Black Child in America: A Guide to Emotional and Psychological Development.* Simon&Schuster, 1975. 408pp. $9.95. 74-28261. ISBN 0-671-21902-2. Index;CIP.

SH-C   This handbook will be useful not only to black parents, but also to their adolescent children and Americans in general. Using an easy-to-read, question-and-answer format, the aim is to help readers cope with the emotions of childhood and racial issues. Answers tend to be a bit shallow and reassuring, but the section on adolescence is excellent.

Dragastin, Sigmund E., and Glen H. Elder, Jr. (Eds.). *Adolescence in the Life Cycle: Psychological Change and Social Context.*(Illus.) Hemisphere (distr.: Halsted/Wiley), 1975. xi + 324pp. $14.95. 74-22002. ISBN 0-470-22155-0. Indexes;CIP.

C-P   These papers are concerned with adolescence as a developmental process, subject to anatomical, psychological, and sociological influences. Problems such as erroneously applying research done on males only to females and influence of socialization and conflicting expectations on females are carefully examined.

Hendin, Herbert. *The Age of Sensation.* Norton, 1975. xxii + 354pp. $9.95. 75-19103. ISBN 0-393-01122-4. Index;CIP.

C   Through interviews with 400 Columbia University students and with emphasis on psychoanalytic interpretation of dreams, Hendin shows how students attempted to cover anger from perceived rejection by parents with a variety of maneuvers including drug abuse, suicide, impotence—even revolution. Most useful as collateral reading in college courses in personality, adolescent and family psychology.

Minton, Lynn. *Growing Into Adolescence: A Sensible Guide for Parents of Children 11 to 14.* (Illus.) Parents', 1972. 288pp. $5.95. 73-190367. ISBN 0-8193-0622-3.

JH-SH-C   A guide to understanding for the concerned parent of pre-adolescents, but it is of value for educators, counselors, and others working with this age group. The book is broad in scope and specific in detail in terms of the important emotional crises at this time of life.

Thornburg, Hershel D., (Ed.). *Contemporary Adolescence: Readings.* (Illus.) Brooks/Cole, 1971. xi + 419pp. $5.75(p). 79-141368. ISBN 0-8185-0007-7.

C   Thornburg presumes that adolescence can best be understood in the context of contemporary biology, psychology and culture, and deals with intellect, sex, family, school, peers, values, delinquency, drugs, activism, work, and personality. One chapter explains the meaning and interpretation of research and statistical terms. Excellent reading for parents, educators and laypersons.

155.937 DYING

See also 170 Ethics.

Grollman, Earl A. *Talking About Death: A Dialogue Between Parent and Child, revised* *ed.* (Illus.) Beacon, 1976. iii + 98pp. $9.95; $3.95(p). 75-37786. ISBN 0-8070-2372-8; 0-8070-2373-6. CIP.

SH-C   An unusually warm and sensitive guide. Contains a section specifically directed to young children; a section of comprehensive guidelines and explanations to the parents; a listing of community, theological and professional organizations available to help mourners; and an age-correlated bibliography of books, films and cassettes.

Kübler-Ross, Elisabeth. *Death: The Final Stage of Growth.* Prentice-Hall, 1975. xxii + 175pp. $7.95; $2.95(p). 75-9688. ISBN 0-13-197012-7; 0-13-196998-6. CIP.

SH-C-P   Kübler-Ross deals with the feelings and philosophies which emerge in those preparing to die and those closely associated with them. The book is a collection of essays which examine dying as a source of growth for the person dying as well as those around him or her. For laypersons and mental or medical health professionals who want to go beyond intellectualizing about dying and death.

Lifton, Robert Jay, and Eric Olson. *Living and Dying.* (Illus.) Praeger, 1974. 156pp. $6.50. 72-85973. ISBN 0-275-25810-3. Index;CIP.

C   The authors present an anecdotal narrative which attempts to identify changing attitudes toward death in American society. They conclude that the same concern for life and death still exists, although specific degrees of concern and its manifestations change as society changes. Recommended for students and health professionals who anticipate encounters with death.

Mills, Gretchen C., et al. *Discussing Death: A Guide to Death Education.* ETC, 1976. iv + 140pp. $8.50; $5.50(p). 75-17885. ISBN 0-88280-026-4; 0-88280-027-2(p). Index;CIP.

SH-C-P   This book provides curricular aid in educating children and young adults about the nature of death and how to deal with it. It contains concise descriptions of 64 different learning experiences which deal with elementary physiology, attitudes and feelings concerning death and old age, etc. Material on autopsies, wills, transplants and causes of death is included.

Schoenberg, Bernard, et al. (Eds.). *Anticipatory Grief.* Columbia Univ. Press, 1974. xvi + 381pp. $12.50. 74-1252. ISBN 0-231-03770-8. Gloss.;index;CIP.

C-P   The process of grieving in anticipation of the demise of self or others is examined in depth in this collection of 41 papers. Topics range from a personal account of intense grief to a 7-stage model of grief. Poorly integrated, but considerable scholarly merit.

Shepard, Martin. *Someone You Love Is Dying: A Guide for Helping and Coping.* (Illus.) Harmony/Crown, 1975. 219pp. $7.59. 75-22088. ISBN 0-517-52375-2. CIP.

C   Shepard's major thesis is that open and honest communication among the dying, the attending physicians, family and close friends can result in much less trauma, fear and anxiety for all concerned. Many chapters include interviews with people who are dying or who have come close to death.

Weisman, Avery D. *On Dying and Denying: A Psychiatric Study of Terminality.* Behavioral Publications, 1972. xvii + 247pp. $9.95. 79-174268. ISBN 87705-068-6.

C-P   Weisman interviewed over 350 dying patients from 1962–65. He identifies unconscious thoughts and feelings which dominate what appear as rational, conscious thoughts, and he reviews basic concepts and common misconceptions.

156 COMPARATIVE PSYCHOLOGY

See also 519.5 and 591.52 Animal Ecology and Behavior.

Alcock, John. *Animal Behavior: An Evolutionary Approach.* (Illus.) Sinauer, 1975. xi + 547pp. $14.00. 74-19892. ISBN 0-87893-022-1. Indexes.

C-P   The author takes an evolutionary approach to animal behavior and presents a sufficient amount of material on the opposing positions for understanding many of the contemporary issues in animal behavior. There is a review of the major points of each chapter. A useful reference book and a fine text.

Amon, Aline. *Reading, Writing, Chattering Chimps.* (Illus.) Atheneum, 1975. 118pp. $7.95. 75-9524. ISBN 0-689-30472-2. CIP.

JH   This is an up-to-date account of efforts to teach chimps various ways of communication. The book is well written, well illustrated, well organized and continuously interesting.

Carr, Donald E. *The Forgotten Senses.* Doubleday, 1972. xi + 347pp. $7.95. 70-157578.

SH-C   Studies by chemists, physicists, neurologists, physiologists, psychologists and naturalists are brought together in this fascinating and provocative book. The basic thesis is that much inexplicable animal behavior may arise because sensory equipment of animals is tuned to signals different from those perceived by humans. Observations concern mainly vision, hearing, chemoreception, touch, temperature and pain.

Chinery, Michael. *Animal Communities.* (Illus.) Watts, 1973. 128pp. $5.95. 72-1891. ISBN 0-00-100167-1. SBN 531-02107-6.

JH   Beginning with animal language and parental care, he develops the concept of an animal community and the evolution of social behavior. The format, the excellent illustrations, and the numerous examples make this an excellent introduction to ethology.

Cohen, Daniel. *Animal Territories.* (Illus.) Hastings House, 1975. 95pp. $5.95. 75-11540. ISBN 0-8038-0368-0. Index;CIP;bib.

JH   Cohen develops the concept of territoriality from an evolutionary perspective, highlighting the text with many fine examples.

Cohen, Daniel. *Watchers in the Wild: The New Science of Ethology.* (Illus.) Little, Brown, 1971. 178pp. $5.95. 78-129908.

JH   An elementary overview of ethology. Sufficient methodological detail is blended with a qualitative treatment of selected data to give the reader a broad, digestible impression of the way in which systematic observations of animal behavior should be conducted. Demonstrates a healthy enthusiasm that could motivate young readers toward a thoughtful and inquiring study of their environment.

Cousteau, Jacques. *Attack and Defense.* (Illus.) World, 1973. 144pp. $7.95. 72-87710. ISBN 0-529-05073-0.

JH-SH   The book is concerned with animal behavior in the sea. The text is descriptive and authentic and correlates well with the many excellent photographs.

Droscher, Vitus B. *The Friendly Beast: Latest Discoveries in Animal Behavior.* (Illus.) Dutton, 1970. 248pp. $8.95. 76-95487. ISBN 0-525-10954.

SH   An excellent and extremely readable summary and analysis of recent work and thought in the field of animal behavior. There is skillful blending of the classic field studies with recent studies by anthropologists, zoologists, comparative and experimental psychologists, ethologists, and neurophysiologists. Although a wide variety of species and behaviors are treated, primary emphasis is on social behavior.

Fichtelius, Karl-Erik, and Sverre Sjölander. *Smarter Than Man? Intelligence in Whales, Dolphins, and Humans.* (Illus.; trans.) Pantheon, 1973. xiv + 205pp. $6.95. 72-3416. IBN 0-394-38149-6.

C-P   This unusual book attacks the notion of the superiority of man and compares humans with the sperm whale. Includes sections on man's place on earth, a discussion of large-brained animals, the history and behavior of the sperm whale, and man and the tree of knowledge. For those interested in animal behavior or environmental studies, or for pleasant reading.

Ford, Barbara. *Can Invertebrates Learn?* (Illus.) Messner, 1972. 96pp. $4.79. 72-1422. ISBN 0-671-32531-0.

JH   An excellent presentation of the scientific method, with the reader doing part of the work. Learning is first defined operationally and then experimental examples for representative members of each phylum are given. Learning is related to complexity of nerve structure and the presence of a central brain with some memory capacity. Student experiments are suggested and described. For either reading alone or as a part of a junior high school science course.

Fox, M.W. *Concepts in Ethology: Animal and Human Behavior.* Univ. of Minnesota Press, 1974. xvii + 139pp. $8.50. 73-93834. ISBN 0-8166-0723-0. Index.

SH-C-P   Provides a current perspective on the interdisciplinary study of human and animal behavior. Fox uses his vast experience in canid behavior to compare changes in animal behavior to the evolutionary development of man. This comparative behavioral approach will be useful to anyone interested in relationships and parallels between animal and human behavior.

Klopfer, Peter H. *On Behavior: Instinct Is a Cheshire Cat.* (Illus.) Lippincott, 1973. 160pp. $5.25. 73-4642. ISBN 0-397-31449-3.

JH-SH   Klopfer considers the concept of instinct and then presents an historical survey of some other ways in which scientists have conceptualized and studied the factors underlying human behavior. He describes numerous experiments with various animals on imprinting, early learning, infant-mother love, habitat preferences and the importance of play.

Lorenz, Konrad. *Studies in Animal and Human Behavior, Vol. I.* (Illus.) Harvard Univ., 1970. xx + 403pp. $10.00. ISBN 0-674-84630-3.

SH-C-P   Six classic papers (1931–1942) by the founder of ethology (the comparative study of the behavior similarities and differences of closely related species) are presented here. Lorenz clarifies the difference between instincts (innate motivational factors) and instinctive behaviors, which are clearly observable, and denies that all human behavior is learned.

Miller, Libuse Lukas. *Knowing, Doing, and Surviving: Cognition in Evolution.* Wiley, 1973. xii + 343pp. $11.50. 73-1198. ISBN 0-471-60512-3.

C-P   Miller incorporates data from many different fields (art, psychology, religion, etc.) and attempts to develop a theory of knowledge which is applicable to animals as well as man. The examples are well chosen. Useful collateral reading in courses on cognition and/or human development.

National Geographic Society *The Marvels of Animal Behavior.* (Illus.) National Geographic Society, 1972. 422pp. $12.65. 72-76734. ISBN 87044-105-1.

SH-C   In this attractive book, with magnificent color photographs and drawings, 22 authorities on animal behavior present outstanding findings which provide an excellent historical review, world-wide in scope, of the development of ethology. The book is authoritative, exceptionally readable and relates animal behavior research to possible applications in understanding human behavior.

Norris, Kenneth S.   *The Porpoise Watcher.* (Illus.) Norton, 1974. 250pp. $7.95. 74-1329. ISBN 0-393-06385-2. CIP.

   SH-C   The author's account of his 20-year work with porpoises includes a summary of recent research in porpoise communication as well as a warning of the dangers posed to the world's porpoise population by commercial tuna fishing activities.

Pryor, Karen.   *Lads Before the Wind: Adventures in Porpoise Training.* (Illus.) Harper&Row, 1975. ix + 278pp. $8.95. 74-15846. ISBN 0-06-013442-9. CIP.

   SH-C-P   Pryor recounts her adventures as a porpoise trainer in Hawaii. She explains the shaping of behavior, criteria, random schedule of reinforcement, stimulus control and extinction in an excellent demonstration of operant conditioning.

Restle, Frank.   *Learning: Animal Behavior and Human Cognition.* (Illus.) McGraw-Hill, 1975. xii + 330pp. $10.95. 74-16006. ISBN 0-07-051910-2. Index;CIP.

   C   Through the use of "behavior flowcharts" Restle integrates the diverse phenomena of human and nonhuman learning. The classic topics are presented and are fitted to an algorithmic structure which also encompasses human choice and decision-making, imagery, grammar and verbal learning. A good text for introductory courses in the psychology of learning.

Rowell, Thelma.   *The Social Behavior of Monkeys.* (Illus.) Penguin, 1973. 203pp. $2.35(p).

   SH-C   Describes special structural and behavioral properties of primates, especially baboons; makes comparisons with other species; and discusses the rationale for most primate studies. Important insights are presented, including the preeminence of visual acuity in the development of complex social groups. The author observed a free-living population of baboons in Uganda for 5 years.

Scientific American.   *Animal Behavior: Readings from* Scientific American. (Illus.) Freeman, 1975. 339pp. $13.00; $7.95(p). 75-2383. ISBN 0-7167-0511-7; 0-7167-0510-9. Index;CIP.

   SH-C   These readings from 1950–1975 focus on the analytical study of whole patterns of behavior in terms of broad nervous system processes and social and environmental interactions. Includes thorough introductory essays, brief abstracts, and excellent schematic diagrams and illustrations.

Scott, John Paul.   *Animal Behavior, 2nd ed., revised.* (Illus.) Univ. of Chicago Press, 1972. xi + 281pp. $12.50; $3.25(p). 57-6989. ISBN 0-226-74334-9. Bib.

   JH-SH-C   This revised edition presents encyclopedic information about all aspects of the subject. The language is simple, often anecdotal. The text is multidisciplinary in approach and thoroughly explains the methodology of studying animal behavior as well as its historical development.

Temerlin, Maurice K.   *Lucy: Growing Up Human: A Chimpanzee Daughter in a Psychotherapist's Family.* (Illus.) Science&Behavior, 1975. xxii + 216pp. $8.95. 75-12455. ISBN 0-8314-0045-5.

   JH-SH-C   This is an account of two species which met, shared, profited and retained their uniqueness. *Lucy* is a fascinating book which is highly recommended for anyone who may wonder what it might be like to live with our closest cousins on the phylogenetic tree.

Van Somers, Peter.   *The Biology of Behaviour.* (Illus.) Wiley, 1972. xii + 184pp. $3.95(p). 72-75661. ISBN 0-471-90045-3; 0-471-89946-1(p).

C The evolutionary basis of behavior provides the central theme of this text for biology and physiological psychology. The book takes both an organismal and a population view of behavior, covering such topics as the genetics of species and natural selection, ecology, homeostasis, sleep, sexual and aggressive behavior, territoriality, dominance, stress and conflict. Novel, direct, readable and useful; highly recommended for both students and teachers.

157 ABNORMAL AND CLINICAL PSYCHOLOGY

See also 616.89 Psychiatry and categories following 361 Social Welfare.

Davids, Anthony. *Abnormal Children and Youth: Therapy and Research.* (Illus.) Wiley, 1972. xii + 211pp. $11.95. 72-6092. ISBN 0-471-19695-9.

    C-P This collection of papers presents the work of psychologists and educators in varied settings devoted to the treatment of deviant children and adolescents. Davids makes no assumptions about the reader's prior knowledge, and so he writes simply, lucidly and to the point. Excellent presentations of the different theoretical approaches and practical applications, excellent integration of theories and research findings, good literature reviews and fascinating case studies.

Fieve, Ronald R., et al. (Eds.). *Genetic Research in Psychiatry.* (Illus.) Johns Hopkins Univ. Press, 1975. xv + 301pp. $16.50. 74-24394. ISBN 0-8018-1660-2. Indexes;CIP

    C-P Data and inference are clearly separated and labeled, and basic applied genetics is given up-to-date treatment in sections on the biochemical basis of genetic disorders and on the genetic studies of schizophrenia, affective and behavioral disorders. Useful as a reference for instructors and students in the social sciences.

Hoch, Erasmus L. *Experimental Contributions to Clinical Psychology.* (Illus.) Brooks/Cole, 1971. viii + 380pp. $12.65. 70-157431. ISBN 0-8185-0011-5.

    C-P A wise and witty discussion of the manipulation of human behavior.

Luria, A.R. *The Man with a Shattered World: The History of a Brain Wound.* (Illus.) Basic, 1972. xx + 165pp. $6.95. 78-174809. SBN 465-04371-2.

    C This case study provides priceless information on the life-struggle of a brain-damaged patient who lost his memory and the ability to speak, read and write. Recommended for all who work with brain-damaged people.

Mausner, Bernard, and Ellen S. Platt (Assisted by Judith S. Mausner). *Smoking: A Behavioral Analysis.* (Illus.) Pergamon, 1971. xiii + 238pp. $14.00. 72-119599. ISBN 0-08-016397-1.

    C Includes a series of psychological investigations into the history of smoking and the results of an experiment on the effect of role-playing on smoking behavior. A scholarly approach to behavior, useful to people trying to change any behavior requiring the sacrifice of present pleasure for future benefit.

Mercer, Jane R. *Labeling the Mentally Retarded: Clinical and Social System Perspectives on Mental Retardation.* (Illus.) Univ. of California Press, 1973. xiii + 319pp. $10.95. 70-182795. ISBN 0-520-02183-5.

    C-P Presents excellent clinical data from a study of individuals with limited skills in coping. Significant findings document the extreme cautiousness with which a diagnosis of mental retardation should be applied to individuals who are apparently mildly retarded but who seem to have no evidence of organic brain disease or of other medical conditions known to produce retardation.

Salzinger, Kurt. *Schizophrenia: Behavioral Aspects.* Wiley, 1974. x + 181pp. $7.25; $4.25(p). 73-1276. ISBN 0471-75091-3; 0-471-75090-5. Index;CIP.

C-P   A categorization of empirical studies of schizophrenia, this book contains a lucid summary of major findings and conclusions. Salzinger discusses such topics as language functions, stimulus control, psychomotor behavior therapy and theories of schizophrenia. For students in the psychological and social sciences.

Suinn, Richard M.   *Fundamentals of Behavior Pathology, 2nd ed.* (Illus.) Wiley, 1975. 595pp. $12.95. 74-30206. ISBN 0-471-83547-1. Glosses.;indexes;CIP;bibs.

SH-C   The exposition is lucid, the treatment sound and the range and breadth of materials encompassed is truly representative. Major strengths lie in the emphasis on possible genetic factors in psychopathology and a discussion of childhood psychopathology and mental retardation, areas lacking in other texts.

Summers, Marcia J., et al.   *Our Chemical Culture: Drug Use and Misuse.* (Illus.) STASH, 1975. xv + 116pp. $5.00(p). 75-15090. ISBN 0-915818-07-8. Gloss.

SH-C   Addressing themselves to their peers, these student authors take the position that drugs are, in themselves, neither good nor bad, but do provide people with the potential for committing constructive and/or destructive acts, and therefore, society really faces a people problem rather than a drug problem.

157.744 SUICIDE

Choron, Jacques.   *Suicide.* Scribner's, 1972. 183pp. $7.95. 75-162757. ISBN 0-684-12577-3.

C   This compact, scholarly book deals with the problem of suicide, its motives, causes and prevention. Considers theological, anthropological, psychological, sociological, and psychiatric aspects. The book is uncluttered by technical jargon.

Haim, Andre.   *Adolescent Suicide.* (Trans.) International Univ., 1975. xiii + 310pp. $15.00. 73-9250. ISBN 0-8236-0090-4. Index.

C-P   The volume's four parts cover (1) classical theories about suicide, definitions, parameters and relevant worldwide statistics; (2) psychoanalytic analysis of attitudes toward suicide; (3) circumstantial, sociological and psychiatric factors which have been advanced as the causes of suicide; (4) the author's hypotheses and conclusions.

McCulloch, J. Wallace, and Alistair E. Philip.   *Suicidal Behaviour.* (Illus.) Pergamon, 1972. x + 123pp. $7.75. 72-188140. ISBN 0-08-016855.

C-P   Covers social, psychiatric and psychological data and ecological correlates of suicidal behavior; sequelae of suicidal behavior; and implications for prevention. Crucial social variables include childhood foster care, multiple school absences, low rate of home ownership, overcrowded living conditions and others.

Shneidman, Edwin S., (Ed.).   *Death and the College Student: A Collection of Brief Essays on Death and Suicide by Harvard Youth.* Behavioral Publications, 1972. xix + 207pp. $9.95; $4.95(p). SBN 87705-083-X; 87705-083-4.

C   Covers contemporary contemplations of death, community studies relating to death and suicide, some sequelae of death, personal reflections about self-destruction, thoughts of contemporary philosophers, and films and books on death. The essays were written by undergraduates enrolled in a course taught by the editor. The editor concludes that what happens in society is much less a cause for suicidal behavior than individual experiences.

158 APPLIED PSYCHOLOGY

Evans, Idella M., and Patricia A. Smith.   *Psychology for a Changing World.* (Illus.) Wiley, 1970. xiii + 444pp. $7.95. 72-110167. ISBN 0-471-24870-3.

SH-GA   Light reading; without technical or highly theoretical discussions. The book is organized around the daily problems of people and how psychologists can help. Not adequate as a college text but good material to introduce the field to high school students or adults.

Foster, Carol. *Developing Self-Control.* (Illus.) Behaviordelia, 1974. viii + 135pp. $4.25(p). ISBN 0-914-47412-X.

SH-C   Foster offers a recipe for the application of behavior control techniques to one's own behavior. Book is designed to teach methods of self-control through the methods of behavior modification in a "programmed learning" fashion.

Freedman, Jonathan L. *Crowding and Behavior.* Freeman, 1975. viii + 177pp. $4.50(p). 75-20217. ISBN 0-7167-0750-0. Index;CIP.

SH-C   The popular conception that people are at their worst in a crowded situation is dispelled here. Freedman presents theory and research on crowding, discusses and effectively dismisses the hypothesis that human behavior is best explained by analogies with animal behavior, and presents some supportive data. Could serve as a text in a wide variety of courses.

Kilpatrick, William. *Identity and Intimacy.* Delacorte, 1975. 262pp. $8.95; $3.25(p). 75-8681. ISBN 0-440-04373-5. Index;CIP.

C-P   Kilpatrick launches a thoughtful and provocative assault on the human potential movement. His notions of identity and its relationship to love, intimacy, responsibility and achievement are deeply rooted in Erik Erikson's ego psychology.

Laird, Donald A., et al. *Psychology: Human Relations and Motivations, 5th ed.* (Illus.) McGraw-Hill, 1975. 449pp. $10.95. 74-22348. ISBN 0-07-036015-4. Index;CIP.

SH-C   The content of this fifth edition (previously titled *Practical Business Psychology*) has not changed; the focus is still on the world of work: orientation, personal efficiencies and performance, job adjustment, human relations and leadership.

Murphy, Gardner, with Morton Leeds. *Outgrowing Self-Deception.* (Illus.) Basic, 1975. xv + 175pp. $8.95. 74-79277. ISBN 0-465-05390-4. Index;CIP.

C   Evidence from diverse fields has been well organized into a stimulating text. Most intriguing is a chapter on everyday ways of coping with self-deception. One drawback is inadequately defined technical terms interspersed in a basically nontechnical work.

Parker, Rolland S. *Emotional Common Sense: How to Avoid Self-Destructiveness.* Harper&Row, 1973. xiv + 219pp. $6.95. 72-9758. ISBN 0-06-013278-7.

SH-GA   The thesis is that minimizing self-destructive reactions is the critical variable in promoting personal and social well-being. Emphasis is given to ways of coping with stress, guilt, anxiety, feelings of worthlessness, loneliness, interpersonal antagonism, irrational anger, depression, inadequate self-assertiveness and sexual inadequacy. A final chapter presents some sound guidance on choosing an appropriate therapist "if all else fails." An eclectic and humane book.

Pietsch, William V. *Human BE-ing: How to have a Creative Relationship Instead of a Power Struggle.* (Illus.) Lawrence Hill, 1974. 254pp. $7.95. 73-20379. ISBN 088208-042-3.

JH-SH-C   This is a fine human relations book which presents a great deal of insightful information in a delightful way. Pietsch offers readers some new insight into why others react the way they do. Recommended for workshops and laboratories of human dynamics.

Skemp, Richard R. *The Psychology of Learning Mathematics.* (Illus.) Penguin, 1971. 319pp. $2.25(p). ISBN 0-14-021210-8.

 SH-C   An interesting attempt to deal with the teaching of mathematical concepts using the psychologists' expertise concerning human learning. The book is well conceived, accurately and carefully written at a level easily comprehensible to the nonpsychologist interested in the problems of teaching.

Tanner, Ira J. *Loneliness: The Fear of Love.* Harper&Row, 1973. xiii + 143pp. $5.95. 72-9158. ISBN 06-014218-9.

 C   The author uses the Transactional Analysis approach to discuss loneliness. He equates loneliness with a fear of love and views it as a disease which has now reached epidemic proportions. The nature of loneliness is traced from early childhood through various stages of life. Useful as ancillary reading for psychology and sociology courses.

Wheelis, Allen. *How People Change.* Harper&Row, 1973. 117pp. $5.00. 72-11245. ISBN 0-06-014558-7.

 SH-C   Wheelis suggests that even given the extreme of circumstances or outside force, we still have choices. Freedom depends upon awareness, and the therapist may assist the patient to extend that awareness. The author offers a knowledgeable and authoritative conviction that, given the insights and will to act, change and freedom for the individual are possible.

170 Ethics

170.202 CONDUCT OF LIFE

Behnke, John, and Sissela Bok (Eds.). *The Dilemmas of Euthanasia.* (Illus.) Anchor/Doubleday, 1975. viii + 187pp. $2.95. 75-5267. ISBN 0-385-09730-1. Index;CIP.

 SH-C-P   The book's seven chapters are separate but complimentary essays on the medical, legal and philosophic problems relating to euthanasia. The six appendixes include a will requesting "death with dignity," a statement on patients' rights, a definition of irreversible coma and legal documents pertaining to the problem of euthanasia.

Dempsey, David. *The Way We Die: An Investigation of Death and Dying in America Today.* Macmillan, 1975. xii + 276pp. $9.95. 75-23128. ISBN 0-02-530750-9. Index;CIP.

 SH-C   Dempsey reviews an extensive literature on such subjects as the use of heroic measures in medicine, transplants and other attempts to increase longevity, euthanasia, preparing for death and ways of caring for the terminally ill. Suitable for layreaders and those interested in death education.

Fox, Renee C., and Judith P. Swazey. *The Courage to Fail: A Social View of Organ Transplants and Dialysis.* Univ. of Chicago Press, 1974. xviii + 395pp. $12.95. 73-89788. ISBN 0-226-25947-1. Index.

 SH-C   Examines sociological aspects and ethical concerns of kidney donors and recipients. Recommended for collateral reading in sociology.

Russell, O. Ruth. *Freedom to Die: Moral and Legal Aspects of Euthanasia.* Human Sciences/Behavioral, 1975. 352pp. $14.95. 74-8946. ISBN 0-87705-216-6. Index;CIP.

 SH-C   This careful book reviews attitudes toward euthanasia before 1930 and in the decades of swift medical technological change thereafter. Traditional arguments as well as legislative proposals to resolve the dilemma faced by doctors and relatives are examined.

Sarvis, Betty, and Hyman Rodman. *The Abortion Controversy.* Columbia Univ. Press, 1973. 222pp. $8.95. 72-12534. ISBN 0-231-03656-6.

   SH-C   This balanced, well-written synopsis of both sides of the abortion controversy illustrates the polemical nature of most discussions of this topic. The authors document how reform laws earlier expanded the grounds for legal abortion but created problems of interpretation in situations involving medical, psychiatric and social factors.

174 Professional Ethics

Etzioni, Amitai. *Genetic Fix.* Macmillan, 1973. 276pp. $7.95. 73-7350.

   SH-C   Asks some pointed questions about genetic engineering and its ethical and sociological implications.

Gorovitz, Samuel, et al. (Eds.). *Moral Problems in Medicine.* Prentice-Hall, 1976. xxiv + 552pp. $11.50. 75-42150. ISBN 0-13-600817-8. Index;CIP.

   C   Sources ranging from Seneca's *Moral Letters* to articles in current medical journals cover most of the ethical problems related to health matters. Organized by broad philosophical topics, the volume could be used for college-level ethics, philosophy and religion courses.

Hilton, Bruce, et al. (Eds.). *Ethical Issues in Human Genetics: Genetic Counseling and the Use of Genetic Knowledge: Proceedings of a Symposium.* (Illus.) Pienum, 1973. xi + 455pp. $14.95. 72-93443. ISBN 0-306-30715-4. Index;gloss.

   C   This symposium dealt with the technical aspects of amniocentesis, analysis of genetic defects, the culture of amniotic fluid cells, the risks of amniocentesis, and the technical, legal and ethical problems that arise during screening for defects in homozygotes and heterozygotes. An important overview by leading professionals in medicine, law, theology, ethics, philosophy and sociology, presented candidly and from many viewpoints.

Hixson, Joseph. *The Patchwork Mouse.* Anchor/Doubleday, 1976. x + 228pp. $7.95. 74-27583. ISBN 0-385-02852-0. Index;CIP.

   SH-C-P   The story of a young dermatologist who was caught misrepresenting data by discoloring patches of skin grafted from one mouse to another. Elucidates some of the very complex relationships between science and public policy.

Jones, Alun, and Walter F. Bodmer. *Our Future Inheritance: Choice or Chance?*(Illus.) Oxford Univ. Press, 1974. ix + 141pp. $13.00; $4.00(p). ISBN 0-19-857384-7;0-19-857390-1. Index.

   SH-C-P   Such topics as artificial insemination, artificial fertilization, genetic screening and selective abortion, organ transplants and their legal and economic aspects are discussed. The authors then attempt to put these topics into perspective and offer some possible solutions to the inherent problems.

Mann, Kenneth W. *Deadline for Survival: A Survey of Moral Issues in Science and Medicine.* Seabury Press, 1970. vii + 147pp. $2.95(p). 77-125835.

   SH-C   Mann argues that there must be a marriage of all forces, from technological to economic to moral to the integrity of man himself. An excellent documentary.

Restak, Richard M. *Premeditated Man: Bioethics and the Control of Future Human Life.* Viking, 1975. xviii + 202pp. $8.95. 75-16171. ISBN 0-670-57333-7. Index;CIP.

   C-P   Restak is concerned with the need to define the allowable limits of scientific power. He raises such questions as, what does one do about the carrier of a defective gene? What price are we willing to pay for a perfect baby? The right of the individual to accept or reject genetic information and its social, moral and psychological consequences are discussed in this thought-provoking book.

179 CRUELTY

Carson, Gerald. *Men, Beasts, and Gods: A History of Cruelty and Kindness to Animals.* (Illus.) Scribner's, 1972. x + 268pp. $8.95. 72-1216. ISBN 684-13039-4.

JH-SH-C   Traces the history of man-animal interactions from prehistory to the present, including the religious and mythological uses of animals. Animal cruelty (domestic, sport, scientific and commercial exploitation) for the benefit of man is well documented and analyzed. Discusses various social pressure groups and legislature developments for the protection of animals.

The Social Sciences

300.07 THE SOCIAL SCIENCES—STUDY AND TEACHING

Bunker, Barbara Benedict, et al. *A Student's Guide to Conducting Social Science Research.* Human Sciences/Behavioral, 1975. 120pp. $3.95(p). 74-11814. ISBN 0-87705-236-0; 0-87705-238-7(p). CIP.

SH-C   This book is an attempt to teach the research methods of the social sciences. The authors deal with the procedures of selecting a research problem, formulating a hypothesis, selecting a method, collecting data, testing the hypothesis and writing the research report. Then they summarize two pieces of research and provide eight research activities that the high school student can do.

Denzin, Norman K., (Ed.). *The Value of Social Science.* Aldine, 1970. 195pp. $5.95. 77-115944.

C . Denzin shows something of the inner workings of social science research: the nature of research in the field, how social scientists view themselves, social research and social action, and the uses and misuses of social science knowledge. A good beginning book.

Weinland, Thomas P., and Donald W. Protheroe. *Social Science Projects You Can Do.* (Illus.) Prentice-Hall, 1973. 143pp. $4.95. 72-8098. ISBN 0-13-818260-4.

JH   A series of 56 short social science projects help develop questions and project objectives and evaluate findings of an investigation. Encourages use of telephone and personal interviews, and the resources of family, library, neighborhood and media to develop a factual basis. Presents interesting ways to examine political accountability, consumer rights, citizen participation and environmental awareness.

301 SOCIOLOGY

Becker, Howard S. *Sociological Work.* Aldine, 1970. x + 378pp. $11.75. 77-115936. ISBN 0-202-30096-X.

SH-C   This lucid book can easily be used as a general introduction to sociology or as collateral reading in the social sciences. Many areas of sociological inquiry, topics, and processes are straightforwardly approached and well discussed. Students will find the topics relevant to their own major life experiences.

Feigelman, William, (Ed.). *Sociology Full Circle: Contemporary Readings on Society.* Praeger, 1972. xi + 371pp. $13.50. 72-75689.

SH-C   An excellent collection of readings from the turn of the century to the end of the '60s. Methodological, theoretical and empirical topics are covered, including the nature of sociology, value-free science, airport art, cultural influence in pain perception, waiting lines as social systems, alienation, teaching troubled children, urban family experience, Christmas card sociology, racial and sexual discrimination, black power, homosexuality, Wallace supporters, population, the hippie movement, and much more.

Gans, Herbert J. *Popular Culture and High Culture: An Analysis and Evaluation of Taste.* Basic, 1975. xii + 179pp. $10.00. 74-79287. ISBN 0-465-06021-8. Index;CIP.

  C A sociological study of popular vis-à-vis high culture and their relations with their publics. Gans comments upon and describes each of the five taste cultures, and goes beyond culture analysis and suggests cues for evaluating other societal phenomena involving axiological questions. Book should interest the general reader and stimulate the professional sociologist.

Gouldner, Alvin W. *The Coming Crisis of Western Sociology.* Basic, 1970. xv + 528pp. $12.50. 77-110771. ISBN 0-465-01278-7.

  SH-C This thought-provoking book is impressionistic and highly philosophic, but almost always stimulating. Topics include a brief history of sociology, sociological positivism, Marxism, classical sociology and Parsonianism; a long analysis of Parson's theories, and discussions on the welfare state, conservative and reflexive sociology, and sociology in the Soviet Union.

Schneider, Louis. *The Sociological Way of Looking at the World.* McGraw-Hill, 1975. xv + 343pp. $12.50. 74-30460. ISBN 0-07-055463-3. Index;CIP.

  C Schneider presents sociology as a perspective rather than a collection of methods and statistics. He focuses on the idea of social structure, culture, society, personality and cultural and social change with illustrative material from medicine, psychiatry and criminology. A good introduction to the sociologist's point of view.

Theodorson, George A., and Achilles G. Theodorson. *A Modern Dictionary of Sociology.* Crowell, 1969. 469pp. $10.00. 69-18672.

  C This comprehensive list of words relating to a special subject is thoroughly cross-referenced, and it includes terms from other related disciplines (psychology, anthropology, social psychology and statistics). It has a few brief etymologies but no pronunciations, no reference list, no index and no illustrations. (Some biographical information is included in the listing as well as some references.)

301.04 THEORIES OF SOCIAL CAUSATION

Bresler, Jack B., (Ed.). *Genetics and Society.* (Illus.) Addison-Wesley, 1973. xv + 280pp. $4.50(p). 72-2650.

  C-P Deals with the interaction between heredity and environment in human development; specific traits that characterize ethnic groups and the observed consequences of intermarriage; and the relation of social status to inherited patterns. Thought provoking and reasonably thorough; an ideal reference for layreaders and students.

Claiborne, Robert. *God or Beast: Evolution and Human Nature.* (Illus.) Norton, 1974. xiii + 259pp. $7.95. 73-20454. ISBN 0-393-06399-2. Index;CIP.

  SH-C This explanation of the author's views on the evolution of human behavior emphasizes the uniqueness of man among other animals becasue of the advantages of speech. Liberal use of provocative terminology restricts this book to libraries serving more sophisticated readers.

Clark, Grahame. *Aspects of Prehistory.* (Illus.) Univ. of Calif. Press, 1970. xiii + 161pp. $5.95. 73-93989. ISBN 0-520-01584-3.

C   An extension of a series of lectures which deals with the interaction of biological and cultural evolution. People do not make themselves, states Clark, but are the result of natural selection, physically as well as culturally. Even conceptual life and social institutions reflect material progress and are products of natural selection.

Dubos, Rene Jules.   *Beast or Angel: Choices That Make Us Human.* Scribner's, 1974. xiii + 226pp. $8.95. 74-10737. ISBN 0-684-13901-4. Index;CIP.

JH-SH-C   Dubos reconciles the recently expounded view that man's "nasty and brutish" traits are genetically structured with the more optimistic view of most anthropologists that culture determines all. Demonstrates that the human potential for good matches its potential for evil.

Etzioni, Amitai.   *Social Problems.* Prentice-Hall, 1976. x + 182pp. $7.50; $3.50(p). 75-38703. ISBN 0-13-817411-3; 0-13-817403-2. Indexes;CIP.

C   Etzioni reviews perspectives offered by structural-functionalism, the conflict-alienation approach, ethnomethodology, and the neoconservative orientation, then presents his own "societal guidance" theory—an analysis of a society's knowledge, goal setting, organization, power and consensus. Valuable but not exciting.

Hagedorn, Robert, and Sanford Labovitz.   *An Introduction into Sociological Orientations.* (Illus.) Wiley, 1973. viii + 136pp. $3.50(p). 72-7193. ISBN 0-471-33862-1. Index;gloss.

C   This is an introductory sociology text concerned with such topics as functionalism, verstehen, positivism, radical sociology, structuralism and social action.

McKern, Sharon S.   *The Many Faces of Man.* (Illus.) Lothrop, Lee & Shepard, 1972. 192pp. $5.95. 72-1100. Index;bib.

JH   Presents the patterns of development of peoples and their societies. General principles of evolution and adaptation are discussed, including the hypotheses of Herodotus, Linnaeus, Darwin, Mendel and Jensen. Stresses both the difficulty of classifying humans into races, and the fact that there are no superior or inferior "races." Valuable as a reference for teachers and more advanced students, and as collateral reading.

Millman, Marcia, and Rosabeth Moss Kanter (Eds.).   *Another Voice: Feminist Perspectives on Social Life and Social Science.* Anchor/Doubleday, 1975. xvii + 382pp. $3.50. 75-8206. ISBN 0-385-04032-6.

C-P   This collection of 12 essays has as its main objective the determination of needed directions in sociological theory and research in order for the discipline to reflect both male and female interests. The work would be most useful to those already familiar with the principal questions and approaches in sociology.

Montagu, Ashley.   *Man's Most Dangerous Myth: The Fallacy of Race, 5th ed.* (Illus.) Oxford Univ. Press, 1974. xvi + 542pp. $15.00 73-92869. ISBN 0-19-501775-7. Index.

SH-C   Montagu points out the dearth of corroborative evidence on the inherent superiority of one race over another. An authoritative book with very relevant ideas.

Murphy, Robert F.   *The Dialectics of Social Life: Alarms and Excursions in Anthropological Theory.* Basic, 1971. x + 261pp. $7.95. 72-147015. ISBN 0-465-01643-X

C-P   From Simmel, Freud, Marx, and Hegel and their interpreters, to Claude Levi-Strauss, Murphy summarizes the development of dialectic social theories and finds them useful in understanding today's issues as well as historical processes. His exposition of the history, meaning, and application of dialectical theory combines erudition and witty informality.

Neaman, Judith S. *Suggestion of the Devil: The Origins of Madness.* Anchor/ Doubleday, 1975. x + 226pp. $2.50(p). 73-9041. ISBN 0-385-08569-9. Index;CIP.

SH-C-P   Neaman contrasts the medieval psychology of mysticism and madness to the current and social views of the supernatural and insanity. The scope encompasses the broad patterns of social understanding about human behavior.

301.1 SOCIAL PSYCHOLOGY

Aronson, Elliot, (Ed.). *Readings about the Social Animal.* (Illus.) Freeman, 1973. xvi + 436pp. $5.95(p). 72-12664. ISBN 0-7167-0833-7.

SH-C-P   The readings include both classical and contemporary examples of research in social psychology. Could be a very useful supplement to most social psychology texts.

Aronson, Elliot. *The Social Animal.* (Illus.) Viking, 1972. viii + 342pp. $7.95. 72-85866. ISBN 0-670-65513-9.

SH-C   An unusual, interesting and relevant introduction to social psychology. Highly selective, and topics are discussed in detail. The nature of the discipline, sensitivity training, ethical questions in research, conformity, mass communication, propaganda, persuasion, self-justification, human aggression, prejudice and attraction are included. Examples from "real life" are frequently and effectively used. A refreshing approach.

Bagdikian, Ben H. *The Information Machines: Their Impact on Men and the Media.* (Illus.) Harper&Row, 1971. vii + 359pp. $8.95. 71-123913. ISBN 06-010198-9.

SH-C-P   Studies the effect that present and projected machines for mass communication can have on the relationship between the citizen and society. Useful for collateral reading in courses in journalism, government, sociology or psychology.

Cherry, Colin. *World Communication: Threat or Promise? A Socio-technical Approach.* (Illus.) Wiley-Interscience, 1971. xiv + 229pp. $9.50. 79-147195. ISBN 0-471-15343-5.

C-P   The author deals with world communication from two points of view: the technological and the social. This amalgamation generates a wealth of new ideas and patterns of thinking about the communication process. Unusual, refreshing, provides new ideas and knowledge.

Gross, Alan E., Barry E. Collins and James H. Bryan. *An Introduction to Research in Social Psychology: Exercises and Examples.* Wiley, 1972. ix + 447pp. $5.95(p). 74-38967. ISBN 0-471-32815-4.

C   Contains everything needed to conduct 14 simple surveys and experiments, including reprints of published paradigms. Techniques range from the use of hidden observers and unobtrusive measures to construction of an attitude scale. Topics include control of conversation via verbal reinforcement, role playing in a group, anxiety, role model effects, etc. Provides a concise review of a five-phased, scientific problem-solving process and discusses random error, experimental design, validity and reliability.

Halacy, D.S., Jr. *Social Man.* (Illus.) Macrae, 1972. 162pp. $4.95. 76-16392. ISBN 0-8255-4041-0.

JH-SH   Halacy presents an overview of a worldwide human society, covering the social organization of animal groups, the human family, and nations and civilizations. A nice balance is achieved between individual human rights and social change, but the approach is somewhat idealistic.

McFeat, Tom. *Small-Group Cultures.* (Illus.) Pergamon, 1974. xii + 209pp. $12.00; $7.00(p). 73-8612. ISBN 0-08-017073-0; 0-08-017770-0. Index;CIP.

C-P   McFeat assesses the fit of small-group principles to ethnographic descriptions of group interaction and to experimental groups. Descriptions include a Nootka whaling crew, a sub-arctic hunting task force, historical development of Pueblo social organization, and a World War II American troop crossing a river in an ungainly craft.

Raven, Bertram H., and Jeffrey Z. Rubin. *Social Psychology: People in Groups.* (Illus.) Wiley, 1976. xx + 591pp. $12.95. 75-32693. ISBN 0-471-70970-0. Indexes;CIP.

SH-C   Depicts historical and current events from the perspective of acquired knowledge of the principles which govern interpersonal behavior and group interactions. A variety of social phenomena are discussed which are relevant to most readers; technical results and empirical findings are presented in the context of the reader's already established frame of reference. (Noticeably absent are comprehensive treatments of attitude formation and attitude change.)

Sampson, Edward E. *Social Psychology and Contemporary Society, 2nd ed.* (Illus.) Wiley, 1976. x + 567pp. $12.95. 75-30225. ISBN 0-471-75116-2. Indexes;CIP.

C   A clearly written, integrated introduction to social psychology. Sampson discusses major psychologically oriented research programs and weaves into the discussion new sociological theory. He provides one of the few coherent integrations of the sociological and psychological underpinnings of social psychology (but with inadequate discussion of methodological-design issues.)

Schiffer, Irvine. *Charisma: A Psychoanalytic Look at Mass Society.* Univ. of Toronto Press, 1973. xiv + 184pp. $8.50. 72-95816. ISBN 0-8020-1958-7.

SH-C   The author, a Canadian psychiatrist and psychoanalyst, addresses himself to political leadership in general and charismatic leadership in particular. Insightful and provocative, but the psychoanalytic theory is largely reductionist. Both the psycho-philosophical and sociological dimensions of charisma are clearly elucidated.

Sherwin, Martin J. *A World Destroyed: The Atomic Bomb and the Grand Alliance.* Knopf, 1975. xvi + 326pp. $10.00. 75-8213. ISBN 0-394-49794-5. Index;CIP.

SH-C   Historian Sherwin traces the political steps in the development of the bomb in the U.S. and Britain during WWII. He illustrates repeatedly the peculiar thinking processes that war induces in males. Almost one-third of the book consists of notes, a bibliographic essay, and selected documents. The writing is clear.

301.2 CULTURE AND CULTURAL PROCESSES

Barnouw, Victor. *An Introduction to Anthropology: Ethnology.* (Illus.) Dorsey, 1971. xii + 358pp. $7.55(p). 72-146732.

C   This introduction covers virtually all of anthropology from the early hominids to astronauts and aquanauts. Cultural variations, psychological and physical aspects of family life, political organizations, law, cultural sanctions, religion and folklore, recreational and artistic expressions of man, as well as the diversity of personality types within subcultures are considered.

Bjerre, Jens. *The Last Cannibals.* (Illus.; trans.) Drake, 1974. 192pp. $3.95(p). 74-5971. ISBN 0-87749-692-7. CIP.

SH-C   Fascinating eye-witness account of daily life among primitive tribes of central Australia and Northern New Guinea. Describes food gathering, weapons, rites, kinship, sex life, spirits and magic. Vivid accounts also document local flora and fauna.

Bleeker, Sonia. *The Zulu of South Africa: Cattlemen, Farmers, and Warriors.* (Illus.) Morrow, 1970. 160pp. $3.95. 71-118059.

JH   A well-written account of the Zulu people—an historical narrative; a full picture of Zulu customs concerning daily life, marriage and cattle raising; and a description of life in modern South Africa. Some fictional characters tell about customs, but historically known leaders are included.

Clarke, Robin, and Geoffrey Hindley. *The Challenge of the Primitives.* McGraw-Hill, 1975. 240pp. $8.95. 75-6907. ISBN 0-07-011234-7. CIP.

C   The authors discuss the ecology of the preindustrial world, emphasizing the positive, adaptive aspects of the life of the hunter and gatherer and the "slash and burn" horticulturist. Social structure is discussed, as are the economic systems of nonmarket societies. The authors call for a move away from the scientific mode of modern social sciences and urge a reunification of humans and nature.

Hanbury-Tenison, Robin. *A Pattern of Peoples: A Journey Among the Tribes of Indonesia's Outer Islands.* (Illus.) Scribner's, 1975. 220pp. $9.95. 75-4089. ISBN 0-684-14363-1. Index.

SH-C   The author describes his tours among several distinctive cultures in the Indonesian archipelago which are feeling the impact of external forces and yet continue to exhibit patterns of behavior which distinguish them from the national norm of Indonesia. Useful as a reference.

Harris, Marvin. *Culture, Man, and Nature: An Introduction to General Anthropology.* (Illus.) Crowell, 1971. xxv + 659pp. $9.95. 78-146065. ISBN 0-690-23034-6.

C-P   Archaeological, linguistic, racial, and ethnographic subdisciplines are well summarized as are other studies ancillary to anthropology. All are integrated effectively, with persuasive continuity between subjects. Gives considerable emphasis to the realms of cultural and social anthropology and finds a useful framework for description and analysis in topical treatments of domestic patterns in different societies.

Latchem, Colin. *Looking at Nigeria.* (Illus.) Lippincott, 1976. 64pp. $5.95. 75-15967. ISBN 0-397-31652-6. Index;CIP.

JH   Introduces four major tribes, the Yoruba, Ibo, Hausa and Fulani, each with its own variation of habitat, religion, kinsmanship and subsistence. Social and cultural information is provided, and comparisons could be made between the Nigerian school systems, marriage customs, puberty rituals and religious traditions and those of the readers.

Leach, Edmund. *Claude Lévi-Strauss, rev. ed.* Viking, 1974. xi + 146pp. $7.50; $2.95(p). 74-1122. SBN 670-22515-0; 670-01980-1. Index.

C-P   Leach's summary of the contribution of Lévi-Strauss to the field of social anthropology constitutes a straightforward, sometimes critical, and concise account of the French scholar's innovative approach to presenting anthropological data in a stimulating, provocative way. Recommended for students of anthropological theory and others interested in social behavior.

Linton, Adelin, and Charles Wagley. *Ralph Linton.* (Illus.) Columbia Univ. Press, 1971. 196pp. $7.50; $2.95(p). 76-1747-8. ISBN 0-231-03355-9; 0-231-03398-2. Bib.

C   One of a series of brief volumes on the lives and contributions of leading anthropologists, this volume provides a brief biographical essay and a selection of writings representative of Linton's point of view and range of interest, with emphasis on articles not readily available.

Marcus, Rebecca B. *Survivors of the Stone Age: Nine Tribes Today.* (Illus.) Hastings House, 1975. 124pp. $6.95. 75-6843. ISBN 0-8038-6726-3. Index;CIP;bib.

JH-SH-C The Ik, Bushman, and Pygmy of Africa; the Jivaro and Kranhacrore of South America; the Arunta of Australia; and the Tasaday, Onge, and Papuan of greater Southeast Asia are presented as nine tribes still maintaining a Stone Age existence. Insights into their cultures, problems and future are provided.

Martin, Robert. *Yesterday's People.* (Illus.) Doubleday, 1970. 158pp. $5.95. 68-22486.

JH The remarkable cultural adaptation that permits the extremely primitive Bushmen of the Kalahari desert to survive is described in detail in the context of ecological principles and cultural evolution.

Massola, Aldo. *The Aborigines of South-Eastern Australia As They Were.* (Illus.) Heinemann (distr.: Scribner's), 1974 (c.1971). x + 166pp. $12.50. 73-95166. ISBN 0-85561-002-6. Index;bib.

JH-SH Details the traditional culture of the nearly extinct aborigines of southeastern Australia, their Asian origins and early campsites, contacts with white settlers and the ultimate loss of the aborigines' way of life and most of their population. Dozens of reproductions of early engravings, photographs and drawings.

Mead, Margaret. *Culture and Commitment: A Study of the Generation Gap.* Doubleday, 1970. 110pp. $5.00. 79-93204.

C Mead deals with child-raising practices in traditional and complex societies and provides many provocative ideas. Three idealized cultural styles are conceptualized: postfigurative cultures, in which children learn primarily from their forebears; cofigurative cultures, in which both children and adults learn from their peers; and prefigurative cultures, in which adults learn also from their children.

Mead, Margaret. *Ruth Benedict.* (Illus.) Columbia Univ. Press, 1974. vii + 180pp. $8.95. 74-6400. ISBN 0-231-03519-5; 0-231-03520-9(p). CIP.

C-P Mead continues Columbia's series on the leaders in modern anthropology in this two-part volume, half biography by Mead, and half selected works by Benedict. Provides only a tantalizing hint of the important aspects of Benedict's life, but does leave the reader with an appreciation and desire for further study of this prominent anthropologist.

Montagu, Ashley, (Ed.). *Culture and Human Development: Insights into Growing Human.* Prentice-Hall, 1974. ix + 181pp. $7.95; $2.95(p). 74-18338. ISBN 0-13-195578-0; 0-13-195560-8. CIP.

C-P Thirteen readings comprise this collection which addresses the question of the effect of culture on human physical and mental development. With such a wealth of controversial and discussable ideas, it will make a nice contribution to sociology and anthropology courses.

Montagu, Ashley, (Ed.). *Frontiers of Anthropology.* (Illus.) Putnam's, 1974. xvii + 617pp. $12.50. 73-78599. SBN 399-11199-9. Bib.

C-P Chronological arrangement of representative readings on humankind's place in nature and society. Subjects range from Greek notions about culture and race to a detailed study of the distribution of sickle cell genes in West Africa as an evolutionary response to changed disease environment.

Murphy, E. Jefferson. *The Bantu Civilization of Southern Africa.* (Illus.) Crowell, 1974. xii + 273pp. $6.95. 73-17194. ISBN 0-690-00399-4. Index;CIP.

JH-SH-C The author presents an excellent in-depth study of the Bantu "nation," emphasizing the nation as a civilization united by deep ties of language, culture and history. The text is exceptionally well written and beautifully illustrated.

Murphy, Robert F. *Robert H. Lowie.* Columbia Univ. Press, 1972. viii + 159pp. $7.50; $2.95(p). 72-1969. ISBN 0-231-03375-3; 0-231-03397-4.

SH-C Half the text is devoted to remembrances and discussion of Lowie's work, and half contains annotated sections from his publications. Sympathetic to Lowie's aims and interests, yet presents a balanced criticism of the products of those aims and interests.

Scientific American. *Biology and Culture in Modern Perspective: Readings from* **Scientific American.** (Illus.) Freeman, 1972. 441pp. $12.00. 72-4237. ISBN 0-7167-0861-2.

SH-C The text is highly accurate; the clarity of exposition is very good; the scope of the work is adequate; and the quality of illustrations is excellent. Included are articles on human origins, genetics and evolution, animal behavior, prehistory of culture and civilization; other traditional concerns in cultural anthropology.

Severin, Timothy. *The Horizon Book of Vanishing Primitive Man.* (Illus.) American Heritage (distr.: McGraw-Hill), 1973. 379pp. $22.00. ISBN 0-07-056348-9; 0-07-056349-7.

SH-GA-C Ten different peoples—including the Ainu, the Australian aborigines and Eskimos—are discussed in historical and anthropological terms. The descriptions provided by outsiders are also examined.

Steward, Julian H. *Alfred Kroeber.* (Illus.) Columbia Univ. Press, 1973. xii + 137pp. $8.00; $2.95(p). 72-8973. ISBN 0-231-03489-X; 0-231-03490-3.

C The first half of the book is a discussion of Kroeber's career and a general sketch of the major patterns of his thought and the second half contains selections from his writings. It constitutes an introduction to one of the most important figures in anthropology theory and provides some material for thought.

Stewart, T.D. *The People of America.* (Illus.) Scribner's, 1973. x + 261pp. $10.00; $3.95(p). 73-1371. SBN 684-13539-6; 684-13570-1.

C-P A major contribution to anthropological literature, this book should be required in any anthropology course. Further, its multifaceted content and humanistic aspects recommends it to students and professionals in other areas such as sociology, political science and urban planning.

301.24 SOCIAL CHANGE

Alexander, Theron. *Human Development in an Urban Age.* (Illus.) Prentice-Hall, 1973. xxi + 336pp. $8.95. 73-506. ISBN 0-13-444786-7.

SH-C-P Alexander traces the development of a human being in society by relating individual development to conditions of urban and modern society. He emphasizes the consequences of social change and deals with development of human beings through socialization, cultural interaction and their ability to modify their environment through mobility and change in "life space" (degree of crowding). Useful for collateral reading for high school courses. A fine reference for advanced students and teachers at all levels.

Berger, Peter L. *Pyramids of Sacrifice: Political Ethics and Social Change.* Basic, 1975. xiv + 242pp. $10.00. 74-78304. ISBN 0-465-06778-6. Index;CIP.

C-P Berger's humanistic study deals primarily with the problems of the Third World and the more general issue of political ethics as applied to social change. He is much concerned with the prices people have to pay for "progress."

Brubaker, Sterling. *In Command of Tomorrow: Resource and Environmental Strategies for Americans.* Resources for the Future/Johns Hopkins Univ. Press, 1975. xii + 177pp. $7.95. 74-24401. ISBN 0-8018-1700-5. Index;CIP.

    C-P Within the confines of current U.S. goals—income growth, supply security, environmental quality—Brubaker attempts to form an interim strategy for future survival. Many problems are considered, while the study focuses on technological improvements. Heavy reading, but useful as a reference.

Carey, Maureen, et al. *Deciding on the Human Use of Power: The Exercise and Control of Power in an Age of Crisis.* (Illus.) Piover (distr.: North Country Pub.), 1974. 73-83110. ISBN 0-8489-017-1.

    SH-C Written in casebook style, this volume presents case sketches in the use of power to control, and to bring about either creative or destructive changes. A provocative approach, and valuable for students and teachers pursuing the lines of John Dewey's inquiry philosophy.

Forman, Shepard. *The Raft Fishermen: Tradition & Change in the Brazilian Peasant Economy.* (Illus.) Indiana Univ. Press, 1970. xv + 158pp. $8.50. 78-126208. ISBN 0-253-39201-2.

    JH-SH This study of technological innovation meets every criterion for teaching material in the social sciences. Readable and useful from junior high school to the planning and execution of research. The book is a delight.

Gerlach, Luther P., and Virginia H. Hine. *Lifeway Leap: The Dynamics of Change in America.* Univ. of Minnesota Press, 1973. 332pp. $12.50; $4.95(p). 72-97762. ISBN 0-8166-0672-2.

    SH-C-P This stimulating, unbiased and entertaining collection of essays on the topic of change in America examines the mechanism of reform, the means a society has to effect desirable ends, and contemporary "movements."

Heilbroner, Robert L. *An Inquiry into the Human Prospect.* Norton, 1974. 150pp. $5.95. 73-21879. ISBN 0-393-05514-0; 0-393-09274-7(p). Index;CIP.

    SH-C-P Concerns the future of man and deals with such topics as overpopulation, the human ability to respond to crises, and the characteristics of capitalistic and socialistic societies now in existence. Heilbroner places very little faith in man's future technological progress or his ability to make necessary changes in society and himself. Humanity must accept the need for change, Heilbroner stresses. Well written and easy to read.

Hoover, Dwight W., and Warren Vander Hill (Eds.). *American Society in the Twentieth Century: Selected Readings.* Wiley, 1972. xiv + 247pp. $9.95; $4.95(p). 77-37172. ISBN 0-471-4089-4; 0-471-40895-6.

    C A collection of essays that deal with some of the changes in social roles in the United States since 1900. The selections present the "flavor" of various facets of society during the period in question.

Inglis, Brian. *The Forbidden Game: A Social History of Drugs.* Scribner's, 1975. 256pp. $8.95. 75-12382. ISBN 0-684-14428-X. Index;CIP.

    C Inglis presents an historical account of the growth of the now-extinct rituals of some primitive peoples into the lucrative drug trade of the 20th century. The author also deals with the political manipulations and profiteering of those nations that modified their priorities to coincide with those of drug suppliers. The book offers the reader rare insight into the introduction of, and changing attitudes toward, such varied drugs as tea, tobacco, coca and opium into Western society.

Mead, Margaret. *World Enough: Rethinking the Future.* (Illus.) Little, Brown, 1975. xxxii + 218pp. $17.50. 75-25680. ISBN 0-316-56470-2. CIP.

SH-C   Mead's masterful presentation explains how earth and its inhabitants have arrived at our present state as a result of industrialization, modernization and urbanization. This is a reasoned and urgent call to action, both in words and pictures.

Pettitt, George A. *Prisoners of Culture.* Scribner's, 1970. xii + 291pp. $8.50. 68-57070.

C-P   A thoughtful, controversial, and thorough book on the problems of youth in contemporary culture, based on more than 20 years of study of human societies, both primitive and civilized, throughout the world. Sharp criticism is levied against compulsory education in particular.

301.243 SOCIAL CHANGE DUE TO SCIENCE AND TECHNOLOGY

Boffey, Phillip M. *The Brain Bank of America: An Inquiry into the Politics of Science.* McGraw-Hill, 1975. xxiii + 312pp. $10.95. 74-23842. ISBN 0-07-006368-0. Index;CIP.

SH-C   This is the report of an investigation of the National Academy of Sciences by Ralph Nader's Center for Study of Responsive Law. Well documented support for the conclusion that most public policy questions are not purely technical but involve value judgments.

De Nevers, Noel, (Ed.). *Technology and Society.* (Illus.) Addison-Wesley, 1972. ix + 307pp. $3.95. 77-172801.

GA-C   These short essays are written in adversary style, and they are concerned with the conflicting roles of humanism and industrialism in American society. The essays supplement seven books that must be acquired separately—one each by Huxley, White, Morison, Ehrlich, Nader, and two by Snow. The importance of this topic today, and the importance of the supplementary books, makes this collection recommended reading.

Dessel, Norman F., et al. *Science and Human Destiny.* (Illus.) McGraw-Hill, 1973. x + 318pp. $6.95(p). 72-7400. ISBN 0-07-016580-7.

SH-C   Deals with the historical development of the models and paradigms of science and introduces the student to the problems of the ecosystem and their influence in human destiny. Discusses the major concepts of science (life and physical sciences), the status of man in the universe, and the mutual dependence of science and technology.

Hellman, Hal. *Technophobia: Getting Out of the Technology Trap.* M. Evans (distr.: Lippincott), 1976. xvi + 307pp. $8.95. 75-44372. ISBN 0-87131-206-9. Index;CIP.

C-P   An objective appraisal of the relationship between technology and society which clearly indicates that technology is a powerful force upon which our successful survival is almost totally dependent. Many suggestions are presented on how to approach this force rationally and thoughtfully in order to define its proper application to the benefit of all humans.

Holton, Gerald, and William A. Blanpied (Eds.). *Science and Its Public: The Changing Relationship.* D. Reidel, 1976. xxv + 289pp. $26.00; $11.00(p). 75-41391. ISBN 90-277-0657-3; 90-277-0658-1. Index;CIP.

C-P   This collection of 17 authoritative papers deals with the role of science and scientists in areas of history, philosophy, politics, public policy, journalism, evolution, objectivity, and ethics, and emphasizes roles, responsibilities, expectations, and misconceptions. Science devotees, political observers, and philosophers will profit from it.

Hoyle, Fred. *The New Face of Science.* (Illus.) New American Library, 1971 (distr.: Norton). xxiii + 132pp. $6.95. 70-142132.

C   These short pieces, half of them originally written in the 1960s, are still absorbing and provocative, and in parts remarkably current and incisive in analyzing
broad social problems. Topics included are the indiscriminate uses of technology, the
importance of physicists, astronomy, assessment of societies by their energy consumption, and the moon landing. Hoyle's conclusions and opinions are distinctive,
unambiguous, and thought-provoking.

Inglis, David Rittenhouse. *Nuclear Energy: Its Physics and Its Social Change.* (Illus.)
Addison-Wesley, 1973. xiv + 395pp. $6.95; $4.95(p). 78-186840.

C   Clear, concise and objective. Basic ideas in physics, the physics of a power
plant and energy sources, nuclear reactors, radioactivity and controls, other
power sources, nuclear explosives, and the arms race are discussed. Provides data on
existing reactors, numbers of nuclear weapons possessed by the U.S. and the USSR,
and excerpts from papers and speeches on nuclear technology and the arms race.

Kahn, Herman, et al. *The Next 200 Years: A Scenario for America and the World.*
(Illus.) Morrow, 1976. xv + 241pp. $8.95. 76-5425. ISBN 0-688-03029-7; 0-688-08029-
4(p). Index;CIP.

GA-C   This book is a painstaking attempt to look positively towards the year 2176
with its world population of 15 billion, world "gross product" of $300
trillion and per capita income of $20,000 (1975 dollars). There is persuasive discus-
sion not only of improvements in conventional agriculture but new triumphs by the
chemists. Economic growth is projected at rates slower than the recent exponential
rate, and the conclusions are optimistic. Eight basically uncertain issues and eight
basically solvable issues are dealt with.

Kranzberg, Melvin, and William H. Davenport (Eds.). *Technology and Culture: An
Anthology.* Shocken, 1972. 364pp. $10.00. 73-185318.

SH-C   This anthology gives an historical perspective as well as insight into the
relationships between technology and society. Places in true perspective
the history of technology and science from primitive times to the present and the
future; points out man's dependence on technology as a factor setting him apart from
the rest of the animal kingdom; provides excellent examples of the uses of technology
in humanistic studies, and pays attention to the need for continuing creativity.

Mesarovic, Mihajlo, and Eduard Pestel. *Mankind at the Turning Point: The Second
Report to the Club of Rome.* (Illus.) Dutton/Reader's Digest, 1974. xii + 210pp. $12.95.
74-16787. ISBN 0-525-15230-X; 0-525-03945-7(p). Index.

C-P   The authors describe a flexible, computer-based planning instrument which
contains a multi-level regionalized model of the world system, developed as
a planning tool for economic and political decision-making. An excellent, well-
documented reference with graphs and briefs on various worldwide situations and
crises.

Muller, Herbert J. *The Children of Frankenstein: A Primer on Modern Technology and
Human Values.* Univ. of Indiana, 1970. xiii + 431pp. $10.00. 76-103926. ISBN 0-253-
11175-7. Index;bib.

SH-C   A disturbing, thought-provoking, problem-illuminating, comprehensive
survey, addressed to the general reader, on the question, "What has modern technology done *to* as well as *for* people?" Muller examines war, science, government, business, language, higher education, natural environment, social environment, the city, mass media, traditional culture, religion, technological development
without human value development, the future, the past, and human nature. Excellent!

Murray, Bruce C. *Navigating the Future.* Harper&Row, 1975. xii + 175pp. $7.95.
74-1839. ISBN 0-06-013122-5. CIP.

C   Murray sought the advice of eight prominent individuals whose expertise
ranges from science to theology and blended their diverse viewpoints into an
erudite and cohesive discourse on humankind's alternatives for surviving a bleak
future. Provocative and profound reading.

Papanek, Victor. *Design for the Real World: Human Ecology and Social Change.*
(Illus.) Pantheon, 1971. xxvi + 339pp. $8.95. 70-154020. ISBN 0-394-47036-2.

SH-C   A comprehensive and free-wheeling analysis of past design failures. The
theory of manipulated taste and planned obsolescence which creates need
instead of eliminating it is roundly criticized. The author finds it hard to show when
and how the industrial design profession has had a positive impact on this nation. An
interesting and thought-provoking book.

Parkman, Ralph. *The Cybernetic Society.* (Illus.) Pergamon, 1972. x + 396pp. $15.00;
$7.50(p). 78-185338. ISBN 0-08-016949-X; 0-08-017185.

C-P   Shows how the computer age has and will continue to affect every aspect of
our lives; discusses the industrial revolution, the advent of mass production,
the development of the computer, post-WWII socioeconomics, the use of technology
in government, the "system approach" to planning, the control and privacy aspects
of computers and communication, and computer technology's relationship to the
human mind, the arts, and educational theory. An excellent basic text for interdisciplinary computer courses and collateral reading for advanced courses in sociology,
education, group psychology, and modern history.

Piel, Gerard. *The Acceleration of History.* Knopf, 1972. 369pp. $8.95. 71-171118.
ISBN 0-394-47312-4.

C   Piel attempts to place science at the very heart of human values. His essay
topics include anthropology, medicine and health care, and he writes with real
concern for scientific values. Highly recommended for anyone concerned with science and its role in society.

Reich, Albert H. (Ed.). *Technology and Man's Future.* St. Martin's, 1972.
xiv + 274pp. $3.95(p). 73-190777.

SH-C-P   Critically questions the values and virtues of the technological society.
Contains selected readings on social, ethical and political implications of
the technological present and future, and a wide range of views and recommendations on technology assessment. Hard-to-obtain original documents are reprinted.

Snow, C.P. *Public Affairs.* Scribner's, 1971. 224pp. $6.95. 77-162752. ISBN 0-684-
12570-6.

C-P   Seven essays from *Two Cultures* (1959), *The State of Siege* (1968), and *The
Case of Leavis* and *The Serious Case* (1970) are reprinted with a prologue
and epilogue. Snow's statements change as his vision of the relation of mankind to
science is altered by experience and as a result of dialogue generated by each lecture.
These warnings of future tragedies are not light reading.

Susskind, Charles. *Understanding Technology.* (Illus.) Johns Hopkins Univ. Press,
1973. x + 163pp. $6.95. 72-12344. ISBN 0-8018-1304-2.

SH-C   An overview of the development of modern technology and of its main
social and political consequences, a detailed description of the various
Nazi methods of euthanasia and genocide, and an excellent treatment of computer
technology, contemporary technology, the challenges to technology and the problems that society is facing or will face as a result of these technological advances.

301.3 ECOLOGY AND COMMUNITY

See also 333.7 –.9 Conservation and Utilization of Natural Resources.

Berry, James W., et al.   *Chemical Villains: A Biology of Pollution.* (Illus.) Mosby, 1974. vii + 189pp. $5.75(p). 73-11099. ISBN 0-8016-0663-2. Index.

   SH-C   This is a true story of occurrences in Borneo which show how the environment is a complex of both living and nonliving interacting factors. Valuable for readers with advanced training in ecology.

Blaustein, Elliott H., et al.   *Your Environment and You: Understanding the Pollution Problem.* (Illus.) Oceana, 1974. 197pp. $8.75. 74-5230. SBN 379-00802-5. Gloss.

   JH   Presents material which will instill environmental awareness in students at all levels of the educational process. This is a welcome introduction to ecology which bridges existing gaps in environmental instruction in school curriculums.

Del Giorno, Bette J., and Millicent E. Tissair.   *Environmental Science Activities: Handbook for Teachers.* (Illus.) Parker, 1974. 245pp. $8.95. 74-20560. ISBN 0-13-283275. Gloss.;index.

   SH-C   This environmental handbook for teachers may be considered a classic. The authors explain why we study the environment, the tools we use to study it, and how environmental study benefits us. The handbook is adaptable to diverse student groups, and sufficient material is presented for different levels of student abilities.

Detweiler, Robert, et al. (Eds.)   *Environmental Decay in Its Historical Context.* Scott, Foresman, 1973. 142pp. $2.50(p). 72-81162. ISBN 0-673-07678-4.

   SH-C   Human attitudes toward the earth habitat are examined by four competent authorities, all of whom point strongly to the lack of sound land ethics adequate for our survival. The roots of environmental decay and blighted urban life are attributed to the population explosion and industrialization. These factors are traced in their historical context. The final section is concerned with development of an environment ethic. *Must* reading.

Ehrlich, Paul R., and Anne H. Ehrlich.   *Population, Resources, Environment: Issues in Human Ecology, 2nd ed.* (Illus.) Freeman, 1972. xiv + 509pp. $9.50. 70-180800. ISBN 0-7167-695-4. Bibs.

   C   Included are sections on forest resources, net reproduction rates and zero population growth, the effects of pollution growth on the environment, heavy metals pollution and the environmental impacts of the war in Indochina and much more. The "Teacher's Guide" contains annotated bibliographies and a more comprehensive index. This is a good general reference and text on this subject; well suited for any course in human ecology.

Ehrlich, Paul R., Anne H. Ehrlich and John P. Holdren.   *Human Ecology: Problems and Solutions.* (Illus.) Freeman, 1973. xi + 304pp. $4.75(p). 72-1282. ISBN 0-7167-0595-8.

   SH   An introduction to the biological and physical aspects of human ecological problems and to possible solutions. Major problems considered concern demography, resource utilization, pollution and ecological systems in developed and underdeveloped countries. Solutions discussed emphasize the importance of population control and of changing human behavior.

Emmel, Thomas C.   *An Introduction to Ecology and Population.* (Illus.) Norton, 1973. x + 196pp. $6.95; $2.95(p). 72-14170. ISBN 0-393-06393-3; 0-393-09371-9.

   SH-C   This concise book explains the basic phenomena and language of ecology and population biology in a coherent, readable fashion. Human presence in the ecosystem—overpopulation, pesticides, pollution, etc.—is treated in a calm, objective manner.

Gildea, Ray Y., Jr. *Arsenic and Old Lead: A Layman's Guide to Pollution and Conservation.* (Illus.) Univ. Press of Mississippi, 1975. 132pp. $5.00(p). 74-17509. ISBN 0-87805-065-5. Index.

JH-SH   Gildea describes in an easy-to-read and understandable manner how humans are killing themselves with a multitude of toxins (in addition to arsenic). He presents the problems associated with air, water, soil, vegetation, wildlife, minerals, population and wastes and recommended solutions for these problems. A useful reference.

Gill, Don, and Penelope Bonnett. *Nature in the Urban Landscape: A Study of City Ecosystems.* York, 1973. xii + 209pp. $12.00. 73-76409. ISBN 0-91275-03-3. Index;gloss.;bib.

SH-C   A holistic approach to the biological environment where most of us live. Provides a very clear explanation of a city ecosystem, with fascinating detailed descriptions of microclimates and wildlife habitats in large cities.

Goldsmith, Edward, and the Editors of *The Ecologist. Blueprint for Survival.* Houghton Mifflin, 1972. xiv + 189pp. $5.95. 72-4132. ISBN 0-395-14098-6.

SH-C   Discusses the interlocking problems facing society and maintains that individuals, banded together in small communities, must control things. Various aspects of the environment are discussed, as well as the interaction of social and ecological systems and suggestions for changes that might ensure survival and enhance our life styles.

Hardin, Garrett. *Exploring New Ethics for Survival: The Voyage of the Spaceship Beagle.* (Illus.) Viking, 1972. xiv + 274pp. $7.95. 78-186737. ISBN 670-30268-6.

SH-C   A stimulating and frightening book about the direction in which we are taking our environment. Explains fully the origins of our present plight and makes projections about current resource utilization practices and their impact on the future quality of human life. Should be read by every concerned citizen.

Harrison, Gordon. *Earthkeeping: The War with Nature and a Proposal for Peace.* Houghton Mifflin, 1971. xii + 276pp. $5.95. 72-108684. ISBN 0-395-12711-4.

GA-C   Treats the development of natural systems, the role of humans within these systems, and the development of human cultures as an abiological process. Recommended for both the general reader and in courses in resource management and related fields.

Hirsch, S. Carl. *Guardians of Tomorrow: Pioneers in Ecology.* (Illus.) Viking, 1971. 192pp. $4.53. 76-136818. ISBN 0-670-35647-6.

JH   The story of the gradual depletion of America's natural resources is told simply and concisely by reference to the lifeworks of eight prominent Americans, from Thoreau to Rachel Carson. The author emphasizes the interrelation of all living things and their nonliving environment and, in outlining the main dangers facing us today, places them in their historical perspective.

Mason, William H., and George W. Folkerts. *Environmental Problems: Principles, Readings, and Comments.* (Illus.) Brown, 1973. x + 399pp. $5.50(p). 72-91813. ISBN 0-697-04700-8.

SH-C   Discusses ecological principles and specific topics, including zero population growth within a very complex human ecosystem, the limitations of finite resources, humans' threat to each other, abuses of land resources, legal ramifications of environmental management, and the motivations underlying the anti-environment movement. Terminology is skillfully presented and most articles are followed by excellent comments by the editors.

Mossman, Archie S. *Conservation.* (Illus.) Intext Educational, 1974. xii + 196pp. $8.50; $4.50(p). 73-16215. ISBN 0-7002-2446-7. CIP.

C-P   Mossman outlines the many phases of living which must be revised before disasters are brought on. The reader is urged to evaluate each problem in its broadest ecological context and in its esthetic impact. Gives discussion material at the end of each chapter.

Passmore, John. *Man's Responsibility for Nature: Ecological Problems and Western Traditions.* Scribner's, 1974. x + 213pp. $7.95. 73-19273. ISBN 0-684-13815-8. Indexes.

C-P   Passmore discusses humankind's plight in its confrontation with nature and analyzes major ecological problems—pollution, conservation, preservation and overpopulation. His thesis is that people have a responsibility to respond to their immediate needs without jeopardizing the freedom of future generations. The solutions to ecological problems cannot be left to scientists.

Pirages, Dennis C., and Paul R. Ehrlich. *Ark II: Social Response to Environmental Imperatives.* Viking, 1974. x + 344pp. $8.95. 73-10555. ISBN 0-670-13282-9. Index;CIP.

SH-C   The authors draw an analogy between the legend of Noah and his ark and our present plight. "We are all now caught in a gigantic tragedy of the commons; each person, each nation is struggling to stay ahead while the whole system is on the verge of collapse," the authors state. A fine text for social science and environmental courses.

Scientific American. *The Human Population.* (Illus.) Freeman, 1974. 147pp. $8.50; $3.95(p). 74-19465. ISBN 0-7167-0515-X; 0-7167-0514-1. Index;CIP.

SH-C-P   This book contains 11 articles on population problems. An appropriate source book for those who are interested in the most fundamental of human endeavors and conditions, from the genetic basis of population diversity in humans to transfer of technology to underdeveloped countries.

Swan, James A., and William B. Strapp (Eds.). *Environmental Education: Strategies Toward a More Livable Future.* (Illus.) Halsted/Wiley, 1974. 349pp. $15.00. 72-98046. ISBN 0-470-83859-0. CIP.

C-P   A collection of papers which exposes the reader to environmental education strategies and serves as a resource for evaluating environmental programs. Although some papers are controversial in their approaches, all give the reader basic knowledge of the problems of the human environment.

Tuan, Yi-Fu. *Topophilia: A Study of Environmental Perception, Attitudes, and Values.* (Illus.) Prentice-Hall, 1974. x + 260pp. $8.95; $4.95(p). 73-8974. ISBN 0-13-925248-7; 0-13-925230-4.

C-P   Tuan ranges widely through psychology, anthropology, philosophy, art and literature to examine the human "love of place." Hundreds of references. A delightful interdisciplinary work in which knowledge and wisdom come together.

Wagner, Kenneth A., P.C. Bailey, and H. Campbell. *Under Siege: Man, Men, and Earth.* (Illus.) Intext, 1973. xii + 386pp. $10.00. 73-166. ISBN 0-7002-2434-3.

SH-C   Biological concepts are applied to problems of present day environment. A broad overview of environmental problems is followed by a discussion of the effects of technology on air, water, soil, and radiation exposure. Ecological and personal stresses between humans and other organisms are discussed in detail. Suggestions are given for individual and community actions, along with addresses of agencies that can help.

Ward, Barbara, et al. *Who Speaks for Earth?* Norton, 1973. 173pp. $6.95; $1.75(p). 72-11726. ISBN 0-393-06392-5.

SH-C-P   Lectures given in Stockholm in 1972 by Barbara Ward, Rene Dubos, Thor Heyerdahl, Gunnar Myrdal, Carmen Miro, Lord Zuckerman, Aurelio Peccei, etc., on human ecology, preservation of our diverse social and cultural characteristics, the vulnerability of the oceans, the economics of maintaining an improved environment, population problems, and the pivotal contributions of science to environmental management and planning. The theme: all nations *must* act in nonselfish unity to preserve the earth. A sober, philosophical presentation.

301.32 DEMOGRAPHY

Brown, Lester R. *In the Human Interest: A Strategy to Stabilize World Population.* Norton, 1974. 190pp. $6.95. 74-6339. ISBN 0-393-05526-4. CIP.

C   Written as part of an on-going collaborative study of the interrelationships of environment, energy, population, and resources, this book eloquently illustrates the many factors involved in the world population problem. The book's theme amply underscores human dependence on the ecosystem as well as our interdependence with each other.

Drummond, A.H., Jr. *The Population Puzzle: Overcrowding and Stress Among Animals and Men.* (Illus.) Addison-Wesley, 1973. 143pp. $4.50. 72-8046. ISBN 0-2-1-01566-8.

McLung, Robert M. *Mice, Moose and Men: How Their Populations Rise and Fall.* (Illus.) Morrow, 1973. 64pp. $4.25. 73-4926. ISBN 0-688-20087-7; 0-688-30087-1.

JH   Both books discuss population patterns, how they change and what the inherent dangers are when conditions develop which permit unlimited population growth.

Frejka, Tomas. *The Future of Population Growth: Alternative Paths to Equilibrium.* (Illus.) Wiley, 1973. xix + 268pp. $9.95. 73-1607. ISBN 0-471-27875-0.

C-P   Frejka estimates future population by assuming that birth rates will decline to a net reproduction rate of 1.0, with an average of two children per family at some time in the future. He discusses current fertility, mortality and reproduction rates and provides a variety of population projections for the world and its various areas. For planners, politicians, academicians and the general reader, there is no other comparable work.

Gastil, Raymond D. *Cultural Regions of the United States.* (Illus.) Univ. of Wash. Press, 1976. xvi + 366pp. $12.95. 75-8933. ISBN 0-295-95426-4. Index;CIP.

C-P   Original settlement and migration patterns and variations in cultural features (religion, politics, housing and settlement styles, dialects, music and food forms) are examined. Then variations in well-being scales, homicide and violence factors, educational performance and infant mortality are shown for 13 regions and their subdivisions. A very useful reference for both students and teachers.

Hart, Harold H., (Ed.). *Population Control: For and Against.* Hart, 1973. 239pp. $7.50; $2.45(p). 70-188205. SBN 8055-1107-5; 8055-0128-2.

SH-C-P   This collection of essays offers one of the better presentations of the issues surrounding population control. The essays are by a fascinating collection of people: Nat Hentoff, Daniel Callahan, Margaret Mead, Nat Glazer, Max Lerner, Amitai Etzioni and others equally famous. Well worth reading.

Loebl, Suzanne. *Conception, Contraception: A New Look.* (Illus.) McGraw-Hill, 1974. ix + 147pp. $6.95. 73-8018. ISBN 0-07-038339-1; 0-07-038340-5. Bib.

SH-C   The author begins with the need to limit births and then presents the story of Margaret Sanger and her work. There is a discussion of the development of oral contraceptives and the redevelopment of IUDs. The material on the reproductive facts is clear and accurate. Will interest students of sociology and population dynamics.

Lowenherz, Robert J.   *Population.* (Illus.) Creative Education Press, 1970. 120pp. $5.95. 74-104928. ISBN 0-8719-042-X.

JH   A meaningful and sympathetic approach to a complex topic. Photos, graphs, diagrams, and tables are well chosen and adequately explained; world problems such as urbanization, pollution and the quality of life are also discussed. Highly relevant.

Moraes, Dom.   *A Matter of People.* Praeger, 1974. vii + 266pp. $7.95; $3.95(p). 73-6497. Index;CIP.

SH-C   Dom Moraes travelled through 12 countries on four continents for the UN Fund for Population Activities to report on the population problem. The material is in diary form, useful for collateral reading in social science courses.

Petersen, William (Ed.).   *Readings in Population.* (Illus.) Macmillan, 1972. xi + 483pp. $5.95. 78-169981.

C-P   An eclectic presentation of the demographic point of view, with most of the articles comprehensible to readers without prior demographic experience. Excellent introductory comments by the editor.

Pringle, Laurence.   *One Earth, Many People: The Challenge of Human Population Growth.* (Illus.) Macmillan, 1971. 83pp. $4.95. 71-133559.

JH-SH   Rarely does a book present so completely in so little space the basic components of population dynamics. Topics range from reproduction and food supply, to attitudes of people, to family size and birth control methods including abortion and vasectomies. Presents both the optimistic and pessimistic sides of the population problem without any polemics and holds the reader's attention.

301.34–.36 COMMUNITIES

See also 711 Area Planning.

Banfield, Edward C.   *The Unheavenly City.* (Illus.) Little, Brown, 1970. viii + 308pp. $6.95. 77-105564.

C   Focuses light from a new direction on the problems of cities—that cities now are in many ways better places to live in than in the past. Stresses the inevitability of city ills as a natural consequence of growth. Suggests that most programs aimed at solutions to city problems are doomed to failure because of human class mores antithetical to upward mobility or environmental modification. His suggestions for treatment of many city problems will probably be unacceptable to society.

Higbee, Edward.   *A Question of Priorities: New Strategies for Our Urbanized World.* Morrow, 1970. xxxiv + 214pp. $6.00. 75-101701.

SH-C   Written with clarity and perception, the book is concerned with technological potential; social resistances toward a theory of the city, of governments and budgets; and social responsibility. Strong stress on the future; a fresh approach to social responsibilities.

Mercer, Charles.   *Living in Cities: Psychology and the Urban Environment.* (Illus.) Penguin, 1975. 240pp. $2.95(p). ISBN 0-14-021895-5. Index.

SH-C   This very readable and provocative book presents some of the issues of content and methodology with which behavioral scientists and planners of the built environment are tussling. Mercer is, however, highly selective in the topics he presents.

Roebuck, Janet. *The Shaping of Urban Society: A History of City Forms and Functions.* (Illus.) Scribner's, 1974. x + 256pp. $8.95. 73-18205. ISBN 0-684-13644-9; 0-684-13700-3(p). CIP.

SH-C-P   Comprehensive review of the socioeconomic factors which have shaped urban society from the age of prehistoric hunters to the modern city. Each chapter deals with a different historical period and provides a valuable point of reference for urban planners.

Schwartz, Alvin. *Central City/Spread City: The Metropolitan Regions Where More and More of Us Spend Our Lives.* (Illus.) Macmillan, 1973. 132pp. $4.95. 72-81068. ISBN 0-02-781320-7. Bib.

JH-SH   Schwartz presents a vivid and accurate picture of the living and working conditions in most central cities and their surrounding suburbs. Many photographs, maps and charts.

Scientific American. *Cities: Their Origin, Growth and Human Impact: Readings from Scientific American.* (Illus.) Freeman, 1973. 297pp. $12.00; $5.50(p). 73-2575. ISBN 0-7167-0870-1; 0-7167-0869-8.

SH-C-P   Discusses problems of population, health, environment, transport and city planning; cities in developing countries; and group relations in cities. The articles, published during the past two decades, are introduced by Kingsley Davis, Director of International Population and Research (U.C., Berkeley). The theme is "the profound contrast between the contributions that cities make to human society . . . and the effect they have on human beings."

Stearns, Forest, and Tom Montag (Eds.). *The Urban Ecosystem: A Holistic Approach.* (Illus.) Dowden, Hutchinson&Ross (distr.: Halsted), 1975. xv + 217pp. $18.00. 75-1001. ISBN 0-470-82079-9. Index;CIP.

SH-C   Stearn and Montag deal with the need to develop a new attitude towards urban life. In this holistic approach, more than 90 specialists from natural and social sciences worked with planning practitioners on a project of the Institute of Ecology to define and determine the "urban ecosystem."

Warren, Donald I. *Black Neighborhoods: An Assessment of Community Power.* (Illus.) Univ. of Michigan Press, 1975. xii + 194pp. $9.00. 73-90888. ISBN 0-472-08960-9. Gloss.

C-P   This ecological approach to the problems of black neighborhoods is based on extensive study of neighborhoods in Detroit. The material is directed away from the usual theoretical models, is not longitudinal, but is rooted in sound data and observation. Useful collateral reading.

301.41 THE SEXES AND THEIR RELATIONS

See also 155.3 Sex Psychology.

Lieberman, Bernhardt (Ed.). *Human Sexual Behavior: A Book of Readings.* (Illus.) Wiley, 1971. ix + 444pp. $11.50; $7.95(p). 70-162422. ISBN 0-471-53424-2; 0-471-53423-4.

C   This introduction to the scholarly literature on sexual behavior consists of selections based on two criteria: "that they deal with what we commonly consider usual heterosexual behavior, and that they meet contemporary standards of rigor and scholarship." A sophisticated look at the best literature in the field.

Lieberman, E. James, and Ellen Peck.   *Sex and Birth Control: A Guide for the Young.* (Illus.) Crowell, 1973. xvi + 299pp. $5.95. 73-7806. ISBN 0-690-72985-5.

JH-SH   The authors stress the element of choice and decision on the part of the girl, both in the decision to begin a sexual relationship and in the use of a particular contraceptive. For collateral reading, reference, or classroom text.

Masters, William H., and Virginia E. Johnson, with Robert J. Levin.   *The Pleasure Bond: A New Look at Sexuality and Commitment.* Little, Brown, 1975. xvi + 268pp. $8.95. 74-18390. ISBN 0-316-54981-9. CIP.

C-P   Dialogues between several groups of couples and the authors comprise about half the text and provide a basis for more abstract discussion. This is a "how-*not*-to" book, offered as preventive medicine. Themes of communication, trust, loyalty, feeling, commitment and growth are emphasized.

Pierson, Elaine C.   *Sex Is Never an Emergency: A Candid Guide for Young Adults, 3rd ed.* (Illus.) Lippincott, 1973. ix + 103pp. $3.95; $1.25(p). 78-148244. ISBN 0-397-47292-7; 0-397-47291-9.

SH-C   Pierson answers the questions which she has most frequently been asked as a college counselor. She discusses the gynecological examination, abortion, and questions relating to orgasm and dyspareunia. May be used as a starter for discussions.

Rimland, Ingrid.   *Psyching Out Sex.* Westminster, 1975. 142pp. $6.00; $3.25(p). 74-26537. ISBN 0-664-20724-3; 0-664-24815-2. CIP.

SH   Topics discussed include individual decision-making regarding sexual relations; how love is felt and perceived by boys and girls; the problem of guilt in our culture; the contradictory pressures on males for sexual achievement, and the females' responses to sexual pressures. Briefly discusses women's liberation and homosexuality.

301.412 WOMEN—SOCIAL ROLE AND FUNCTION

Bernard, Jessie.   *The Future of Motherhood.* Dial, 1974. xiii + 426pp. $10.00. 74-13391. ISBN 0-8037-2747-X. Index;CIP.

C   A provocative and fair look at such important issues as population problems, personal choice, and individual responsibility as they relate to motherhood. Discusses advantages and disadvantages of motherhood in a straightforward and nonjudgmental manner. Good reading for those seeking to know more about feminine identity.

Bird, Caroline.   *Enterprising Women.* Norton, 1976. 256pp. $8.95. 75-33662. ISBN 0-393-08724-7. Index;CIP.

SH-C   An exercise in reinterpreting history by increasing the attention given to those women who played an important economic role in U.S. history. The book is a series of brief biographies of "enterprising women," with commentaries about the conditions which enabled them to succeed in a man's world.

Bullough, Vern L., with Bonnie Bullough.   *The Subordinate Sex: A History of Attitudes Toward Women.* Univ. of Illinois Press, 1973. viii + 375pp. $10.95. 72-91079. ISBN 0-252-00320-9.

SH-C   The history of what men have said, thought and felt about women as a class. That women have occupied (and still occupy) second-class status in every culture is amply documented. The primary reason is said to be biological: unlimited fertility and its debilitating consequences, lack of physical strength, and the consequent rationalizations of society transmitted to each new generation of girls through the socializing process.

Carlson, Dale. *Girls Are Equal Too: The Women's Movement for Teenagers.* (Illus.) Atheneum, 1973. xiii + 146pp. $6.25. 73-76333. ISBN 0-689-30106-5.

   JH-SH   The author draws from recent feminist writings, providing enough infor-
         mation to raise everyone's level of awareness. A book which should not
be limited to teenagers.

Group For The Advancement of Psychiatry. *The Educated Woman: Prospects and Problems.* Scribner's, 1975. 188pp. $7.95. 75-2372. ISBN 0-684-14211-2. Index;CIP.

   C-P   Questions every assumption previously accepted by psychiatrists, behav-
         ioral scientists and laypersons regarding the psychology of women. Topics
include development of the female sex role, the college experience and expectations
of women regarding college and career, resolution of conflicts between need for
personal accomplishment and autonomy and need to develop relationships with men,
etc. Appropriate for women's studies, developmental psychology or education.

Huber, Joan (Ed.). *Changing Women in a Changing Society.* Univ. of Chicago Press, 1973. 295pp. $7.95. 72-96342. ISBN 0-226-35644-2.

   C-P   Diverse, provocative essays about the past, current and changing status of
         women; this anthology offers something for everyone.

Kundsin, Ruth B. (Ed.). *Women and Success: The Anatomy of Achievement.* Morrow, 1974. 256pp. $7.95. 73-11350. ISBN 0-688-00229-3.

   SH-C-P   A symposium volume designed to relate the characteristics of individual
         women to the larger patterns of society. Most interesting chapters are
those devoted to the influence of husbands and fathers on the success and happiness
of career-oriented women. Wide-ranging in scope.

Lloyd, Cynthia B. (Ed.). *Sex, Discrimination, and the Division of Labor.* (Illus.) Columbia Univ. Press, 1975. xiv + 431pp. $17.50; $6.00(p). 74-32175. ISBN 0-231-03750-3; 0-231-03751-1. CIP.

   C   This collection of essays challenges the existing division of labor between the
         sexes. Empirical data on female labor force participation, male/female earnings
differentials and occupational segregation is followed by articles reporting new economic research. In Lloyd's introduction, the economic implications of political, social and psychological attributes of sexual status are discussed, and almost half the
book deals with unpaid employment.

Martin, M. Kay, and Barbara Voorhies. *Female of the Species.* (Illus.) Columbia Univ. Press, 1975. x + 432pp. $15.00; $6.50(p). 74-23965. ISBN 0-231-03875-5; 0-231-03876-3. Gloss.;index;CIP.

   C-P   Two anthropologists examine the roles played by men and women in forag-
         ing, horticultural, agricultural, pastoral and industrial societies. Biological
and psychological aspects of sex differences are reviewed, and central concepts,
research evidence and controversies are accurately presented.

Murphy, Yolanda, and Robert F. Murphy. *Women of the Forest.* Columbia Univ. Press, 1974. xii + 236pp. $10.00; $3.45(p). 74-9912. ISBN 0-231-03682; 0-231-03881-X. Index;CIP.

   SH-C-P   The authors trace the effect that encroaching civilization has had on the
         relationships between men and women in villages inhabited by the Mun-
durucu Indians of Amazonian Brazil. Some comparisons with European societies are
made.

Oakley, Ann. *The Sociology of Housework.* Pantheon, 1974. x + 242pp. $10.00. 75-4668. ISBN 0-394-49774-0; 0-394-73088-7(p). Index;CIP.

C-P   Oakley treats housework as work rather than as an aspect of the feminine role in marriage. The research is based upon interviews conducted in the homes of 40 London housewives. Will be of interest to sociologists and lay readers.

Peck, Ellen, and Judith Senderowitz (Eds.).   *Pronatalism: The Myth of Mom and Apple Pie.* (Illus.) Crowell, 1974. xii + 332pp. $5.95. 74-6087. ISBN 0-690-00498-2. CIP.

SH-C-P   *Pronatalism* reviews the social pressures and attitudes which encourage reproduction and exalt the role of parenthood. Delves into the overlaps between sexism and the pressures exerted on women to reproduce. This book is the first to bring together writings on pronatalism and attempts to provide incentives for further research.

Pomeroy, Sarah B.   *Goddesses, Whores, Wives, and Slaves: Women in Classical Antiquity.* (Illus.) Schocken, 1975. xiii + 264pp. $8.95. 74-8782. ISBN 0-8052-3562-0. Index;CIP.

SH-C-P   This is a scholarly and objective study of women in the Greek and Roman world. The differing views and roles of women in Greek mythology, in the Homeric period, in fifth century Athens, in the Hellenistic age, in the Roman Republic and in the Early Empire are examined.

Raphael, Dana, (Ed.).   *Being Female: Reproduction, Power, and Change.* Mouton (distr.: Aldine), 1975. xiii + 293pp. $14.50. ISBN 0-202-01151-8. Indexes;bibs.

C-P   This is one of two volumes devoted specifically to women of the more than 50 volumes in the World Anthropology Series. The book is an excellent starter for further reading; there are sections on biomedical and sociocultural aspects of reproduction; women's domestic and economic power; and worldwide social trends in women's roles.

Reed, Evelyn.   *Woman's Evolution: From Matriarchal Clan to Patriarchal Family.* Pathfinder, 1975. xviii + 491pp. $15.00; $4.95(p). 74-26236. ISBN 0-87348-421-5; 0-87348-422-3. Gloss.;index.

SH-C-P   Reed details a controversial theory in which family structure began with maternal clans or hordes, developed into complex tribes with matrilinial descent, then into the matriarchal family and eventually evolved into the patriarchal family (as the concept of private property developed). Useful in sociology and anthropology courses.

Reiter, Rayna, (Ed.).   *Toward an Anthropology of Women.* Monthly Review, 1975. 416pp. $15.00. 74-21476. ISBN 0-85345-372-1. CIP.

SH-C   The authors, most of them Marxists, expose the traditional male biases in anthropological theory. Power is examined as it relates to control over critical food nutrients; poverty as a dependency-creating phenomenon; consensual bonds as a strategy for survival under slum conditions; and differential access to work and schooling; and Western (capitalistic) economy as a controlling factor in role assignment. Provides suggestions for new guidelines for anthropologists and those interested in women's studies.

Rosaldo, Michelle Zimbalist, et al. (Eds.).   *Woman, Culture, and Society.* Stanford Univ. Press, 1974. xi + 352pp. $12.50; $3.95(p). 73-89861. ISBN 0-8047-0850-9; 0-8047-0851-7. Index.

C-P   These 16 articles examine women's position in culture and society. Theoretical papers explore the relationship between women's participation in domestic life and their public activity, the limitations and implications of power women may hold inside and outside the home, and the way cultural ideologies hold women in "place." Provocative, scholarly, and well written, this collection is essential reading for any student of human culture.

301.42 MARRIAGE AND FAMILY RELATIONSHIPS

Gerzon, Mark. *A Childhood for Every Child: The Politics of Parenthood.* Outerbridge &Lazard, 1973. xii + 270pp. 72-97986. $7.95.

   C-P Gerzon presents interesting ideas on the effect of the Vietnam war, of alternative life styles and of the youth culture on the way Americans now raise their children.

Gilbert, Sara D. *What's a Father For? A Father's Guide to the Pleasures and Problems of Parenthood with Advice from the Experts.* (Illus.) Parents', 1975. xxiii + 231pp. $6.95. 74-20992. ISBN 0-8193-0793-9. Index;CIP.

   SH A well-written report on the common vicissitudes of different models of fatherhood discussed from many points of view. The author is realistic in her exploration of how cultural differences, changes in society and big city or small town life may affect the father's role in the family. While no stand is taken, readers will respond to the pleasant style.

Goodman, Elaine, and Walter Goodman. *The Family: Yesterday, Today, Tomorrow.* Farrar, Straus & Giroux, 1975. xii + 114pp. $5.95. 74-32069. ISBN 0-374-32260-0. Index;CIP.

   SH A generally clear, sound, readable and relevant book that defends the family as a useful institution. Seven chapters cover the primitive family, cross-cultural practices, state and church influence, and the American family. A more refreshing approach than most traditional family-life texts.

Grollman, Earl A. *Talking about Divorce: A Dialogue Between Parent and Child.* (Illus.) Beacon, 1975. 87pp. $7.50; $2.95(p). 75-5289. ISBN 0-8070-2374-4; 0-8070-2375-2. CIP;bib.

   C-P This book is a timely and sensitive aid to divorcing parents and their children, and a useful reference for those professionals who counsel and advise distressed couples. A carefully researched and eclectic listing of counseling and self-help services available to families is included.

Rohner, Ronald P. *They Love Me, They Love Me Not: A Worldwide Study of the Effects of Parental Acceptance and Rejection.* (Illus.) HRAF (Box 2015 Yale Sta.), 1975. 300pp. $12.00; $6.00(p). 75-17092. ISBN 0-87536-332-6. Index.

   C-P Rohner blends the fields of anthropology and psychology and addresses himself to the issue of prediction in the behavioral sciences. Almost half the book consists of appendixes, which supply exhaustive anthropological data, statistical tables, references and explanatory notes.

Stuart, Irving R., and Lawrence Edwin Abt (Eds.). *Interracial Marriage: Expectations and Realities.* Grossman, 1973. xiv + 336pp. $12.50. 72-77705. SBN 670-40014-9.

   C-P This wide-ranging collection of original papers by psychologists, sociologists, anthropologists, social workers, and others emphasizes comparative or cross-cultural aspects of interracial marriage. Contents include interracial marriage in Puerto Rico, Cuba, Brazil, Japan, and Hawaii; interviews of students discussing intergroup relations; an analysis of the Moynihan Report; a review of existing empirical studies; Murstein's theory of marital choice; interracial dating, adoption, marital role conflict, etc.

Yorburg, Betty. *The Changing Family.* Columbia Univ. Press, 1973. viii + 230pp. $2.95. 7284. ISBN 0-231-03461-X.

   SH-C Provides students and general readers with a short yet comprehensive view of the American family in terms of ethnic contrasts and class differences in the past, present and future. Exceptionally well written and referenced, with good integration of theory and factual material.

Young, Michael, and Peter Willmott. *The Symmetrical Family.* (Illus.) Pantheon, 1974. xxvii + 398pp. $10.00. 73-7009. ISBN 0-394-48727-3. CIP.

    C Although the study upon which this book is based was centered in London, readers may draw parallels with American culture as the authors explore the progression of familial relationships which culminate in the "symmetrical family." Historical analysis is integrated with 1970 survey material to establish comparisons of social life in historical and contemporary London and to trace the relationship between the individual, family and society.

301.435 AGED IN SOCIETY

See also categories following 155 Differential and Genetic Psychology.

Benet, Sula. *How to Live to be 100: The Life-Style of the People of the Caucasus.* (Illus.) Dial, 1976. xvi + 201pp. $8.95. 75-40471. ISBN 0-8037-3834-X. Index;CIP.

    GA A detailed study of the people of the Caucasus, where to live to be over 100 years old in a hearty state is apparently not rare and where such elders are revered. It gives a fascinating view of these extraordinarily robust people through the eyes of an accomplished anthropologist who details the striking value differences between their society and the industrialized West.

Davies, David. *The Centenarians of the Andes.* Anchor/Doubleday, 1975. xvi + 150pp. $6.95. 75-6270. ISBN 0-385-09914-2.

    C-P This is a short but detailed account of the four visits Davies made to southern Ecuador to study the oldest living people in the world. Possible causes of longevity are discussed at length and the author's conclusions are based on geographical, dietary, genetic and psychological factors. A fresh approach to a topic of universal interest.

Jonas, Doris G., and David J. Jonas. *Young Till We Die.* Coward, McCann&Geoghegan, 1973. 316pp. $6.95. 72-94128. SBN 698-10516-8.

    SH-C-P Using zoological, anthropological and historical evidence, the authors attempt to explain the complex reasons for the lowered status of the elderly and present an imaginative series of proposals aimed at restoring vigor, purpose and usefulness to the later years. Chapters dealing with causative factors suffer from inadequate documentation.

Kent, Donald, et al. (Eds.). *Research Planning and Action for the Elderly: The Power and Potential of Social Science.* Behavioral Publications, 1972. xix + 569pp. $22.95. 72-140049. ISBN 87705-056-2.

    C-P Shows how research planning can be translated into action. Contains diverse essays by some outstanding researchers and practitioners in gerontology. Emphasizes the difficulties inherent in behavioral research.

Knopf, Olga. *Successful Aging.* Viking, 1975. xii + 229pp. $8.95. 74-19250. ISBN 0-670-68081-8. Index;CIP;bib.

    SH-C Addressed specifically to the aged themselves, this book deals with a wide range of topics, including retirement planning, the use of leisure time, the pursuit of a second career, sexual adaptation in later life, health problems and adjustments, the aged parent and family relationships, making out the will, and "death with dignity."

301.451 AGGREGATES OF SPECIFIC NATIONAL, RACIAL, AND ETHNIC ORIGINS

Balikci, Asen. *The Netsilik Eskimo.* (Illus.) The Natural History Press, 1970. xxiv + 264pp. $8.95. 71-114660.

C-P Balikci's summary of the culture and social life of these Eskimos draws upon his own careful field work and research and other studies by early explorers. Topics include Netsilik technology, family and kinship patterns, social tensions and religious concepts and practices.

Burt, Jesse, and Robert B. Ferguson. *Indians of the Southeast: Then and Now.* (Illus.) Abingdon, 1973. 304pp. $7.95. 72-4695.

JH-SH-C This is a history of the destruction of the Indians and their culture by white diseases, military power and political subjugation. There are details about Indian leaders, past and present. Indian history up to the present decade and current Indian political and cultural affairs are discussed.

Chamberlin, J.E. *The Harrowing of Eden: White Attitudes Toward Native Americans.* Seabury, 1975. 248pp. $8.95. 75-9941. ISBN 0-8164-9251-4. Index;CIP.

SH-C-P This is an in-depth probing into the history of the way whites have perceived and treated native Americans, from earliest contacts to the present. A good work for reference or classroom use.

Coleman, James S. *Resources for Social Change: Race in the United States.* Wiley-Interscience, 1971. xii + 119pp. $7.95. 77-152494. ISBN 0-471-16493-3.

Coleman develops a general theory for social change based on the development of a systematic multivariate theory that sees change as the result of definition of goals and examination and application of available resources and change mechanisms. The theory incorporates social variables, world views, religion, economics and other factors. A considerably more sophisticated approach then revolutionary or simplistic economic models.

Davis, George A., and O. Fred Donaldson. *Blacks in the United States: A Geographic Perspective.* (Illus.) Houghton Mifflin, 1975. xi + 270pp. $5.95(p). 74-14362. ISBN 0-395-14066-8. Index.

SH-C A provocative book that analyzes sociospatial patterns of human communities in terms of definite socioeconomic causes. The authors deal with the factors affecting the distribution of blacks in the United States from the days of slavery to the present time. Emphasis is on how black people have perceived their own situation.

Dutton, Bertha P. *Indians of the American Southwest.* (Illus.) Prentice-Hall, 1975. xxix + 298pp. $14.95. 74-30311. ISBN 0-13-456897-4. Index;CIP.

C Dutton has attempted an ambitious synthesis of the current state of anthropological knowledge of the southwestern Indians. She treats each language group separately and provides an astonishing amount of detail along with short historical and contemporary commentaries.

Erdoes, Richard. *The Rain Dance People: The Pueblo Indians, Their Past and Present.* (Illus.) Knopf, 1976. 280pp. $7.95. 74-157. ISBN 0-394-82394-X; 0-394-92394-4. Index;CIP.

JH-SH Journalist-photographer-artist Erdoes relates the story of the survival of Pueblo peoples and their ancestors, who faced challenges ranging from prehistoric climatic-environmental changes to modern business interests. Their perseverance is explained in terms of cultural values. Useful for amplifying and illustrating points in history, current affairs and other social science courses.

Erny, Pierre. *Childhood and Cosmos: The Social Psychology of the Black African Child.* (Illus.; trans.) Black Orpheus, 1973 (hardcover); New Perspectives (distr.: Independent Publishers Group), 1973(p). 232pp. $12.50; $3.45(p). 71-172332. ISBN 0-87953-004-9; 0-87953-303-X(p). Index;CIP.

   JH-SH-C   Erny presents a structural study of the social psychology of the African child as reared in the various tribal cultures. This is a sensitive portrayal of child development within the cultural matrix. Recommended for collateral reading for any student of culture or personality.

Glazer, Nathan, Daniel P. Moynihan, et al. (Eds.). *Ethnicity: Theory and Experience.* Harvard Univ. Press, 1975. ix + 531pp. $15.00. 74-21230. ISBN 0-674-26855-5. Index.

   C-P   This product of a 1972 conference consists of 16 articles divided into three sections: theory, established nations and new states. How ethnic groups are established and related to other groups is explained. A definitive work about intergroup relationships at the macro level.

Goldstein, Rhoda L., (Ed.). *Black Life and Culture in the United States.* (Illus.) Crowell, 1971. xiii + 400pp. $6.95. 74-146281. ISBN 0-690-14598-5.

   SH-C   These readings cover a wide variety of topics concerning black history, social status, education, art, slavery, music, theater, political ideology, and organized prejudice groups in black culture encapsuled within the context of white culture. The articles are scholarly and well documented. The book would be most desirable as a supplement in courses dealing with black culture.

Greer, Colin, (Ed.). *Divided Society: The Ethnic Experience in America.* Basic, 1974. x + 405pp. $12.50. 73-88029. SBN 465-01679-0; 465-01680-4(p). Index.

   C   The articles deal with the problems of the immigrant in America. The concept of America as a great melting pot where all nationalities are successfully absorbed is challenged.

Gridley, Marion E. *Indian Tribes of America.* (Illus.) Hubbard, 1973. 63pp. $4.95. 72-91788. ISBN 0-8331-0017-3.

   JH   A good brief introduction to North American Indian tribal life, although the Comanches and Paiutes are omitted. Chapters cover tribes of the forest, plains, desert, southland, and western coast, and are preceded by convenient reference tables showing the names of tribes and original and current location. No bibliography or list of definitions of Indian terms.

Hays, H.R. *Children of the Raven: The Seven Indian Nations of the Northwest Coast.* (Illus.) McGraw-Hill, 1975. xii + 314pp. $12.95. 75-6668. ISBN 0-07-027372-3. Index;CIP.

   SH-C   The author deals with the native cultures of the Northwest coast of North America. The typical problems associated with colonialism—native exploitation by the fur trade in this area and social disorganization brought about by missionary and governmental activity—are described along with the more recent and continuing problems of education and economics. Excellent collateral reading and a good basic introduction for the general reader.

Henri, Florette. *Black Migration: Movement North, 1900-1920.* Anchor/Doubleday, 1975. xi + 419pp. $9.95. 74-9453. ISBN 0-385-04030-X. Index;CIP.

   GA-C-P   Henri deals with both the origins and the consequences of the mass migrations of blacks away from the old South. An excellent analysis.

Kluckhohn, Clyde, W.W. Hill, and Lucy Wales Kluckhohn. *Navaho Material Culture.* (Illus.) Harvard Univ., 1971. xiv + 488pp. $25.00. 78-122217. ISBN 0-674-60620-5.

SH-C-P   The Navajo are probably the best studied American Indian group, yet
this volume brings before the reader a whole new perspective on Navajo
life and their material culture. Information is presented on 263 items (subsistence,
shelter, clothing, ritual, and recreation etc.) taken from 25 years of field notes by
several anthropologists. The general presentation is excellent, and the content constitutes an outstanding description of the Navajo way of life in terms of the goods he
manufactures to sustain himself and his family.

Kuper, Leo, (Ed.).   *Race, Science and Society, rev. ed.* Columbia Univ. Press, 1975.
370pp. $12.00; $4.95(p). 74-11278. ISBN 0-231-03908-5; 0-231-03910-7. Index;CIP.

SH-C-P   This collection of papers could fit nicely into physical and cultural anthropology courses and social science courses dealing with race as a
social issue. Should be used with rather than instead of the first edition.

Levitan, Sar A., et al.   *Minorities in the United States: Problems, Progress, and Prospects.* (Illus.) Public Affairs, 1975. 106pp. $3.50(p). 75-21685. ISBN 0-8183-0242-9.

SH-C-P   The book's central aim is to assess the socioeconomic gains won for
blacks, Mexican-Americans, Puerto Ricans and Indians in the 1960s.
Income, employment and educational situations for each group are described. An
excellent supplement for college courses.

Marriott, Alice, and Carol K. Rachlin.   *Peyote.* Crowell, 1971. 111pp. $6.95. 75-
146284. ISBN 0-690-61697-X. Index;bib.

SH   An excellent introduction to the Native American Church. This Indian religion, developed around 1900, centers around the use of peyote, a hallucinogenic portion of the flower of a small cactus growing in the American Southwest. The spread of this church (sometimes underground) has affected the life of
American Indians and others today. This highly readable book is for anyone interested in the 20th-century American Indian. Excellent bibliography.

Mooney, James.   *Historical Sketch of the Cherokee.* (Illus.) Aldine, 1975. xiv + 272pp.
$12.50; $4.95(p). 75-20706. ISBN 0-202-1136-4; 0-202-01137-2. Index.

Royce, Charles C.   *The Cherokee Nation of Indians.* (Illus.) Aldine, 1975. xiv + 272pp.
$12.50; $4.95(p). 75-20708. ISBN 0-202-01138-0; 0-202-01139-9. Index.

SH-C-P   Mooney and Royce were staff members of the Bureau of American
Ethnology during the late decades of the 19th century. Their publications
are both classic studies. Royce's presentation of land cessions and treaties from
1785–1900 includes official text and historical background discussions. His work is
academic, straightforward and limited in scope. Mooney's broad sketch on contact
ethnohistory of Cherokee groups from Spanish days to 1900 covers contact by Euro-
peans, Americans and other Indian peoples of the east central United States. Both
include original maps and photographs, and introductory remarks by the elected
principal chief of the Cherokee Nation and by the president of Tulsa's Tsa-La-Gi-Ya
Cherokee Community. These well-done works would require some additional references and familiarity for proper use.

Nelson, Richard K.   *Hunters of the Northern Forest: Designs for Survival Among the
Alaskan Kutchin.* (Illus.) Univ. of Chicago Press, 1973. xv + 399pp. $10.50. 72-97941.
ISBN 0-226-5177-7.

SH-C   A collection of information on how to hunt, cook, and use various animals
of the North Country, particularly in sub-Arctic Alaska, this book provides
directions and descriptions of river boats, fish nets and netting, ice fishing, tracking
and snaring. Valuable for students of anthropology or researchers in the Alaskan
culture.

Oswalt, Wendell H.   *This Land Was Theirs: A Study of the North American Indian, 2nd
ed.* (Illus.) Wiley, 1973. xx + 617pp. $13.00. 72-11973. ISBN 0-471-65717-4.

C   Contains discussions of twelve Native American groups in the United States
and Canada. Each chapter gives the history of one group since European
contact, a summary of the way of life at a specific time, and the contemporary
situation of the group about 1970. The final chapter deals with policies and attitudes
toward Native Americans.

Pratson, Frederick John. *Land of the Four Directions.* (Illus.) Chatham Press (distr.:
Viking Press.), 1970. 131pp. $7.95. 75-122759. ISBN 0-85699-016-7.

JH-SH   This attractive volume presents 150 photographs with a brief text describing Indian peoples and their communities in Eastern Maine (Pas-
samaquoddis) and New Brunswick (Micmacs, Maliseets). It is clear that these Indians retain and are proud of their separate cultural origins.

Ryan, Joseph, (Ed.). *White Ethnics: Their Life in Working Class America.* Prentice-
Hall, 1974. 184pp. $7.95; $2.95(p). 74-22326. ISBN 0-13-957712-2; 0-13-957704-1.
CIP.

C   A collection of readings about the working lives of European-Catholic ethnic
Americans. Provides an interesting contrast to the Andrew Greeley and Studs
Terkel approaches to Catholic values and the status of working Americans. For
classroom or collateral reading.

Titiev, Mischa. *The Hopi Indians of Old Oraibi: Change and Continuity.* (Illus.) Univ.
of Michigan Press, 1972. xii + 379pp. $12.50. 79-142590. ISBN 0-472-08900-5.

C-P   This is an annotated diary covering Titiev's eight-month stay in Old Oraibi
from August 1933 to March 1934. The annotations of more recently gathered
data illustrate areas of continuity or change which have occurred in the Pueblo
between 1934 and 1967. For collateral reading in almost any anthropology or American Indian course.

Wagner, Nathaniel N., and Marsha J. Haug (Eds.). *Chicanos: Social and Psychological Perspectives.* (Illus.) Mosby, 1971. xxvii + 303pp. $5.75. ISBN 0-8016-5315-0.

C   This collection of 32 articles, largely gathered from publications of the last few
years, is psychologically oriented and includes articles written by Americans of
Mexican descent. The volume will provide an understanding of the predicament of
Chicanos in the United States.

Wetmore, Ruth Y. *First on the Land: The North Carolina Indians.* (Illus.) Blair, 1975.
xiii + 196pp. $8.95. 74-84151. ISBN 0-910244-80-4. Index.

JH-SH   Through archeological data and historical documentation, the past life
ways of these people are reconstructed. The present condition of North
Carolina's Indians is also described. Comprehensive but not very deep.

301.5 SOCIAL INSTITUTIONS

See also 001.6 Data Processing.

Chamberlain, Neil W. *The Place of Business in America's Future: A Study in Social
Values.* Basic, 1973. vii + 338pp. $12.50. 72-89186. SBN 465-05778-0. Bib.

C-P   The book synthesizes America's intellectual heritage in a welcome and readable fashion, with good explanations of the thoughts of others. Chamberlain
finds America's social values to be basically those of its business-industrial class, and
presents futuristic scenarios.

Dunnette, Marvin D. *Work and Nonwork in the Year 2001.* (Illus.) Brooks/Cole, 1973.
x + 212pp. $5.95(p). 72-94643. ISBN 0-8185-0080-8. Bib.;index.

SH-C-P   Ten essays discuss "work" (activities normally performed for pay) and
"nonwork" (productive leisure activities) from historical, cultural, and

institutional perspectives. For the general reader and for courses in industrial and organizational psychology.

Fischer, John.   *Vital Signs, U.S.A.* Harper&Row, 1975. viii + 197pp. $8.95. 74-15823. ISBN 0-06-011247-6. Index;CIP.

   SH   Fischer writes about progress made by little people in dealing creatively with both centralizing and decentralizing processes in our complex industrial society. He includes personal reflections about the New Deal's Farm Security Administration and valuable historical summaries. An inventory of solid social achievements.

Helmer, John.   *The Deadly Simple Mechanics of Society.* Seabury, 1974. 313pp. $9.95. 73-6417. ISBN 0-8164-9162-3. Index;CIP.

   C-P   This is simultaneously a well-documented exposé of the current state of affairs in sociology and a needed defense of workers against prevalent stereotypes employed by sociological writers in their treatment of the working class.

Kranzberg, Melvin, and Joseph Gies.   *By the Sweat of Thy Brow: Work in the Western World.* Capricorn/Putnam's, 1975. 248pp. $6.95. 74-79654. SBN 399-11312-6. Index.

   SH-C   A great deal of material has been condensed to produce this history of the nature of work from prehistoric times to the present. Descriptions of ancient agricultural and industrial technology are followed by the growth of the factory system and mass production. Recommended for those who want factual information without much intellectual stimulation.

Rule, James B.   *Private Lives and Public Surveillance: Social Control in the Computer Age.* Schocken, 1974. 382pp. $10.00. 73-90685. ISBN 0-8052-3542-6. Index.

   C-P   This exercise in empirical sociology uncovers the relationship between computer assembly of personal documentation and the growth of modern social forms. Investigates five bureaucratic systems which rely heavily on the use of personal information, and reveals the social forces shaping the continued development of surveillance practices. Provides a balanced consideration of the present capabilities and future directions of mass surveillance.

301.6 SOCIAL CONFLICT

Finifter, Ada W. (Ed.).   *Alienation and the Social System.* Wiley, 1972. x + 367pp. $9.95; $5.95(p). 70-180242. ISBN 0-471-25887-3; 0-471-25889-X.

   C-P   The articles comprising this book have been carefully selected to give the reader a deep understanding of the concept of alienation and its relation to a wide range of social concerns. Problems needing further examination are identified, and viewpoints from political philosophy, political science, sociology and psychology are represented. Thoughtfully organized.

Hirsch, S. Carl.   *The Riddle of Racism.* Viking, 1972. 222pp. $5.50. 75-185348. SBN 670-59791-0.

   JH-SH-C   This lucid, dramatically informative intellectual and social history is a decade-by-decade account of racism from the 1880s to the 1970s. It describes how the racism of the present developed out of the racism of the past, how racism imbues every institution of American life, how it functions, and how it has been made use of and fought against.

Karlins, Marvin, (Ed.).   *Psychology in the Service of Man: A Book of Readings.* Wiley, 1973. ix + 390pp. $8.95; $5.95(p). 72-7095. ISBN 0-471-45866-X; 0-471-45867-8.

   C-P   These 31 research reports are on major social problems such as education, crime and violence, drug abuse, urban overcrowding, poverty, ghettos, malnutrition, and youth in revolt. For each article, the editor has written an incisive introduction. Of particular merit for students seeking reasonable solutions to social issues.

Pasternack, Stefan A., (Ed.). *Violence and Victims.* Spectrum (distr.: Halsted), 1975. xiii + 215pp. $14.95. 74-14971. ISBN 0-470-66921-7. Index;CIP.

C-P   This collection of papers brings together the little we know, psychiatrically and psychologically, about those who commit violence and their victims. Among the topics covered are prediction of aggression, impact of legal procedures on victims, child abuse, Vietnam violence and theories of anger.

310 Statistical Collections

McHale, John. *World Facts and Trends, 2nd ed.* (Illus.) Macmillan, 1972. 96pp. $6.94; $2.95(p). 70-186436.

SH-C   A compendium of graphs and figures related to the environmental crisis, culled from a variety of sources and connected with sufficient text to make the whole effort cohesive.

Showers, Victor. *The World in Figures.* (Illus.) Wiley, 1973. xii + 585pp. $14.95. 73-9. ISBN 0-471-78859-7.

JH-SH-C   There are tables relating to physical geography, demography, climate, engineering and education for some 1,635 countries and cities. Numeric data are given in both English and metric units, and there is a cross-referenced index of more than 6,800 names of the individual entries. Has the flavor of the Guinness Book of Records.

333.7 CONSERVATION OF RESOURCES

See also 301.3 Ecology and Community and 639.9 Conservation of Biological Resources.

Andrews, William A., Donna K. Moore and Alex C. LeRoy. *A Guide to the Study of Environmental Pollution.* (Illus.) Prentice-Hall, 1972. x + 260pp. $5.95; $3.00(p). 72-1128. ISBN 0-13-370858-6; 0-13-370833-0.

SH   An effective tool for building an awareness of environment. Describes components of an undisturbed ecosystem and the air and water pollution problem. Challenging and practical questions for discussion and investigation and laboratory and field exercises covering the more important experimental and observational techniques are provided.

Barnes, Peter, (Ed.). *The People's Land: A Reader on Land Reform in the United States.* (Illus.) Rodale, 1975. xii + 260pp. $9.95; $6.95(p). 75-1028. ISBN 0-87857-091-8; 0-87857-093-4. Index;CIP.

SH   This collection of essays excerpted from congressional hearings, conference proceedings and other literature presents important historical and regional perspectives and covers technology and resource exploitation, the role of tax policy, rural poverty and an agenda of reform.

Blaustein, Elliott H. *Anti-Pollution Lab.* (Illus.) Sentinel, 1972. 128pp. $1.50(p). SBN 911360-54-9. Bib.

JH-SH   One of the few sources available to young people concerned with environmental problems that provides many completely detailed procedures for measuring quantitatively the air, water and solid matter pollution levels in communities.

Commoner, Barry. *The Closing Circle: Nature, Man, and Technology.* Knopf, 1971. 326pp. $6.95. 76-127092. ISBN 0-394-42350-X.

SH-C-P   An excellent overview, in simple language, of what the current crisis of our physical environment means. The concept of the ecosphere is explained, and the four laws of ecology are developed: everything is connected to everything else; everything must go somewhere; nature knows best; and there is no such thing as a free lunch (there are always costs and benefits). One of the best books available on the environmental crisis.

Davies, J. Clarence, III. *The Politics of Pollution.* Pegasus, 1970. xii + 231pp. $6.00; $1.95(p); 72-114174.

SH-C   An elementary, instructive and readable account of pollution control by a political scientist, which discusses setting standards, enacting legislation, allocating funds, enforcing laws, and the interrelation of government, special interest groups, and public opinion.

Elliott, Sarah M. *Our Dirty Land.* (Illus.) Messner, 1976. 64pp. $6.25. 76-184. ISBN 0-671-32770-4; 0-671-32771-2. Index;CIP.

JH   Litter, solid waste disposal, the debate over using the timber and minerals in natural parks and forests, pesticides, urban congestion, and land-use planning are covered. Frankly pro-environmentalist, focused on current issues.

Halacy, D. S., Jr. *Now or Never: The Fight against Pollution.* (Illus.) Four Winds, 1971. 208pp. $5.95. 71-161018.

JH   This readable book about the alarming proportions of the pollution problem includes background information on our current pollution difficulties, sections on aspects such as air and water pollution, solid waste, pesticides, radioactive contaminants, and noise and concludes with recommendations for action and a prognosis.

Hellman, Hal. *Energy in the World of the Future.* (Illus.) Evans (distr.: Lippincott), 1973. 240pp. $5.95. 72-90980. ISBN 0-87131-123-2.

SH-C   Hellman brings the reader from consideration of our present energy uses and supplies to the problem of pollution, and then to a consideration of future sources, either being developed or more visionary. There is information on coal, oil, boiling water, nuclear reactors, fast breeder reactors, geothermal energy, hydroelectric power, magneto-hydrodynamics, wind power, etc.

Hollon, W. Eugene. *The Great American Desert: Then and Now.* (Illus.) Univ. of Nebraska Press, 1975 (c.1966). xxvi + 284pp. $3.95(p). 75-5512. ISBN 0-8032-5806-2. Index;CIP.

SH-C   Hollon writes about that area which extends west of the Mississippi River and from southern California into northern Mexico. The author emphasizes the need for long-range planning, noting that the region demands a more useful human-land relationship. Useful, entertaining and thought-provoking.

Laycock, George. *Air Pollution.* (Illus.) Grosset&Dunlap, 1972. 81pp. 71-182014. ISBN 0-448-21440-7; 0-448-26207-X.

JH-SH   A brief, straightforward and elementary discussion, ranging from a historical and physical perspective on the earth's atmosphere, through a current inventory of pollutants, and ending with the pertinent issues of health effects, possible remedies and testing procedures.

Marx, Wesley. *Man and His Environment: Waste.* (Illus). Harper&Row, 1971. x + 179pp. $6.00; $3.25(p). 79-137805.

JH-SH-C   Much broader than its title suggests, and covers most of the environmental problems of today—solid waste, air and water pollution, pesticides, oil, radioactivity, noise pollution, etc. Lucid discussions include economic, social and political interactions and provide a general introduction to today's environmental problems.

Miller, C. Tyler, Jr.   *Living in the Environment: Concepts, Problems, and Alternatives.* (Illus.) Wadsworth, 1975. 592pp. $13.95. 74-79032. ISBN 0-534-00347-8. Gloss;index.

C   The world population problem, the crises in energy, food and mineral resources, land-use problems and air and water pollution are dealt with in detail.

Murdoch, William W., (Ed.).   *Environment: Resources, Pollution and Society, 2nd ed.* (Illus.) Sinauer, 1975. viii + 488pp. $10.95. 74-24361. ISBN 0-87893-503-7. Index.

SH-GA-C   This revised, updated edition retains an approach balanced between technical and economic, political and social solutions to environmental problems. An excellent introduction to ecological systems is followed by sections on such topics as population and resource abuse.

Potter, Jeffrey.   *Disaster by Oil: Oil Spills: Why They Happen, What They Do, How We Can End Them.* (Illus.) Macmillan, 1973. xii + 301pp. $7.95. 72-85183. Bib.

JH-SH-C   Describes four major oil spill disasters (the tankers *Torrey Canyon, Ocean Eagle,* and *General Colocotronis* and the platform spill at Santa Barbara) in a factual, fascinating and informative manner. The conclusions reflect the complexity of the problem, but do not spare the oil industry. There are only a few photographs and maps, no index and the many references are not annotated.

Sewell, Granville H.   *Environmental Quality Management.* (Illus.) Prentice-Hall, 1975. xii + 299pp. $9.95; $6.95(p). 75-4647. ISBN 0-13-283127-9; 0-13-283168-6. Index;CIP.

SH-C   Each component of the human environment is introduced by defining an environmental factor and explaining the basic aspects of that factor. The current status of and problems created by environmental management are examined, and a realistic view of the near future is presented. A clearly written introduction to a broad and complex subject.

Utgard, Russell O., and Garry D. McKenzie (Eds.).   *Man's Finite Earth.* (Illus.) Burgess, 1974. xi + 368pp. $4.95(p). 73-90807. ISBN 8087-2104-6. Gloss.

C-P   Collection of articles intended to familiarize the reader with the scientific disciplines impacting on the conflict between environmental preservation and social and economic growth. Recommended for first and second year college science students as well as environmental engineers and scientists.

Wagner, Richard H.   *Environment and Man, 2nd ed.* (Illus.) Norton, 1974. xiii + 528pp. $7.95. ISBN 0-393-09317-4. Index.

C   Wagner attempts to separate fact and opinion in his assessment of our impact on the environment. Although controversial in its treatment of dryland ecosystems, the book is of particular value for human ecology courses and for general reading.

Watson, Jane Werner.   *Toward a Better Environment for Our World Tomorrow.* (Illus.) Golden/Western, 1973. 140pp. $6.95. 73-81962.

JH-SH   Environmental problems in energy production, chemical production and use, foreign materials in air, water and land are covered in brief case studies. History, data collection, solutions, and opportunities and challenges for young scientists are presented factually but (necessarily) superficially.

333.72 CONSERVATION IN GENERAL

Allsopp, Bruce. *The Garden Earth: The Case for Ecological Morality.* Morrow, 1972. viii + 117pp. $5.00. 77-188179.

> **SH-C** Shows that Earth's resources are not inexhaustible and that they must not be exploited but propagated. Written in a simple yet eloquent and persuasive style.

Clepper, Henry, (Ed.). *Leaders of American Conservation.* Ronald, 1971. vii + 353pp. $10.00. 75-155206.

> **SH-C-P** This unique reference book includes more than 300 one-page biographies of American conservation leaders, each summarizing the individual's education, career, and principal achievements. An especially good record of early leaders.

Curry-Lindahl, Kai. *Conservation for Survival: An Ecological Strategy.* (Illus.) Morrow, 1972. xiv + 335pp. $6.95. 78-142417.

> **SH-C** This is not a "gloom and doom" type of conservation book, but the author does not gloss over the fact that we are now in an ecological crisis which will worsen as world population increases. The factual foundations on which plant and animal ecology must be based are so well presented that this book could be used as a text in a nonspecialist course in ecology.

Dasmann, Raymond F. *Environmental Conservation, 3rd ed.* (Illus.) Wiley, 1972. xi + 473pp. $10.50; $6.95(p). 71-37168. ISBN 0-471-19601-0.

> **SH-C** Covers a wide range of topics, including a description of the nature of our environment and a contrast of the major biotic regions of the earth. Traces the development of culture from prehistory to the present showing how the human impact on resources has increased with time. Describes the problems involved in utilization of major renewable and nonrenewable resources, such as water, timber, wildlife and minerals. Well illustrated; a good text.

Fritsch, Albert J. *The Contrasumers: A Citizen's Guide to Resource Conservation.* (Illus.) Praeger, 1974. ix + 182pp. $7.95; $3.50(p). 74-6726. ISBN 0-275-10140-1; 0-275-63540-6. Index;CIP.

> **SH-C** Written with an almost revolutionary fervor, this book calls for immediate action to reduce personal energy consumption and conserve diminishing energy resources. This useful tract should provide ammunition for any public speaker.

National Academy of Sciences. *The Earth and Human Affairs.* (Illus.) Canfield, 1972. xii + 142pp. $3.95; $1.95(p). 72-6804. ISBN 0-06-385491-0; 0-06-385490-2.

> **SH** This carefully documented plea to the general public to make more effective use of the world's limited resources and to reduce pollution is based on two major hypotheses: that Earth is the only suitable habitat we have, and that the Earth's resources—living space, energy, and materials—are limited. These axioms are controversial, but it is nevertheless a readable and comprehensive description of the problem.

Owen, Oliver S. *Natural Resource Conservation: An Ecological Approach.* (Illus.) Macmillan, 1971. xii + 593pp. $9.95. 74-119125.

> **C** An outstanding textbook, possibly the best available. Brings out the value of natural resources, the principles of ecology which underlie natural resource management, the techniques and policies by which resources can be managed effectively, the urgency of our conservation crisis, and the extent to which the public can participate in solving it.

Robertson, James, and John Lewallen (Eds.). *The Grass Roots Primer.* (Illus.) Sierra Club, 1975. 288pp. $7.95(p). 75-10017. ISBN 0-87156-142-5. Index;CIP.

SH-C-P *Primer* is mostly a compilation of environmental battles won by citizens throughout the United States and Canada. Valuable as a guidebook for people who are concerned about environmental degradation and want to take action.

Smith, Guy-Harold. *Conservation of Natural Resources, 4th ed.* (Illus.) Wiley, 1971. xiii + 685pp. $11.95. 71-129661. ISBN 0-471-80192-5.

SH-C A compendium of immensely broad subject matter woven into an integral, applicable, and understandable unit by 21 specialists. The major works are prefaced by a concise history of the early concerns for our resources by Gifford Pinchot, John Muir and Theodore Roosevelt. Land, water, air, and plant, animal and human communities are dealt with as related parts of a unified whole.

333.75 FOREST LANDS

Clawson, Marion. *Forests for Whom and for What?* Resources for the Future/Johns Hopkins Univ. Press, 1975. xi + 175pp. $11.95; $3.65(p). 74-24399. ISBN 0-8018-1698-X. Index;CIP.

SH-C-P Clawson, a member of the President's Advisory Panel on Timber and the Environment, has written a book with policy implications. While many of his conclusions are controversial, his arguments are logical and convincing. Of interest to the nonspecialist, specialist and anyone concerned with our natural resources.

Jones, Dewitt, and T. H. Watkins. *John Muir's America.* (Illus.) American West (distr.: Crown), 1976. 159pp. $20.00. 75-30536. ISBN 0-517-526387. Index;CIP.

GA A biographical narrative interlaced with insights about Muir's psychological beliefs. Drawings by Muir are interspersed throughout the text along with Dewitt Jones's versatile color photography. The book provides a good background and reference on landscape conservation; nature and history buffs will be delighted and inspired; outdoor photographers will be envious.

Minckler, Leon S. *Woodland Ecology: Environmental Forestry for the Small Owner.* (Illus.) Syracuse Univ. Press, 1975. xvi + 229pp. $9.95. 74-21909. ISBN 0-8156-0109-4. Index;CIP.

SH-C Minckler addresses conservation and woodland philosophy, ethics and ecological concepts; woodland economics; social values; timber production; wildlife and fish habitat; recreation and aesthetics; watershed values; and miscellaneous management investment and protection problems. This belongs in the library of every woodland owner.

Randall, Janet. *To Save a Tree: The Story of the Coast Redwoods.* (Illus.) David McKay, 1971. 115pp. $4.95. 72-165079.

JH-SH A well-balanced discussion of the redwood forests, their uses, and the need for preservation. Various aspects of botany, prehistory, discovery and history, lumbering and forest conservation are considered. Includes informative plates.

Robinson, Glen O. *The Forest Service: A Study in Public Land Management.* Resources for the Future/Johns Hopkins Univ. Press, 1975. ix + 337pp. $16.95; $4.95(p). 75-11352. ISBN 0-8018-1723-4; 0-8018-1768-4. CIP.

GA-C Robinson describes the history, organization, decision-making processes and major management responsibilities of the Forest Service. A critical analysis of some of the major land-management problems, controversies and policies are presented as well.

Silverberg, Robert. *John Muir: Prophet Among the Glaciers.* Putnam's, 1972. 256pp. $4.69. 74-183391.

JH-SH   An authoritative biography of a man devoted to the conservation of America's natural areas. Muir was a crusader for conservation from 1875 until his death in 1914, and he is here portrayed vividly as a heroic figure who was, nevertheless, a misfit in his own time. Provides a good introduction to the turn-of-the-century conservation movement.

Stephens, Rockwell R. *One Man's Forest.* (Illus.) Stephen Greene, 1974. 159pp. $7.95; $4.50(p). 74-23687. ISBN 0-8289-0224-0; 0-8289-0225-9. Index;CIP.

SH-C   This is a practical, clearly written account of how to buy, manage and profit from a woodlot. Based on the experiences of a retired couple who forsook city life for rural Vermont, this will be valuable for others who need to learn basic silvicultural principles and techniques as they apply to the Northeast.

Stone, Christopher D. *Should Trees Have Standing? Toward Legal Rights for Natural Objects.* Kaufmann, 1974. xvii + 102pp. $6.95; $2.95(p). 73-19535. ISBN 0-913232-09-2; 0-913232-08-4. Index;CIP.

C-P   Stone's thesis is that natural resources should be granted legal standing in American jurisprudence as aggrieved parties in their own right. Full of legal doctrine and footnoted legal citations, this volume would be of most interest to the student of environmental law.

Wood, Frances, and Dorothy Wood. *Forests Are for People: The Heritage of Our National Forests.* (Illus.) Dodd, Mead, 1970. xi + 210pp. $4.95. 74-128863. ISBN 0-396-06227-X.

JH-SH   The description of the national forest is well conveyed, and questions about forest management and policy are answered. The book is well written and provides entertaining and pleasant reading. For junior and senior high school reference for natural resources and biological sciences assignments and of interest to campers.

333.78 RECREATIONAL LANDS

Kirk, Ruth. *Yellowstone: The First National Park.* (Illus.) Atheneum, 1974. 103pp. $6.25. 74-76273. ISBN 0-689-50006-8. Index.

JH-SH   Kirk presents a number of interesting items for those who have never visited the park. Also includes a good discussion of wildlife policies and what is meant by the concept of a "National Park."

Udall, Stewart L. *America's Natural Treasures: National Nature Monuments and Seashores.* (Illus.) Country Beautiful (distr.: Rand McNally), 1971. 226pp. $14.95. 73-161502. ISBN 0-87294-029-2.

JH-SH-C   The geology, biology, folklore, and exploration of the 73 National Nature Monuments and Seashores are considered; 96 color and 130 black-and-white photographs of excellent quality are included; information is listed alphabetically by Monument. Useful collateral reading in intermediate biology and earth-science courses.

Watkins, T. H., and Charles S. Watson, Jr. *The Lands No One Knows: America and the Public Domain.* (Illus.) Sierra Club Books, 1975. 256pp. $9.95. 75-1276. ISBN 0-87156-130-1. Index;CIP.

C   Outlines the give-away and squandering of the public domain—land held in trust by the federal government for the people of the U.S. Emphasis is on public land laws as they evolved and are evolving in response to continuous demands on and damage to these lands. A fine classroom reference and basic source book.

Wayne, Bennett, (Ed.). *They Loved the Land.* (Illus.) Garrard, 1974. 168pp. $3.98. 74-915. ISBN 0-8116-4908-3. Index;CIP.

JH   Contains brief biographies of John Audubon, John Muir, Luther Burbank, and Rachel Carson, each of whom made significant contributions to the preservation and enhancement of the environment and the scenic beauty of the United States.

333.8 UTILIZATION OF MINERAL RESOURCES

See also 338.2 Mineral Production.

Caudill, Harry M. *My Land Is Dying.* (Illus.) Dutton, 1971. 144pp. $6.50. 76-158582. ISBN 0-525-16230-5.

SH-C   Caudill is a lawyer, and he includes references to many of the key legal battles over regulation of strip mining. The history of strip mining in Kentucky is developed fully; the environmental impact is described and lamented. Most of the text is devoted to political and social aspects.

Gordon, Richard L. *U.S. Coal and the Electric Power Industry.* Resources for the Future/Johns Hopkins Univ. Press, 1975. xiii + 213pp. $12.50. 74-24403. ISBN 0-8018-1697-1. Index;CIP.

C-P   A significant and timely book which includes an historical survey of the coal market, its sources and price structure, and the present situation with regard to coal use, geographical distribution and pricing. Provides useful background information for concerned citizens, lawmakers, industrialists.

Toole, K. Ross. *The Rape of the Great Plains: Northwest America, Cattle and Coal.* Atlantic Monthly/Little, Brown, 1976. ix + 271pp. $8.95. 75-31927. ISBN 0-316-84990-1. Index;CIP.

SH-C   This admittedly biased but mainly accurate view of the exploitation of the Northern Great Plains, chiefly Montana, maintains that the great coal and energy companies are using "inexhaustible resources . . . for blatant propaganda all over America." Toole discusses the region itself, the Cheyenne and the Crow Indians, cattlemen, community development, land reclamation, water use, the consequences of transmission lines, and the environmental movement in Montana.

333.82 FUELS AND ENERGY

See also 621.4 Heat Engineering and Prime Movers.

Abelson, Philip H. *Energy for Tomorrow.* (Illus.) Univ. of Wash. Press, 1975. ix + 78pp. $5.95; $2.50(p). 75-1368. ISBN 0-295-95413-2; 0-295-95414-X. CIP.

C   This book provides a concise and accurate summary of the current energy situation in the United States. Abelson stresses the highly complex nature and long duration of the energy problem and states that no simple or cost-free solutions are likely to emerge.

Clark, Wilson. *Energy for Survival: the Alternative to Extinction.* (Illus.) Anchor/Doubleday, 1974. xvi + 652pp. $12.50. 72-89297. ISBN 0-385-3501-2. Index;CIP.

SH-C   Essential background material is provided for understanding the broad problems associated with energy transformation, conservation and use through 1973. An extensive literature survey is included for each chapter. Useful collateral reading for both students and teachers.

Commoner, Barry, et al. (Eds.). *Energy and Human Welfare—A Critical Analysis; The Social Costs of Power Production (Vol. 1).* (Illus.) Macmillan Information/

Macmillan, 1975. xx + 217pp. $14.95 ($40.00; 3 vol. set). 75-8986. ISBN 0-02-468420-1. Index;CIP.

C-P   The environmental and societal impact of the two major power generation
       systems are presented by experts in a variety of technologies. About half the
book deals with fossil fuel power generation and the other half with nuclear power.
Primarily useful to professionals.

Commoner, Barry, et al. (Eds.).   *Energy and Human Welfare—A Critical Analysis;*
Human Welfare: The End Use for Power (Vol. 3). (Illus.) xvi + 185pp. $14.95. 75-8992.
ISBN 0-02-468440-6. Index.

C   This volume focuses on the effect of energy utilization on human welfare.
     Tables and graphs abound. Recommended for courses in technology impact.
(For Vol. 2 in this set, see *621.4 Heat Engineering and Prime Movers.*)

Commoner, Barry.   *The Poverty of Power: Energy and the Economic Crisis.* Knopf,
1976. 314pp. $10.00. 75-36798. ISBN 0-394-40371-1. Index;CIP.

SH-C-P   Commoner examines the politics, principles and practice of power con-
          sumption in the world today, starting with an entertaining and relatively
understandable account of thermodynamics and how energy is accounted for. The
general lack of courage in setting energy policy in the world today is highlighted. For
all economists, biologists and students of the political process.

Ford Foundation.   *A Time to Choose: America's Energy Future.* (Illus.) Ballinger/
Lippincott, 1974. xii + 511pp. $10.95; $3.95(p). 74-14787. ISBN 0-88410-023-5;
0-88410-024-3. CIP.

C   This think-tank study provides a choice of three "futures" scenarios with
     estimates of the energy required for each. The major conclusion is that
America can reduce its energy growth to zero percent with no adverse economic
effects. Gives a rational response to oil price increases and OPEC's energy politics.

Freeman, S. David.   *Energy: The New Era.* Walker, 1974. 386pp. $14.50. 74-77980.
ISBN 0-8027-0460-3. Index.

SH-C   Background information on the energy crisis is followed by a history of
        energy development and patterns of energy consumption. Problems of
government policy, foreign policy, domestic policies, conservation and better utilization of existing resources are discussed and interrelated.

Gabel, Medard, et al.   *Energy, Earth and Everyone: A Global Energy Strategy for*
Spaceship Earth. (Illus.) Straight Arrow (distr.: Simon&Schuster), 1975. 160pp.
$4.95(p). 75-9429. ISBN 0-87932-095-8. Gloss.

SH-C   The authors treat the whole earth and all its energy needs as one functional
        unit and identify steps that could and should be taken between 1975 and
1985 in order to realize a global strategy for energy development to meet humanity's
needs.

Marshall, James.   *Going, Going, Gone? The Waste of Our Energy Resources.* (Illus.)
Coward, McCann&Geoghegan, 1976. 94pp. $5.49. 75-44014. SBN GB-698-30607-4;
TR-698-20355-0. Index;CIP.

JH-SH   Contains a wealth of information presented clearly, comprehensively and
         in detail, including the energy crisis, fossil fuels, nuclear energy, steps
that can be taken towards energy conservation, solar, wind and geothermal energy,
ocean gradient, and fusion. Useful as a beginning reference or a textbook.

National Academy of Sciences.   *Energy: Future Alternatives and Risks: Academy*

Forum. (Illus.) Ballinger, 1974. ix + 227pp. $11.50. 74-13084. ISBN 0-88410-025-1. Index.

C-P   A multidisciplinary analysis of the scope and possible solutions to the energy problem by leaders in the sciences, social sciences, humanities, government, industry and public interest groups. Presents lively discussions that illuminate controversial issues. The book is a transcript of a public forum.

Pringle, Laurence. *Energy: Power for People.* (Illus.) Macmillan, 1975. 147pp. $6.95. 74-19033. ISBN 0-02-775330-1. Gloss;index;CIP.

JH-SH-GA   A lucid and interesting book that hits hard at the wasteful use of fossil fuels in the United States. Following a discussion of the 1973–74 energy crisis, the author describes new methods of fuel extraction and possible alternative sources.

Szulc, Tad. *The Energy Crisis.* Watts, 1974. 133pp. $6.95. 74-17335. ISBN 0-531-02752-X. Index;CIP.

SH-C   The author examines the energy sources available, particularly oil, coal, natural gas and electricity. The questions of how and why the crisis developed are addressed. The interplay among government, the oil industry, developing nations and international oil conglomerates and the resulting effect on the development of the energy shortage is clearly discussed.

333.9 UTILIZATION OF OTHER NATURAL RESOURCES

Baxter, William F. *People or Penguins: The Case for Optimal Pollution.* Columbia Univ. Press, 1974. 110pp. $5.95; $1.95(p). 74-6102. ISBN 0-231-03820-8; 0-231-03821-6. CIP.

SH-C   A short exposition of elementary economic theory as applied to the management of pollution control technologies. Concludes with a carefully constructed argument for the adoption of a tax to penalize industrial polluters in proportion to the amount of damage caused. A fine springboard for discussions of the economics of clean air and water.

Frisken, William R. *The Atmospheric Environment.* (Illus.) Resources for the Future (distr.: Johns Hopkins Univ. Press), 1973. x + 68pp. $3.50(p). 73-8139. ISBN 0-8018-1530-4. Bib.

C-P   This book reviews man's interaction with his atmospheric environment, focusing chiefly on atmospheric pollution. There is also a brief review of our present state of knowledge concerning climate change and its cause. For the scientist, but also recommended as a text in an introductory meteorology course.

Lavaroni, Charles W., and Patrick A. O'Donnell. *Air Pollution.* (Illus., teacher's ed.) Addison-Wesley, 1971. 94pp. $1.68.

Lavaroni, Charles W., Patrick O'Donnell, and Lawrence A. Lindberg. *Water Pollution.* 94pp. $1.68.

JH   Two of a series of three (the other deals with noise) for a junior high school unit on pollution. Each book gives the problem, additional background information, student investigations, and questions and suggestions for additional investigations.

Tannenbaum, Beulah, and Myra Stillman. *Clean Air.* (Illus.) McGraw-Hill, 1974. 64pp. $4.33. 73-3332. ISBN 0-07-062892-0.

JH   An informative book on air pollution covering smog alert, what's in the air, burning, soot, sulfur dioxide, hydrocarbons, food and water, winds, and today's and tomorrow's cities.

333.91 WATER AND SHORE CONSERVATION

Adamson, Wendy Wriston. *Saving Lake Superior: A Story of Environmental Action.* (Illus.) Dillon, 1974. 75pp. $5.95. 74-17351. ISBN 0-87518-083-3. Gloss.;CIP.

JH-SH   Adamson presents the geology and history of the largest of our Great
Lakes, as well as information on environmental political action. A good reference on environmental study.

Berkman, Richard L., and W. Kip Viscusi. *Damming the West: Ralph Nader's Study Group Report on the Bureau of Reclamation.* Grossman, 1973. xiv + 272pp. $7.95. 72-77707. SBN 670-25460-6.

SH-C   This extensively documented report accuses the Bureau of Reclamation of
promoting projects (mostly irrigation dams in Western states) which consume billions of taxpayers' dollars and serve only to assure the continuing existence of the Bureau. Several case studies expose the use of distorted technical criteria and faulty data. Illustrates the disastrous environmental consequences of irresponsible planning.

Bradford, Peter Amory. *Fragile Structures: A Story of Oil Refineries, National Security, and the Coast of Maine.* Harper's Magazine Press/Harper&Row, 1975. xiii + 392pp. $12.95. 72-12096. ISBN 0-06-120450-1. Index;CIP.

C   This is an historical account of the efforts (1961–1974) of small-time pro-
moters, politicians, and big-time oil company executives to install oil depots and refineries at deep water harbors of Maine. Facts and events are carefully documented. A useful reference.

Braun, Ernest, and David Cavagnaro. *Living Water.* (Illus.) American West, 1971. 184pp. $17.50. 75-142443. ISBN 0-910118-20-5.

GA   In this excellent book, each photograph can stand alone as an unique work of
art. The text makes a strong plea for restructuring our life styles so that humans can enter into an environmentally balanced relationship with the rest of nature.

Brokaw, Dennis, and Wesley Marx. *The Pacific Shore: Meeting Place of Man and Nature.* (Illus.) Dutton, 1974. 144pp. $20.00. 74-10667. ISBN 0-525-17438-9. Index;CIP.

SH-C   A comprehensive geographic and historic summary of a magnificent,
resource-rich coast now in jeopardy from exploitation and wastefulness. Presents a strong case for modifying our demands on coastal resources. An interesting, fact-filled reference.

Carter, Luther J. *The Florida Experience: Land and Water Policy in a Growth State.* (Illus.) Resources for the Future/Johns Hopkins Univ. Press, 1975. xvi + 355pp. $15.00. 74-6816. ISBN 0-8018-1646-7. Index;CIP.

C-P   One of the nation's most knowledgeable writers on environmental and con-
servation issues has written a history of Florida's land and water. Political aspects of key decisions are analyzed and regulatory policies are suggested. City, county and state planners, representatives and leaders take note.

Moorcraft, Colin. *Must the Seas Die?* (Illus.) Gambit, 1973. x + 194pp. $6.95. 72-91818. ISBN 0-87645-069-9.

SH-C   A thoroughly holistic explanation of what is happening to the oceans from
an international viewpoint. Overfishing, chemical and radioactive pollution and other specific problems are discussed and solutions suggested.

Neal, Harry Edward.   *The People's Giant: The Story of TVA.* (Illus.) Messner, 1970. 96pp. $3.95. 74-124296. ISBN 0-671-32363-6.

JH   This book is a historical-geographical model of a river system, the Tennessee, in the humid, well-watered, southeastern U.S. The book tells of the social, economic and geographic conditions of the Tennessee Valley as they existed before TVA and the changes that ensued with the building of TVA. The major part of this book is given to the construction and benefits of the TVA.

Owen, Marguerite.   *Tennessee Valley Authority.* (Illus.) Praeger, 1973. ix + 275pp. $9.50. 72-85985.

SH-C   Some statistics but much more about people and politics and struggle between public and private power interests over TVA. Marguerite Owen was the Washington representative for TVA for 33 years, and her well-written history will be enjoyed by anyone interested in U.S. history, public vs. private power, conservation and ecology.

Rondiere, Pierre.   *Purity or Pollution: The Struggle for Water.* (Illus.) Watts, 1971. 128pp. $4.95. 74-153828. ISBN 0-531-02103-3.

JH-SH   Reviews concisely the early development and use of water in various parts of the world, particularly for irrigation and the growing of food crops. The systematic presentation of water in the air, ground, and surface, and of the hydrologic cycle also provides a vivid insight into our capacity to pollute.

Steinhart, Carol E., and John S. Steinhart.   *Blowout: A Case Study of the Santa Barbara Oil Spill.* (Illus.) Duxbury, 1972. xvi + 138pp. $3.00(p). 72-075102.

SH-C-P   An account of the Santa Barbara oil spill, this book is also a primer of wisdom for the knee-jerk environmentalist, an introductory text in vicissitudes for inadvertent polluters, an object lesson for bureaucrats, a paradigm of objective historical reporting and comment, and a pleasure to read.

Voigt, William, Jr.   *The Susquehanna Compact: Guardian of the River's Future.* (Illus.) Rutgers, 1972. x + 337pp. $15.00. 71-185398. ISBN 0-8135-0722-7.

C-P   The *Compact* created a coordinating and regulatory Commission to aid in the conservation and management of the water resources of a watershed located within three sovereign states. The book is an interpretive history of the *Compact* — its origins, purposes, and promise. It is mandatory for professionals and citizens to understand effective resource management on a regional basis.

333.95 BIOLOGICAL RESOURCES

See 639.9 Conservation of Biological Resources.

338.1 AGRICULTURAL PRODUCTION

Aykroyd, W.R.   *The Conquest of Famine.* (Illus.) Reader's Digest Press (distr.: Dutton), 1975. 216pp. $7.95. 74-32653. ISBN 0-88349-054-4. Index.

SH-C-P   The author defines famine and discusses the different reasons for its occurrence. The history of major famines in China, Egypt, India, Ireland, Netherlands, Soviet Union and other parts of Europe is well described. Recommended to nutrition and population specialists, students, teachers, and the general public.

Bickel, Lennard.   *Facing Starvation: Norman Borlaug and the Fight Against Hunger.* Reader's Digest Press (distr.: Dutton), 1974. 376pp. $8.95. 74-1078. ISBN 0-88349-015-3. CIP

SH-C   Challenges the environmentalists who oppose the use of fertilizers and insecticides in food production, predicting that, unless action is taken now, fifty million people will starve in the next few years. Must reading for those who want both sides of the story.

Brown, Lester R., with Erik P. Eckholm. *By Bread Alone.* ((Illus.) Praeger, 1974. xi + 272pp. $8.95. 74-16477. ISBN 0-275-33540-2; 0-275-63640-2(p). Index;CIP.

   SH-C   The authors describe and defend two policy recommendations regarding overpopulation and food shortages: cooperative efforts to promote agricultural production in poorer areas, and an immediate dampening of population growth.

Brown, Lester R., and Gail W. Finsterbusch. *Man and His Environment: Food.* (Illus.) Harper&Row, 1972. x + 208pp. $3.25(p). 76-178104. ISBN 06-040984-3; 06-040983-5.

   SH-C   The kinds of foods which were and are available, their sources and the recent historical changes in agricultural techniques are all discussed. The authors also consider the change from hunting to farming, and future food production and needs.

Brown, Walter R., and Norman D. Anderson. *Historical Catastrophes: Famines.* (Illus.) Addison-Wesley, 1976. 191pp. $5.95. 75-22297. ISBN 0-201-00827-0. Index;CIP.

   JH   A dozen famines of the past century are described, including the recent one in Bangladesh; the plight of a particular family or group is sometimes emphasized. Historical background, consequences, efforts to avoid famines, and some research into causes are summarized.

Eckholm, Erik P. *Losing Ground: Environmental Stress and World Food Prospects.* Norton, 1976. 223pp. $7.95. 75-41397. ISBN 0-393-06410-7; 0-393-09167-8(p). Index;CIP.

   C   A valuable contribution to the understanding of basic ecology and the causes of current or impending droughts and famines. Emphasizes the continuous cause/effect cycle of population, environmental degradation and poverty. Eckholm recommends increased education, changing land tenure, ecological planning and changes in technology.

Edwin, Ed. *Feast or Famine: Food, Farming, and Farm Politics in America.* Charterhouse, 1974. xii + 365pp. $10.95. 74-82976. ISBN 0-88327-038-2. Index.

   SH-C   This timely book relates aspects of farming and food production to the politics of American agriculture. Suitable for both biologists and political scientists.

Green, Daniel. *The Politics of Food.* Cremonesi (distr.: Atheneum), 1975. xi + 220pp. $14.95. ISBN 0-86033-000-1. Index.

   C   Deals with food policies, statistics, consumer and farm pressure groups, and problems arising from the history, ideology and laws of the various national groups. Five countries are used to illustrate national and international politics and land reform measures. More for readers versed in economics than the general reader.

Idyll, C.P. *The Sea Against Hunger: Harvesting the Oceans to Feed a Hungry World.* (Illus.) Crowell, 1970. xii + 221pp. $7.95. 72-113859. ISBN 0-690-72264-8.

   C   Reviews the hunger problem, the contribution of the sea in providing food, and the technical difficulties of increasing such production. Fish, their present and potential contributions, and the social customs and laws which hinder the exploitation of the sea are also discussed.

National Research Council. *World Food and Nutrition Study: Interim Report.* Nat'l. Academy of Sciences, 1975. xix + 85pp. $5.50(p). 75-37120. ISBN 0-309-02436-6.

   SH-C   Discusses what steps the U.S. should and can take toward assisting other countries in developing and increasing their food production. An excellent overview but somewhat shallow; includes recommendations on nutrition research.

Slack, A.V. *Defense against Famine: The Role of the Fertilizer Industry.* Doubleday, 1970. 232pp. $5.95. 76-104982. Index.

SH-C The history of the fertilizer industry and its ability to produce food products of greater quantity and quality is traced, and the task of producing and distributing plant nutrients to assure adequate worldwide food production is discussed.

Vicker, Ray. *This Hungry World.* Scribner's, 1975. ix + 270pp. $9.95. 75-11920. ISBN 0-684-14383-6. Index;CIP.

SH-C The author reviews the complexities of the hunger crisis—population growth, technological development, weather, environmental protection, land situation and political priorities in both rich and poor countries.

338.2 MINERAL PRODUCTION

See also 333.8 Utilization of Mineral Resources.

Allvine, Fred C., and James M. Patterson. *Highway Robbery: An Analysis of the Gasoline Crisis.* (Illus.) Indiana Univ. Press, 1974. xii + 261pp. $10.00. 74-1598. ISBN 0-253-13750-0. Index;CIP.

C-P The authors provide a detailed account of the development, practices and economic base of the petroleum industry, which they consider responsible for the 1972 gas shortage. A valuable book for both energy policymakers and economists.

Inglis, K.A.D., (Ed.). *Energy: From Surplus to Scarcity?* (Illus.) Halsted/Wiley, 1974. ix + 242pp. $21.95. 73-22112. ISBN 0-470-42731-0. Indexes;CIP.

C-P The wide-ranging scope of this energy analysis focuses on problems of competition, energy planning, growth of markets, outlook for oil producers, and alternative sources of energy. The required level of understanding and depth of analysis gear this book toward more advanced students and graduate courses emphasizing energy.

Miller, Roger LeRoy. *The Economics of Energy: What Went Wrong?* Morrow, 1974. 131pp. $4.95. 74-172. ISBN 0-688-00302-8.

SH Miller advocates letting market prices allocate existing supplies of petroleum products among present users, and existing stocks of crude oil reserves among present and future users. Bureaucrats and politicians are blamed for the energy crisis.

Ridgeway, James, and Bettina Conner. *New Energy: Understanding the Crisis and a Guide to an Alternative Energy System.* Beacon, 1975. xiii + 224pp. $7.95. 74-16669. ISBN 0-8070-0504-5. Index;CIP.

SH-C-P The energy crisis of 1973–74 is viewed as part of the process to reorganize industry-government relations. Histories of the energy industries and case studies of congressional reform efforts document the trend towards centralization. An alternative energy system is suggested, based on new sources and local control.

Sampson, Anthony. *The Seven Sisters: The Great Oil Companies and the World They Made.* (Illus.). Viking, 1975. xv + 334pp. $10.00. 75-20268. ISBN 0-670-63591-X. Index;CIP.

SH-C Sampson describes the way the image of the oil companies has changed for the worse. The "seven sisters" are Exxon, Texaco, Shell, Gulf, Mobil, Socal (Chevron) and British Petroleum (BP). Many insights into the future of the "petroleum age" are provided.

350.855 SCIENCE AND GOVERNMENT

Primack, Joel, and Frank von Hippel.   *Advice and Dissent: Scientists in the Political Arena.* Basic, 1974. xi + 299pp. $12.95. 73-90136. SBN 465-00090-8. Index.

C-P   A report on and analysis of the politics of technology development; presents case studies involving public acceptance or rejection of issues such as banning DDT and the supersonic transport. Provides valuable insights into the bureaucratic manipulation of scientific advice and data to serve special interest groups.

Scott, David L.   *Pollution in the Electric Power Industry: Its Control and Costs.* (Illus.) Lexington/Heath, 1973. xvi + 104pp. $10.00. 73-6559. ISBN 0-699-89219-X.

C-P   Scott believes there must be a balance between the costs of reducing pollution and the economic benefits that such reduction might achieve. He urges that a national energy policy be adopted. Useful as collateral reading for students of economics, professional managers, and public officials.

Taubenfeld, Howard J.   *Controlling the Weather: A Study of Law and Regulatory Procedures.* Dunellen, 1970. xvi + 275pp. $10.00. 76-132981. ISBN 0-8424-0017-6.

SH-C   This book is the first report of a task group on the legal problems of weather modification activities organized in 1968 by Southern Methodist University Law School. Deals with legislation enacted in 29 states.

Thomas, William A., (Ed.).   *Scientists in the Legal System: Tolerated Meddlers or Essential Contributors?* Ann Arbor Science, 1974. vii + 141pp. $14.50. 73-90414. ISBN 0-250-40050-2. Index.

C-P   This collection of papers by lawyers who have had extensive experience with the lawyer-scientist relationship is addressed to scientists and underlines scientific inputs in the legal system. Contributes to a better understanding of the problems and benefits of collaboration between the two systems.

361 SOCIAL WELFARE

Demone, Harold W., Jr., and Dwight Harshbarger (Eds.).   *A Handbook of Human Service Organizations.* Behavioral, 1974. xx + 600pp. $22.95. 73-12280. ISBN 0-87705-120-8.

C-P   The 39 contributors, from the areas of social and psychological sciences, administration, and health planning, present guidelines for new directions for personnel in human service organizations. Considers mental health, hospitalization, education, relationship to politics, National Health Insurance, organizational alternatives, etc.

Pilisuk, Marc, and Phyllis Pilisuk (Eds.).   *How We Lost the War on Poverty.* Transaction (distr.: Dutton), 1973. 338pp. $7.95; $2.95(p). 72-91471. ISBN 0-87855-079-8; 0-87855-574-9.

C-P   The editors have compiled the works of writers affiliated with social programs of the 1960s and have provided an explanation of why social reform so often fails. There are interesting insights into the future of social and health care legislation. Highly useful and readable.

Schulberg, Herbert C., and Frank Baker (Eds.).   *Developments in Human Services, Vol. II.* Behavioral, 1975. x + 404pp. $17.95; $13.95(p). 73-6840. ISBN 0-87705-168-2. Index.

C   This volume consists of three comprehensive, nontechnical monographs, useful to students, legislators, planners, administrators and general readers concerned with the burgeoning human services sector of the evolving American socioeconomic system. A well-referenced review.

362.1 MEDICAL CARE POLICIES AND SERVICES

See also specific services and problems.

Alford, Robert R.   *Health Care Politics: Ideological and Interest Group Barriers to Reform.* Univ. of Chicago Press, 1975. xiv + 294pp. $12.50. 74-75611. ISBN 0-226-01379-0.

   C   Alford's qualitative analysis of health-care policies and politics suggests that "crises" arise out of battles among interest groups fighting for control of key health-care resources and institutions. He focuses on New York City, but his analyses may be applied to national problems. An important book.

Friedman, Marcia.   *The Story of Josh.* Praeger, 1974. 281pp. $7.95. 73-21463. ISBN 0-275-19960-6. CIP.

   SH-C   An excruciatingly vivid account by a mother of her son's unsuccessful battle with a malignant brain tumor. Chronicles the attitudes of the medical staff ranging from stupid and unfeeling to sympathetic. A penetrating and powerful commentary on the contemporary social scene.

Fuchs, Victor R.   *Who Shall Live? Health, Economics, and Social Choice.* Basic, 1975. 168pp. $8.95. 74-79283. ISBN 0-465-09185-7. Index;CIP.

   GA-C-P   A clear and concise summary of health problems and the related economic choices currently facing America. Necessary reading for health professionals as well as the public.

Ginzberg, Eli, and Miriam Ostow.   *Men, Money & Medicine.* Columbia Univ. Press, 1969. 291pp. $8.50. 79-101134. ISBN 0-231-03366-4.

   C-P   Presents the important changes in the structure and functioning of our health services since the close of World War II, with major attention to the Medicare-Medicaid legislation of 1965. The political and economic aspects of health services are presented, along with warnings of the need for careful planning of medical care.

Illich, Ivan.   *Medical Nemesis: The Expropriation of Health.* Pantheon, 1976. viii + 294pp. $8.95. 75-38118. ISBN 0-394-40225-1. Index.

   C   Concerns the causation of disease by physicians or by the whole medical complex, through negligence, incompetence, unnecessary surgery, drugs, diagnostic testing, etc. Illich writes with brilliant insight and logic, resulting in a well-documented major critique of the continuing industrialization of our medical-care complex.

Johnson, G. Timothy.   *Doctor! What You Should Know About Health Care Before You Call a Physician.* McGraw-Hill, 1975. vi + 424pp. $8.95; $3.95(p). 74-31459. ISBN 0-07-032664-9; 0-07-032663-0. Index;CIP.

   SH-C   The author describes several important facets of American medicine today as they relate to consumers and provides basic information on common medical problems. Very good as collateral reading in health courses.

Klaw, Spencer.   *The Great American Medicine Show: The Unhealthy State of U.S. Medical Care, and What Can Be Done About It.* Viking, 1975. xvii + 316pp. $11.95. 75-22040. ISBN 0-670-34836-8. Index;CIP.

   SH-C   A comprehensive overview of many of the problems of contemporary American medical care, with substantial documentation through quantitative data, case illustrations and quotes from experts.

Levin, Arthur.   *Talk Back to Your Doctor: How to Demand (and Recognize) High-*

Quality Health Care. Doubleday, 1975. xv + 245pp. $7.95. 74-12696. ISBN 0-385-03455-5. Index;CIP.

GA-C   Using this guide, patients will be able to choose physicians and services more wisely, judge the quality of the care they receive, and talk back if the care is substandard.

Magnuson, Warren G., and Elliot A. Segal. *How Much for Health?* Luce (distr.: McKay), 1974. xii + 210pp. $7.95. 74-7834. ISBN 0-88331-068-6.

C   Authors point out problems in the national health-care system and detail the weaknesses of existing legislation on health research and medical care. Ample use of case histories (particularly regarding health expenditures).

Motto, Jerome A., et al. *Standards for Suicide Prevention and Crisis Centers.* Behavioral, 1974. xiv + 114pp. $8.95. 73-17029. ISBN 0-87705-105-4. Indexes;CIP.

SH-C-P   This small, comprehensive manual is a must for anyone involved in crisis intervention and suicide prevention. Presents previously nonexistent standards and criteria which will aid in the establishment of crisis centers as integral components of the community health system.

Rushmer, Robert F. *Humanizing Health Care: Alternative Futures for Medicine.* (Illus.) MIT Press, 1975. xii + 210pp. $13.50. 75-1399. ISBN 0-262-18075-8. Gloss.;index;CIP.

C-P   In this provocative essay on issues surrounding medical care now and in the immediate future, Rushmer provides an historical note on major issues and then sets forth significant questions as well as alternatives and solutions.

Scientific American. *Life and Death and Medicine.* (Illus.) Freeman, 1973. 147pp. $7.50; $3.75(p). 73-16097. ISBN 0-7167-0892-2; 0-7167-0891-4. Index;CIP.

C-P   An in-depth review of the socioeconomic impact of health care which recognizes that medical care has advanced beyond acute illnesses and that the patient must be treated as a whole person within the context of the individual's socioeconomic environment.

Smith, David B., and Arnold D. Kaluzny. *The White Labyrinth: Understanding the Organization of Health Care.* (Illus.) McCutchan, 1975. xiv + 379pp. $14.00. 75-7012. ISBN 0-8211-1854-4. Index.

C-P   An imaginative, eclectic description of how the American health-care system works (and doesn't work) at the grass roots, particularly in hospitals.

Stevens, Robert, and Rosemary Stevens. *Welfare Medicine in America: A Case Study of Medicaid.* Free Press/Macmillan, 1974. xxii + 386pp. $13.95. 74-2870. ISBN 0-02-931520-4. Index;CIP.

C-P   The authors explain the purpose and need of our present Medicaid program. Their approach is both historical and philosophical as they analyze the nation's need for planned health insurance.

Wood, Madelyn, and Science Book Associates. *Medicine and Health Care in Tomorrow's World.* (Illus.) Messner, 1974. 157pp. $6.25; $5.79(p). 74-7588. ISBN 0-671-32679-1; 0-671-32680-5. Index;CIP.

SH-GA-C   Geared toward the general reader, this book discusses problems of medical specialization, maldistribution of physicians, new health professions, health care costs, and the social aspects of medicine.

362.2042 SOCIAL PROBLEMS—ALCOHOL AND DRUG ABUSE AND TREATMENT

See also 613.8 Drugs, Alcohol and Tobacco.

Addeo, Edmond G., and Jovita Reichling Addeo. *Why Our Children Drink.*

Prentice-Hall, 1975. xvi + 191pp. $7.95. 75-20326. ISBN 0-13-959460-4. CIP;bib.

SH-C   The authors discuss the opinions of different experts about alcohol education, describe some programs that seem to be working, and assess the role
of the mass media in alcohol education. Chapters on drinks, drinking and being
drunk; defining personal attitudes about drinking; and how to help your children
handle alcohol.

Burgess, Louise Bailey. *Alcohol and Your Health.* (Illus.) Charles, 1973. 243pp.
$12.50. 72-83315. ISBN 0-912880-01-5.

SH-C   Primary concerns of this book are programs for the prevention and treatment of alcohol abuse. The HEW report, *Alcohol and Health* (1971), is
contained in this volume, as are lists of national organizations dealing with alcohol-
related problems.

Einstein, Stanley. *Beyond Drugs.* (Illus.) Pergamon, 1975. xiv + 290pp. $12.00;
$7.00(p). 73-7940. ISBN 0-08-017767-0; 0-08-017768-9. Gloss.;index;CIP.

SH-GA-C   This comprehensive review of drugs and their effects attempts to elicit
critical and analytical thinking on the part of the reader. Lists many
information resources.

Engel, Madeline H. *The Drug Scene: A Sociological Perspective.* (Illus.) Hayden,
1974. 179pp. $4.06(p). 74-7464. ISBN 0-8104-5527-7. Index;CIP.

SH-C   A sociological aproach to drug consumption which exposes some of the
myths of drug use and offers a new perspective on drug abuse.

Englebardt, Stanley L. *Kids and Alcohol, The Deadliest Drug.* Lothrop,
Lee&Shepard, 1975. 64pp. $4.50. 75-20327. ISBN 0-688-41717-5; 0-688-51717-X. In-
dex;CIP.

SH   A considerable amount of information on alcohol, drinking behavior and
alcoholism is provided. The author emphasizes the recent increase of drinking and problem drinking among teenagers.

Fleming, Alice. *Alcohol: The Delightful Poison: A History.* (Illus.) Delacorte, 1975.
v + 138pp. $5.95. 74-22629. ISBN 0-440-01796-3; 0-440-02524-9. Index;CIP.

JH-SH   The author discusses the uses, abuses and history of alcohol, as well as
the social impact, causes and treatment of alcoholism. The role of alcohol
in traffic fatalities is illustrated with pictures.

Fort, Joel. *The Pleasure Seekers.* Bobbs-Merrill, 1969. 255pp. $6.50. 69-13090.

SH-C   Drugs from a pharmacological, sociological and legal point of view; the
history of drug use; international drug trade; legal solutions to drug problems; and offenses against civil liberties are covered in this attack upon irrational and
hypocritical approaches to the drug scene. Much data, but somewhat dated and
biased (especially in favor of marijuana).

Hochman, Joel Simon. *Marijuana and Social Evolution.* Prentice-Hall, 1972.
viii + 184pp. $5.95. $2.45(p). 72-8952. ISBN 0-13-556217-1; 0-13-556209-0.

SH-C-P   This study of the psychosociology of marijuana users, written by a psychiatrist, also examines the middle-class urban society in which chemical
modification of behavior and conscious processes is increasingly common and apparently becoming relatively respectable.

Hyde, Margaret Oldroyd.   *Alcohol: Drink or Drug?* McGraw-Hill, 1974. xii + 150pp.
$4.72. 73-17760. ISBN 0-07-031635-X. Index;CIP;bib.

JH-SH   Provides a simple and clearly written introduction to alcoholic beverages
and to medical and social problems which derive from them. A descrip-

tion of the manufacture of alcohol, a brief history, and suggestions for the prevention and treatment of alcoholism are provided.

Lee, Essie E. *Alcohol—Proof of What?* (Illus.) Messner, 1976. 191pp. $6.25. 75-45149. ISBN 0-671-32788-7; 0-671-32789-5. Index;CIP;bib.

    JH-SH   Lee and her colleagues evaluated 27,000 questionnaires to determine how young people use and abuse alcohol, and their conclusions and a sampling of responses are included. Their evaluation is from a sociological rather than a psychological perspective. Includes a chapter on drinking customs in various cultures.

Marin, Peter, and Allan Y. Cohen. *Understanding Drug Use: An Adult's Guide to Drugs and the Young.* Harper&Row, 1971. 163pp. $5.95. 69-15318.

    C-P   The emphasis is on adolescence rather than on drugs—people and the world they live in are much more important to understand than the drugs themselves. Also included are easily understandable and comprehensive descriptions of the common drugs used.

Nelkin, Dorothy. *Methadone Maintenance: A Technological Fix.* Braziller, 1973. ix + 164pp. $6.95; $1.95(p). 72-96071. ISBN 0-8076-0681-2; 0-8076-0680-4.

    SH-C-P   A lucid examination of the ethics and mechanics of readdicting the addicted to a cosmetically more acceptable drug. Tracing the use and abuse of heroin gives a detailed and objective background to a poignant study of a methadone treatment program. The issue is, is methadone a rational alternative to heroin addiction when little or nothing is done about treating the basic human needs that result in addiction in the first place? For counselors, teachers, politicians, bureaucrats and concerned students.

Rublowsky, John. *The Stoned Age: A History of Drugs in America.* Capricorn/Putnam's, 1974. 218pp. $6.95. 73-93744. SBN 399-50321-8; 399-11306-1. Index.

    SH   The principal hypothesis is that prohibition has caused the increase in drug usage in America and that users of marijuana, heroin, and hallucinogens have now become America's scapegoats.

Silverstein, Alvin, and Virginia B. Silverstein. *Alcoholism.* Lippincott, 1975. 128pp. $5.50; $2.25(p). 75-17938. ISBN 0-397-31648-8; 0-397-31649-6. Index;CIP.

    JH-SH   This short, simple book is an excellent overview of the use and abuse of alcohol. Summarizes its production and consumption throughout history, the reasons for drinking, its social milieu, and the physical, psychological and social consequences of drinking.

Stamford Connecticut Board of Education. *Stamford Guide to Drug Abuse Education.* Ferguson, 1971. xvii + 96pp. $4.00. 70-134585. Gloss.;bib.

    C-P   A comprehensive curriculum (4th to 12th grade) that could be a useful reference. A policy statement and procedure for psychologist and social worker to use in dealing with students under the influence of drugs is included. Lists films and other resource materials.

Szasz, Thomas Stephen. *Ceremonial Chemistry: The Ritual Persecution of Drugs, Addicts, and Pushers.* Anchor/Doubleday, 1974. xvii + 243pp. $6.95. 74-2834. ISBN 0-385-06627-9. Index;CIP.

    C-P   Highlights the mythological and ceremonial aspects of drug usage and society's attitudes toward the drug dependent. Szasz sees modern drug laws as quasi-religious: a condemnation of deviance will somehow make us holy. He does not, however, give carte blanche to drug usage, but instead urges a demythologization of drugs.

Tracy, Don. *What You Should Know About Alcoholism.* Dodd, Mead, 1975. 157pp. $6.95. 74-10007. ISBN 0-396-06997-5.

JH-SH-C   Presented in a question-and-answer format, this lucid book offers practical help to the alcoholic through his or her family, friends, associates
and employer. Emphasizes ways of recognizing and helping those suffering from
alcoholism. Alcoholics Anonymous oriented, but not an abstinence tract. Valuable
collateral reading for health courses.

Wise, Francis H. *Youth and Drugs: Prevention, Detection and Cure.* Association Press, 1971. 191pp. $4.95. 78-132392. ISBN 0-8069-1781-1.

SH-C   This extremely lucid book begins with a discussion of drugs from marijuana
to heroin. There are thought-provoking chapters for parents on the role of
the family in setting standards for their children and the need for understanding and
counseling so that the children maintain self-esteem and learn to live with authority.

362.3 MENTAL RETARDATION

See also 155.45 Exceptional Children and 371.9 Special Education.

De La Cruz, Felix F., and Gerald D. LaVeck (Eds.). *Human Sexuality and the Mentally Retarded: Proceedings of a Conference on Human Sexuality and the Mentally Retarded.* Brunner/Mazel, 1973. xviii + 347pp. 72-90257. SBN 87630-063-8.

SH-C-P   Specialists discuss psychosexual development and sex education, physical and biological aspects, institutional and community attitudes and
practices, directions for research, and sexuality in a broader context. One of the most
comprehensive, advanced and objective treatments of the subject.

Koch, Richard, and James C. Dobson (Eds.). *The Mentally Retarded Child and His Family: A Multidisciplinary Handbook, rev. ed.* (Illus.) Brunner/Mazel, 1976. xiii + 546pp. $15.00. 76-122743. SBN 87630-121-9. Gloss.;index;CIP.

C-P   For students in education, medical care, nutrition, psychology, social work,
speech and hearing, community organization, rehabilitation, physical and
occupational therapy, law, architecture, and recreation. Describes mental retardation, its causes and psycho-social aspects; education and training; community services; and multidisciplinary services.

Nichtern, Sol. *Helping the Retarded Child.* Grosset&Dunlap, 1974. xii + 289pp. $8.95. 73-11395. ISBN 0-448-01305-3. Index.

C   Nichtern approaches the subject from an interventionist's point of view. His
advice ranges from preventive genetic counseling to means for socialization,
self-care and specific types of therapy. Culminating sections dealing with community
organizations and the legal rights of the retarded make this an authoritative handbook
for both parents and the medical community.

362.7 PROBLEMS OF YOUNG PEOPLE AND THEIR ALLEVIATION

See also 155.5 Psychology of Adolescents and 362.2042 Social Problems—Drug and Alcohol Abuse and Treatment.

Fredericksen, Hazel, and R.A. Mulligan. *The Child and His Welfare, 3rd ed.* (Illus.) Freeman, 1972. x + 434pp. $8.50. 70-172242. ISBN 0-7167-0905-8.

C-P   Authoritative, professional and comprehensive; covers the spectrum of services subsumed by public assistance and child welfare programs. Sociological
considerations are interlaced with a wealth of factual information. Discussions of
legislation and Supreme Court decisions are included.

Lieberman, Florence, et al. *Before Addiction: How to Help Youth.* Behavioral Publications, 1973. ix + 131pp. $7.95. 73-7803. SBN 87705-112-7.

C-P   This book contains a broad base of information concerning the psycho-social needs of adolescents and their parents. The importance of parents as a resource in the prevention and treatment of problems is stressed. An excellent book for pediatricians, parents, students, and workers in helping professions.

Richette, Lisa Aversa. *The Throwaway Children.* Lippincott, 1969. 342pp. $6.95. 71-77864.

C-P   The case histories reported here are those of culturally deprived, neglected, and criminally abused and exploited children of both poverty-ridden and affluent parents. The deficiencies in the traditional methods of counseling, housing, correcting, and rehabilitating juvenile delinquents and law-breakers are clearly described.

Roby, Pamela, (Ed.). *Child Care—Who Cares? Foreign and Domestic Infant and Early Childhood Development Policies.* Basic, 1973. xviii + 456pp. $16.00. 72-89179. SBN 465-00983-2.

C-P   This advocate's book and guide to needed legislation is well written, packed with information, and analytic. The book, distinguished by the direct quality of the reports, is in the best tradition of investigative reporting.

Schorr, Alvin L., (Ed.). *Children and Decent People.* Basic, 1974. xvii + 222pp. $7.95. 73-82894. SBN 465-01041-5. Index.

SH-C-P   The thesis is that all children are receiving poor treatment—by the welfare system, the day-care foster family system, in the institutions where they may be placed, in the courts and by the health-care system. Workable solutions are, unfortunately, missing.

364 CRIME

Hyde, Margaret O. *Speak Out on Rape!* McGraw-Hill, 1976. xi + 145pp. $5.72. 75-10916. ISBN 0-07-031636-8. Index;CIP.

JH-SH   Hyde examines what is and what is not known about rape, hoping to "encourage understanding, both of the need for research and for action." Treats rape not as a political action but as a violent crime of assault involving all of society. Rape prevention methods are reviewed.

Kennedy, Daniel B., and August Kerber. *Resocialization: An American Experiment.* Behavioral Publications, 1973. x + 191pp. $9.95; $4.95(p). 72-10326. ISBN 0-87705-033-3; 0-87705-091-0. Index;bib.

C-P   Examines programs aimed at altering human behavior and accomplishments and summarizes theory and its practical applicability. For students in education, psychology, and sociology and for professionals working with the chronic offender, the hardcore unemployed, or those in long-term psychotherapy programs.

370 Education

370.1 EDUCATION—PHILOSOPHY, THEORIES, PRINCIPLES

Barnard, John, and David Burner (Eds.). *The American Experience in Education.* New Viewpoints/Watts, 1975. xx + 268pp. $5.95(p). 74-13455. ISBN 0-531-05361-X; 0-531-05569-8. Index;CIP.

SH-C-P   These 15 essays trace some prominent themes and explore some of the continuing problems of American education. Particularly interesting are the essays on minority education, parochial schools and testing.

Becker, Gary S.   *Human Capital: A Theoretical and Empirical Analysis, With Special Reference to Education, 2nd ed.* (Illus.) Nat'l Bureau of Economic Research (distr.: Columbia Univ. Press), 1975. xvii + 268pp. $10.00. 74-83469. ISBN 0-87014-513-4.

C-P   In the standard reference in this area, Becker presents theoretical and cost/benefit analyses of activities such as education, on-the-job training, and migration, that increase the "capital" embodied in individuals. Emphasis is on education; contains a statistical analysis of the costs and benefits of college education to various population groups. Useful both to economists and educators. Index.

Flescher, Irwin.   *Children in the Learning Factory: The Search for a Humanizing Teacher.* Chilton, 1972. ix + 180pp. $5.95(p). 72-10268. ISBN 0-8019-5751-6.

C   Thoughts and suggestions for interpreting what goes on inside the "factory," urging dialogue among parents, teachers and school administrators. Most of the ideas can be found elsewhere, but the worth of this book is that it speaks to parents and teachers in a charming and persuasive style.

Goodwin, Dwight L., and Thomas J. Coates.   *Helping Students Help Themselves: How You Can Put Behavior Analysis into Action in Your Classroom.* (Illus.) Prentice-Hall, 1976. xvi + 205pp. $9.95; $5.95(p). 75-33541. ISBN 0-13-386490-1; 0-13-386482-0. Index;CIP.

C-P   The authors present logical, step-by-step approaches to behavioral analysis, with a rationale for why it works. The precedures are useful for teachers of "normal kids." Observation and recording techniques are included.

Jencks, Christopher, et al.   *Inequality: A Reassessment of the Effect of Family and Schooling in America.* Basic, 1972. xii + 400pp. $12.50. 72-89172. SBN 465-03264-8.

C-P   An argument against the long-standing belief that good schools produce more successful, achieving adults than do bad schools. The authors conclude that "success" must be the result, finally, of "luck." An excellent book which should be read by all school administrators and public policy makers.

Middleton, John, (Ed.).   *From Child to Adult: Studies in the Anthropology of Education.* Natural History Press, 1970. xx + 355pp. $3.95(p).

C-P   These 15 classic selections deal with small-scale societies experiencing little rapid change and reflecting education for socialization rather than for the acquisition of skills. Should stimulate educational researchers to enlarge their perspectives.

Naylor, F.D.   *Personality and Educational Achievement.* (Illus.) Wiley, 1972. viii + 162pp. $6.95. 72-1570. ISBN 0-471-63074-8; 0-471-63075-6(p).

C-P   This introductory work gives a fairly comprehensive overview of the correlational literature in the personality-achievement domain. Fundamental personality theory and correlational statistics are incorporated. Can be easily understood by readers with a fundamental grasp of specific methods.

371.4 GUIDANCE AND COUNSELING

Ancona, George.   *And What Do You Do? A Book About People and Their Work.* (Illus.) Dutton, 1976. 47pp. $6.95. 75-40285. ISBN 0-525-25605-9. CIP.

JH   Describes 21 jobs which do not require college (e.g., computer operator), how to obtain them and how to advance in various fields. Nonstereotyped.

Chandler, Caroline A., and Sharon Kempf. *Nursing as a Career.* (Illus.) Dodd, Mead, 1970. vii + 157pp. $3.75. 74-123501. ISBN 0-0396-06230-X.

   SH   A more complete cross-section of the nursing profession than in many previous monographs. The sections describing the medical environment and deep involvement in the adventures of life and death are true and practical.

Clarke, Emerson, and Vernon Root. *Your Future in Technical and Science Writing.* (Illus.) Richards Rosen, 1972. 162pp. $3.99. 75-158563. ISBN 0-8239-0242-0.

   JH-SH   Clarity, completeness, organization, and vocabulary all contribute to the excellence of this career summary. Interest is sustained by examples and illustrations.

Dowdell, Dorothy, and Joseph Dowdell. *Careers in Horticultural Sciences.* (Illus.) Messner, 1975. 222pp. $6.95. 75-9612. ISBN 0-671-32738-0; 0671-32739-9. Index;CIP.

   JH-SH   The coverage is good and the appendixes list academic institutions offering various kinds of horticultural training. The style is somewhat condescending.

Fanning, Odom. *Opportunities in Environmental Careers, rev. ed.* (Illus.) Vocational Guidance Manuals, 1975. xviii + 251pp. $6.95; $3.95(p). 74-25902. ISBN 0-89022-198-7; 0-89022-008-5. Index; bib.

   SH-C   The best book on this subject. Gives background on environment and environmental management, on factors influencing employment, and on the necessary education and experience. Describes a number of programs and includes places to write for further information.

Fenten, D.X. *Ms.—M.D.* (Illus.) Westminster, 1973. 144pp. $4.95. 72-11999. ISBN 0-664-32524-6. Index.

   JH-SH-C   The book guides the student from contemplation of a career in medicine through the preparatory high school and college courses, to applications and testing for medical school entrance to the general medical curriculum and finally into the choice of a specialty. The author focuses on medicine as a job choice for women and discusses the difficulty of maintaining a medical practice, a marriage and a family.

Greenebaum, Louise G. *Looking Forward to a Career: Electronics.* (Illus.) Dillon, 1975. 123pp. $5.95. 75-12620. ISBN 0-87518-102-3. Index;CIP;bib.

   JH-SH   An excellent analysis of opportunities in a wide variety of military, space, industrial, business and consumer occupations. Tells the reader how to evaluate him/herself for potential success in these careers and outlines the steps necessary to prepare for entry into them.

Harmon, Margaret. *The Engineering Medicine Man—The New Pioneer.* (Illus.) Westminster, 1975. 224pp. $6.95. 75-25690. ISBN 0-664-32581-5. Index;CIP.

   SH   This timely look at a very important technical career opportunity discusses job opportunities and academic prógrams in communications, medicine, bioresearch, computers, and such areas as acupuncture. Uses an historical approach, and narratives by people in the field.

Lee, Essie E. *Careers in the Health Field.* (Illus.) Messner, 1972. 191pp. $4.79. 78-176381. ISBN 0-671-32502-7; LB 0-671-32513-2.

   JH-SH   This overview discusses requirements, duties, salaries, potential for advancement, and sources of further information. Conveys the interest and excitement of this field.

McCall, Virginia, and Joseph R. McCall. *Your Career in Parks and Recreation.* (Illus.) Messner, 1974. 191pp. $6.25. 74-8191, ISBN 0-671-32687-2; 0-671-32688-0. Index;CIP.

JH-SH   Illustrates the numerous opportunities in parks maintenance and recreation management. Includes educational requirements and sample programs in colleges offering appropriate degrees, as well as a survey of 1973-74 salaries. Provides a good balance between the glamor and challenge and the sometimes monotonous routine of recreations work.

McLeod, Sterling, et al. *Careers in Consumer Protection.* (Illus.) Messner, 1974. 191pp. $6.25. 74-7591. ISBN 0-671-32691-0; 0-671-32692-9. Index;CIP.

JH-SH   The authors put into perspective the interplay between the dangers faced by the consumer and the various careers relating to protecting the consumer.

Metz, L. Daniel, and Richard E. Klein. *Man and the Technological Society.* (Illus.) Prentice-Hall, 1973. xi + 180pp. $5.00(p). 72-12177. ISBN 0-13-550996-3.

SH-C   The authors address themselves to the growing alienation of students from technology. With great dash and elan, they have constructed illustrations, problems and anecdotes, and have done a thorough job of showing potential engineering students what engineering is like in the real world.

Millard, Reed, and Science Book Associates. *Careers in Environmental Protection.* (Illus.) Messner, 1974. viii + 188pp. $5.79. 73-19231. ISBN 0-671-32665-1; 0-671-32666-X. Index;CIP;bib.

JH-SH-C   This is a guidebook to possible vocations in world resource protection. A comprehensive picture of the various jobs, the necessary education, the possible salaries and the working conditions of each career is presented.

Rettig, Jack L. *Careers: Exploration and Decision.* (Illus.) Prentice-Hall, 1974. 120pp. $3.52. 74-611. ISBN 0-13-114686-6; 0-13-114678-5. Index;CIP.

SH-C-P   A valuable resource for aiding students in establishing a basis for career exploration and the realization of personal goals. Its organization helps readers individualize and interact with the material.

Robinson, Alice, and Mary Reres. *Your Future in Nursing Careers.* (Illus.) Arco, 1974 (c.1972). xii + 113pp. $1.95(p). 72-75218. ISBN 0-688-03429-7.

JH-SH-C   A useful guide to nursing careers, personal and educational prerequisites, job opportunities, and the nursing environment. Recommended for both male and female students as well as their teachers and career counselors.

Splaver, Sarah. *Nontraditional Careers for Women.* Messner, 1973. 224pp. $5.64. 73-5384. ISBN 0-671-32619-8; 0-671-32620-1.

SH-C-P   This introductory text discusses traditional and nontraditional careers, and provides statistics on the employment situation for women throughout the '70s.

Walker, Greta. *Women Today: Ten Profiles.* (Illus.) Hawthorn, 1975. 174pp. $6.95. 74-22925. ISBN 0-8015-7760-8. Index.

JH-SH   Each profile contains descriptions of hectic work schedules, as well as descriptions of the types of people who have helped these women carry out their tasks and achieve their goals. Useful for the girl or woman searching for an identity and life goal.

Whitney, Leon F., and George Whitney. *Animal Doctor: The History and Practice of Veterinary Medicine.* (Illus.) McKay, 1973. 104pp. $4.95. 72-83038.

JH   Contains descriptions of the practice of the profession; job possibilities in research, teaching, etc.; three typical days in the life of the veterinarian; the required education, State Board examinations and establishing a practice, and a listing of all 18 U.S. schools of veterinary medicine.

90 SPECIAL EDUCATION

371.9 SPECIAL EDUCATION

See also 155.45 Exceptional Children and 362.3 Mental Retardation.

Barnard, Kathryn E., and Marcene L. Powell. *Teaching the Mentally Retarded Child: A Family Care Approach.* (Illus.) Mosby, 1972. xi + 158pp. $4.50. ISBN 0-8016-0485-0.

    SH-C-P Extensive and psychologically sound lists of parent-child activities appropriate to developmental and competence goals are presented for seven age groupings, ranging from 1–3 months to 49–52 months. Special education teachers, nurses-in-training and parents will benefit from this.

Blake, Kathryn A. *The Mentally Retarded: An Educational Psychology.* (Illus.) Prentice-Hall, 1976. x + 403pp. $11.95. 75-38702. ISBN 0-13-576280-4. Indexes;CIP.

    C-P Proceeds from background knowledge to application, through literature on both research and practice. Covers classification and delivery systems, growth and development, language, education and vocational-civic characteristics, psychometrics, verbal and motor learning, problem solving, and basic learning processes, based predominantly on Blake's own research. Many examples given. A remarkable book.

Bleck, Eugene E., and Donald A. Nagel (Eds.). *Physically Handicapped Children: A Medical Atlas for Teachers.* Grune&Stratton (distr.: Academic), 1975. xiv + 304pp. $16.50. 75-4827. ISBN 0-8089-0863-4. Gloss.;index;CIP.

    C-P An atlas of facts and practical suggestions for teachers working with disabled children in special or regular schools and for college educators preparing those teachers. Covers a variety of handicapping conditions puzzling to teachers. Includes material on anatomy and sequences of developmental expectancies in childhood.

Hallahan, Daniel P., and James M. Kauffman. *Introduction to Learning Disabilities: A Psycho-Behavioral Approach.* (Illus.) Prentice-Hall, 1976. x + 310pp. $10.95. 75-29498. ISBN 0-13-485524-8. Index;CIP.

    C-P Describes the various learning disabilities and discusses standard assessment and some treatment programs for each. Applied behavior analysis is a major focus. Contains an excellent historical overview and a chapter on diagnostic-prescriptive teaching. Useful for paraprofessionals, students, and for in-service training.

389 METROLOGY AND STANDARDIZATION

Branley, Franklyn M. *Think Metric!* (Illus.) Crowell, 1973. 53pp. $4.50. 72-78279. ISBN 0-690-81861-0; 0-690-81862-9.

    JH-SH An excellent and very simple overview of the metric system which will familiarize the reader with new terms. Discusses conversions and equivalents and suggests measurement activities. Emphasizes the ease of changing from one metric unit to another.

Cunningham, James B. *Teaching Metrics Simplified.* (Illus.) Prentice-Hall, 1976. vii-i + 184pp. $10.95; $5.95(p). 75-44162. ISBN 0-13-893883-0; 0-13-893875-X. Index;CIP.

    C-P A sophisticated treatment for teachers who have had little or no experience with the metric system. The material includes the customary (English) system; the basic metric units; the use of prefixes; the derived units of area, volume, pressure, etc.; scientific notation; and the conversion between units. Includes a large set of classroom activities.

Gallant, Roy A. *Man the Measurer: Our Units of Measure and How They Grew.* (Illus.) Doubleday, 1972. 111pp. $4.95. 73-160879.

JH   A selective history of measurement units, measuring, and the metric system. Each section includes several experiments that students can carry out.

Hahn, James, and Lynn Hahn. *The Metric System.* (Illus.) Watts, 1975. 63pp. $3.90. 74-31386. ISBN 0-531-00834-7. Index;CIP.

SH   This basic introduction to the metric system describes its history and spread
        throughout the world. Former measurement systems and their uses are also
covered. Teachers will find this a useful reference.

Hirsch, S. Carl. *Meter Means Measure: The Story of the Metric System.* Viking, 1973. 126pp. $4.95. 72-91398. ISBN 0-670-47365-0.

JH-SH   The author details the development of metric measure, the growing world
            commitment to it and the attempts since the 18th century to have it
adopted in the U.S.

Hopkins, Robert A. *The International (SI) Metric System and How It Works.* (Illus.) AMJ Publishing/Polymetric Services, 1973. 281pp. $12.95(p).

SH-C   The author reviews the history and current status of SI with a brief discus-
          sion of the cost/benefits to society and to industry of using metric meas-
urements. Basic and derived units of the SI system and conversion factors, tables and scales between the SI and inch-pound systems are presented.

Klein, H. Arthur. *The World of Measurements: Masterpieces, Mysteries and Muddles of Metrology.* (Illus.) Simon&Schuster, 1974. 736pp. $14.95. 74-7656. SBN 671-21565-5. Index.

JH-SH-C   Dozens of units are included: length, weight, time, the SI units, tem-
              perature scales, electromagnetic, pressure, density, sound, flow, proof,
and measures of radioactivity and its effects. An excellent addition to any library.

Kurtz, V. Ray. *Teaching Metric Awareness.* (Illus.) Mosby, 1976. vii +83pp. $4.95. 75-22174. ISBN 0-8016-2811-3. CIP.

GA-C-P   The author avoids the conversion approach, shows that experiences are
            necessary to learn the metric system, provides activities in game-like
format, gives backgrounds for each concept, and provides logical arguments about why we should change at all. Includes a list of metric suppliers. Intended for elementary teachers, but other adults will find it useful.

400   Language

410–419   LINGUISTICS

Bosmajian, Haig A. *The Language of Oppression.* Public Affairs, 1975. 156pp. $4.50(p). 75-24984. ISBN 8183-136.

SH-C-P   An excellent exposition of the role that language may play and has played
            in furthering the oppression of sociological minorities, including women.

Charlip, Remy, et al. *Handtalk: An ABC of Finger Spelling and Sign Language.* (Illus.) Parents', 1974. 38pp. $4.95. 73-10199. ISBN 0-8193-0705-X; 0-8193-0706-8.

JH-SH   Introduces the two modes of manual communication: signs (gestures
            which represent words) and fingerspelling (words are spelled using the
letters of the manual alphabet), with scenes of persons making various signs. Captures the liveliness and sparkle of this method of communication.

Claiborne, Robert, et al. *The Birth of Writing.* (Illus.) Time-Life, 1975. 160pp. $5.95. 74-83646. Index.

JH-SH-C   A compelling introduction to the origin and development of the written word. The early civilizations of Mesopotamia and Egypt where scripts were first produced are recreated with lively illustrations.

Hogben, Lancelot. *The Vocabulary of Science.* Stein&Day, 1970. $6.95. 184pp. 77-108314. ISBN 0-8128-1287-5.

C   Covers Latin and Greek origins of modern scientific terms and provides insight into the rationale of scientific nomenclature. Rules are given for the invention of new terms. For reference and browsing.

Howell, Richard W., and Harold J. Vetter. *Language in Behavior.* Human Sciences/Behavioral, 1976. ii + 395pp. $14.95. 74-8363. ISBN 0-87705-157-7. Index;CIP.

GA   A nontechnical, informative, and entertaining introduction to linguistics. Basic concepts and ideas of contemporary interest are presented with clear and appealing examples.

Kavanagh, James F., and James E. Cutting (Eds.). *The Role of Speech in Language.* (Illus.). MIT Press, 1975. xiii + 335pp. $15.00. 75-12561. ISBN 0-262-11059-8. Indexes;CIP.

SH-C-P   This book, aimed primarily at the specialist in linguistic studies, would also be of interest to lay readers. Covers the origins and evolution of speech, sign language, primate studies, etc. Presents novel and creative approaches to linguistics.

Liles, Bruce L. *An Introduction to Linguistics.* (Illus.) Prentice-Hall, 1975. xiii + 336pp. $12.00; $7.50(p). 74-23249. ISBN 0-13-486134-5; 0-13-486126-4. Index;CIP.

C   First introduces Chomsky's transformational grammar via numerous examples of rules, tree diagrams and exercises. The remaining chapters cover such topics as nonverbal communications, language acquisition and universals, change in language and other contemporary variations. Detailed presentation of syntax and phonology.

Norman, James. *Ancestral Voices: Decoding Ancient Languages.* (Illus.) Four Winds/Scholastic, 1975. 242pp. $7.95. 75-14426. ISBN 0-590-17333-2. Index;CIP;bib.

SH-C   This is an account of how various ancient languages are and have been deciphered, with biographical information on the principal figures involved.

Ogg, Oscar. *The 26 Letters, rev. ed.* (Illus.) Crowell, 1971. 294pp. $6.95. 70-140646.

SH-C   Tends more to calligraphy than to the historical origins of the 26 letters. The text is simple and readable, the illustrations are well done and exceptionally clear. Useful for collateral reading for students of linguistics.

Scott, Joseph, and Lenore Scott. *Hieroglyphs for Fun: Your Own Secret Code Language.* (Illus.) Van Nostrand Reinhold, 1974. 79pp. $6.95. 72-12443. ISBN 0-442-27523-4. CIP.

JH   Egyptian numbers and mathematics, storytelling, games, and similar aspects of culture are presented in a fascinating, easy to understand manner and are delightfully illustrated. Characteristics of Egyptian writing are presented.

Pure Sciences

500.1 NATURAL SCIENCES

Asimov, Isaac. *Science Past—Science Future.* Doubleday, 1975. xiv + 346pp. $8.95. 74-25092. ISBN 0-385-09923-1. CIP.

> **JH-SH-C**   Asimov articulates the significance of science and technology for the future of humankind. Using historical examples of the dependence of humanity upon technology, he argues that science and technology are the answers to our problems.

Bova, Ben. *Through Eyes of Wonder: Science Fiction and Science.* (Illus.) Addison-Wesley, 1975. 127pp. $5.75. 74-13893. ISBN 0-201-09206-9. Index;CIP.

> **JH-SH**   Bova's thesis is that science fiction frequently points the way for real science and also contributes to society's understanding of rapid technological and social change. May stimulate the reader to undertake a deeper study of science.

Clarke, Robin, (Ed.). *Notes for the Future: An Alternative History of the Past Decade.* Universe, 1976. 238pp. $10.00; $4.50(p). 75-37295. ISBN 0-87663-255-X; 0-87663-929-5.

> **C**   Articles by Wald, Ehrlich, Meadows, Hardin, Schumacher, Commoner, and Roszak et al. argue that a profound shift in public awareness is needed and can occur to alter fundamentally the relationship of science to the rest of society. Useful for the concerned citizen-scientist and excellent for courses on science, technology and public policy, environment or alternative futures.

Cottrell, Alan. *Portrait of Nature: The World as Seen by Modern Science.* (Illus.) Scribner's, 1975. 236pp. $9.95. 75-837. ISBN-0-684-14355-0. Index;CIP.

> **SH-C**   Describes astronomy, physics, chemistry, and biology and touches on evaluation and behavior. Such topics as space travel, gravity, origin of the universe, energy, electric charges, atomic structure and the nature and origin of life are explained. Excellent for liberal arts students.

Günter, Altner, (Ed.). *The Human Creature.* (Illus.) Anchor/Doubleday, 1974. viii + 467pp. $3.50. 73-11640. ISBN 0-385-04947-1. Gloss.;index.

> **SH-C**   This collection establishes a dialog between the social and natural sciences and provides the reader with a unique view of evolving humanity, its biology, culture and future. Papers vary in degree of sophistication.

Jueneman, Frederic B. *Limits of Uncertainty: Essays in Scientific Speculation.* (Illus.) Industrial Research, 1975. viii + 229pp. $5.95.

> **SH-C**   An intellectually stimulating collection of 49 speculative, scientific vignettes by the author from *Industrial Research* magazine (1971–1974) examines fields as diverse as astronautics, geologic history, number theory, the Velikovsky phenomenon, magnetohydrodynamics, carbon dating, the ether, Kirlian photography and others.

Rinehart, Kenneth L., Jr., et al. (Eds.). *Wednesday Night at the Lab: Antibiotics, Bioengineering, Contraceptives, Drugs and Ethics.* (Illus.) Harper&Row, 1973. xii + 226pp. $2.95(p). 73-281. SBN 06-045409-1.

> **GA-C**   In this collection of lectures, science is shown to form an interface with the humanities—both deal with basic human problems.

Young, Louise B., (Ed.). *Exploring the Universe, 2nd ed.* (Prepared by Amer. Foundation for Continuing Education.) (Illus.) Oxford, 1971. xii + 731pp. $12.50; $6.50(p). 79-151188.

    SH-C Designed to meet the needs of adult education in science, the book consists of 100 popular essays or excerpts on the methods, values, and limitations of science; the universe and our place in it; and cosmology, astronomy, and space. Authors range from Aristotle through Thomas Paine, Administrator of NASA (1970). Some are outstanding, all are intellectually stimulating.

500.2 PHYSICAL SCIENCES

Bonner, Francis T., Melba Phillips, and Jane Raymond. *Principles of Physical Science, 2nd ed.* (Illus.) Addison-Wesley, 1971. x + 398pp. $10.75. 73-137833.

    C Aimed at the liberal arts major, the text is still classical and encyclopedic. Covers topics from the whole of astronomy, physics, and chemistry.

Booth, Verne H. *Elements of Physical Science: The Nature of Matter and Energy.* (Illus.) Macmillan, 1970. xii + 468pp. $8.95. 74-85791.

    C Designed for the nonscience major, this text conveys the message that everyone must be aware of basic scientific facts in order to evaluate science's social impact. Includes reminders of the importance of the ethical responsibilities of the scientist to society.

Communication Research Machines, Inc. *Concepts in Physics.* (Illus.) CRM Books, 1973. 396pp. $10.95(p). 72-92897. Index;gloss.;bib.
Physical Science Today. (Illus.) 639pp. $12.95. 72-85255. SBN 87665-136-8. Index;gloss;bib.

    JH-SH-C These texts are intended for liberal arts students and are among the finest available in physics or the physical sciences. (*Physical Science Today* adds 12 chapters on the earth sciences and chemistry.) No end-of-chapter problems or review material. The contributors are a "Who's Who" of the sciences.

Dobbs, Frank W., Albert Forslev and Robert L. Gilbert. *The Physical Sciences.* (Illus.) Allyn & Bacon, 1972. xviii + 604pp. $11.95. 72-075065.

    SH-C A clearly written, thoughtfully illustrated introduction to physics, chemistry, and the earth and planetary sciences. Lucid, contemporary examples are accompanied by uncomplicated and sometimes novel illustrations of ideas discussed in the text. Skimpy treatment of global tectonics, however.

Holton, Gerald, and Stephen G. Brush. *Introduction to Concepts and Theories in Physical Science, rev. 2nd ed.* (Illus.) Addison-Wesley, 1973. xix + 589pp. $13.95. 72-2787.

    C A standard work for the nonscience major, with good revisions and new material. It shows how historical developments have played key roles in the present-day understanding of scientific laws and theories. Additions to this edition include a brief discussion of quantum mechanics and of Einstein's Special Theory of Relativity. A solid, intellectually honest treatment of the basic concepts of physical science; especially useful for high school physics teachers.

Miles, Vaden W., et al. *College Physical Science, 3rd ed.* (Illus.) Harper&Row, 1974. xi + 655pp. $12.95. 73-13200. ISBN 0-06-044443-6. Index;CIP.

    SH-C A textbook designed for nonscience majors at the freshman and sophmore level of college. Subjects covered are classical astronomy, principles of motion, energy, fluids, waves, electricity, the structure of matter, the fundamentals of chemical stoichiometry, bonding and organic matter, earth science, and the geology of the moon.

Payne, Charles A., and William R. Falls. *Modern Physical Science.* (Illus.) Brown, 1974. xvii + 542pp. $11.95. 73-86809. ISBN 0-697-05755-0. Index;gloss.

JH-SH-C   The authors present an interrelated approach to astronomy, chemistry, earth science and physics which includes excellent historical introductions, descriptions of historic experiments, chapter summaries, and review problems and questions. Very readable and enjoyable.

Solomon, Joan. *The Structure of Matter: The Growth of Man's Ideas on the Nature of Matter.* (Illus.) Halsted/Wiley, 1974. 179pp. $9.95. 73-8541. ISBN 0-470-81222-2. Index.

SH-GA-C   A very readable introduction to a wide range of topics in the history and philosophy of science. Half of the book deals with science since the discovery of the electron. Radioactivity, atoms-or-waves, antimatter, war and fission are also discussed, bringing the reader up to the "new pieces of matter." Especially for the liberal arts student.

Swartz, Clifford E., and Theodore D. Goldfarb. *A Search for Order in the Physical Universe.* (Illus.) Freeman, 1974. 315pp. $10.50. 73-19743. ISBN 0-7167-0345-9. Index;CIP.

SH-C   The authors have made an admirable effort to make the physical world understandable and useful to nonscience, nontechnical readers. Emphasis is on physics, and numerical as well as philosophical questions are included.

Townsend, Ronald D. *Energy, Matter, and Change: Excursions into Physical Science.* (Illus.) Scott, Foresman, 1973. xv + 561pp. $6.21. 72-87831. ISBN 0-673-01972-1.

JH   There are five major units: light, electricity, motion, heat and chemistry. Illustrations are excellent; good accuracy, scope, clarity.

Wiggins, Arthur W. *Physical Science: With Environmental Applications.* (Illus.) Houghton Mifflin, 1974. xiii + 305pp. $8.76. 73-9406. ISBN 0-395-17072-9. Gloss.;index.

JH-SH-C   This introduction emphasizes modern theory and topics. Student interest is enhanced through friendly writing, humor and problem-solving.

500.9 NATURAL HISTORY

Borgese, Elisabeth Mann. *The Drama of the Oceans.* (Illus.) Abrams, 1975. 258pp. $25.00. 74-16165. ISBN 0-8109-0337-7. Index;CIP;bib.

GA-C   This beautifully done book, with excellent color plates, is not a textbook or reference, but a "prose poem." Geophysical, biological, ecological, economic, medical, political, geographical and historical aspects are covered. Accurate; uses an advanced vocabulary.

Bronowski, J. *The Ascent of Man.* (Illus.) Little, Brown, 1974. 448pp. $15.00. 73-20446. ISBN 0-316-10930-4. Index;CIP.

SH-C-P   The book deals with all of humankind's activities, from the rituals of our society to our curious preoccupation with nature. Bronowski's major premise is that humans by nature have produced artifacts, history and knowledge, and "science." A well-written book with some questionable analogies.

Cahalane, Victor H., (Ed.). *Alive in the Wild.* (Illus.) Prentice-Hall, 1970. ix + 244pp. $9.95. 77-81581. ISBN 0-13-022160-0.

JH-SH   Some 35 essays on birds, reptiles, and mammals, each authored by a notable in the field, make up this natural history. Black-and-white drawings are pleasing and accurate, but there are many undefined terms.

Cousteau, Jacques-Yves, with Philippe Diolé. *Life and Death in a Coral Sea: The*

Undersea Discoveries of Jacques-Yves Cousteau. (Trans.; illus.) Doubleday, 1971. 302pp. $8.95. 69-13003.

JH-SH-C-P   This is an excellent introduction to coral reefs and islands. Beautifully illustrated and informative, with much geography, oceanography, and natural history in an exciting and palatable form.

Heckman, Hazel.   *Island Year.* (Illus.) Univ. of Wash. Press, 1972. xi + 255pp. $7.95. 74-17801. ISBN 0-295-95171-0.

JH-SH-C   The author has recorded her observations of several years on Anderson Island. The ecologically oriented student and those who appreciate nature will find this an interesting, readable, and accurate account of the seasonal changes of plant and animal life.

Nutting, William B.   *Basic Natural History, A Procedural Approach.* (Illus.) Macmillan, 1972. xii + 386pp. $6.50(p). 76-156988.

SH-C   A study guide for a one-semester course in natural history. Its approach is field-oriented and includes equipment lists and sections showing in detail how experimental data should be handled. The main thrust is biological and ecological. Well organized, with a broad range of content.

Palmer, E. Laurence.   *Fieldbook of Natural History.* (Illus.;rev. by H. Seymour Fowler.) McGraw-Hill, 1975. xviii + 779pp. $19.95. 73-18290. ISBN 0-07-048425-2; 0-07-048196-2(text ed.). Index;CIP.

JH-SH-C   A unique book that gives concise information about the natural world. This edition covers rocks, stars, planets, plants and animals.

Platt, Rutherford.   *Water: The Wonder of Life.* (Illus.) Prentice-Hall, 1971. x + 274pp. $9.95. 73-146045. ISBN 0-13-945808-5.

JH-SH-C   Discusses water in the universe, the discovery of the chemical nature of water, its chemical, physical and biological properties, the amount of water available to Americans, and the necessity of keeping it clean and unpolluted. An excellent book.

Sage, Bryan L.   *Alaska and Its Wildlife.* (Illus.) Viking Press, 1973. 128pp. $14.00. 72-76836. SBN 670-11170-8.

JH-SH-C-P   A fresh, full-scale introduction to Alaska, one of the world's last sanctuaries of wildness. This perceptive English geologist and field naturalist views each of the regions, first from a helicopter, then from a hiker's viewpoint. Autobiographical.

Sanger, Marjory Bartlett.   *Billy Bartram and His Green World.* (Illus.) Farrar, Straus&Giroux, 1972. 205pp. $6.50. 78-175822. ISBN 0-374-30707-5.

SH-C   Bartram (1739–1823) described 215 native birds, and his bird paintings influenced Wilson and Audubon. This fascinating volume tells the life of this great early naturalist and the story of science in America when our country was young.

Steele, Mary Q.   *The Living Year: An Almanac for My Survivors.* Viking Press, 1972. xi + 109pp. $6.50. 71-188531. ISBN 670-43594-5.

SH-C   The author gives a very personal sense of her identification with and involvement in natural history in Tennessee in vignette-like essays which will appeal to teachers of English, conservationists, bird watchers, biology teachers, and sensitive students.

Stevens, Peter S.   *Patterns in Nature.* (Illus.) Atlantic Monthly/Little, Brown, 1974. 240pp. $10.00. 73-19720. ISBN 0-316-81328-1. Index;CIP.

JH-SH-C   Will please the biologist, mathematician or artist who appreciates the basic unity and diversity of patterns in nature. Topics include scale, spacefilling, topology, networks, fluid flow, spirals and related forms, branching and growth patterns. Excellent illustrations.

Ternes, Alan, (Ed.).   *Ants, Indians, and Little Dinosaurs: Selections from Natural History.* Scribner's, 1975. 391pp. $8.95. 75-4212. ISBN 0-684-14312-7. Index.

JH-SH-C-P   A collection of essays on the natural history of humans and other animals.

Zwinger, Ann.   *Run, River, Run: A Naturalist's Journey Down One of the Great Rivers of the West.* (Illus.) Harper&Row, 1975. xiii + 317pp. $10.95. 74-1874. ISBN 0-06-014824-1. Index;CIP.

SH-C   Zwinger writes of the natural history of the Green River, from its source in the Wind River Mountains of Wyoming to its mouth at the Colorado River of Utah. A rich, delightful book.

501   Philosophy and Theory

Eiseley, Loren.   *The Invisible Pyramid.* (Illus.) Scribner's, 1970. 164pp. $6.95. 71-123826.

SH-C   Eiseley speculates that the "afterlife" is the possibility of humanity's successful establishment on another planet. Eiseley's analogy is to the slime mold and its means of continuance through spore dispersal after aggregation and formation of capsules to encase the spores, which he likens to our forming urban complexes and pouring vast resources and energy into our space program.

Hanson, Norwood Russell.   *Observation and Explanation: A Guide to Philosophy of Science.* Harper&Row, 1971. ix + 84pp. $1.50 76-154654. ISBN 0-06-131575-3.

SH-C   This well-written introduction to the philosophy of science covers observation, facts, measurement, induction, experiment, causality, explanation, theories, laws, hypothetico-deduction, retroduction, theoretical entities, Craig's theorem, verification, falsification, and models.

Harre, Rom, (Ed.).   *Problems of Scientific Revolution: Progress and Obstacles to Progress in the Sciences.* Clarendon/Oxford Univ. Press, 1975. vi + 104pp. $11.25. ISBN 0-19-58211-0. Index.

C-P   Four (of six) essays question the linkage of progress in science with the inevitable, steady improvement of the human condition. The nature of scientific progress is itself questioned. Will be useful to students interested in the interrelationships of science, technology and social progress.

Holton, Gerald.   *Thematic Origins of Scientific Thought: Kepler to Einstein.* Harvard Univ. Press, 1973. 495pp. $10.00; $3.95(p). 72-88126. SBN 674-87745-4.

C-P   Interesting, articulate and sometimes profound essays on the general history and philosophy of science. Physicists from Kepler through Einstein are discussed, as is the place of science in the armamentarium of a healthy society or an educated person. A fascinating (if difficult) test for a course in the history of modern physics.

Malville, Kim.   *A Feather for Daedalus: Explorations in Science and Myth.* (Illus.) Cummings, 1975. 152pp. $3.95(p). 75-1955. ISBN 0-8465-4335-4. Index.

C   The role of the individual in science serves as the focal point. Exciting and well-drawn ideas and concepts are presented in a way that will inform and stimulate the well-read college student, lay person and professional. Explores the humanistic implications of science.

Medawar, Peter B.   *The Hope of Progress: A Scientist Looks at Problems in Philosophy, Literature and Science.* Anchor/Doubleday, 1973. 145pp. $1.95(p). 73-81423. ISBN 0-395-07615-0.

> **SH-C**   The author presents his opinions on such subjects as psychoanalysis, genetics, the sanctity of life and the future of the human race. The reactions of critics and writers to Medawar's essay on science and literature are also included.

Morgenthau, Hans J.   *Science: Servant or Master?* New American Library (distr.: Norton), 1972. xxi + 153pp. $6.95. 70-155726.

> **C**   In this erudite yet passionate little book Morgenthau argues that science has lost its transcendent meaning that once reflected concern for humanity and its future and now gropes aimlessly in a moral vacuum. As a result we have entered an era of nuclear complacency where scientists make scientifically acceptable the possibility of nuclear conflict. It is still *must* reading.

Rosenbluth, Arturo.   *Mind and Brain: A Philosophy of Science.* (Illus.) MIT Press, 1970. x + 128pp. $5.95. 70-95287. ISBN 0-262-18041-3.

> **C**   The contributions that neurophysiology can offer for the formulation of a philosophy of science are stressed. The author's central theme concerns the relation between personal mental states (conscious experiences, feelings, doubts, beliefs) and events which occur in the material universe. He introduces language and concepts from both neurophysiology and philosophy.

Schlegel, Richard.   *Inquiry into Science: Its Domain and Limits.* (Illus.) Doubleday, 1972. x + 128pp. $4.95. 74-144295.

> **GA-C**   Defines and describes the domain and limits of science. Topics range from quantum theory to cosmology. The book is quite readable and provides some easily understood concepts in the philosophy of science. The author's commentary on the relationship of science and the humanities will be of interest to both scientists and laypersons.

Sklar, Lawrence.   *Space, Time, and Spacetime.* (Illus.) Univ. of Calif. Press, 1974. xii + 423pp. $15.00 73-76096. ISBN 0-520-02433-8. Index.

> **C-P**   This excellent treatise bridges the gap between physics and philosophy and shows how philosophical presuppositions color decisions to adopt particular scientific theories.

502   SCIENCE—MISCELLANY

Brown, Bob.   *Science Circus #3.* (Illus.) Fleet, 1972. 112pp. $5.50. 60-7506. ISBN 8303-0122-4.

> **JH-SH**   More than 160 experiments are included on air and air pressure, light, heat, water, sound, electricity, magnetism, biology, mechanics, chemistry, etc. The equipment is easy to obtain. A good guide for a teacher who is willing to demonstrate physical principles and explain the results more fully than the author does.

Dellow, E. L.   *Methods of Science: An Introduction to Measuring and Testing for Laymen and Students.* (Illus.) Universe Books, 1970. 268pp. $8.95. 71-121793. ISBN 0-87663-129-4.

> **JH-SH-C**   A worthwhile attempt to arouse interest in and awareness of science and technology and their accomplishments.

Houwink, R.   *Data: Mirrors of Science.* (Illus.) American Elsevier, 1970. 210pp. $9.50. 76-100399. ISBN 0-444-00068-2.

SH-C   Scientific data are converted into images to help the reader see clearly and remember mutual relationships between quantities, sizes and capabilities. Illustrates the relation of data presentation to the comprehension of the reality behind the data.

Houwink, R.   *Sizing Up Science*. (Illus.) Murray (distr.: Transatlantic Arts), 1975. x + 164pp. $12.50. ISBN 0-7195-3101-2. Index.

SH-C   The author successfully translates scientific data into numerical relationships that can be visualized. His facts and analogies cover every field of science, including energy, strength and human hair, tornadoes and pollution from automobiles. An excellent reference for students and teachers.

Ritchie, Carson I.A.   *Making Scientific Toys*. (Illus.) Nelson, 1976. 169pp. $6.50. 75-23110. ISBN 0-8407-64331-2. CIP.

JH-SH   The toys are classified in nine categories: optical, acoustic, flying, gravity and balance, heat, weather and climate, chemical, electric and miscellaneous. The instructions are clear and specific.

Schwartz, Julius.   *It's Fun to Know Why: Experiments with Things Around Us, 2nd ed.* (Illus.) McGraw-Hill, 1973. 159pp. $5.72. 73-3306. ISBN 0-07-055733-0.

JH   Various experiments and observations that students can make using mainly their eyes and hands are described. Covered are such subjects as how wood is changed to coal, why bread is full of holes, and how to make plastic from milk. Each experiment is preceded by a list of materials, most of which can be found around the house.

Stoffer, Janet Y. (Ed.).   *Science Fair Project Index, 1960–1972*. Scarecrow, 1975. vi + 728pp. $25.00. 74-30269. ISBN 0-8108-0783-1. Bib;CIP.

JH-SH   A very useful reference for running school science fairs. The projects, listed in alphabetical order, are suitable for illustrating particular scientific principles.

Stone, A. Harris.   *Science Project Puzzlers: Starter Ideas for the Curious*. (Illus.) Prentice-Hall, 1969. 63pp. $3.95. 68-57675. ISBN 0-13-795369-0.

JH   Describes the materials and equipment needed, asks questions to guide the experimenter, and lists scientific ideas that can be studied or demonstrated. Hazardous procedures are carefully marked. Sources of other experimental methods and ideas are listed, and an appendix offers suggestions on how scientific observations are tabulated.

Webster, David.   *More Brain Boosters*. (Illus.) Doubleday, 1975. 176pp. $5.95. 73-78092. ISBN 0-385-02091-0; 0-385-02497-5.

JH-SH   This is an interesting collection of science brain teasers recommended to any science teacher. Webster includes puzzles about structures which relate to physical laws, many experiments with air and water, mystery photographs of shadows and other unusual patterns, kitchen chemistry, electricity experiments, and some number puzzles.

White, Laurence B., Jr.   *Investigating Science with Coins*. (Illus.) Addison-Wesley, 1970. 95pp. $3.50. 69-15796.

JH   A considerable number of investigations are described using coins and a few other common items. The effects are directly related to such ideas as surface tension, sensory perception, chemical reactions of metals, energy, momentum, friction, refraction, and various mathematical concepts.

White, Laurence B., Jr. *Investigating Science with Paper.* (Illus.) Addison-Wesley, 1970. 123pp. $3.75. 79-105874. ISBN 0-201-086581-1.

JH   Describes many interesting activities through which students can learn about paper and explore important ideas of science and engineering using paper and a few other household items. Experiments in papermaking, absorption, chromatography, filtration, invisible ink, and topology are included.

502.8 MICROSCOPY

Klein, Aaron E. *The Electron Microscope: A Tool of Discovery.* (Illus.) McGraw-Hill, 1974. 86pp. $5.72. ISBN 0-07-035029-9. Index;CIP.

JH-SH-C   Covers the theory and development of the electron microscope and indicates future uses for it. Highly recommended as collateral reading in any of the natural sciences.

Moellring, F.K. *Beginning with the Microscope.* (Illus.) Sterling, 1972. 72pp. $4.95. 72-180465. ISBN 0-8069-3044-6; 0-8069-3045-4.

C   A practical guide on how to use the microscope, which combines informality of style and good illustrations. The text is devoted only to use; specimen preparation techniques are not discussed.

503 DICTIONARIES AND ENCYCLOPEDIAS

See also 030 General Encyclopedic Works.

Asimov, Issac. *More Words of Science.* (Illus.) Houghton Mifflin, 1972. iv + 267pp. $5.95. 79-187422. ISBN 0-395-13722-5. Index.

JH-SH   Takes up where the earlier volume left off and provides the reader with 250 clearly and interestingly written explanations drawn from all fields of science. So well written that many people will want to read it cover to cover.

Ballentyne, D.W.G., and D.R. Lovett. *A Dictionary of Named Effects and Laws in Chemistry, Physics, and Mathematics.* Barnes&Noble, 1970. viii + 335pp. $9.50. ISBN 0-412-09600-5.

C   These definitions are clear and easy to understand, and illustrations of processes, chemical formulas and mathematical equations are included when necessary. Recommended for all science reference collections.

Bennett, H. (Ed.). *Concise Chemical and Technical Dictionary, 3rd ed.* Chemical Publishing, 1974. xxxix + 1175pp. $35.00.

SH-C-P   This edition contains 75,000 definitions and has been updated with new trademark products, chemicals, drugs and other technical terms. A valuable reference.

Gillispie, Charles Coulston, (Editor-in-chief). *Dictionary of Scientific Biography, Vols. I-XIV.* Scribner's, 1970–1976. 8935pp.(total). $35.00ea. 69-18090.

SH-C-P   A useful reference for scientists, historians, journalists, and students interested in the development of scientific ideas and concepts throughout recorded history (living scientists are omitted). The sketches provide factual information on the scientists' lives and discussions of their work in relation to that of their predecessors, contemporaries, and successors. A selected bibliography follows each article, and many of the listed items are not available in other references. A must for all libraries.

Lapedes, Daniel N., (Ed.). *McGraw-Hill Dictionary of Scientific and Technical Terms.* (Illus.) McGraw-Hill, 1974. xv + 1634pp. $39.50. 74-16193. ISBN 0-07-045257-1. CIP.

SH-C-P This dictionary provides an extensive listing of terms for approximately 100 well-defined fields of science and technology. Should be in all public, secondary school and college libraries.

McGraw-Hill. *McGraw-Hill Yearbook of Science and Technology: Comprehensive Coverage of the Important Events of the Year 1972.* (Illus.) McGraw-Hill, 1973. 468pp. $27.50. 62-12028. ISBN 07-045340-3. Index; CIP.
1973. McGraw-Hill, 1974. 465pp. $28.00. ISBN 07-045341-1.
1974. McGraw-Hill, 1975. 460pp. $28.50. ISBN 0-07-045342-X.
1975. McGraw-Hill, 1976. 437pp. $29.50. ISBN 0-07-045343-8.
1976. McGraw-Hill, 1977. 454pp. $29.50. ISBN 0-07-045344-6.

SH-C-P In each of these volumes the articles are written by recognized authorities, and the very broad range of topics accurately reflects the areas of major scientific and technological importance in each year. All major libraries should have these volumes, although some of the articles are too technical for popular understanding.

Tressler, Arthur G., et al. (Eds.). *Science Year: The World Book Science Annual, 1976.* (Illus.) Field Enterprises, 1975. 432pp. $8.95. 65-21776. ISBN 0-7166-0576-7. Index.
Science Year: The World Book Science Annual, 1977. (Illus.) 431pp. $7.95. 65-21776. ISBN 0-7166-0577-5. Index.

SH-GA-C These and earlier volumes in the series provide information on current accomplishments of science and technology and offer "special reports"—in depth treatments of significant and timely subjects—and a "science file"—articles and close-up reports on the year's developments. The authors are specialists, but they write for the literate nonscience major.

507 STUDY AND TEACHING

Bobrowsky, Kenneth, and the Editors of *Science World.* *Science/Search (Series).* (Illus.) Scholastic, 1971. Approx. 80pp. ea. $1.00ea. *You: A Body of Science; The Living Scene; The Air Above the Ground Below; Power and Man. Teaching Guide.* 96pp. $1.95.

JH-SH This program involves students in their own investigations by using everyday materials and experiences. It is written for a fifth-grade reading level, in a clear, concise manner geared to attract the students' interest. The units include physiology, earth science, biology, and physical science. Should be used in every school where science is taught and reading deficiency is a problem.

Brandwein, Paul F., et al. *Matter: Its Forms and Changes, 2nd ed.* (Illus.) Harcourt Brace, 1972. xii+564pp. $5.40. ISBN 0-15-366340-5. *Teacher's Manual.* ISBN 0-15-366341-3.

SH An excellent text which is concise, readable and holds the interest of the students. The color photographs are stunning. Encourages the student to think and probe the methods and operations of science. The teacher's manual has excellent references and suggestions.

Brandwein, Paul F., and Hy Ruchlis. *Invitations to Investigate: An Introduction to Scientific Exploration.* (Illus.) Harcourt Brace Jovanovich, 1970. 158pp. $4.95. 77-102440. ISBN 0-15-238835-4.

JH-SH A wide variety of areas to investigate is suggested, although biological topics predominate. The importance of careful measurement, recording observations and setting up controls is stressed.

Daish, C.B.   *Learn Science Through Ball Games*. (Illus.) Sterling, 1972. 128pp. $3.95. 72-81047. ISBN 0-8069-3050-0.

SH-C   Relates physical principles to sports activities. Daish's physics is authentic; his knowledge of various sports nuances is striking; his descriptions of physical phenomena are unusually clear. Most of text deals with the impact, flight, roll, spin and bounce of balls. There is something here for both participants and spectators.

Marean, John H., and Elaine W. Ledbetter.   *Physical Science: Investigating Matter and Energy, rev. ed. (A Laboratory Approach)*. (Illus.) Addison-Wesley, 1972. xi + 308pp. $6.40. *Teacher's Guide*. iv + 116pp. $2.60.

JH-SH   Basic principles of physical science are developed through a series of laboratory investigations integrated into the narrative of the text. Includes processes of science, basic chemistry and physics and an introduction to chemical kinetics and equilibrium. The *Teacher's Guide* includes suggestions for lab preparation, answers to questions, and detailed instructions for constructing an inexpensive vacuum tube electroscope. Useful for students who take no courses in chemistry or physics.

Vrana, Ralph.   *Junior High School Science Activities*. (Illus.) Parker, 1969. x + 213pp. $8.95. 69-18744.

JH   Presents experiences and techniques in chemistry, life sciences, earth sciences, electricity, mechanics, optics and light, heat, and mathematics for science. A good number are presented with a "new twist." There is a helpful chapter on materials and equipment for the lab program.

Wong, Harry K., and Malvin S. Dolmatz.   *Ideas and Investigations in Science*. Vol. 1: *Biology*. Prentice-Hall, 1971. 240pp. $7.96ea. *Teacher's Manual*. 243pp. $7.96ea. Student exercise books, $1.12ea.: *Inquiry; Evolution; Genetics; Homeostasis; Ecology*. Vol. 2: *Physical Science*. 248pp. *Teacher's Manual*. 240pp. Exercise books: *Predicting; Matter; Energy; Interaction; Technology*.

SH   Directed toward the average or below-average student who is not college-bound and not interested in traditional academic activities, and written in an informal, slangy style illustrated with humorous and useful drawings and photographs, these programs seek to develop an understanding of several major ideas of science; to promote skill in using the processes of science; and to establish the relevance of science to the individual and to society.

Wood, W.G., and D.G. Martin.   *Experimental Method: A Guide to the Art of Experiment for Students of Science and Engineering*. Athlone (distr.: Humanities Press), 1974. 106pp. $4.25(p). ISBN 0-485-12022-4. Index.

C-P   The authors present the characteristics of theoretical and experimental studies and show how various techniques are applied to the two studies. Discusses the use of the logbook, consideration of variables, planning the experiment, treating the results and preparing the final report.

508 COLLECTIONS AND SURVEYS

Asimov, Isaac.   *The Left Hand of the Electron*. Doubleday, 1972. xii + 225pp. $6.95. 75-181479.

SH-C   Science teachers can learn from these essays ideas, anecdotes, and how to simplify complicated ideas without distorting or falsifying them. Essays cover a spectrum of topics from chemistry and mathematics to history, biology and the population problem.

Kone, Eugene H., and Helene J. Jordan (Eds.). *The Greatest Adventure: Basic Research that Shapes Our Lives.* (Illus.) Rockefeller Univ. Press, 1973. x + 294pp. $9.80. 73-83747. SBN 87470-018-3.

SH-GA-C These essays by eminent scientists demonstrate uses of the "pure" sciences and argue for public appreciation and support of scientific research. Topics include cosmology, water, air and earth, the neurosciences, population growth, nutrition, immunology and genetics, polymers, optical instruments, computers, resources, power and communication.

Metzger, Norman. *Men and Molecules.* (Illus.) Crown, 1972. x + 246pp. $5.95. 71-185075.

SH-GA-C Eleven unrelated topics are presented to the intelligent lay reader in an entertaining and thoughtful manner. Genetics, aging, the universe, superheavy elements, neutrinos, ancient glass, ribonuclease, nitrogen fixation, insects, poly(?)water and pollution are the themes. Conveys not only the thrill of discovery but the need for convincing proof.

Smithsonian Institution. *Frontiers of Knowledge.* (Illus.) Doubleday, 1975. xii + 399pp. $15.00. 74-18793. ISBN 0-385-03151-3. CIP.

C This volume containing the Frank Nelson Doubleday lectures at the Smithsonian Institution includes "Technology and the Frontiers of Knowledge," "Creativity and Collaboration," and "The Modern Explorers." The essays are well written, serious and stimulating.

509 HISTORY OF SCIENCE

See also individual disciplines and 509.2 Science—Biographies.

Basalla, George, William Coleman and Robert H. Kargon (Eds.). *Victorian Science.* (Illus.) Doubleday, 1970. x + 510pp. $2.45(p). 77-89137.

SH-C A critical, questioning analysis of the development of science training and facilities and of the general progress of science during one of the most productive periods of human history. The main materials are drawn from the presidential addresses of the British Association for the Advancement of Science. A must as supplementary material in history of science courses.

Bernal, J.D. *Science in History.* Vol. 1: *The Emergence of Science.* Vol. 2: *The Scientific and Industrial Revolution.* Vol. 3: *The Natural Sciences in our Time.* Vol. 4: *The Social Sciences: Conclusion.* (Illus.) MIT Press, 1971. xxv + 362pp. $3.95ea.(p). 78-136489. ISBN 0-262-52020-6. Bibs.

SH-C A breath-taking view of the place of science in history. Vol. 1 documents scientific developments from the first neolithic toolmaking traditions to the end of feudal systems in the late Middle Ages. Vol. 2 shows scientific advances from the Renaissance and the Reformation to the end of the 19th century. Vol. 3 deals with physical and biological sciences of the 20th century, and Vol. 4 contains a remarkable, well-referenced synthesis. Many photographs with detailed captions, comprehensive tables, and informative notes throughout.

Goran, Morris. *Science and Anti-Science.* Ann Arbor Science, 1974. xi + 128pp. $10.00. 73-90416. ISBN 0-250-40049-9. Index.

SH-C Provides a brief treatment of the human frailties and human achievements of science. Goran retains a sense of proportion and is even-handed in his treatment of science and its adversaries.

Kohlstedt, Sally Gregory. *The Formation of the American Scientific Community: The*

American Association for the Advancement of Science 1848–60. (Illus.) Univ. of Illinois Press, 1976. xiii + 264pp. $10.95. 75-37748. ISBN 0-252-00419-1. Index;CIP.

C   A history of the establishment and the first years of AAAS, when it gained stature as the leader of the scientific community. This accurate and well-written record will serve as a reference for courses in the history of science or the intellectual development of this period.

Neyman, Jerzy, (Ed.). *The Heritage of Copernicus: Theories "Pleasing to the Mind."* (Illus.) MIT Press, 1974. ix + 542pp. $25.00. 74-6415. ISBN 0-262-14021-7. Index;CIP.

SH-GA-C   This volume in honor of Copernicus's 500th birthday describes major scientific revolutions in many fields over the last two centuries. Each contribution is of uniformly high quality, intellectually stimulating and authoritative. Intended for the general reader with some science background, this is one of the best science books in years and a key addition to any library.

Parry, Albert. *The Russian Scientist.* (Illus.) Macmillan, 1973. 196pp. $5.95. 72-92454. Bib.

JH-SH   Includes a history of science in prerevolutionary Russia; brief biographical sketches of scientists such as Lomonosov, Lobachevsky, Mendeleyev and Pavlov; a look at the Soviet Academy in modern times; and a brief discussion of the new Soviet scientist city (Akademgorodok in Siberia). Briefly examines Russia's lack of science and technology before the 1700s.

Price, Derek de Solla. *Science Since Babylon, enlarged ed.* (Illus.) Yale Univ. Press, 1975. xvi + 215pp. $15.00; $3.45(p). 74-79976. ISBN 0-300-01797-9; 0-300-01798-7. Index.

C   Although this classic has been generally revised, it does not supplant the earlier edition nor the companion classic *Little Science, Big Science.* New additions include the history of automata and geometrical amulets, and an essay on the relation between science and technology. Of the highest professional caliber, but easily read by college or high school teachers and their students.

Ronan, Colin. *Lost Discoveries: The Forgotten Science of the Ancient World.* (Illus.) McGraw-Hill, 1973. 125pp. $10.95. 73-4968. ISBN 0-07-053597-3.

SH-C   Ronan concentrates on the science and technology in Egypt, Mesopotamia, Greece and Rome, but includes discussions of developments in China and India, of the accomplishments of prehistoric inhabitants of northern Europe, pre-Columbian Americans and seafaring peoples of the Pacific.

Russell, C.A., and D.C. Goodman (Eds.). *Science and the Rise of Technology Since 1800.* (Illus.) England: John Wright, 1972. xii + 338pp. £3.95. ISBN 0-7236-0336-7.

SH-C   An anthology of literate declarations of scientific milestones; includes original papers by Davy, Lister, Pasteur, Diesel, Einstein, Faraday and others. Well executed, with adequate editorial inclusions; originally compiled for the Open University in Britain.

Sambursky, Samuel, (Ed.). *Physical Thought from the Presocratics to the Quantum Physicists: An Anthology.* (Illus.) Pica (distr.: Universe), 1974. xv + 584pp. $20.00. 74-12946. ISBN 0-87663-712-8. Indexes.

SH-C   The six parts of this anthology are "Antiquity," "The Middle Ages," "Copernicus to Pascal," "The Royal Society to LaPlace," "Dalton to Mach" and "Planck to Pauli." There are 311 selections spanning 2500 years.

Sarton, George. *A History of Science.* Vol. 1: *Ancient Science Through the Golden Age of Greece.* Norton, 1970. xvi + 646pp. $3.25ea.(p). ISBN 0-393-00525-9. Vol. 2: *Hellenistic Science and Culture in the Last Three Centuries B.C.* 554pp. ISBN 0-393-00526-7.

SH-C   This is the paperbound edition of the most detailed and scholarly account available of the beginnings and development of science in pre-Christian eras. Can be used for either reference or collateral reading.

Taylor, James. *The Scientific Community.* Oxford Univ. Press, 1973. viii + 79pp. $3.50(p). ISBN 0-19-858314-1. Index.

C-P   Traces the growth of science through the Ancient World, the Greek impact, the "New Learning" and the British Royal Society, the Industrial Revolution, and two world wars to the present time. Presents a contemporary overview of the relation of science and society and cogent, eloquent thoughts on how science, along with society, should approach contemporary problems in order to improve the quality of life.

Westfall, Richard S. *The Construction of Modern Science: Mechanisms and Mechanics.* (Illus.) Wiley, 1971. xiii + 171pp. $7.50; $3.95(p). 72-151730. ISBN 0-471-93530-1; 0-471-93531-X.

C-P   A readable and unified account of the revolutionary changes in the sciences, theology and philosophy in the 17th century. Contrasts the different philosophical positions held and elucidates the assumptions and logic behind various scientific accomplishments, including those in chemistry, biology and mechanics. A sense of continuity, almost of logical necessity, is developed which substantially enriches contemporary scientific perspectives.

Woodcroft, Bennet, (Trans. and Ed.).   *The Pneumatics of Hero of Alexandria: From the Original Greek.* (Illus.) American Elsevier, 1971. xix + 141pp. $10.25. 70-132638. ISBN 0-444-19623-4.

C-P   This is one of the first translations into English of an early Greek manuscript on technology. *The Pneumatics* (c. 150 B.C.) is an illustrated description of 78 devices which Hero had observed, improved upon or invented, many of which are ingenious innovations still in use today. Students of physics and historians of science will welcome this fascinating revival.

Wynter, Harriet, and Anthony Turner. *Scientific Instruments.* (Illus.) Scribner's, 1976. 239pp. $27.50. 75-14941. ISBN 0-684-14471-9. Index; gloss.

C-P   This book covers telescopes and other astronomical instruments, navigation and surveying instruments, microscopes and accessories and sundials in the period from the mid-16th to the mid-19th centuries. Aimed primarily at collectors, it includes brief comments on the collectibility of some instruments.

509.2 SCIENCE—BIOGRAPHIES

See also individual disciplines.

Agassi, Joseph. *Faraday as a Natural Philosopher.* (Illus.) Univ. of Chicago, 1971. xiv + 359pp. $12.50. 73-151130. ISBN 0-226-01046-5.

C   This interesting book combines a biography of Faraday with a historical and philosophical discussion of his work and ideas. The author has drawn from Faraday's extensive lecture notes, scientific writings, and correspondence as well as the writings of his scientific contemporaries.

Bedini, Silvio A. *The Life of Benjamin Banneker.* (Illus.) Scribner's, 1972. xvii + 435pp. $14.95. 78-162755. ISBN 0-684-12574-9.

C   The definitive biography of Benjamin Banneker: copiously annotated and with an exhaustive bibliography and extensive excerpts from documents privately owned and from other primary sources not readily available. An extensive preface contains Bedini's evaluation of Banneker and observations on the prodigious research necessary to uncover the facts about a man who lived in obscurity but became a legendary figure after his death.

Bedini, Silvio A. *Thinkers and Tinkers: Early American Men of Science.* (Illus.) Scribner's, 1975. xix + 520pp. $17.50. 72-1174. ISBN 0-684-14268-6. Index;CIP.

SH-C   Bedini discusses practical science in terms of the technology required and its application in the New World, from the first settlement to the 19th century. Immediate needs included charting the wilderness, navigating waterways, establishing territorial boundaries, making the necessary instruments and training people. There is a particularly useful technical glossary.

Cohen, I. Bernard. *Benjamin Franklin: Scientist and Statesman.* (Illus.) Scribner's, 1975 (c.1972). 95pp. $6.95; $2.65(p). 75-7595. ISBN 0-684-14251-1; 0-684-14252-X. Index.

JH-SH   Cohen clearly establishes the magnitude of Franklin's contributions. A portfolio of pictures of equipment, sketches, writings, portraits and title pages is included.

Conant, James B. *My Several Lives: Memoirs of a Social Inventor.* (Illus.) Harper &Row, 1970. xvi + 701pp. $12.50. 72-83590.

SH-C   An engrossing, rich commentary on our times which provides a potent argument that creative people should deliberately and radically alter their careers to retain their youthful creativity.

Drake, Stillman. *Galileo Studies: Personality, Tradition, and Revolution.* Univ. of Michigan, 1970. 289pp. $8.50. 73-124427. ISBN 0-472-08283-3.

SH-C-P   Thirteen essays dealing with Galileo's personality, relationship with his colleagues, and scientific achievements: theory of tides, problems of free fall and acceleration, the concept of inertia, sunspot observation, and problems connected with bodies floating in water.

Eakin, Richard M. *Great Scientists Speak Again.* (Illus.) Univ. of Calif. Press, 1975. viii + 119pp. $6.95. 74-22960.

SH-C-P   Eakin selected six biological and medical scientists of the past and dramatized his class presentations by having each of the six serve as "guest lecturers." Harvey, Mendel, Beaumont, Speman, Darwin and Pasteur were the scientists; the lectures are reproduced verbatim.

Fleming, Thomas. *The Man Who Dared the Lightning: A New Look at Benjamin Franklin.* (Illus.) Morrow, 1971. 532pp. $12.50. 79-133289.

SH-C   This fascinating biography provides insights into the character of the man and the nature of his times. Franklin's science is placed in proper perspective; its influence on his personal relationships, its use as an entry into powerful political circles, and its significance as a way of providing public service without political involvement are all noted.

Hayden, Robert C. *Seven Black American Scientists.* (Illus.) Addison-Wesley, 1970. 172pp. $4.75. 77-118997.

SH   The scientists are Charles Drew, Daniel Hale Williams, Benjamin Banneker, Charles H. Turner, Ernest E. Just, Matthew Henson, George W. Carver. The contribution of each to his scientific field is described without technical jargon, and the question of racial barriers is treated objectively.

Hoyle, Fred. *Nicolaus Copernicus: An Essay on His Life and Work.* (Illus.) Harper &Row, 1973. xi + 94pp. $5.95. 73-4092. ISBN 0-06-011971-3.

C-P   A brief, personal essay considers the problems of establishing the accurate story of Copernicus' life and work. Hoyle then mathematically compares the theories of Ptolemy and Copernicus with modern theory and shows that the primary importance of Copernicus' theory was that it contributed to a new world view which made theoretical scientific progress easier.

Ronan, Colin A. *Galileo.* (Illus.) Putnam's, 1974. 264pp. $14.95. 74-76233. SBN 399-11364-9. Index.

   SH-C   After developing the historical background, attention is focused on
         Galileo—his life, work and recovery from trial by the Inquisition. The book is very well written, the story told is interesting, and the reader's interest is maintained.

Schultz, Pearle, and Harry Schultz. *Isaac Newton: Scientific Genius.* (Illus.) Garrard, 1972. 144pp. $2.79. 70-177892. ISBN 0-8116-4514-2.

   JH   A good account of Newton's major discoveries concerning light and the gravi-
         tational field, his encounters with other scientists, and his involvement in the 17th century English political scene.

Wilson, Mitchell. *Passion to Know: The World's Scientists.* Doubleday, 1972. xiii + 409pp. $10.00. 79-171329.

   SH   Destroys many long-standing popular myths about the character of successful
         scientists, but the reader is left with the impression that they, like many famous artists, are slaves of an all-consuming passion to know—a passion that forces them to endure years of disappointment, frustration, and physical stress for the rare ecstasy of discovery.

510 Mathematics

Adams, William J. *Finite Mathematics for Business and Social Science.* (Illus.) Xerox Coll., 1975. xiv + 354pp. $10.95. 73-84448. ISBN 0-536-00986-4. Index.

   C   An impressive, no-nonsense book written without cluttering jargon. Examples
         are well chosen and realistic, explanations are lucid, and the text is organized for flexibility. The second half could also be used as a text for an elementary probability course.

Burns, Marilyn. *The I Hate Mathematics! Book.* (Illus.) Little, Brown, 1975. 127pp. $6.95; $3.95(p). 75-6707. ISBN 0-316-11741-2. Index;CIP.

   JH-SH   This book has an informal format and a large variety of mathematical
         ideas presented in an interesting way.

Dowdy, S.M. *Mathematics: Art and Science.* (Illus.) Wiley, 1971. xviii + 282pp. $8.95. 70-140516. ISBN 0-471-22020-5. Bibs.;index.

   C   Shows what mathematicians do and demonstrates the creativity and beauty in
         mathematical thinking. Touches on number theory, algebra, geometry, lattices, logic, set theory, infinite series, probability, and computer sciences. Excellent features include varied topics, good exercises, and important guidance notes for the instructor.

Hildebrand, Francis H., and Cheryl G. Johnson. *Finite Mathematics, 2nd ed.* (Illus.) Prindle, Weber&Schmidt, 1975. vii + 535pp. $12.95. 74-34287. ISBN 0-87150-169-4. Index;CIP.

   SH-C   A fine text for a finite mathematics course, this book covers logic sets,
         vectors and matrices, probability and statistics, linear programming and game theory.

Holt, Michael, and D.T.E. Marjoram. *Mathematics in a Changing World.* (Illus.) Walker, 1973. vii + 293pp. $10.00. 72-95756. ISBN 0-8027-0450-0.

SH-C Describes the growing use of mathematics in many fields: psychology, physiology, medicine, economics, commerce, cartography, and demography. Diverse concepts are simplified and described, including models and specific applications. Some knowledge of mathematics is helpful but not necessary.

Linn, Charles F. *The Golden Mean: Mathematics and the Fine Arts.* (Illus.) Doubleday, 1974. xvii + 131pp. $4.95. 73-15480. ISBN 0-385-04110-1.

JH-SH-C Linn bridges the gap between mathematics and aesthetics. His well-illustrated treatment ranges from the golden mean and the Greek musical scale to the "form and pattern" rules of computer decision-making. A valuable addition to any math or art teacher's library.

Mizrahi, Abe, and Michael Sullivan. *Finite Mathematics with Applications for Business and Social Science, 2nd ed.* (Illus.) Wiley, 1976. xii + 586pp. $13.95. 75-15681. ISBN 0-471-61193-X. Index;CIP.

SH-C The authors develop, with excellent examples, logic, sets, elementary counting, mathematical models, linear equations and inequalities, matrix algebra, linear programming, mathematics of finance and probability and statistics. They also introduce the simplex method, directed graphs, Markov chains and game theory. For students with minimal mathematical background.

Volker, Benjamin W., and Andrew S. Wargo. *Fundamentals of Finite Mathematics.* (Illus.) Intext, 1972. xi + 368pp. $9.95. 79-177306. ISBN 0-7002-2355-X.

C This book illustrates principles and theory with practical examples. Its broad coverage provides suitable introductory background for both the applications- and the theory-oriented student. Clear exposition, good exercises.

510.2 MATHEMATICS—MISCELLANY

Barr, George. *Entertaining with Number Tricks.* (Illus.) McGraw-Hill, 1971. 144pp. $4.33. 71-157478. ISBN 0-07-003842-2.

JH Over 50 easily understood tricks, such as finding a person's favorite number, guessing age, mental arithmetic and magic squares, are described. Of special interest to the teacher are the clear explanations of why the tricks work.

Benson, William H., and Oswald Jacoby. *New Recreations with Magic Squares.* (Illus.) Dover, 1976. viii + 198pp. $4.00(p). 74-28909. ISBN 0-486-23236-0.

SH-C Gives prescriptions, with proofs, which enable the reader to write magic squares of various sizes and with additional interesting properties.

Emmet, E.R. *Puzzles for Pleasure.* (Illus.) Emerson, 1972. x + 310pp. $6.95. 71-189618. ISBN 87523-178-0.

SH-C One hundred two new logical puzzles, most of which require finding an answer based on what appears to be insufficient data (e.g., correcting a long division problem in which every single digit is wrong). Requires an understanding of long division and such terms as prime, square and cube.

Eves, Howard W. *In Mathematical Circles: A Selection of Mathematical Stories and Anecdotes, 2 vols.* Prindle, Weber&Schmidt, 1969. xi + 145pp. & xvii + 136pp. $12.50set. 70-94459. ISBN 0-87150-056-8.

Mathematical Circles Revisited: A Second Collection of Mathematical Stories and Anecdotes. Prindle, Weber&Schmidt, 1971. xii + 186pp. $10.00. 71-155300. ISBN 0-87150-121-X.

Mathematical Circles Squared: A Third Collection of Mathematical Stories and Anecdotes.
(Illus.) Prindle, Weber&Schmidt, 1972. xxi + 186pp. $15.00. 76-175520. ISBN
87150-154-6.

> **SH-C** Anecdotes, often unconnected except by a rough chronology, comprise a
> splendid variety: in the first two volumes, biographical material, problems,
> puzzles, constructions, controversies and folklore; in the second collection, anec-
> dotes concerned with numbers and numerals, pi, counting boards and tally sticks,
> computers, metrology, symbols and terminology (with advice on classroom tactics).
> In the third collection, there are varied selections, from the time of Saint Jerome (400
> A.D.) to space-age mathematicians, from absent-minded professor stories to magic
> squares and unsolved problems. Heavier reading, but the suggestions for writing
> projects and mathematical models make the book more than just a collection of
> curiosities.

Gardner, Martin. *Martin Gardner's Sixth Book of Mathematical Games from* **Scientific
American.** (Illus.) Freeman, 1971. x + 262pp. $9.95. 75-157436. ISBN 0-7167-0944-9.
Bib.

> **SH-C** The author's format is that of a clear and concise discussion of the mathe-
> matical subject matter, followed by challenging problems using ideas pre-
> sented, and their solutions. While the book lacks an overall unity, each chapter
> stands on its own.

Gardner, Martin. *Mathematical Carnival: A New Round-Up of Tantalizers from* **Scientific American.** (Illus.) Knopf, 1975. xii + 272pp. $8.95. 75-8208. ISBN 0-394-49406-7.
CIP.

> **SH-C** The content includes coin puzzles, card shuffles, infinities, physics puzzles
> and lightning calculation tricks. A very fine book for anyone with intellec-
> tual curiosity.

Jefimenko, Oleg D. *How to Entertain with Your Pocket Calculator: Pastimes, Diversions, Games and Magic Tricks.* (Illus.) Electret Scientific, 1975. xv + 189pp. $7.00;
$3.00(p). 74-84864. Index.

> **JH-GA** The author has collected a considerable quantity and variety of entertain-
> ing problems for pocket calculators. Most are carefully explained alge-
> braically and provide a lighthearted introduction to number theory.

Kordemsky, Boris A. *The Moscow Puzzles: 359 Mathematical Recreations.* (Illus.;
trans.; edited by Martin Gardner) Scribner's, 1972. 309pp. $10.00. 74-162770. ISBN
0-684-12586-2.

> **SH** The varied "puzzles" cover a wide range of difficulty, and some are intrigu-
> ing games rather than problems. They are often presented in the context of
> stories about Russian life and customs and thus provide interesting, nonmathematical
> insights. Useful supplementary mathematics study material.

510.7 MATHEMATICS—STUDY AND TEACHING

Bassler, Otto C., and John R. Kolb (John E. Searles, Ed.). *Learning to Teach Secondary School Mathematics.* Intext Educational, 1971. xiii + 434pp. $6.50. 70-151637.
ISBN 0-7002-2320-7.

> **C** A truly innovative approach, this collection of exercises simulates aspects of
> the teaching process and identifies recurring problems of content. Teachers
> learn by doing, and an introductory section defines terms and provides background.
> Recommended for all teachers of mathematics and as a text for courses in mathemat-
> ics teaching.

Behr, Merlyn J., and Dale G. Jungst. *Fundamentals of Elementary Mathematics: Number Systems and Algebra.* (Illus.) Academic, 1971. xx + 419pp. $9.00. 77-137609.

SH-C   If this remarkable book were used in a required course for all junior and senior high school math teachers, an educational revolution would result. A combination of deep treatment and intuitive motivation, a direct effort to develop the student's skill in writing proofs, and very extensive problems are the distinguishing features of this excellent book.

Cooley, James A., and Ralph Mansfield. *Basic Mathematics Review.* (Illus.) Vol. 1. *Arithmetic* 178pp. Vol. 2. *Elementary Algebra.* Macmillan, 1970. 235pp. $3.95ea.; $6.95set. 75-89927.

SH-C   A comprehensive text-workbook (with perforated pages) for students of diverse mathematical background. Provides lead material with examples, concise summaries, abundant exercises, and compares most favorably with other basic reviews. Should have a large audience even outside of the regular classrooms.

Greenes, Carole E., et al. *Problem Solving in the Mathematics Laboratory: How to Do It.* (Illus.) Prindle, Weber&Schmidt, 1972. 158pp. $5.95(p). 72-88526.

C-P   This handbook provides a collection of activities using attribute blocks, geoboards, multibase blocks and cuisenaire rods. Specific applications are described and illustrated. The materials are used to play games, find patterns, create designs, test algorithms, and make constructions. Adaptable for grades K through 12.

Jacobs, Harold R. *Mathematics, A Human Endeavor.* (Illus.) Freeman, 1970. xvii + 529pp. $8.50. 70-116898. ISBN 0-7167-0439-0. *Teacher's Guide.* 1971. x + 211pp. $5.00(p).

JH-SH   Deftly presents such items as sequences, graphs, logarithms, curves and probability so that they are challenging and yet easily understood. Comic strips illustrate concepts. Bright but disinterested junior high students and disinterested 11th or 12th grade low achievers would profit. The *Teacher's Guide* includes answers to exercises, and solid explanation and guidance.

Kadesch, Robert R. *Math Menagerie.* (Illus.) Harper&Row, 1970. 112pp. $4.11. 66-11498.

JH-SH   A collection of mathematical experiments on probability, binary and unusual numbers, shapes, mappings, transformations, etc. Some discussion of the rationale is included along with explicit instructions on how to construct the apparatus or perform the experiment.

Kulm, Gerald. *Laboratory Activities for Teachers of Secondary Mathematics.* (Illus.) Prindle, Weber&Schmidt, 1976. iv + 124pp. $6.50(p). 75-44036. ISBN 0-87150-211-9. CIP.

C   Introduces teachers to the laboratory technique of simplifying abstract topics for students. The topics include radioactive decay, figurate numbers, quadratic roots, surface area and volume, square function and construction of products, quotients and square roots. A simulating supplement, encouraging both group cooperation and independent thinking.

Meserve, Bruce E., and Max A. Sobel. *Introduction to Mathematics.* (Illus.) Prentice-Hall, 1969. xi + 420pp. $7.95. 75-77849. ISBN 0-13-487322-X.

C   Intended for a liberal arts course or for prospective elementary teachers. Extremely broad in scope, and the coverage is almost superficial. Number bases and theory, Egyptian numerals, multiplication system, algebraic properties, modular arithmetic, set theory and logic, linear equations and graphs, discrete probability, symbolic logic, Euclidean, non-Euclidean and metric geometry are treated.

Scandura, Joseph M. *Mathematics: Concrete Behavioral Foundations.* (Illus.) Harper &Row, 1971. xx + 459pp. $11.95. 70-144235. ISBN 0-06-045757-0.

 C   Primarily intended for elementary school teachers of mathematics and for persons who have taken mathematics without comprehending it or its relation to the physical world. A description of mathematics skills and of basic competencies is followed by exposition of significant ideas in contemporary mathematics. In addition to the exercises, a workbook is available.

Wheeler, Ruric E. *Modern Mathematics: An Elementary Approach, 3rd ed.* (Illus.) Brooks/Cole, 1973. 667pp. $10.95. 72-86159. ISBN 0-8185-0070-0. Bibs.

 C   Designed for a one or two semester final course in mathematics for prospective elementary teachers or liberal arts majors. Useful lists of symbols used and definitions, and suggested course outlines. Principal topics are simple logic and mathematical reasoning, sets and relations, whole, rational and real numbers, geometry, some algebra and basic statistics.

510.83 MATHEMATICAL TABLES

Bartsch, Hans-Jochen. *Handbook of Mathematical Formulas.* (Illus.; trans.) Academic, 1974. 525pp. $9.50. 73-2088. ISBN 0-12-080050-0. Index.

 SH-C-P   This book is loaded with formulas and general facts from many areas of elementary mathematics. Particularly complete are the sections on arithmetic, synthetic and analytic geometry, algebra, vectors, calculus and differential equations.

Selby, Samuel M., (Ed.) *CRC Handbook of Tables for Mathematics, 4th ed.* Chemical Rubber Co., 1970. 1120pp. $24.50. 62-15661.

 C-P   This fourth edition has almost double the pagination and many new and improved features. Will serve well anyone who needs mathematical tables—students, scientists, engineers, architects, economists, sociologists, hobbyists, etc.

510.9 MATHEMATICS—HISTORICAL, GEOGRAPHICAL, AND BIOGRAPHICAL TREATMENTS

Dodge, Clayton W. *Numbers and Mathematics, 2nd ed.* (Illus.) Prindle, Weber &Schmidt, 1975. xvi + 557pp. $12.95. 74-31133. ISBN 0-87150-180-5. Index;CIP.

 SH-C   Dodge delightfully comments on the topics and on the humanistic, historical and sociological background of the people who have made mathematics what it is today.

Dubbey, J.M. *Development of Modern Mathematics.* (Illus.) Crane, Russak, 1972 (c.1970). 145pp. $6.75. 72-88125. ISBN 0-8448-0115-1. Bib.

 C-P   A survey of mathematics from the Sumerians' contributions (3000–2000 B.C.) to Cohen's 1964 proof of the independence of the continuum hypothesis and the axiom of choice. Achievements and errors of major mathematicians are presented together with descriptions of their intellectual and technological milieu. The three great crises in mathematics (the Pythagorean discovery of irrational numbers, the imprecise use of infinitesimals in early calculus, and the consistency of axiomatic systems) are major themes.

Gies, Joseph, and Frances Gies. *Leonard of Pisa and the New Mathematics of the Middle Ages.* (Illus.) Crowell, 1969. 127pp. $3.95. 71-81952. Bib.

 SH-C   The known facts concerning Leonard of Pisa are placed in their historic setting. Artistic illustrations and careful documentation. Style is not exciting but content is both historically and mathematically interesting.

Gillings, Richard J. *Mathematics in the Time of the Pharaohs.* (Illus.) MIT, 1972. x + 286pp. $25.00. 74-137469. ISBN 0-262-07045-6.

   C-P   The history and philosophy of Egyptian mathematics has not been presented with greater clarity, vitality and thoroughness. This introduction presents hieroglyphic and hieratic writing and numbers, the Egyptian approach to arithmetic operations, fractions and the "G-Rule," the Rhind Mathematical Papyrus, problems in completion, the "red Auxiliaries," and much more.

Mahoney, Michael Sean. *The Mathematical Career of Pierre De Fermat (1601–1665).* (Illus.) Princeton Univ. Press, 1973. xviii + 419pp. $20.00. 72-733. ISBN 0-691-08119-0.

   SH-C   Mahoney re-creates the atmosphere of mathematics and the sciences in the 17th century in this scholarly and entertaining account of Fermat's mathematical life and times. Includes strong circumstantial evidence that Fermat never really worked on a proof of his "last" theorem and suggests new historico-mathematical themes to be explored.

Osen, Lynn M. *Women in Mathematics.* (Illus.) MIT Press, 1975. xii + 185pp. $4.95(p). 73-19506. ISBN 0-262-15014-X. Index; CIP.

   SH-GA   This welcome addition to the history of women in science contains biographical sketches of the lives and work of Hypatia, C. Herschel, S. Kovalevsky and other mathematician-scientists. The style is lucid and lively.

Reid, Constance. *Hilbert.* (Illus.) Springer-Verlag, 1970. xi + 290pp. $8.80. 76-97989.

   C-P   As the author carries us through Hilbert's life we are easily swept up into the enthusiasm of pure research and the personal relationships existing among people so involved. An appendix is devoted to Hermann Weyl's summary of Hilbert's work.

Ulam, S.M. *Adventures of a Mathematician.* (Illus.) Scribner's, 1976. xi + 317pp. $14.95. 75-20133. ISBN 0-684-14391-7. Index;CIP.

   C   Ulam's autobiography is a magnum opus of the first order. His childhood, intellectual development, ontogeny of interests in fields outside of mathematics (literature and classical languages), anecdotes of professional and social contacts, and work on nuclear weapons are all melded into a perfectly charming book.

Zaslavsky, Claudia. *Africa Counts: Number and Pattern in African Culture.* (Illus.) Prindle, Weber&Schmidt, 1973. x + 328pp. $12.50. 72-91248. ISBN 87150-160-0. Index.

   SH-C-P   Zaslavsky reveals the contrasts, mysteries and accomplishments of people native to the African continent south of the Sahara. She concerns herself with "sociomathematics" in this fascinating book.

511 Mathematical Generalities

Bower, Julia Wells. *Mathematics: A Creative Art.* (Illus.) Holden-Day, 1973. xvi + 315pp. $7.50. 72-83241. ISBN 0-8162-1074-8. *Instructor's Manual with Answers and Hints.* 62pp. ISBN 0-8162-1084-5. Bib.

   C   An introduction to modern mathematical thought for the liberal arts student. Strategically selected, basic topics from abstract algebra, geometry (including non-Euclidean), and set theory point to the power and pay-off of the postulational method.

Dilson, Jesse. *Curves and Automation: The Scientists' Plot.* (Illus.) Lippincott, 1971. 141pp. $5.50. 72-141448.

JH-SH   In addition to the diverse fields in which curve plotting is used, the author
discusses historical background, devices for plotting curves automatically
and their applications, and the shorthand notation and equations used by mathematicians to describe curves.

Katzenberg, Arlene Chmil. *How to Draw Graphs.* (Illus.) Behaviordelia, 1974.
ix + 150pp. $3.95(p). ISBN 0-914-47413-8.

SH-C-P   Using a programmed instruction format, this book teaches graphing to
meet the requirements of the American Psychological Association.
However, anyone interested in presenting data by means of graphs would benefit
from the excellent series of examples and exercises presented.

Khazanie, Ramakant, and Daniel Saltz. *Introduction to Mathematics.* (Illus.)
Goodyear, 1974. viii + 465pp. 74-11741. ISBN 0-87620-469-8. Index;CIP.

SH-C   This outstanding text will help liberal arts students understand concepts
and acquire skills in basic mathematical areas. Four sections cover background, number systems, geometry, and probability and statistics. A superior text for
a high school senior honors program or a college level 1-year survey course.

Lancaster, Peter. *Mathematics: Models of the Real World.* (Illus.) Prentice-Hall, 1976.
xi + 164pp. $9.95. 75-17716. ISBN 0-13-564708-8. Index;CIP.

C   Focusing on mathematical models, Lancaster treats linear programming and
population genetics together with population growth and optimization models
that can be attacked by discrete methods. He explains the mechanics of particle
collision by examining a sequence of increasingly refined models. Good understanding of mathematics required.

Lang, Serge. *Basic Mathematics.* (Illus.) Addison-Wesley, 1971. xv + 431 + 44pp.
$9.95. 75-132055.

SH-C   The theoretical emphasis is in two areas: (1) a thorough understanding of
the proofs of a few of the more important and later-used theorems and the
relations of these theorems to solving computational problems; and (2) the relations
between fundamental intuitive notions and axioms or assumptions upon which mathematical machinery can be built. Fun and easy to read.

Malkevitch, Joseph, and Walter Meyer. *Graphs, Models, and Finite Mathematics.*
(Illus.) Prentice-Hall, 1974. x + 515pp. $10.95. 73-7580. ISBN 0-13-363465-5. Index;CIP.

SH-C   A stimulating and interesting finite mathematics text for liberal arts students. Exercises and examples are drawn from a wide variety of fields,
including social and life sciences.

Maxfield, John E., and Margaret W. Maxfield. *Keys to Mathematics.* (Illus.) Saunders, 1973. ix + 328pp. $8.25. 72-86450. ISBN 0-7216-6193-9.

SH-C   Unusual as well as standard topics are covered, including networks and
graphs, modular arithmetic, logic, Boolean algebra, matrices, groups, analytic geometry, topology, statistics, and computers. Many of the exercises are not
just routine computations. Good for a general education math course.

Ohmer, Merlin M. *Mathematics for a Liberal Education.* (Illus.) Addison-Wesley,
1971. vi + 330pp. $10.00. 79-119669.

SH-C   A strong presentation of many topics in secondary school mathematics,
this text may serve as a reference for teachers and honors students. In
addition to brief historical notes, topics include logic, sets, numeration, algebraic
systems, intuitive calculus, and probability.

Smith, Karl J. *Introduction to Symbolic Logic.* (Illus.) Brooks/Cole, 1974. viii + 119pp. $3.95(p). 73-87250. ISBN 0-8185-0115-4. Index.

SH-C   This brief work examines truth tables, tautologies, and fallacies of reasoning. Useful for enrichment topics or independent study programs at the high school or junior college level.

Spector, Lawrence. *Liberal Arts Mathematics.* (Illus.) Addison-Wesley, 1971. xv + 560pp. $9.95. 70-133895.

C   This lively and unconventional book deals with the foundations of mathematical logic and its applications to modern arithmetic. Covers the fundamentals of mathematical logic, the properties of numbers and various number systems, and basic number theory. Reflects the spirit of modern mathematics.

Stolyar, Abram Aronovich. *Introduction to Elementary Mathematical Logic.* (Illus.;trans.; edited by Elliott Mendelson.) MIT, 1970. vii + 209pp. $5.95. 69-12759. ISBN 0-262-19054-0. Indexes;bib.

SH-C   A highly readable introduction to elementary mathematical logic, concerned with propositional logic, propositional calculus, and predicate logic, with appendixes for certain proofs. Numerous helpful examples.

Vilenkin, N. Ya. *Combinatories.* (Illus.; trans.) Academic, 1971. xii + 296pp. $12.00. 77-154369.

SH-C   A discussion of general rules of combinations; samples, permutations; and combinations, distributions and partitions, recurrence relations, and series. A strong feature is the inclusion of over 400 solved examples ranging from simple to difficult. The exposition is informal and intuitive.

Willerding, Margaret F., and Ruth H. Hayward. *Mathematics: The Alphabet of Science, 2nd ed.* (Illus.) Wiley, 1972. xvi + 516pp. $9.95. 70-180274. ISBN 0-471-94661-3.

C   Includes topics from logic, number theory, mathematical systems, finite geometry, the Pythagorean theorem, permutations and combinations, matrices, computers, sets and relations, and analytic geometry. A book upon which a capable teacher can build a rich mathematics survey course.

512 ALGEBRA

Anton, Howard. *Elementary Linear Algebra.* (Illus.) Wiley, 1973. xii + 296pp. $10.25. 72-5511. ISBN 0-471-03247-6.

C   Matrices are introduced as devices for solving systems of linear equations, and the reader learns to manipulate matrices before proceeding to vector spaces and linear transformations. Thorough preparation for each new level of abstraction includes convincing the reader of the desirability of reaching for the higher levels. A sensible introductory text for an average class.

Bradley, Gerald L. *A Primer of Linear Algebra.* (Illus.) Prentice-Hall, 1974. xiv + 382pp. $11.95. 74-2025. ISBN 0-13-700328-5. Index;CIP.

C   A comprehensive text designed for an introductory course in linear algebra. Well-developed concepts and problem-solving techniques are related to previously acquired knowledge, although concepts do become more complex (and unexplained) in later chapters.

Brauer, Fred, John A. Nohel and Hans Schneider. *Linear Mathematics: An Introduction to Linear Algebra and Linear Differential Equations.* Benjamin, 1970. xiii + 347pp. $13.95. 74-102268. ISBN 0-8053-1206-4.

C   The first part of the book can be read by able high school seniors to learn the essential details of linear algebra and vector spaces. The exposition is concise but quite clear, the choice of examples and exercises is exceptional. An excellent choice for a first course in linear algebra.

Brinkmann, Heinrich W., and Eugene A. Klotz. *Linear Algebra and Analytic Geometry.* (Illus.) Addison-Wesley, 1971. xvi + 535pp. $11.50. 79-132056.

    C   A truly unusual linear algebra text. The computational development (including flow charts) and the development of analytic geometry are outstanding. Very well written, with applications to the behavioral sciences as examples to broaden its usefulness.

Gillett, Philip. *Linear Mathematics.* (Illus.) Prindle, Weber&Schmidt, 1970. x + 373pp. $10.95. 71-143529. ISBN 0-87150-115-5.

    C   This text offers some fresh approaches to the abstractions of mathematics beyond calculus. It begins with a treatment of vector spaces, covers differential equations, linear maps, matrices, linear systems, and systems of differential equations. The writing is clear and concise.

Goldstein, Larry Joel. *Abstract Algebra: A First Course.* (Illus.) Prentice-Hall, 1973. xii + 385pp. $11.95. 72-12790. ISBN 0-13-000851-6.

    C   Presents the solution of polynomial equations from the perspective of abstract algebra. Historical references are provided, as are the prerequisites of set theory and a brief exposition of number theory. High school teachers will find this very useful for reference and review.

Johnson, Richard E. *Elementary Linear Algebra.* (Illus.) Prindle, Weber&Schmidt, 1971. viii + 291pp. $9.95. 72-169130. ISBN 0-87150-134-1.

    C   A straightforward presentation of linear algebra for a first course. The content is a conventional study of finite-dimensional vector spaces, linear mappings on such spaces, and associated matrices with appropriate applications (systems for solving linear equations, analyses of graphs and quadratic equations, and the solution of certain linear differential equations).

Knopp, Paul J. *Linear Algebra: An Introduction.* (Illus.) Hamilton/Wiley, 1974. xvi + 435pp. $10.95. 73-10499. ISBN 0-471-49550-6.

    SH-C   Presents the essential ideas of linear algebra: the real, n-dimensional vector spaces and their linear transformations; real matrices; the determinant function; distance; change of base; equivalence; eigenvalues and quadratic functions. Numerical examples illustrate concepts before discussions of generalities; abstract set terminology is avoided. A pleasure to read.

Larney, Violet Hachmeister. *Abstract Algebra: A First Course.* (Illus.) Prindle, Weber &Schmidt, 1975. xiii + 367pp. $14.95. 75-28495. ISBN 0-87150-209-7. Index;CIP.

    C   One of the better books of its type, covering basic properties of groups, rings and fields, and elementary number theory. Can be read by someone who has studied no mathematics beyond high school algebra. Includes many well-chosen examples and exercises, and sketches of mathematicians.

Murdoch, D.C. *Linear Algebra.* (Illus.) Wiley, 1970. xi + 312pp. $9.95. 72-121911. ISBN 0-471-62500-0.

    C   This text is an excellent follow-up to *Linear Algebra for Undergraduates.* The material has been rewritten to give a more abstract approach to linear algebra. The book exhibits original approaches to similarity and diagonalization and is written from a mature standpoint.

Painter, Richard J., and Richard P. Yantis. *Elementary Matrix Algebra with Linear*

Programming. (Illus.) Prindle, Weber&Schmidt, 1971. vii + 372pp. $8.95. 70-136199. ISBN 0-87150-121-X.

C Applications to linear programming and Markov chains complete a text which can be most useful for a 1-semester course. Uses a behavioral method of presentation, and the text is suitable only in a course whose objective is to "teach for use."

Weiss, Edwin. *First Course in Algebra and Number Theory.* Academic, 1971. xi + 547pp. $12.95. 73-158813.

C Designed for the true beginner in modern algebra, this gentle introduction to abstract algebra covers four broad topics (elementary number theory, rings and domains, congruences and polynomials, and groups), subdivided so that the instructor has some leeway in selection. Although designed for math classes, the book is also useful for self-study by others in the sciences.

Zelinsky, Daniel. *A First Course in Linear Algebra, 2nd ed.* (Illus.) Academic, 1973. x + 260pp. $9.95. 72-88368.

C For a lower level course in the calculus sequence following one-variable functions but preceding nonlinear functions and differential equations. The emphasis is more geometric than usual, and the book contains some classical vector analysis. Linear programming, vectors, planes and lines, linear functions, solution of equations, dimension, determinants and transposes, eigenvalues, quadratic forms and change of bases are covered.

512.7 NUMBER THEORY

Adams, William W., and Larry Joel Goldstein. *Introduction to Number Theory.* (Illus.) Prentice-Hall, 1976. xiii + 362pp. $13.95. 75-12686. ISBN 0-13-491282-9. Index;CIP.

C For a 2-semester course. The first half contains the usual material for a 1-semester elementary course, and requires high school algebra. The second half introduces the algebraic theory of numbers and presupposes an undergraduate course in abstract algebra. The theory of Diophantine equations is used as a unifying theme.

Barnett, I.A. *Elements of Number Theory, rev. ed.* (Illus.) Prindle, Weber&Schmidt, 1972. x + 213pp. $9.95. 69-12290. ISBN 87150-157-0. Bib.

C Very well developed coverage of prime factorization, divisibility rules, the Euclidean algorithm, linear Diophantine equations, congruences, Euler's function, simultaneous congruences, Pythagorean equations, Fermat's last theorem, Gaussian integers, and construction with ruler and compass. Intended for prospective junior and senior high school math teachers.

Lankford, Francis G., Jr., Donald D. Heikkinen and Ina M. Silvey. *Numbers and Operations.* (Illus.) Harcourt, Brace&World, 1970. x + 461pp. $4.50.

SH An excellent book which prepares students for the rigors of higher mathematics. The teacher's manual is well done, with additional information for presentation to the class. Interesting and enjoyable for a secondary school course in mathematics, or as a source book.

St. John, Glory. *How to Count Like a Martian.* (Illus.) Walck, 1975. 66pp. $6.95. 74-19714. ISBN 0-8098-3125-2. CIP.

JH A four-symbol message from Mars is the unifying thread in this investigation of central number concepts and systems. The style is conversational, and the treatment is brief and pleasant. Heartily recommended.

Youse, Bevan K.   *Arithmetic: An Introduction to Mathematics.* (Illus.) Canfield, 1971. ix + 294pp. $8.95. 76-143695. ISBN 0-06-389650-8.

    SH-GA   Elementary number theory and introductory set theory without use of axioma tics are the principal concerns of this handy monograph for the intelligent lay reader who is interested in a sketch of "new math." This monograph, with its numerous graded exercises and answers, should find considerable use as supplementary reading.

512.9 BASIC SECONDARY AND COLLEGE ALGEBRA

Armstrong, James W.   *Elements of Mathematics.* (Illus.) Macmillan, 1970. xi + 306pp. $9.95. 72-80304.

    SH-C   Nonmathematicians are the audience for this book on logic, mathematical systems, whole numbers and number concepts, set theory, two geometries, analysis, probability and limits of sequences. Lucid, with enough historical background to generate some enthusiasm for the subject.

Brudner, Harvey J.   *Algebra and Trigonometry: A Programmed Course with Applications.* (Illus.) McGraw-Hill, 1971. xiv + 522pp. $8.95. 78-115151. ISBN 0-07-046381-6.

    SH-C   This programmed text covers algebraic operations and expressions, exponents, complex numbers, trigonometric identities and equations, triangles, etc., and is particularly useful for self-instruction by students in engineering or in technically oriented programs at community colleges or institutes.

Denbow, Carl H.   *College Algebra.* (Illus.) Harper&Row, 1970. xi + 434pp. $8.95. 79-98199.

    SH-C   One of the best. Topical coverage is conventional; however, the topics are in a modern format and treated in an integrated manner. Symbols and concepts are not confused; both are clearly explained. Each topic is well motivated and developed. Copious use of examples and counter-examples; fine prose and interesting anecdotal material.

Dolciani, Mary P., and Robert H. Sorgenfrey.   *Elementary Algebra for College Students.* (Illus.) Houghton Mifflin, 1971. x + 398pp. $7.95. 78-146720. ISBN 0-395-12069-1. *Instructor's Guide and Solutions.* 179pp. ISBN 0-395-12071-3.

    SH-C   Incorporates several innovative and interesting features, such as the quantity and variety of exercises. Included are the typical topics in elementary algebra (numbers and sets through quadratic equations). Flow charts introduce an organized, step-by-step method of problem analysis and synthesis. The instructor's guide is particularly valuable, listing the behavioral objectives and typical test questions which could be used directly or as models.

Dolciani, Mary P., and William Wooton.   *Book 1: Modern Algebra, Structures and Method, rev. ed.* (Illus.) Houghton Mifflin, 1970. 655pp. $6.60. *Book 2: Modern Algebra and Trigonometry, Structure and Method, rev. ed.* (with Simon L. Berman). (Illus.) 674pp. $6.50.

    JH-SH   Book 1 is basic and introductory algebra; Book 2 is advanced algebra leading into quadratic equations, trigonometric functions, identities and formulas, complex numbers, circular functions and inverses, matrices and determinants, permutations, combinations, and probability. The teacher's editions explain the philosophy and objectives of each book and provide detailed directions for teaching and answer supplements. The texts are clean and direct, with frequent historical notes and practical applications.

Gobran, Alfonse. *Introductory Algebra.* (Illus.) Prindle, Weber&Schmidt, 1970. xii + 378pp. $7.95. 79-132048. ISBN 0-87150-102-3.

SH-C   This excellent textbook presents algebra as a logical mathematical structure, successfully develops an understanding of basic mathematical principles, and emphasizes comprehension rather than rules and techniques. There is ample opportunity to apply mathematical principles in solving problems. Contains numerous exercises and problems of varying difficulty.

Groza, Vivian Shaw. *Elementary Algebra: A Worktext.* (Illus.) Saunders, 1975. xiii + 728pp. $10.95. 74-4567. ISBN 0-7216-4321-3. Index;CIP. *Instructor's Guide.* 59pp.

SH-C   Although largely intended for individualized use, this "worktext" is adaptable to the classroom and provides a leisurely, highly detailed introduction to algebraic techniques. A large number of problems and examples are included in this highly structured approach.

Hart, William L. *Algebra and the Elementary Functions: With Included Instructor's Guide.* (Illus.) Goodyear, 1974. xv + 455pp. $13.95. ISBN 0-87620-033-1. Index.

SH-C   Clear and concise review of algebraic, exponential, logrithmic, and trigonometric functions. Valuable for any mathematics teacher. The many examples and exercises are different and novel; the entire approach is unique.

Henderson, George L., et al. *Algebra: A Personalized Approach.* (Illus.) Prindle, Weber&Schmidt, 1976. xiii + 578pp. $9.95(p). 75-31597. ISBN 0-87150-201-1. Index;CIP.

Introduction to Algebra: A Personalized Approach. (Illus.) Prindle, Weber&Schmidt, 1975. xi + 387pp. $9.95(p). 75-31632. ISBN 0-87150-200-3. Index;CIP.

SH-C   These two texts present modular approaches to intermediate and introductory algebra, respectively. They cover the standard topics in a simple and pointed manner. Particularly suitable for self-study.

Janowitz, Melvin F. *Intermediate Algebra.* (Illus.) Prentice-Hall, 1976. 436pp. $10.50. 75-29258. ISBN 0-13-469528-3. Index;CIP.

SH-C   The material is clearly presented with a wealth of illustrative examples and exercises. Some essentials of analytic geometry are also covered. Strongly recommended.

Keedy, Mervin L., and Marvin L. Bittinger. *Vol. 1: College Algebra: A Functions Approach.* (Illus.) Addison-Wesley, 1974. vi + 513pp. $8.95(p). 73-22729. ISBN 0-201-03718-4. Index.

Vol. 2: Algebra and Trigonometry: A Functions Approach. (Illus.) viii + 694pp. $9.95. 73-8240. ISBN 0-201-03659-2(p). Index.

SH-C   This two-volume set presents conventional algebraic material in an attractive format. Volume 1 contains review material followed by sections on functions, systems of linear equations, exponential and logarithmic functions, complex numbers, etc. The combined volume adds trigonometric functions, identities, and the trigonometry of the triangle. Valuable for learning techniques to solve standard types of mathematical problems.

Keedy, Mervin L., and Marvin L. Bittinger. *Essential Mathematics: A Modern Approach.* (Illus.) Addison-Wesley, 1972. xvi + 641pp. $7.50(p).

C   Not just a textbook but an instructional package complete with objectives, tests, developmental and practice exercises, answers to tests and problem sets, and an instructor's manual. The content selected from arithmetic, algebra, numerical trigonometry, and sliderule usage covers practically all areas in which a student might be deficient upon entering college.

Minnick, John H., and Raymond C. Strauss. *Beginning Algebra, 2nd ed.* (Illus.) Prentice-Hall, 1976. xiii + 335pp. $12.50. 75-25881. ISBN 0-13-073791-7. Gloss.;index;CIP.

C   For the college freshman who has no previous experience with algebra or who needs a review. Limited to the topics essential for further study in calculus or science, with well-designed exercises.

Mueller, Francis J. *Elements of Algebra, 2nd ed.* Prentice-Hall, 1975. xii + 436pp. $9.95. 73-22199. ISBN 0-13-262436-2. Index;CIP.

JH-SH   Problems and worked examples abound in this teachable text which progresses from arithmetic to algebra and concludes with an indepth treatment of logarithms.

Munem, Mustafa. *College Algebra.* (Illus.) Worth, 1974. ix + 518pp. $9.95. 73-85130. ISBN 0-87901-026-6. *Study Guide.* (Illus.) 346pp. $3.95(p). ISBN 0-87901-027-4.

C   An introductory, precalculus algebra text which covers the ground adequately, from sets to matrices via various elementary functions. The semi-programmed study guide is adaptable to self-study or "individualized" instruction.

Munem, Mustafa A., and James P. Yizze. *Functional Approach to Precalculus,, 2nd rev. ed.* (Illus.) Worth, 1974. 569pp. $11.95. ISBN 0-87901-030-4. Index.

SH-C   Chapters on sets, numbers, functions, trigonometry, and analytic geometry. Appendixes of logarithms, functions, powers and roots.

Peterson, John M., and Floyd E. Haupt. *College Algebra.* (Illus.) Prindle, Weber &Schmidt, 1974. ix + 346pp. $11.50. 74-1309. ISBN 0-87150-167-8. Index;CIP.

C   This book is unusual because it stresses mechanics and methodology while dealing minimally with derivations. The student is continually directed to the applicability of the subject matter in real world situations. There are excellent examples, accompanied by well-chosen graphs, charts, and other illustrations.

Plachy, Jon M., and Orason L. Brinker. *Elements of Algebra: A Worktext, 2nd ed.* Prindle, Weber&Schmidt, 1973. x + 274pp. $5.95(p). 68-31628. SBN 87150-071-1.

JH-SH-C   A concise but excellent overview of the elements of traditional algebra. Written in the format of a workbook, each subsection is preceded by an explanation of the fundamental ideas involved. Useful as a rapid introduction to the subject, as a review medium and as a supplementary text for teachers.

Salas, Saturnino, and Charles G. Salas. *Precalculus: A Short Course.* (Illus.) Xerox Coll., 1975. xi + 289pp. $9.95. 74-83347. ISBN 0-536-01049-8. Index.

SH-C   Five chapters provide background for calculus: algebra, analytic geometry, functions, trigonometry, and a final chapter on special topics—mathematical induction, upper bound theorem, logarithms, inequalities, polar coordinates and complex roots of quadratic equations.

Swokowski, Earl W. *Elementary Functions with Coordinate Geometry.* (Illus.) Prindle, Weber&Schmidt, 1971. ix + 371pp. $9.50. 76-155292. ISBN 0-87150-117-1.

SH-C   Topics include functions, polynomials, exponential and logarithmic functions, trigonometry, and coordinate geometry. Combines modern and traditional approaches while covering what is essential. The approach is mainly intuitive; the style lucid but pedestrian. A successful classroom text, but less satisfactory for independent study.

Swokowski, Earl W. *Fundamentals of Algebra and Trigonometry, 2nd ed.* Prindle, Weber&Schmidt, 1971. viii + 455pp. $2.95.(p). 68-13118. ISBN 0-87150-126-0.

SH-C   A strong, well-written textbook for a course preparatory to calculus. In-. cludes the standard topics, from set language to complex numbers, polynomials, and sequences through the "standard route." Ample problems plus handy review sections.

Swokowski, Earl W.   *Fundamentals of College Algebra, 2nd ed.* (Illus.) Prindle, Weber &Schmidt, 1971. ix + 340pp. $10.00(p). 68-13117. ISBN 0-87150-123-4.

   C   The fundamentals are well covered in compact style well suited to practical application. The author is consistent in his development of the principles of algebra based on the set concept. An excellent text in the hands of a good instructor capable of amplifying some of the theoretical developments.

Vance, Elbridge P.   *Modern Algebra and Trigonometry, 3rd ed.* (Illus.) Addison-Wesley, 1973. 436pp. $10.75. 72-11471. ISBN 0-201-08036-2.

   SH-C   This conceptual approach to the study of mathematics incorporates many of the recommendations of groups such as SMSG in the organization and development of the mathematical concepts. Covers sets, algebra of numbers, algebraic expressions, geometry of real numbers, integrated algebra and trigonometry, and applications of the circular functions. Many problems provided.

Whitesitt, J. Eldon.   *Principles of Modern Algebra, 2nd ed.* Addison-Wesley, 1973. viii + 243pp. $10.75. 74-184163.

   C   The topics in this text—groups, rings and fields—have been carefully selected and well developed. Two appropriate examples are worked out in detail: the group of isometrics in the plane and ring of polynomials. For a student who had studied rigid motions in geometry, the first of these topics would be an excellent extension. Intended for high school teachers.

Willerding, Margaret F.   *Modern Intermediate Algebra, 2nd ed.* (Illus.) Wiley, 1975. xi + 411pp. $11.95. 74-16160. ISBN 0-471-94667-2. Index;CIP.

   SH-C   Suitable as a text or reference for college students with a weak mathematical background. Covers real numbers, exponents and radicals, polynomials, rational expressions, first degree equations and inequalities, relations and functions, quadratic equations and inequalities over the real numbers, quadratic functions, systems of linear equations, complex numbers, sequences and series and exponential and logarithmic functions.

513 ARITHMETIC

Anderson, John G.   *Technical Shop Mathematics.* (Illus.) Industrial, 1974. viii + 504pp. $12.00. 74-16115. ISBN 0-8311-1085-6. Index;CIP.

   SH   Algebra fundamentals are discussed sufficiently for the student to develop and interpret shop equations and expressions and derive specific relationships and answers. Considerable coverage is given to gear ratios and tapers. Plane geometry and right- and oblique-angle trigonometry are also covered.

Jonas, Harry H.   *Pre-Algebra.* (Illus.) Wiley, 1972. xi + 299pp. $8.50. 70-166315. ISBN 0-471-44702-1.

   C   Intending to bridge the gap between arithmetic and algebra, the author emphasizes numerical computation and algorithmic processes that have analogies in algebra. (There is no geometry, measurement, graphing in the plane, or work with the function concept.) Carefully developed concepts and processes.

Kaufmann, Jerome E., and William C. Lowry.   *The Many Facets of Mathematics.* Prindle, Weber&Schmidt, 1971. 264pp. $8.95. 77-136198. ISBN 0-87150-116-3.

   SH-C   Contains chapters on logic, probability, number theory, deductive mathematics, modular arithmetic and geometry via arithmetic and algebra. These last two show how mathematics is used in other sciences. Much mathematical history is included, and it presents a different discussion of many of the topics and ideas currently included in mathematics texts.

Keedy, Mervin L., and Marvin L. Bittinger. *Arithmetic: A Modern Approach.* (Illus.) Addison-Wesley, 1971. xii + 420pp. $5.75(p).

  JH-SH-C   As a class text, workbook, or for independent study, this book should find wide use in the study of basic arithmetic. Each page contains the lesson objective and supplementary observations, diagrams, and questions. There are exercise sets and tests.

Meserve, Bruce E., and Max Sobel. *Foundations of Number Systems.* (Illus.) Prentice-Hall, 1973. x + 292pp. $8.95. 73-2021. ISBN 0-13-329094-8.

  C   This is a teachable and provocative book with a lively and interesting tone. The "pedagogical considerations," spaced at close intervals throughout the text, embed the mathematics solidly in teaching strategies.

Munem, Mustafa A., and William Tschirhart. *Beginning Algebra.* (Illus.) Worth, 1972. ix + 460pp. $7.95. 76-188910. ISBN 0-97901-017-7.

  SH-C   Much of this book is concerned with the basic operations of arithmetic. The set approach to the delineation of the various kinds of numbers is well done. There is an extended treatment of absolute values of numbers, and many examples are cited.

Peterson, John A., and Joseph Hashisaki. *Theory of Arithmetic, 3rd ed.* (Illus.) Wiley, 1971. xi + 339pp. $8.95. 70-132855. ISBN 0-471-68320-5.

  C   Covers the development of the real number system from counting or whole numbers through integers to rational and irrational numbers. There are introductory chapters on numeration systems, sets, relations and geometry. The book is comprehensive, with the proper balance of informality. Emphasis is on understanding and using concepts and principles. Prospective elementary school teachers should find this text enjoyable and challenging.

Smithsi, Thomas G. *Basic Mathematical Skills.* Prentice-Hall, 1974. 332pp. $6.95(p). 73-18462. ISBN 0-13-063420-4. CIP.

  JH-SH   An exercise-oriented remedial text in arithmetic. Motivational material stresses basic operations and practice and sets the stage for the student's success in learning.

Stokes, William T. *Notable Numbers.* (Illus.) Creative Publications, 1972. 77pp. $4.00(p).

  C   Topics include Venn diagrams, number charts of various kinds, a variety of number lines, lists or diagrams of perfect, amicable, factorial, palindromic, Fibonacci and figurate numbers, number patterns, Pascal's triangle, modular arithmetic tables (and blank answer sheets). Will be a great help to the busy teacher.

Willerding, Margaret F. *A First Course in College Mathematics.* Prindle, Weber &Schmidt, 1975. $3.95ea.(p). 72-14172. CIP. *Module 1: The Decimal System of Numeration; Addition and Subtraction of Whole Numbers; Multiplication and Division of Whole Numbers.* (Illus.) xii + 183pp. ISBN 87150-185-6. *Module 2: Factors and Factoring; Rational Numbers and Fractions; Decimals.* xii + 212pp. ISBN 87150-187-2. *Module 3: Per Cent; Measurement; Geometric Figures and their Measures.* xii + 215pp. ISBN 87150-188-0.

  JH-SH-C   These work-text modules are "intended for the student who needs to review the basic concepts of arithmetic and to improve computation and problem-solving skills." Includes examples, exercises, reviews, tests, and answers. Each page is perforated.

514 TOPOLOGY

Flegg, H. Graham. *From Geometry to Topology*. (Illus.) Crane, Russak, 1975. xii + 186pp. $10.50. 74-78155. ISBN 0-8448-0364-2. Index.

    C Flegg combines a geometrical, expository approach to topography with comprehensible proofs of theorems. Topics covered include graph theory, map coloring, classification of surfaces and the Jordan curve. Highly recommended as additional reading in undergraduate topology and geometry courses and will also appeal to high school students and teachers of geometry.

Griffiths, H.B. *Surfaces*. Cambridge Univ. Press, 1976. xi + 120pp. $12.50. 74-25660. ISBN 0-521-20696-0. Indexes;CIP.

    C Develops the combinatorial topology of surfaces. Also discusses invariance of Euler characteristic and Morse theory of surfaces. High school algebra and geometric intuition are needed, but not formal topology. Intended for students who wish to teach high school mathematics and recommended for high school teachers who want to learn surface topology.

Struble, Mitch. *Stretching a Point*. (Illus.) Westminster, 1971. 128pp. $4.95; $2.95(p). 74-155901. ISBN 0-664-32499; 0-664-34002-4.

    JH-SH This book approaches the subject of network relations from the point of view of mazes, Moebius strips and Klein bottles. The topics are developed in an orderly fashion and are easy to understand and interesting to study. The puzzles are clever and to the point.

515 CALCULUS

Baxter, Willard E., and Clifford W. Sloyer. *Calculus with Probability: For the Life and Management Sciences*. (Illus.) Addison-Wesley, 1973. xiii + 648pp. $13.95. 72-1937.

    C Offers all the usual elements of basic calculus, in addition to mathematical models in the life and management sciences and integration techniques.

Campbell, Howard E., and Paul F. Dierker. *Calculus*. (Illus.) Prindle, Weber &Schmidt, 1975. xii + 752pp. $16.95. 74-32491. ISBN 0-87150-181-3. Index;CIP.

    C A conventional text for a 3- or 4-semester course in calculus, this book includes relevant exercises of varying levels of difficulty. The language is clear and concise. Excellent for its type.

Ceder, Jack G., and David L. Outcalt. *A Short Course in Calculus*. Worth, 1971. xv + 341pp. $8.95. 68-55362.

    C The text assumes only algebra and geometry and covers how to apply the methods of elementary analysis, through separation of variables in differential equations, in the social sciences. An intuitive, geometrically motivated approach, yet definitions and theorems are kept as mathematically precise as possible. Many complete examples, and exercises are categorized for methods, applications, or theory.

Cruse, Allan B., and Millianne Granberg. *Lectures on Freshman Calculus*. (Illus.) Addison-Wesley, 1971. xi + 641pp. $10.50. 79-136118.

    SH-C An outstanding selection of exercises; clear, detailed examples illustrate the concepts; an inductive approach illuminates the concepts of continuity and limit. A useful reference for advanced-placement high school physics and math classes, a basal text for liberal arts majors, and a suitable first-year calculus text for engineers and scientists.

Duren, William L., Jr. *Calculus and Analytic Geometry*. (Illus.) Xerox Coll., 1972. xvi + 773pp. $14.50. 75-162426. ISBN 0-536-00665-2; 0-536-00869-8 (internat'l.)

C   An outstanding book, providing an arrangement of material with a strong technical emphasis. The three parts are calculus of elementary functions, multivariable calculus, and analytic geometry and calculus. Several effective teaching
methods are employed: meaningful examples, reviews, sufficient motivation and
introduction of new subjects, and relevant exercises and problems.

Fleenor, Charles R., et al. *The Elementary Functions, 2nd ed.* (Illus.) Addison-
Wesley, 1973. xii + 367pp. $9.95. 75-178266.

SH   The book has an adequate number of exercises (with answers to selected
exercises), and explanations are clear and complete. An average student with
adequate background should have no difficulty completing the material in one semester. An excellent text.

Gemignani, Michael. *Calculus: A Short Course.* (Illus.) Saunders, 1972. viii + 269pp.
$8.50. 76-173333. ISBN 0-7216-4097-4.

C   This well-written text contains the appropriate topics—such as exponential
and logarithmic functions—to prepare students of the management and social
sciences for further work in statistics and operations research, while eliminating
those topics which waste their time.

Gray, Mary W. *Calculus with Finite Mathematics for Social Sciences.* (Illus.)
Addison-Wesley, 1972. xi + 593pp. $12.95. 74-174335.

C   Eclectic harmony between classical integro-differential calculus and linear
algebra in the context of social-humanistic applications is the principal theme.
Intended as a first calculus course for both mathematics majors and students in the
social or management sciences. Very carefully written and quite clear.

Herstein, I. N., and Reuben Sandler. *Introduction to the Calculus.* (Illus.) Harper
&Row, 1971. viii + 309pp. $9.95. 74-129477.

C   The text will introduce the student to the major concepts and tools of calculus
without losing him or her in the detail of mathematical method and unnecessary
rigor. Plausible argument is substituted for formal proof. Well-behaved functions are
employed to make first acquaintance with calculus a gratifying, positive experience.
Should be required reading for all calculus teachers!

Lang, Serge. *A First Course in Calculus, 3rd ed.* (Illus.) Addison-Wesley, 1973.
xii + 469 + 29pp. $10.95. 70-183670.

SH-C   An excellent text with sound explanations, adequate problem coverage,
and a sufficient range of basic topics.

Leithold, Louis. *The Calculus with Analytic Geometry, 2nd ed.* (Illus.) Harper&Row,
1972. xvi + 1064pp. $15.00. 74-168364. ISBN 0-06-043959-9.

SH-C   All the usual topics (including vectors) are covered in this book, with
emphasis on applications. The excellent format utilizes one extra color
with clear figures and diagrams and a generous number of examples and exercises,
including review exercises at the end of each chapter. Quite satisfactory for the usual
courses which combine calculus and analytic geometry.

Lial, Margaret L., and Charles D. Miller. *Essential Calculus: With Applications in
Business, Biology and Behavioral Sciences.* (Illus.) Scott, Foresman, 1975. 346pp.
$11.95. 74-82771. ISBN 0-673-07959-7. Index.

C   Intended for students of business, social science and biology, this practical
applications approach is adequate to develop facility in general calculus. Best
used for advanced courses after exposure to the authors' books on beginning and
advanced algebra.

Loewen, Kenneth. *Calculus for Management, Economics and the Life Sciences.* (Illus.)

Prindle, Weber&Schmidt, 1975. xii + 388pp. $12.50. 74-16393. ISBN 0-87150-170-8. Index;CIP.

C   The use of calculus in economics and the management sciences is shown.
Beginning with the concept of the function, the reader is introduced to the derivative and differential, to the partial derivative and the integral, and is taught to apply the derivative to implicit, exponential and logarithmic functions.

Loewen, Kenneth. *A Short Calculus for Management, Economics, and the Life Sciences.* (Illus.) Prindle, Weber&Schmidt, 1975. xi + 318pp. $11.95. 74-17217. ISBN 0-81750-186-4. Index:CIP.

C   For courses which emphasize the nonphysical science applications of calculus.
Each chapter begins with a problem drawn from the areas listed in the title. Exercises involve practice in both computation and applications. Any gaps occurring in proofs are duly noted.

Loomis, Lynn. *Calculus.* (Illus.) Addison-Wesley, 1974. xvi + 1024pp. $15.50. 73-8241. ISBN 0-201-04307-6. Index.

C   This introductory text relies on the intuitive approach and emphasizes calcula-
tion to allow for maximum flexibility to accommodate various audiences, time schedules, and course aims. With proper direction, most students should easily progress from this text to more advanced work.

Loomis, Lynn. *Introduction to Calculus.* (Illus.) Addison-Wesley, 1975. xiv + 772pp. $12.95. 74-30700. ISBN 0-201-04306-8. Index.

SH-C   A valuable text by a prominent scholar for a basic calculus course. Includes
respectable treatments of power series and functions of two variables but not multiple integration. The highly intuitive approach is designed to motivate the student and to avoid introducing excessive rigor prematurely.

Lowengrub, Morton, and Joseph G. Stampfli. *Topics in Calculus, 2nd ed.* (Illus.) Xerox Coll., 1975. xiii + 442pp. $11.95. 74-78064. ISBN 0-536-01088-9. Index.

C   This book presents the material of one- and two-variable calculus for a 1-year
course with emphasis on economic and business applications. All the basic concepts, good chapter summaries and many exercises are included. A good text for students who haven't enjoyed mathematics previously.

Mizrahi, Abe, and Michael Sullivan. *Calculus with Applications to Business and Life Sciences.* (Illus.) Wiley, 1976. xiii + 414pp. $12.95. 75-17964. ISBN 0-471-61192-1. Index;CIP.

SH-C   This elementary exposition of calculus, which can be read by anyone com-
petent in high school algebra, emphasizes simple concepts with a minimum of mathematical terms. The business applications are numerous and interesting; the life sciences ones are few and of little practical value. However, the life scientist can still learn calculus from this book, and it could provide a refresher for a non-mathematician.

Salas, Saturnino L., and Einar Hille. *Calculus: One and Several Variables.* (Illus.) Xerox Coll., 1971. xi + 800pp. $13.95. 74-141896.

C   The approach is modern and sound, and the topics are familiar. The recom-
mended 2-semester course includes two chapters on elementary mathematics. No exotic material and only occasional historical references. The figures, examples and problems are clear, and an adequate number of proofs is included.

Simon, Arthur B. *Calculus.* (Illus.) Macmillan, 1970. xiii + 626pp. $11.95. 71-77489.

SH-C   In this text for a 2- to 3-semester course, rigor is espoused, leading to a long
delay in the introduction of differentiation and integration; however, Chap-

ter 2, "Computational Calculus," provides straightforward cookbook calculus. Elementary rules of differentiation and integration and many exercises are helpful. Well written, with sufficient examples.

Swokowski, Earl W. *Calculus with Analytic Geometry.* (Illus.) Prindle, Weber &Schmidt, 1975. ix + 854pp. 74-30065. ISBN 87150-179-1. Index;CIP.

C   May be used as a text for a 3-semester calculus course or as a reference book. There are over 450 solved examples and more than 4000 exercises. Approximation methods are brought in early, and there are applications to areas outside mathematics and physics.

Whipkey, Kenneth L., and Mary N. Whipkey. *The Power of Calculus.* (Illus.) Wiley, 1972. xv + 297pp. $9.95. 76-172955. ISBN 0-471-93777-0.

SH-C   A calculus text for the nonmathematician which has a chance of succeeding in its aims. Most individuals with at least 2 years of high school algebra will find it a source of ideas, concepts and methods being used in many graduate programs. The extent of ideas mentioned and reasonably explained is considerable. Many clear applications, practically all of which are *not* in the physical sciences.

515.3 DIFFERENTIAL EQUATIONS

Bellman, Richard, and Kenneth L. Cooke. *Modern Elementary Differential Equations, 2nd ed.* (Illus.) Addison-Wesley, 1971. xii + 228pp. $9.50. 70-100854.

C   An excellent book which stresses those principles that are really important to the average science and mathematics student in a first-year course in differential equations. Topics include how differential equations arise, power series methods, numerical solutions, the nature of iterative processes, the Laplace transform, and the uniqueness of theorems.

Boyce, William E., and Richard C. DiPrima. *Introduction to Differential Equations.* (Illus.) Wiley, 1970. 310pp. $7.50. 72-96049. ISBN 0-471-09338-6.

C   An extraordinary, flexible, teachable text. Three chapters consider first-order differential and second-order linear equations, with excellent illustrations of theory and difficulties. Applications to physics, biology and the social sciences include discussions of the construction of mathematical models, series solutions of second-order linear equations, higher-order linear equations, and systems of linear equations, power series and matrices, and a comparison of different numerical techniques. The problems are *all* answered.

Dickinson, Alice B. *Differential Equations: Theory and Use in Time and Motion.* (Illus.) Addison-Wesley, 1972. ix + 271pp. $8.95. 77-136120.

C   A differential equations book which would serve as an introduction to applied mathematics as well as to the theorem-proof style. Gives particular attention to the transition from the physical situation to the mathematical model. The derivation of the mathematical models of the gravitational field, oscillatory motion of the vibrating string and the wave equation are lucid and logical.

Hagin, Frank G. *A First Course in Differential Equations.* (Illus.) Prentice-Hall, 1975. ix + 342pp. $12.95. 74-22396. ISBN 0-13-318394-7. Index;CIP.

C   Hagin treats numerical methods in a manner that includes FORTRAN and BASIC programs and allows the use of small calculators. A number of applied exercises and examples are provided, but the approach is theoretical. First-order differential equations, linear equations of order two or more, numerical techniques, power series, linear systems and Laplace transforms are covered.

Plaat, Otto. *Ordinary Differential Equations.* Holden-Day, 1971. xi + 295pp. $10.95. 70-156869. ISBN 0-8162-6844-4.

C-P   An excellent introduction to the central ideas and methods of ordinary linear and nonlinear differential equations, using what the author describes as "the qualitative point of view . . .which not only dominates contemporary research, but is ideally suited to make the subject accessible and meaningful to the beginner. To enable the student to think about differential equations, rather than merely to manipulate them."

Rabenstein, Albert L. *Introduction to Ordinary Differential Equations, 2nd ed.* (Illus.) Academic, 1972. x + 526pp. $12.75. 78-185031. ISBN 0-12-573957-5.

C-P   The stress is on the mathematical techniques, but the book also treats special topics useful in applications. Assumes a knowledge of elementary calculus. A well-written, understandable reference or text for a traditional course.

515.7 FUNCTIONAL ANALYSIS

Apostol, Tom M. *Mathematical Analysis, 2nd ed.* (Illus.) Addison-Wesley, 1974. xvii + 492pp. $14.95. 72-11473. ISBN 0-201-00288-4. Indexes.

C   Consisting mainly of "advanced calculus," this volume also contains the background material for additional study of function theory. Topics include real and complex number systems, point set theory and topology, limits and continuity, series, sequences and various advanced areas of both differential and integral calculus. Suitable for teaching purposes and a fine reference book.

Kolman, Bernard, and William F. Trench. *Elementary Multivariable Calculus.* (Illus.) Academic, 1971. vii + 505pp. $12.50. 73-156267.

C   The main merits of this book are its clear presentation and its profuse examples. Intended for sophomores, but a good background is needed. Exercises are both applications and theory-oriented problems. The topics treated are a selection from those found in engineering mathematics and in advanced calculus.

Merritt, Frederick S. *Applied Mathematics in Engineering Practice.* (Illus.) McGraw-Hill, 1970. vii + 289pp. $12.50. 73-114449.

C   Designed for practicing engineers to learn applied mathematical methods. Differential, integral and operational calculus; ordinary differential equations, their nonelementary solutions and numerical integration; partial differential equations; complex variables and conformal mapping; probability and statistics are covered. Clear, concise, easy to understand.

Riley, K. F. *Mathematical Methods for the Physical Sciences: An Informal Treatment for Students of Physics and Engineering.* (Illus.) Cambridge Univ. Press, 1974. xvi + 533pp. $26.00; $8.95(p). 73-89765. ISBN 0-521-20390-2; 0-521-09839-4. Index.

C-P   This informal treatment is strongly classical in tone and well organized in its discussion of abstract mathematics. Useful as a text for science and engineering students and for those desiring to broaden their knowledge of currently useful mathematics.

516 GEOMETRY

Behr, Merlyn J., and Dale G. Jungst. *Fundamentals of Elementary Mathematics: Geometry.* (Illus.) Academic, 1972. xix + 326pp. $9.50. 77-182621. ISBN 0-12-184740-X.

C   Intended for inservice teacher training, the choice of topics is good, and treatment is generally well done. Coverage includes transformations and groups, measure, congruence and distance. The exercises in some sections are designed for classroom use by the teachers.

Earle, James H. *Descriptive Geometry.* (Illus.) Addison-Wesley, 1971. vii + 344pp. $8.95. 72-139161.

C   Traditional topics—revolution, intersections, developments, vectors, and graphic solutions—are developed in a significantly different way: emphasis is placed on the design and creative aspects of engineering geometry; most of the problems emphasize applications; and step-by-step solutions and demonstrations of fundamental principles are used throughout.

Ellison, Elsie C. *Fun With Lines and Curves.* (Illus.) Lothrop, Lee&Shepard, 1972. 95pp. $4.25. 72-1095. ISBN 0-688-40012-44.

JH   Opens with several very attractive designs, then a short chapter describes the tools one needs to make them—a ruler, compass and protractor. With its many illustrations and clear directions, this is a delightful book.

Hemmerling, Edwin M. *Fundamentals of College Geometry, 2nd ed.* (Illus.) Wiley, 1970. viii + 464pp. $8.95. 75-82969.

C   A carefully organized, well-written development of Euclidean geometry. Whenever the extension is a natural one, the author has introduced 3–D concepts. Model exercises are plentiful; the writing style is clear, straightforward and to the point; and the book is appropriate for self-study.

Holden, Alan. *Shapes, Space, and Symmetry.* (Illus.) Columbia Univ., 1971. 200pp. $11.00. 71-158459. ISBN 0-231-03549-7.

SH-C   Holden explores solids and produces many variations of the nine basic solid shapes. Three-dimensional visualization is constantly required, and the method of "learning by discovery" is used. Excellent photographs and detailed construction directions are provided to show the results of the operations of symmetry, reflection, rotation and decomposition. Fascinating!

Jacobs, Harold R. *Geometry.* (Illus.) Freeman, 1974. xii + 701pp. $9.00. 73-20024. ISBN 0-7167-0456-0. Index;CIP.

SH-C   Jacobs provides excellent chapters on deductive reasoning, transformations, inequalities, right triangles, regular polygons and circles, geometric solids, and non-Euclidean geometries. Useful in either a high school academic or college liberal arts curriculum.

Keedy, Mervin L., and Charles W. Nelson. *Geometry: A Modern Introduction, 2nd ed.* (Illus.) Addison-Wesley, 1973. xiii + 369pp. $10.95. 79-178267.

C   This revision omits conic sections and adds a chapter on transformations. Intended as a text for prospective elementary teachers, it follows the CUPM recommendations for Level 1. More than a high school math background is necessary for some sections.

Loeb, Arthur L. *Space Structures: Their Harmony and Counterpoint.* (Illus.) Addison-Wesley, 1976. xviii + 169pp. $19.50; $9.50(p). 75-16198. ISBN 0-201-04650-4; 0-201-04651-2. Index;CIP.

C-P   A precise mathematical exposition focusing on the concept of valency. Begins with the very simple ideas of statistical symmetry, moves through hyperpolyhedra, lattices, space-filling polyhedra, and concludes with a 16-page photographic essay, "Unwrapping the Cube." A book for scientists, architects and sculptors.

Maxwell, E. A. *Geometry by Transformations.* (Illus.) Cambridge Univ. Press, 1975. xii + 276pp. $19.50. 74-76568. ISBN 0-521-20405-4. Index.

C-P   The author introduces the concept of reflections by paper folding; then continues with parallelism and area concepts, isometries, similarity transformations and matrices. Proofs are well done and easy to follow; notation and definitions are presented as needed. The symbolism and vocabulary are British, but all teachers of high school or college geometry should own this book.

Nowlan, Robert A., and Robert M. Washburn. *Geometry for Teachers.* (Illus.) Harper&Row, 1975. xvii + 391pp. $11.95. 74-7098. ISBN 0-06-044866-0. Index;CIP.

C   The authors provide a nice introduction to transformations and their use to prove theorems. Simple constructions are explained and illustrated in great detail. There are ample exercises but no answers. The text is arranged for a 2-semester course, but an outline for only a 1-semester course is supplied.

Ogilvy, C. Stanley. *Excursions in Geometry.* (Illus.) Oxford Univ., 1969. vi + 178pp. $6.00. 78-83014.

SH-C   Gives brief insights into inversive, projective and advanced Euclidean geometries, carefully designed to take the reader rapidly into exciting theorems—often unexpected but always beautiful—and even to unsolved problems.

Stubblefield, Beauregard. *An Intuitive Approach to Elementary Geometry.* (Illus.) Brooks/Cole, 1969. xi + 254pp. $7.95. 69-11122.

SH-C   Vivid and picturesque analogies encourage intuitive development of fundamental geometric concepts. Imaginative exercises lead into many varied ideas and show their interrelations. The remarkably complete survey develops Euclidean, non-Euclidean, coordinate, vector and finite geometries as well as the postulational approach to creating mathematical systems.

Wenninger, Magnus J. *Polyhedron Models.* (Illus.) Cambridge Univ. Press, 1971. xii + 208pp. $14.50. 69-10200. ISBN 0-521-06917-3.

JH-SH-C   The author describes each polyhedron and gives clear, simple directions for its construction, including the necessary patterns and a photograph of his own completed model. Elaborate directions and hints regarding the use of templates, scoring, gluing, and color arrangements are given.

Young, John E., and Grace A. Bush. *Geometry for Elementary Teachers.* (Illus.) Holden-Day, 1971. xii + 273pp. $9.95. 77-155559. ISBN 0-8162-9984-6.

C   The aim of this excellent text is to reinforce the knowledge of geometric concepts held by elementary education majors. Ideas and language used relate geometry to experiences of the reader. Emphasis is on experimentation rather than traditional rigorous proof. The authors integrate contemporary and classical mathematics with originality and skill.

516.3 ANALYTIC GEOMETRY AND TRIGONOMETRY

Holton, Jean Laity. *Geometry: A New Way of Looking at Space.* (Illus.) Weybright&Talley, 1971. x + 70pp. $4.50. 70-128093.

SH-C   Starting with a light introduction to geometry, Holton presents the ideas of axiom structures, emphasizing projective geometry. The book is suitable for teachers or individuals who have already studied geometry.

Keedy, Mervin L., and Marvin L. Bittinger. *Trigonometry: A Functions Approach.* (Illus.) Addison-Wesley, 1974. viii + 321 + 32pp. $7.95(p). 73-22731. ISBN 0-201-03719-X. Index.

SH-C Contains the usual topics in trigonometry as well as detailed discussions of relations, functions, transformations, and exponential and logarithmic functions. Presupposes eleventh-grade math. A valuable text for students continuing with calculus where material of this type may be taken for granted.

Morrill, W.K., S.M. Selby and W.G. Johnson. *Modern Analytic Geometry, 3rd ed.* (Illus.) Intext, 1972. xii + 481pp. $9.50. 76-185821. ISBN 0-7002-2413-0.

SH-C This traditional approach using vector methods gives an extensive treatment in two- and three-space. Includes minimum theory, but many examples and numerous problems. Manipulative techniques are explained for operating with determinants which are used throughout. Includes an appendix on matrix theory.

Moser, James M. *Modern Elementary Geometry.* (Illus.) Prentice-Hall, 1971. x + 333pp. $8.95. 78-137660.

SH-C Shows the relation of geometry to other branches of mathematics and includes the usual topics of plane and solid geometry, set theory, elementary logic, simple plane trigonometry, and an introduction to analytic geometry. The clearly written text gives broad coverage of the foundations of mathematics and explains the relevance of geometry to familiar situations.

Preston, Gerald C., and Anthony R. Lovaglia. *Modern Analytic Geometry.* (Illus.) Harper&Row, 1971. x + 319pp. $7.95. 74-127342. ISBN 0-06-045256-0.

C The concepts of analytic geometry are developed on a purely algebraic and set theory base. All geometric elements are developed algebraically or in terms of their properties as sets or as a part of the real number systems. The logical sequence is good, and the illustrations are clear and sufficient to make the text a good, compact package.

Robinson, Thomas J. *Analytical Trigonometry.* (Illus.) Harper&Row, 1972. xiv + 287pp. $7.95. 70-170620. ISBN 06-045506-3.

C Reviews modern algebra, trig functions and properties, graphs of functions, angles and applications, inverse functions and trigonometric equations, complex numbers, logarithms and exponents, linear interpolation and the laws of tangents and tangent half-angles. Emphasis is theoretical, but some applications (such as vectors) are included.

Selby, Peter H. *Geometry and Trigonometry for Calculus.* (Illus.) Wiley, 1975. viii + 424pp. $5.95(p). ISBN 0-471-77558-4. Index;CIP.

SH-C The first four chapters are devoted to plane geometry, followed by trigonometry. Analytic geometry is summarized in two chapters, and a final chapter concerns intuitive limits. Excellent for self-study or review.

Swokowski, Earl W. *Fundamentals of Trigonometry, 3rd ed.* (Illus.) Prindle, Weber &Schmidt, 1975. viii + 243pp. $10.50. 75-8898. ISBN 0-87150-193-7. Index;CIP.

SH-C A good standard text for a one-semester course in trigonometry, this book covers trigonometric functions, analytical trigonometry, exponentials and logarithms, solutions of triangles, complex numbers and some vectors.

519.2–.3 Probabilities and Game Theory

Barton, Richard F. *A Primer on Simulation and Gaming.* (Illus.) Prentice-Hall, 1970. x + 239pp. $8.95. 79-110489.

C An introduction to simulation and gaming for administrators, behavioral scientists, and educators which reviews basic concepts, research, and some future possibilities in a lucid, nontechnical, and well-organized way. Simulation models are explained, as well as computer systems, languages, and Monte Carlo techniques. Strong emphasis on applications and concrete examples.

Chung, Kai Lai. *Elementary Probability Theory with Stochastic Processes.* (Illus.) Springer-Verlag, 1974. x + 325pp. $12.00. 73-21210. ISBN 3-387-90096-9. Index.

C-P   Chung's lively, conversational style fulfills his promise that the book will provide a "thorough and deliberate discussion of basic concepts and techniques of elementary probability with few frills and technical complications."

Davis, Morton D. *Game Theory: A Nontechnical Introduction.* (Illus.) Basic, 1970. xii + 209pp. $6.95. 79-94295. ISBN 0-465-02626-5.

SH-C   The best nontechnical introduction to game theory since Anatol Rapoport's *Fights, Games and Debates* (Univ. of Michigan, 1960). It covers the fundamentals of one, two, and n-person games, zero- and nonzero-sum games and utility measurement. It has no exercises, but it is well written and organized and recommended for self-study.

Owen, D.B., (Ed.). *On the History of Statistics and Probability.* (Illus.) Dekker, 1976. xiv + 466pp. $29.50. 75-32473. ISBN 0-8247-6390-4. Index.

C-P   Among the authors are some of the pioneers of probability theory and statistics, who often describe the motivation behind fundamental work, thus contributing to understanding of current theory as well as its historical evolution. Useful for all teachers and practitioners; some articles could be read profitably by students.

Rényi, Alfréd. *Letters on Probability.* (Trans.) Wayne State Univ. Press, 1973. 86pp. $2.95. 74-179559. ISBN 0-8143-1465-1.

SH-C   Contains four later "letters" of Pascal—actually a witty literary device used by the late, distinguished Hungarian mathematician Alfréd Rényi to explain the fundamental concepts of probability. Useful to introduce probability to students who have an elementary knowledge of combinatorial methods.

519.5 STATISTICAL MATHEMATICS

Aitchison, John. *Choice Against Chance: An Introduction to Statistical Decision Theory.* (Illus.) Addison-Wesley, 1970. ix + 284pp. $9.50. 70-109505.

C-P   Presupposes a minimum of mathematical information (since it considers only finite sets and develops needed probability theory) but much sophistication (since it develops difficult ideas). A number of decision problems are stated; then statistical decision theory is developed in terms of solutions to those problems. Not easy to read because of the nature of the subject.

Alder, Henry L., and Edward B. Roessler. *Introduction to Probability and Statistics,* *5th ed.* (Illus.) Freeman, 1972. xii + 373pp. $9.00. 78-188959. ISBN 0-7167-0444-7.

SH-C   Provides access to statistical analysis of experimental data with a minimum of prerequisite mathematics. Introduces summation notation, graphing, sample means and variances (with some motivation), and provides the necessary background in probability theory and various statistical techniques. Calculus is not necessary, but many computations will require a good proficiency in algebra.

Balaam, L.N. *Fundamentals of Biometry.* (Illus.) Wiley, 1972. xiv + 259pp. $12.95. 72-4170. ISBN 0-470-04571-X.

C   Proceeds in a slow, methodical manner from variables and variation, probability and normal distribution, to the analysis of variance, simple linear regression and correlation and tests for goodness of fit. No calculus needed. Very well written, excellent problems, valuable mainly for students in the biological and agricultural sciences needing a theoretical and mathematic rather than an intuitive treatment.

Bauer, Edward L. *A Statistical Manual for Chemists, 2nd ed.* (Illus.) Academic, 1971. vii + 193pp. $9.50. 73-154404. ISBN 0-12-082756-5.

   C   Outlines in an easily usable form the statistical methods with which every experimentalist should be familiar. Topics covered include various types of distributions, precision versus accuracy, analysis of variance, experimental design, control charts, data correlation, and sampling techniques.

Bradley, James V. *Probability; Decision; Statistics.* (Illus.) Prentice-Hall, 1976. xv + 596pp. $18.50. 75-1083. ISBN 0-13-711556-3. Index;CIP.

   SH-C   Using only high school algebra, Bradley presents commonly used concepts in probability, decision theory and statistics. A well-written, entertaining, but rapid introduction for students in education, psychology and related fields.

Ellis, Richard B. *Statistical Inference: Basic Concepts.* (Illus.) Prentice-Hall, 1975. xiv + 258pp. $10.50. 74-6289. ISBN 0-13-844621-0. Index;CIP.

   SH-C   This textbook, based on the author's lectures, guides the liberal arts student through the standard fare of statistics. Peppery, instructive quotes as well as active problem-solving exercises instruct and encourage students to delve deeper into statistics.

Guttman, Irwin, S.S. Wilks, and Stuart Hunter. *Introductory Engineering Statistics, 2nd ed.* (Illus.) Wiley, 1971. xix + 549pp. $13.95. 72-160214. ISBN 0-471-33770-6.

   C   Includes elements of probability, discrete distributions, acceptance sampling, estimation of population parameters, the Bayesian approach, statistical tests, regression analysis, analysis of variance, and some experimental designs. Only an introductory calculus course is required.

Hoel, Paul G. *Elementary Statistics, 3rd ed.* (Illus.) Wiley, 1971. x + 309pp. $9.50. 78-152497. ISBN 0-471-40300-8.

   C   An admirable presentation of basic statistical theory and practice. A wide range of examples and exercises illustrates situations requiring a probabilistic treatment for making decisions. The theory is developed through the technique of creating distributions by repeated sampling, and no calculus is used. The text covers the traditional topics and should be useful for high school science teachers.

Lawson, Charles L., and Richard J. Hanson. *Solving Least Squares Problems.* Prentice-Hall, 1974. xii + 340pp. $14.50. 73-20119. ISBN 0-13-822585-0. Index;CIP;bib.

   C-P   A good treatise for scientists, engineers and students whose work requires the analysis and solution of linear algebraic equations.

Lehmann, E.L., with H.J.M. D'Abrera. *Nonparametrics: Statistical Methods Based on Ranks.* (Illus.) Holden-Day/McGraw-Hill, 1975. xvi + 457pp. $22.95. 73-94384. ISBN 0-8162-4996-6. Indexes.

   C   A nontheoretical approach which introduces material through examples and includes many exercises and an 80-page appendix of background information. Only minimal mathematical background needed.

Lentner, Marvin. *Introduction to Applied Statistics.* (Illus.) Prindle, Weber&Schmidt, 1975. xi + 388pp. $13.95. 75-16250. ISBN 0-87150-204-6. Index;gloss.;CIP.

   SH-C   The book is devoted to such topics as sampling, randomness, graphs, bar charts, frequency distributions, probability, point and interval estimation, inferences about the means and standard deviation of two samples, and regression.

Leonard, J.M. *Understanding Statistics.* English Universities Press, 1974. 216pp. 60 pence(p). ISBN 0-340-18259-8. Index.

   SH-C   Supplies basic guidelines to aid in the layperson's understanding of numerical information. Final chapters include examples of statistics from diverse areas such as the popularity of records, medicine, voting, and crime. Statistical information and examples relate to Britain, but the book's theme and lively style of writing are also appropriate for American readers.

Lukacs, Eugene. *Probability and Mathematical Statistics: An Introduction.* (Illus.) Academic, 1972. x + 242pp. $8.50. 74-154395. ISBN 0-12-459850-1.

C   Each chapter contains problems in applied mathematics and the book covers probability spaces, random variables, expectation values, limit theorems, sampling, estimation and testing of hypotheses.

Meddis, Ray. *Elementary Analysis of Variance for the Behavioral Sciences.* (Illus.) Halsted/Wiley, 1973. ix + 129pp. $5.75(p). 72-12493. ISBN 0-470-59007-6.

C   Covers methods of calculating various statistics, including the analysis of variance of differences between means and statistical tests used in conjunction with or after obtaining significant F-ratios.

Ostle, Bernard, and Richard W. Mensing. *Statistics in Research; Basic Concepts and Techniques for Research Workers, 3rd ed.* (Illus.) Iowa State Univ. Press, 1975. xiv + 596pp. $18.95. 75-15528. ISBN 0-8138-1570-3. Index;CIP.

C-P   The new edition of this popular textbook retains many of its excellent features, especially its emphasis on theory. There is a good balance between the "how" and the "when and where." Useful to students in engineering and physical sciences.

Remington, Richard D., and M. Anthony Schork. *Statistics with Applications to the Biological and Health Sciences.* (Illus.) Prentice-Hall, 1970. xii + 418pp. $11.95. 71-100588.

SH-C   Standard topics are covered in a lucid style. Ideal for self-instruction in biology, education and the social sciences. Good examples and problems.

Runyon, Richard P., and Audrey Haber. *Fundamentals of Behavioral Statistics, 2nd ed.* Addison-Wesley, 1971. xiii + 351pp. $8.95. 79-136126.

C   Features a balance between simplification and sophistication, and incorporates the recent small-sample statistics while eliminating seldom-used methods. Includes excellent explanation of proportions of area under the normal curve and basic mathematics review. A useful text and reference.

Tanur, Judith M., Frederick Mosteller, et al. (Eds.). *Statistics: A Guide to the Unknown.* (Illus.) Holden-Day, 1972. xxiii + 430pp. $9.95; $4.95(p). 77-188128. ISBN 0-8162-8604-3; 0-8162-8594-2(p).

C   An engagingly written, highly readable and extremely interesting volume about the contributions of statistics to society. Essays cover the use of statistics in public health and pharmacology; in market research, accounting, quality control; and in economics, meteorology, geology, psychology and sociology. Particularly suitable as a supplement at the introductory level.

Welkowitz, Joan, Robert B. Ewen, and Jacob Cohen. *Introductory Statistics for the Behavioral Sciences.* (Illus.) Academic, 1971. xvi + 271pp. $8.50. 76-152747. *Workbook* (by Robert B. Ewen). xiv + 155pp. $2.95.

C   Covers the standard topics such as measures of central tendency and dispersion, transformed scores, the normal curve, hypothesis testing, the *t*-test, linear correlation, power analysis, and one- and two-way analyses of variance. The well-prepared workbook includes computation exercises and problems which require synthesis of the subject matter, interpretation of experimental objectives, and evaluation of results.

Zehna, Peter W. *Introductory Statistics.* (Illus.) Prindle, Weber&Schmidt, 1974. x + 470pp. $12.50. 74-1467. ISBN 87150-171-6. Index;CIP.

> **SH-C**   Designed for high school students with a background in algebra, the book deals mainly with statistical inference but also provides a good introduction to probability and random variables.

519.7–.8 PROGRAMMING AND SPECIAL TOPICS

See also 001.6 Data Analysis.

Ashley, Ruth, with Nancy B. Stern. *Background Math for a Computer World.* (Illus.) Wiley, 1973. xi + 286pp. $3.95(p). 72-8948. ISBN 0-471-03506-8.

> **SH-C**   Treats both basic computer mathematics and elementary computer application topics in business, statistics, mathematics and game theory, but programming itself is not touched on. Very little prior knowledge is assumed, and the treatment is relatively superficial.

Clifford, Jerrold R., and Martin Clifford. *Modern Electronics Math.* (Illus.) Tab, 1976. 684pp. $12.95; $9.95(p). 73-86767. ISBN 0-8306-6655-9; 0-8306-5655-3. Index.

> **SH-C**   An excellent, thorough treatment of mathematics combined with applications to electronics. Coverage ranges from basic math through calculus and computer science. Many examples and figures.

Conrad, Clifford L., et al. *Computer Mathematics.* (Illus.) Hayden, 1975. 210pp. $13.95. 75-6849. ISBN 0-8104-5095-X. Index;CIP.

> **SH-C**   Numeric calculus, operations in any number system, arithmetic algorithms, base conversion, logic calculus, formula and list functions, arithmetic and programming practice, array calculus and matrix multiplication and division are covered, with emphasis on the development of algorithms for performing arithmetic and logical calculations in machine or assembly language programs.

Dorn, William S., et al. *Mathematical Logic and Probability with Basic Programming.* (Illus.) Prindle, Weber&Schmidt, 1973. viii + 216pp. $7.50(p). 72-79073. ISBN 87150-155-4.

> **SH-C**   The authors offer readers a feel for computers, using BASIC, Boolean algebra and the calculus of probability. A supplementary reader for introductory courses.

Emshoff, James R., and Roger L. Sisson. *Design and Use of Computer Simulation Models.* Macmillan, 1970. xvii + 302pp. $11.95. 72-96739.

> **C**   Takes the reader step-by-step through examples to an understanding of skills required to learn and use simulation. Only a basic knowledge of statistics and FORTRAN is needed. Methodology is discussed, as are development, model design, languages, and analysis of simulation.

Gilbert, Jack. *Advanced Applications for Pocket Calculators.* (Illus.) Tab, 1975. 304pp. $8.95; $5.95(p). 74-33620. ISBN 0-8306-5824-6; 0-8306-4824-0. Index;gloss.

> **C**   The author provides a creditable general view of how to extend the use of pocket calculators, with specific reference to some 14 individual machine models from some 12 manufacturers. Presumes a knowledge of college algebra and trigonometry.

Gill, Arthur. *Applied Algebra for the Computer Sciences.* (Illus.) Prentice-Hall, 1976. xv + 432pp. $16.50. 75-2110. ISBN 0-13-039222-7. Indexes;CIP.

> **C**   An excellent, comprehensive study for beginning computer science majors. Describes a variety of concepts and techniques required for more advanced work in computer science. No math prerequisites except high school algebra.

Johnson, Rodney D., and Bernard R. Siskin. *Quantitative Techniques for Business Decisions.* (Illus.) Prentice-Hall, 1976. ix + 485pp. $14.50. 75-31845. ISBN 0-13-746990-X. Index;CIP.

C-P   Provides potential managers with quantitative decision-making tools and emphasizes the way to select an appropriate technique to fit problems. Explains basic probability and decision theory.

Lee, Sang M., and Laurence J. Moore. *Introduction to Decision Science.* (Illus.) Petrocelli/Charter, 1975. xii + 589pp. $14.95. 75-16143. ISBN 0-88405-310-5. Index;CIP.

C-P   An introduction to the quantitative techniques used in decision science, including network models (PERT-CPM), transportation and assignment methods, queuing theory, game theory, simulation analysis and decision theory.

McCuen, Richard H. *FORTRAN Programming for Civil Engineers.* Prentice-Hall, 1975. xv + 448pp. $8.95(p). 75-2401. ISBN 0-13-329417-X. Index;CIP.

C   Topics include FORTRAN quantities, FORTRAN statements, input/output, program structure, transfer of control, flowcharting and iterative operations. McCuen describes the general principles, then uses examples to illustrate them. Could be used as a freshman or sophomore text.

Peckham, Herbert D. *Computers, BASIC and Physics.* (Illus.) Addison-Wesley, 1971. 320pp. $5.50(p). 76-149814.

C   The author's program is very ambitious, but his work is so detailed and clearly done that students will learn a great deal of mathematics even before reaching the material in their math courses. The only prerequisite is a semester of calculus, and this book could provide the best understood semester of mathematics most students will ever encounter.

Snell, J. Laurie. *Introduction to Probability Theory with Computing.* Prentice-Hall, 1975. vii + 294pp. $9.95(p). ISBN 0-13-493445-8. Index.

C   The author presents probability through coin-tossing, fun and games. There are numerous problems, and the emphasis is on computing and computer programs. Refreshing and stimulating.

Wardle, M.E. *SMP Computing in Mathematics: From Problem to Program.* (Illus.) Cambridge Univ., 1972. x + 128pp. $8.75; $4.45(p). ISBN 0-521-08301-X; 0-521-09684-7(p).

SH-C   Teaches programming through simulating computer memory by squares on plastic (or actual matchboxes) and develops ideas from storage and replacement through to an unconditional jump instruction. Examples are numerous and well graded.

520 Astronomy

Alter, Dinsmore, et al. *Pictorial Astronomy, 4th rev. ed.* (Illus.) Crowell, 1974. 328pp. $10.00. 73-15577. ISBN 0-690-00095-2. Index;gloss.

C   A basic book for every library. Its 60 chapters contain something about the sun, earth, moon, eclipses, planets, comets and meteors, stars and nebulae and the space sciences. Profusely illustrated with photographs, line drawings and paintings.

Brandt, John C., and Stephen P. Maran. *New Horizons in Astronomy.* (Illus.) Freeman, 1972. xii + 496pp. $12.50. 74-178298. ISBN 0-7167-0338-6.

C   Addressed to college-level, liberal arts students who are not majoring in science, this book attempts to survey some of the questions that have excited contemporary astronomers and for which they are actively seeking answers. The information content is enormous; the flavor is that of a TV documentary.

Friedman, Herbert. *The Amazing Universe.* (Illus.) National Geographic Soc., 1975. 200pp. $4.25. 74-28806. ISBN 0-87044-179-5. Index;CIP.

JH-SH-C   This highly readable account of astronomical research from ancient times to the present is an outstanding introduction to the subject.

Hopkins, Jeanne. *Glossary of Astronomy and Astrophysics.* Univ. of Chicago Press, 1976. vii + 169pp. $10.95. 75-14799. ISBN 0-226-35172-6.

C   This is a complete, concise and up-to-date dictionary. A fair knowledge of astronomy is required for its use, and it is most appropriate for university libraries.

Jastrow, Robert. *Red Giants and White Dwarfs: Man's Descent from the Stars, rev. ed.* (Illus.) Harper&Row, 1971. xvi + 190pp. $6.95. 79-108939. ISBN 0-06-012181-5.

SH-C   A work that describes in a highly readable manner the origins of the various elements of the universe, including man. It includes information obtained by the space probes to Venus and Mars and the moon flights. Among other topics covered are the birth, evolution, and death of stars; the beginning of the universe; the origin of the solar system and its components; the beginning of life on earth; and questions concerning extraterrestrial life. Good science and good writing.

Jones, Harold Spencer, et al. *The New Space Encyclopedia: A Guide to Astronomy and Space Exploration, rev. ed.* (Illus.) Dutton, 1974. 326pp. $14.95. 73-12348. ISBN 0-87690-108-9.

C   B. Lovell, H.S. Jones, Z. Kopal, H. Newell, P. Moore, J.G. Porter, plus a less famous but very able staff, provide a cogent, comprehensive and remarkable up-to-date compendium. Essays on virtually every topic in modern general astronomy and space science are included. The choice of illustrations is good, but the quality of the photographic reproductions is often inadequate.

Moore, Patrick. *Concise Atlas of the Universe.* (Illus.) Rand McNally, 1974. 192pp. $19.95. 74-421. ISBN 528-83031-7. Index;gloss.

JH-SH-C   Consists of atlases of the earth, from space, the moon, the solar system, and the stars. Useful tables are provided. For readers of all ages.

Moore, Patrick, (Ed.). *The Atlas of the Universe.* (Illus.) Rand McNally, 1970. 272pp. $35.00. 77-653619. Index;gloss.

SH-C   Perhaps the most complete astronomical atlas and reference work of all time. The first section, an illustrated review of exploration of space from the beginnings of naked-eye astronomy to man in space, is followed by atlases of the earth from space, the moon, the solar system and the stars. The work concludes with a catalog of stellar objects.

Roth, G.D., (Ed.). *Astronomy: A Handbook, rev. ed.* (Illus.; trans. and rev. by Arthur Beer.) Springer-Verlag, 1975. xvii + 567pp. $21.40. 74-11408. ISBN 0-387-06503-2. Index;CIP;bib.

SH-C   Indepth articles describing the most popular aspects of astronomy are contained herein; there is much material dealing with observations, physical features and mathematical concepts of astronomical phenomena. This is a skillfully assembled, highly recommended compendium.

Van Der Waerden, Bartel L., et al. *Science Awakening, II: The Birth of Astronomy.*

(Illus.) Oxford Univ. Press, 1974. xiv + 347pp. $35.00. A54-7774. ISBN 0-19-519753-4. Index.

SH-C-P   This distinguished Swiss historian of science presents a critical analysis of the development of humanity's awareness of the greater universe. Includes a fascinating account of the links between our awareness of celestial movements and our religious beliefs.

520.7 ASTRONOMY—STUDY AND TEACHING

Beet, E.A.   *Mathematical Astronomy for Amateurs.* (Illus.) Norton, 1972. 143pp. $7.95. ISBN 393-063887.

JH-SH   May aid the amateur already familiar with descriptive astronomy to become an active, serious astronomer or could serve as a text in those grades where astronomy is included in the science curriculum. With celestial mechanics as the theme, the major subjects are earth, moon and their orbits; time; the celestial sphere; and the solar system and stellar topics.

Brown, Peter Lancaster.   *What Star is That?* (Illus.) Viking, 1971. 224pp. $12.95. 73-149587. ISBN 0-670-75865-5.

JH-SH-C   Brings a new freshness and excitement to the ancient art of astronomy. Beginning with the origin of the constellations, the book continues with instrumentation; the north circumpolar stars; 53 major constellations; meteor showers; asteroids; comets; and novas. Numerous charts and illustrations complement the text. Excellent tables include solar and naked-eye planet positions for 1971 to 1981, and finding charts for 75 variable stars of special interest to the more advanced observer.

Howard, Neale E.   *The Telescope Handbook and Star Atlas, rev. ed.* (Illus.) Crowell, 1975. ix + 226pp. $14.95. 75-6601. ISBN 0-690-00686-1. Index;gloss;CIP.

SH   This is an excellent book for amateur astronomers and high school students with a strong interest in astronomy and some background in the physical sciences. The description and details of the physical optics of the telescopes are very good. An introduction to and explanation of sky charts is included.

Joseph, Joseph Maron, and Sarah Lee Lippincott.   *Point to the Stars, 2nd ed.* (Illus.) McGraw-Hill, 1972. 96pp. $4.72. 71-39765. Index;gloss.

JH   Excellent elementary introduction to sky observing. The earth's motions, time, and the position of the observer are shown to determine the appearance of the sky. Introduces the better-known telescopic objects and discusses celestial coordinate systems and the zodiac. Includes a table of positions.

Moore, Patrick, (Ed.).   *Astronomical Telescopes and Observatories for Amateurs.* (Illus.) Norton, 1973. 256pp. $7.95. ISBN 0-393-06395-X.

SH-C   This is a compilation of articles by 15 different contributors, each dealing with a particular aspect of the design, fabrication or use of telescopes. An excellent introduction for the beginning astronomer or amateur.

Moore, Patrick, (Ed.).   *1976 Yearbook of Astronomy.* (Illus.) Norton, 1976. 215pp. $9.95. ISBN 0-393-6404-2. Bib.

JH-SH-C   This is an excellent handbook (issued yearly) for the amateur astronomer, which contains information and articles on events seen during 1976, from the movement of planets and the phases of the moon to meteors, eclipses, and the minor planets. An excellent set of star maps is included.

Muirden, James. *The Amateur Astronomer's Handbook, rev. ed.* (Illus.) Crowell, 1974. xiii + 404pp. $8.95. 74-5411. ISBN 0-690-00505-9. Index;CIP.

SH-C A basic book for the serious amateur who wants to learn the techniques of observation. Includes numerous photographs taken by beginning observers as well as sections on equipment, the solar system, stars and nebulae, and optical work for amateurs.

Muirden, James. *Beginner's Guide to Astronomical Telescope Making.* (Illus.) Pelham (distr.: Transatlantic Arts), 1975. 201pp. $12.50. ISBN 0-7207-0822-2. Index;bib.

SH-C A thorough, lucid guide by a British expert who writes from a wealth of experience. Discusses making a simple refractor and an f/10, 15cm reflector; grinding, polishing and testing, and figuring the mirror; tubes, mountings, optical alignment and accessories; the design of a two-element object glass and larger reflectors; and making a Maksutov telescope.

Neely, Henry M. *A Primer for Star-Gazers.* (Illus.) Harper&Row, 1970. xiv + 334pp. $6.95. 72-120090.

JH-SH An ingenious scheme, clearly described, permits the reader to relate 96 charts of the sky directly to the appropriate parts of the sky. A calendar and 26 supplemental diagrams are used to orient the book to the sky, area by area. You can either learn the whole sky by using the book at the same time of night in 24 biweekly sessions or use it at all times of the night over a shorter period.

Neely, Henry M. *The Stars by Clock and Fist, rev. ed.* (Illus.) Viking, 1972. 192pp. $5.95. 73-176310. ISBN 670-66826-5.

JH-SH Written in easy-to-follow language, this simplified approach to learning the heavens is useful as a collateral text for elementary science and for anyone interested in learning the stars rapidly, easily and in a delightfully unique manner.

Nourse, Alan E. *The Backyard Astronomer.* (Illus.) Watts, 1973. 118pp. $7.95. 73-4644. ISBN 0-531-02568-3.

JH-SH An excellent book for those who like to engage in astronomy which can be pursued with the naked eye or a good pair of binoculars. Directions are given for location and identification of constellations. Excellent illustrations.

Paul, Henry E. *Telescopes for Skygazing, 3rd ed.* (Illus.) AMPHOTO, 1976. 160pp. $9.95. 65-26425. ISBN 0-8174-2408-3. Index.

SH-C This well-illustrated book for the layperson and astronomy student provides concise and readable explanations of a variety of telescopes, including lenses, mirrors, accessories, and their care, use and mounting. Includes brief expositions of our solar system, galaxies, etc., stellar photography, and the use of binoculars in stargazing.

Peltier, Leslie C. *Guideposts to the Stars: Exploring the Skies Throughout the Year.* (Illus.) Macmillan, 1972, 176pp. $6.95. 72-187797.

JH-SH Includes brief descriptions of planets, comets, meteors, aurorae, etc. Pictures and diagrams show constellations. Too much material covered in too little space, but well done and infused with the author's devotion to the subject.

Robinson, J. Hedley. *Astronomy Data Book.* (Illus.) Wiley, 1973. 271pp. $10.95. 72-9496. ISBN 0-470-72801-9.

SH-C The author, an amateur astronomer, has produced a highly commendable compilation directed to the needs of students, amateur astronomers and teachers of the earth sciences.

Worvill, Roy. *Telescope Making for Beginners.* (Illus.) Orbiting Book Service, 1974. 79pp. $5.95. 73-91740. ISBN 0-914326-00-7. Bib.

JH-SH   Worvill supplies instructions for the construction of simple refracting and reflecting telescopes with accessories, sundials, star clocks and quadrants. An extensive bibliography on general astronomy, instruments, star atlases, moon maps, observational techniques and telescopic projects is included.

520.9 HISTORY OF ASTRONOMY

Adamczewski, Jan, et al. *Nicolaus Copernicus and His Epoch.* (Illus.) Copernicus Society/Scribner's, 1974. 160pp. $7.95.

JH-SH-C   This book, published on the 500th anniversary of Copernicus' birth, weaves together facts about his life, and the political and cultural setting which shaped his career. A valuable addition to school library collections.

Aiton, E.J. *The Vortex Theory of Planetary Motions.* (Illus.) Elsevier, 1972. ix + 282pp. $18.50. ISBN 0444-19595-5.

C   The long conflict between the Cartesians and the Newtonians over the mechanism that would explain the motion of the planets is explained. As a case study in the evolving epistemology and methodology of science, this volume is superb.

Buttman, Gunther. *The Shadow of the Telescope: A Biography of John Herschel.* (Illus.; pub. in German, 1965). Scribner's, 1970. xiv + 219pp. $7.95. 72-85256.

SH-C   Sir John Herschel, the son of the renowned astronomer Sir William Herschel, worked in astronomy, especially with double star systems, and with apparently equal facility in mathematics, geology, paleontology, photography, photochemistry and optical physics. An interesting biography.

Moore, Patrick. *Watchers of the Stars: The Scientific Revolution.* (Illus.) Putnam's, 1974. 239pp. $15.95. 74-78640. SBN 399-11374-6. Index.

SH-C   The achievements, problems, skills, misfortunes and good fortunes that made possible the tremendous strides of such men as Copernicus, Tycho, Kepler, Galileo and Newton are examined. The author has carefully researched his subjects and has provided anecdotal matter which makes fascinating reading for science history buffs.

Ronan, Colin A. *Discovering the Universe: A History of Astronomy.* (Illus.) Basic Books, 1971. 248pp. $6.95. 72-135556. ISBN 0-465-01670-7.

SH-C   This exposition of the entire history of astronomy, from the ancients to the present, for a lay audience, is both lucid and concise—an admirable and difficult combination.

Whitney, Charles A. *The Discovery of Our Galaxy.* (Illus.) Knopf, 1971. xv + 308 + viiipp. $10.00 76-154942. ISBN 0-394-46068-5.

SH-C   Beginning with Greek Epicureanism and proceeding to modern-day discussions of astronomical phenomena, the book mixes descriptions of the personalities of several astronomers with the religion, beliefs, and psychology of their times. The author's opinions are evident, and his comments are always interesting.

Wright, Helen, Joan N. Warnow and Charles Weiner (Eds.). *The Legacy of George Ellery Hale: Evolution of Astronomy and Scientific Institutions in Pictures and Documents.* (Illus.) MIT Press, 1972. 293pp. $17.50. 74-148854. ISBN 0-262-23049-6.

SH-C   The life and scientific achievements of George Ellery Hale (1868–1938)— research astronomer, designer and builder of observatories and instruments and advocate for scientific organizations and the education of scientists—are related. Parts 1 and 2 are a biographical essay and five selections from Hale's writings. Part 3 consists of essays on large telescopes and other instruments, on solar research, and on scientific organizations.

522 OBSERVATIONAL ASTRONOMY

Asimov, Isaac. *Eyes on the Universe: A History of the Telescope.* (Illus.) Houghton Mifflin, 1975. ix + 274pp. $8.95. 75-15830. ISBN 0-395-20716-9. Index;CIP.

SH-C   This is a history of astronomy and cosmology, rather than just a history of
        telescopes. Asimov discusses the development of astronomical thought, techniques, and observations from about 5000 years ago to the space program, radio and radar astronomy, quasars, pulsars, black holes and the proposed Large Space Telescope.

Forbes, Eric G. *Greenwich Observatory: The Royal Observatory at Greenwich and Herstmonceux, 1675–1975; Vol. 1: Origins and Early History (1675–1835).* (Illus.) Taylor&Francis (distr.: Scribner's), 1975. xv + 204pp. $60.00, set. ISBN 0-85066-093-9. Index.
Meadows, A.J. *Vol. 2: Recent History (1836–1975).* (Illus.) xi + 135pp. ISBN 0-85066-094-7. Index.
Howse, Derek. *Vol. 3: The Buildings and Instruments.* (Illus.) xix + 178pp. ISBN 0-85066-095-5. Index.

SH-C-P   This 3-volume history commemorates the Tercentinary of the Greenwich
          Obervatory. The first two volumes emphasize people and ideas and the characterizations are excellent. The third volume, which deals with instruments, contains a physical description of each instrument as well as notes about its maker.

Hey, J.S. *The Radio Universe, 2nd rev. ed.* (Illus.) Pergamon, 1976. viii + 264pp. $15.00; $9.50(p). 75-23134. ISBN 0-08-018760-9; 0-08-018761-7. Indexes;CIP.

C-P   Covers thoroughly the techniques, observations and theory of modern radio
        astronomy, also explaining its full role in the context of our optical view of the heavens and current cosmological models. Well-balanced, accurate, clearly written and illustrated and current through 1973. No other book of this scope and level presently exists.

Knight, David C. *Eavesdropping on Space: The Quest of Radio Astronomy.* (Illus.) Morrow, 1975. 128pp. $4.95. 74-19285. ISBN 0-688-22019-3; 0-688-32019-8. Gloss;index;CIP.

JH-SH   This compact, nicely organized little book is a brief overview of funda-
         mental concepts, simply explained, and such topics as quasars and black holes. Well suited for collateral reading in any beginning science course including astronomy.

Woodbury, David O. *The Glass Giant of Palomar.* (Illus.) Dodd, Mead, 1970. viii + 390pp. $7.50. 77-135210. ISBN 0-396-01919-6.

JH-SH   First published in 1929, the book's main theme is the design and construc-
         tion of the 200-inch Hale reflecting telescope. Photographs have been inserted in this edition to show some of the new discoveries, but much of the substance of modern astronomy is not covered.

523 DESCRIPTIVE ASTRONOMY

Asimov, Isaac. *To the Ends of the Universe, rev. ed.* (Illus.) Walker, 1976. 141pp. $6.50. 75-10524. ISBN 0-8027-6236-0; 0-8027-6235-2. Index.

JH-SH-GA   This fairly elementary but well-organized treatment deals with the
            planets and the solar system, the Milky Way, globular clusters, the

Magellanic clouds, nebulas, novas and supernovas, the sun, red giants, red and white dwarfs, cosmic distances, x-ray and radio sources, exploding galaxies, and quasars. For supplementary reading only.

Hoyle, Fred. *Highlights in Astronomy.* (Illus.) Freeman, 1975. 179pp. $10.00; $5.50(p). 75-1300. ISBN 0-7167-0355-6; 0-7167-0354-8. Index;CIP.

SH-GA-C   Some of the material contained in this book is rather advanced, while much of the remainder is elementary. More than an elementary knowledge of trigonometry is needed to understand the more abstruse concepts presented. Nevertheless, this is a pleasant book, magnificently illustrated, which should interest laypersons.

Jastrow, Robert and Malcolm H. Thompson. *Astronomy: Fundamentals and Frontiers.* (Illus.) Wiley, 1972. xiv + 439pp. $12.50. 78-174770. ISBN 0-471-440752.

JH-SH-C   Suited to the precocious youngster and inquisitive oldster, this book gives a lucid description of exciting discoveries in recent research. Conjecture is carefully separated from fact. The book ends with a splendid, cohesive summary of life in the universe, not only in space, but in time and creation.

Kopal, Zdenek. *Man and His Universe.* (Illus.) Morrow, 1972. 313pp. $7.95. 74-166343.

JH-SH-GA   A very readable book covering many relatively recent advances in astronomy. No mathematical equations are used, but the reader is assumed to be familiar with powers of ten, metric units, and logarithms. Suitable for auxiliary reading; contains excellent photographs.

Levitt, I.M. *Beyond the Known Universe: From Dwarf Stars to Quasars.* (Illus.) Viking, 1974. xi + 131pp. $10.00. 73-5232. ISBN 670-16107-1. Index.

SH-C   An easy-to-read and dramatic text leads the reader from the theory of astrophysics to techniques being used to study and measure astrophysical phenomena. Excellent illustrations highlight the author's understanding of the layman's point of view. The book can be understood by high school students and could be used to illustrate even advanced courses in astrophysics.

Menzel, Donald H. *Astronomy.* (Illus.) Random House, 1970. 320pp. $17.50. 70-127542.

JH-SH   This magnificently illustrated book by a skillful writer and astronomer is thoroughly readable and a mine of information. The author traces the beginning of astronomy to the evolution of stars, galaxies and the universe. No mathematics is involved.

Nicolson, Iain. *Simple Astronomy.* (Illus.) Scribner's, 1974. 64pp. $6.95. 73-11573. ISBN 0-684-13640-6.

JH-SH-C   Nicolson reviews basic astronomy, from the aspects of the sky to instruments, the solar system and on to cosmology. This volume resembles an expanded and illustrated glossary; the treatment is classical. The excitement of modern astrophysics does not come across, but the format and language are readily understandable by a novice and the diagrams genuinely instruct.

Page, Lou Williams. *Ideas from Astronomy.* (Illus.) Addison-Wesley, 1973. vi + 250pp. $3.60(p).

JH-SH-C   Intended as a text, this lucid and lively book describes the sky as seen by the ancients, the Copernican revolution, the mechanics of the solar system, the discovery of Uranus, Neptune, and Pluto; stellar energy sources, stellar evolution, the birth and death of stars; our galaxy, the discovery of other distant galaxies, the expanding universe, the origin of the universe; and quasars. An excellent, nonmathematical treatment for the general reader.

Pananides, Nicholas A. *Introductory Astronomy.* (Illus.) Addison-Wesley, 1973. xv + 344pp. $12.50. 72-1942. ISBN 0201-5675-5. Index;gloss.

    SH-C Includes historical material; covers tools of the astronomer; origin, structure and evolution of the solar system, stars, and galaxies; interrelationships among astronomical bodies; and information on constellations, units, scales, conversions, physical constants and physical and orbital data for the planets, star lists, and the Messier Catalogue of Nebulae and Star Clusters. Useful as a basic text or collateral reading.

Rohr, Hans. *The Beauty of the Universe.* (Illus.; trans.) Viking, 1972. 87pp. $10.00. 77-164990. ISBN 670-15340-0.

    JH-SH-C Primarily a book of exquisite astronomical photographs with an accompanying text. Should be of particular interest to young readers experiencing their first taste of astronomy.

Scientific American. *New Frontiers in Astronomy: Readings from* **Scientific American.** (Illus.) Freeman, 1975. 369pp. $13.00; $7.50(p). 75-8902. ISBN 0-7167-0520-6; 0-7167-0519-2. Index:CIP.

    SH-C The articles—arranged in the categories planetary system, sun, stellar evolution, milky way, galaxies, high energy astrophysics and cosmology—are introduced by Owen Gingerich who shows how exciting and fast-moving events in astronomy are related. A comprehensive course in astronomy of the 1970s.

Smith, Elske v. P., and Kenneth C. Jacobs. *Introductory Astronomy and Astrophysics.* (Illus.) Saunders, 1973. xii + 564pp. $15.95. 72-88853. ISBN 0-7216-8387-8. Bib.

    C Detailed and up-to-date material presented in traditional format. Topics include an introduction to the solar system and celestial and planetary mechanics, basic stellar characteristics and classification, structure and content of our galaxy and of the universe. Many tables, etc., and a review of mathematical methods. Useful as a text or general reference.

Stoy, R.H., (Ed.). *Everyman's Astronomy.* (Illus.) St. Martin's, 1974. 493pp. $10.00. 74-81460. Gloss;indexes.

    SH-C Although the terminology is decidedly British, this book contains a tremendous amount of easily located information. Various authors have contributed, and the chapter on an astronomer's tools is of particular interest. Useful as a reference.

Woods, John A., et al. (Eds.). *The Science of Astronomy.* (Illus.) Harper&Row, 1974. x + 466pp. $8.50(p). 73-10684. ISBN 0-06-041446-4. Index;CIP.

    SH-C This elementary textbook covers the essentials of astronomy. The constellations are identified; there are chapters on the moon, planets, sun, stars, and galaxies, and a brief discussion of cosmology. A well-executed book.

Wyatt, Stanley P. *Principles of Astronomy, 2nd ed.* (Illus.) Allyn&Bacon, 1971. xii + 686pp. $14.95. 70-131204. ISBN 0-205-03116-1.

    SH-GA-C The text is straightforward and a pleasure to read. The arrangement is conventional; topics begin with the earth and move outward through the solar system to stars, interstellar matter, galaxies and the universe at large. Explanatory diagrams and illustrations make up about one-third of the book. The content, composition, typography, and accuracy of information are all meritorious. (A shorter version, c.1974, is also available.)

523.1 COSMOLOGY (UNIVERSE)

Clayton, Donald D. *The Dark Night Sky: A Personal Adventure in Cosmology.* (Illus.) Quadrangle/N.Y. Times, 1975. xii + 206pp. $9.95. 75-9213. ISBN 0-8129-0585-7. CIP.

    C   Clayton interprets the writings, logic and conjectures of scientists and recon-
        structs their observational techniques. Of greatest value is Clayton's discus-
sion of the evolution of his own intellectual development, from child to student to
contributing scientist, and his statement of his hopes for the future of science and
humankind.

Hinkelbein, Albert. *Origins of the Universe.* (Illus.) Watts, 1973. 128pp. $5.95. 72-4195. SBN 531-02112-2.

    SH-C   Hinkelbein offers a wealth of fascinating facts, ideas, and explanations for
        many astronomical phenomena. Opposing theories are dismissed rapidly,
and important recent observations and theories are ignored. Nearly half the book is
striking, full-color illustrations. Appropriate for collateral and pleasure reading but
not for reference.

Hoyle, Fred. *Astronomy and Cosmology: A Modern Course.* (Illus.) Freeman, 1975. xiv + 711pp. $15.95. 74-28441. ISBN 0-7167-0351-3. Index;CIP;bib.

    C   Hoyle keeps mathematics to a minimum by presenting complex matters in
        quantum mechanics, relativity, nuclear physics and cosmology in an imagina-
tive, remarkably clear and understandable way. An exceptionally fine text for an
introductory course.

Hoyle, Fred. *From Stonehenge to Modern Cosmology.* (Illus.) Freeman, 1972. 96pp. $4.95. 72-10836. ISBN 0-7167-0341-6.

    C-P   The author introduces "cosmical constants," and then considers the evi-
        dence provided by practically all sciences to substantiate his personal obser-
vations and thoughts about the evolution, current state and future of the universe.
Extremely interesting.

Kaufmann, William J., III. *Relativity and Cosmology.* (Illus.) Harper&Row, 1973. 134pp. $2.95(p). 72-12002. SBN 06-043568-2.

    JH-SH-C   A delightful presentation, without mathematics, of the origin and impli-
        cations of the general theory of relativity and its current implications
for the universe as a whole. Deals with gravitational collapse, white dwarfs and
neutron stars, black holes, and white holes. Authoritative, balanced, and dependable.

Lovell, Bernard. *Man's Relation to the Universe.* (Illus.) Freeman, 1975. vi + 118pp. $5.95. 75-14096. ISBN 0-7167-0356-4. Index;CIP.

    SH-C-P   A compelling view of the dramatic recent expansion of our knowledge of
        the solar system, galaxy and universe, focusing on modern observational
techniques. Observations are interwoven with theoretical developments. A strong
technical background is needed; however, the clarity is such that a science-minded
high school student could benefit from reading it.

Motz, Lloyd. *The Universe: Its Beginning and End.* (Illus.) Scribner's, 1976. xiv + 343pp. $14.95. 75-6635. ISBN 0-684-14239-2. Index;CIP.

    GA-C   Beautifully combines the descriptive and technical aspects of astronomy.
        Lucidly explains basic concepts, with many of the explanations taken from
everyday life. An excellent book for general readers.

Sciama, D.W. *Modern Cosmology.* (Illus.) Cambridge Univ., 1971. vii + 212pp. $8.95. 73-142961. ISBN 0-521-08069-X.

C-P Devoted to the universe as a whole—its contents, structure, and history. Great care is taken to clarify the limits of confidence in current explanations and theories and the possibility of alternative explanations. The book should provide a dependable background for future new discoveries and expansion of knowledge. Background in mathematics and physics is needed.

Shipman, Harry L. *Black Holes, Quasars, and the Universe.* (Illus.) Houghton Mifflin, 1976. x + 309pp. $12.95. 75-19535. ISBN 0-395-24374-2; 0-395-20615-4(p). Gloss.;index;CIP.

C-P An excellent job of presenting the current ideas and questions about the observable universe and most of the entities therein, blending good observational facts with current theories. Examines late evolution of single stars, black holes, galaxies, quasars, the redshift controversy, and the observable universe and its evolution. Ideally suited for students and laypersons with a deep interest in astronomy and some understanding of the basic concepts of physics.

523.2 SOLAR SYSTEM

Butler, S.T., and Robert Raymond. *The Family of the Sun.* (Illus.) Anchor/Doubleday, 1975. 84pp. $2.50(p). 74-33084. ISBN 0-385-09827-8.

JH-SH The authors discuss the various mysteries surrounding the sun and solar system and speculate about living and working in other areas of the solar system. The illustrations are first-rate and the narrative is well written. A good reference for a science classroom.

Clarke, Arthur C. *Beyond Jupiter: The Worlds of Tomorrow.* (Illus.) Little, Brown, 1972. xvii + 89pp. $12.95. 72-6440. ISBN 0-316-14699-4.

JH-SH-C Reviews current and potential explorations of the planetary system. Includes 27 full-page illustrations, 15 in color. There are also excellent "space-age" photographs and outstanding typography.

Cornell, James, and E. Nelson Hayes (Eds.). *Man and Cosmos: Nine Guggenheim Lectures on the Solar System.* (Illus.) Norton, 1975. 191pp. $8.95. 75-6687. ISBN 0-393-06402-6. CIP.

SH-C The nine lectures cover the history of the solar system, the sun, moon, planets, planetary atmospheres, outer planets, asteroids, comets, and "Perspectives." Treatment is nonmathematical and deals with problems such as the "lost" neutrinos, the evolution of Pluto, the magnetic field of Venus, and the "waterways" of Mars.

Cousins, Frank W. *The Solar System.* (Illus.) Pica, 1972. 300pp. $20.00. 70-175855. ISBN 0-87663-704-7.

C-P An aesthetically pleasing book for the serious student of the solar system which will also appeal to the general reader seeking a straightforward presentation of both historical and recent astronomical information. Includes discussions of aspects of the solar system known in classical and Renaissance times, models of the solar system, planets and satellites, the sun, the earth-moon system, comets, meteors, tektites, interplanetary dust and gases, and the Van Allen belts.

Gardner, Martin. *Space Puzzles: Curious Questions and Answers about the Solar System.* (Illus.) Simon&Schuster, 1971. 95pp. $4.95. 78-144777. ISBN 0-671-65182-X.

JH-SH-GA Findings about the sun, the planets, their atmospheres and moons, meteors, and comets are given in background discussions before and with the questions, as well as in the answers. Myths and naive suppositions are demolished; a lucid and light style supports an elegant structure of solid science.

523.3 MOON

Alter, Dinsmore. *Pictorial Guide to the Moon, 3rd rev. ed.* (Illus.; rev. by Joseph H. Jackson.) Crowell, 1973. 216pp. $8.95. 73-9869. ISBN 0-690-00096-0. Index;gloss.

JH-SH-C Most of the 163 photographs in this book are excellent and are described with long captions. Although the emphasis is somewhat old-fashioned, the book is a good guide for the amateur telescope user. Includes historical facts on early maps of the moon.

Bedini, Silvio A., Wernher Von Braun, and Fred I. Whipple. *Moon: Man's Greatest Adventure.* (Ed. by David Thomas) Abrams, 1970. 268pp. $45.00. 73-121338. ISBN 0-8109-0327-X.

JH-SH-C The moon, in religion, superstition, myths, in the evolution of the reckoning of time and of the calendar, and in the work of the astrologers and astronomers is richly revealed in this history and its accompanying illustrations of deities, embellished artifacts, paintings, sculpture, early scientific instruments, observatories, methods, and U.S. space ventures, including the Apollo 11 Mission. A collector's item.

Branley, Franklyn M. *The Moon: Earth's Natural Satellite, rev. ed.* (Illus.) Crowell, 1972. 117pp. $4.50. 76-146279. ISBN 0-690-55415-X; 0-690-55416-8.

JH-SH A comprehensive discussion of the moon's motions; its surface properties; its gravitational interaction with the earth and sun; and other related topics. Current theories concerning the early history of the earth-moon system are missing, however, and it is difficult to identify particular surface features on drawings of lunar charts. Nevertheless, an excellent reference.

Gamow, George, and Harry C. Stubbs. *The Moon, rev. ed.* (Illus.) Abelard-Schuman, 1971. 126pp. $4.95. 73-137588. ISBN 0-200-71761-8. Index;bib.

JH-SH Among the topics covered are the origin and constitution of the moon, gravity, eclipses, rockets, telescopes, tides, and parallax. Brief diversions into science fiction of the past and the effects of eclipses on some historical figures and events are presented. There are numerous illustrations and photographs.

Kopal, Zdeněk. *A New Photographic Atlas of the Moon.* (Illus.) Taplinger, 1971. vii + 311pp. $20.00. 72-125480. ISBN 0-8008-5515-9.

SH-C A discussion of the moon-earth relationship, the solar system, the moon's varied surface features and probable history and structure, accompanied by more than 200 excellent plates with ample explanatory material for each. The striking photographs sample the far and near sides.

523.4–.6 PLANETS, ASTEROIDS, COMETS

Asimov, Isaac. *How Did We Find Out about Comets?* (Illus.) Walker, 1975. 64pp. $4.95. 74-78115. ISBN 0-8027-6203-4; 0-8027-6204-2. Index.

JH-SH An excellent introduction to the nature of comets written to appeal to a wide audience. Technical data on these intriguing objects are presented so the average layperson can understand. Suitable also as a classroom reference.

Asimov, Isaac. *Jupiter: The Largest Planet.* (Illus.) Lothrop, Lee&Shepard, 1973. 224pp. $5.95. 72-9359. ISBN 0-688-40044-2; 0-688-50044-7.

JH-SH Asimov discusses Jupiter and the planets, the Jovian satellites, Jupiter's gravitational effects, its atmosphere and changing colors, and he notes the unanswered questions: the nature of its Red Spot, internal structure, radio wave sources, etc. Historical approach but data (54 tables) current to 1972.

Baum, Richard. *The Planets: Some Myths and Realities.* (Illus.) Halsted/Wiley, 1973. 200pp. $8.95. 73-7583. ISBN 0-470-05930-3. Bib.

C   This is a well-researched collection of astronomical mystery tales which should fascinate readers familiar with observational astronomy at the college introductory level.

Branley, Franklyn M. *Comets, Meteoroids, and Asteroids: Mavericks of the Solar System.* (Illus.) Crowell, 1974. 115pp. $5.50. 73-16043. ISBN 0-690-20176-1. Index;CIP.

JH-SH   Seventh in a series, this volume complements the author's earlier works on the solar system. Especially recommended for young readers.

Davies, Merton E., and Bruce C. Murray. *The View from Space: Photographic Exploration of the Planets.* (Illus.) Columbia Univ., 1971. xii + 163pp. $14.95. 75-168867. ISBN 0-231-03557-8. Bib.

SH-C   Catchy subtitles and lucid text, together with easily interpreted factual tables and diagrams and abundant good illustrations, tell the story of photographic exploration from space of the earth, moon, and planets. The role of the USSR in the overall achievement is particularly well described. For the technically oriented reader, references and appendixes are provided.

Knight, David C. *The Tiny Planets: Asteroids of Our Solar System.* (Illus.) Morrow, 1973. 95pp. $3.95. 72-12946. ISBN 0-668-20072-9; 0-688-30072-3.

JH   Discusses the discovery, composition, origin, and possible uses of asteroids. The chapter entitled "Physical Nature" is particularly good; density, mass, weight, orbital configuration, and distribution are treated. Deals briefly with "Kirkwood's gaps" and Lagrangian points.

Moore, Patrick, and Charles A. Cross. *Mars.* (Illus.) Crown, 1973. 48pp. $7.95. 73-78847. ISBN 0-517-50527-4.

SH-C-P   A superb account of the red planet which provides a fascinating treatment of physical and topographic characteristics, with the explanations representing the consensus of the scientific community in 1973.

Nourse, Alan E. *The Asteroids.* (Illus.) Watts, 1975. 59pp. $3.90. 74-12020. ISBN 0-531-00822-3. Index;CIP.

JH   Nourse traces the discovery, nature, possible origins and reasons for further study of the asteroids. Fascinating photographs and detailed drawings from NASA and Yerkes Observatory make a book which will engage any star-struck reader.

Nourse, Alan E. *The Giant Planets.* (Illus.) Watts, 1974. 62pp. $3.95. 73-14515. ISBN 0-531-00816-9. Index.

JH   An imaginary space voyage to the giant planets introduces the vocabulary, describes the various planets, and discusses what could be learned on fly-by missions. Time and speed comparisons illustrate possibilities of space travel, the telescope, radio-radar telescope and rockets are explained, and atmospheres, surface temperatures and cloud covers are compared with earth's.

Wetterer, Margaret K. *The Moons of Jupiter.* (Illus.) Simon&Schuster, 1971. 96pp. $4.95. 76-163493. ISBN 0-671-65179-X.

JH-SH-GA   This small volume skillfully and competently recounts the discovery and the events preceding and immediately following the discovery of the moons of Jupiter with the first astronomical use of the telescope. The book is well written for both the beginner and the layperson.

523.7–.8 SUN AND STARS

Bova, Ben. *In Quest of Quasars: An Introduction to Stars and Starlike Objects.* (Illus.) Crowell-Collier, 1970. viii + 198pp. $5.95. 77-83062.

SH-C   Modern astronomy for the general reader, with chapters on quasars; the physical constitution, energy production and evolution of the sun and other stars; galaxies of all types; and various cosmologies and their bases. Contains an excellent collection of astronomical photographs.

Glasby, John S. *The Dwarf Novae.* (Illus.) American Elsevier, 1970. 293pp. $7.50. 74-125628. ISBN 0-444-19633-1.

C-P   A detailed description of the observational evidence on this special class of close stellar binaries. The subject matter as presented is not too technical, but some background in mechanics and spectroscopy is helpful.

Golden, Frederic. *Quasars, Pulsars, and Black Holes.* (Illus.) Scribner's, 1976. xv + 205pp. $7.95. 75-37646. ISBN 0-684-14501-4. Gloss.;index;CIP;bib.

JH-SH-GA-C   Concerns the history of astronomy from its ancient beginnings through 1974, and gives evidence for and against the various theories on the nature of the universe. The last third of the book covers quasars, pulsars and black holes. Intended for the layperson.

Richardson, Robert S. *The Stars & Serendipity.* (Illus.) Pantheon, 1971. 129pp. $5.95. 70-77435. ISBN 0-394-82022-3; 0-394-92022-3(p).

JH-SH   An account of some 13 examples of astronomical serendipity. The author describes such examples as the discoveries of Uranus, the sun spot cycle, the white dwarf companion of Sirius, the red shift law, galactic radio radiation, and pulsars. The accounts are very well written and highlight the human side of the stories.

Taylor, R.J. *The Stars: Their Structure and Evolution.* (Illus.) London: Wykeham, 1970. xi + 207pp. $3.60. ISBN 0-85109-110-5.

C-P   One of the best basic books on the subject. The extended work descriptions along with the mathematical treatment help to clarify what is often a very messy subject. The diagrams are simple and convey the information concisely.

Weart, Spencer. *How to Build a Sun.* (Illus.) Coward-McCann, 1970. 95pp. $4.29. 125326.

JH-SH   Conceptual models help answer such questions as, what is the sun? How was it born? How old is it? What is it made of? This book is written with enthusiasm for experimental research, appreciation of the role of the theorist, and of the indebtedness of both to the technician.

Zim, Herbert S. *The Sun, rev. ed.* (Illus.) Morrow, 1975. 64pp. $4.59. 74-34461. ISBN 0-688-22033-4; 0-688-32033-3. Index;CIP.

JH   Zim introduces the metric system in linear measurements, the Celsius temperature scale, and replaces British Thermal Units with calories in this edition. Other information has been revised in the light of new scientific discoveries.

525 EARTH (ASTRONOMICAL GEOGRAPHY)

Allison, Linda. *The Reasons for Seasons: The Great Cosmic Megagalactic Trip Without Moving from Your Chair.* (Illus.) Little, Brown, 1975. 124pp. $6.95; $3.95(p). 75-5930. ISBN 0-316-03440-1. CIP.

JH   Allison first supplies a background presentation of historical and scientific information and then invites the reader to try related experiments. When interest is aroused, she provides whys, wherefores and further information on the title topic.

Asimov, Isaac. *How Did We Find the Earth Is Round?* (Illus.) Walker, 1973. 64pp. $4.50. 72-81378. ISBN 0-8027-6121-6; 0-8027-6122-4.

JH   A chronological description of the intellectual voyage from the days of the flat earth to the final awareness of sphericity. Unfortunately, some explanations are incomplete and will leave even the most intelligent student puzzled. Charmingly and skillfully written.

Bodechtel, Johann, et al. *The Earth from Space.* (Illus.; trans.) Arco, 1974 (c. 1969). 176pp. $16.00. 72-97584. ISBN 0-668-02960-9. Index;gloss.

SH-C-P   Color and black-and-white photographs taken from space of earth, a few pictures of the moon and an interpretive map for each make up this extraordinary book. Photographs are identified by date, satellite, camera, lens, film, altitude of exposure, axis and approximate scale; techniques (IR, radar, visible photography microwave radiometry and absorption spectroscopy, etc.) and uses of results (mapping, structure recognition, crop management, etc.) are described.

Nicks, Oran W., (Ed.). *This Island Earth.* (Illus.) NASA Sp-250(distr.: Supt. of Docs., GPO, Wash., DC), 1970. 182pp. $6.00. 73-608969.

JH-SH-C   These dramatic and outstanding satellite photographs show the earth's lands, mountains, plains, deserts, and oceans in clear and accurate detail. Very useful for introductory geography classes.

Smith, Anthony. *The Seasons: Life and Its Rhythms.* (Illus.) Harcourt Brace Jovanovich, 1970. 318pp. $12.50. ISBN 0-15-179924-5. Index.

SH-C   This beautifully illustrated book explains how the rotation of the earth on its axis once in every 24 hours and the rotation of the earth in an orbit around the sun every year are responsible for nights, days, and the orderly progress of the four seasons. The very complete text considers the seasons terrestrially, seasonal conditions on other planets, in the past through geological time, and at present. Includes a geological calendar.

526 GEODESY AND CARTOGRAPHY

Breed, Charles B. *Surveying, 3rd ed.* (Illus., rev. by Alexander J. Bone & B. Austin Barry.) Wiley, 1971. xvii + 495pp. $8.95.

C   Largely a condensation of the two-volume classic, *The Principles and Practice of Surveying* by Breed and George L. Hosmer, the text has been updated to include instruments developed since 1957. Modern surveying, computing and drafting methods and materials are discussed, and logarithmic solutions are retained. The emphasis is on principles and use of instruments, and the text continues to be a valuable exposition.

Brindze, Ruth. *Charting the Oceans.* (Illus.) Vanguard, 1972. xii + 108pp. $5.95. 77-134674. ISBN 0-8149-0715-6.

JH-SH   This lively account of our efforts to chart the oceans covers map making, wreck hunting, charting the ocean floor and currents, and icebergs. Includes photographs, historical maps, and charts.

Herubin, Charles A. *Principles of Surveying.* (Illus.) Reston, 1974. x + 288pp. $12.95. 73-15796. ISBN 0-87909-618-7. CIP.

C   Designed to provide a self-study unit for surveying, this book will serve as an excellent accompanying text in a technician's field course. The author supplies clear definitions and excellent illustrations. He describes horizontal and vertical measurement and the use of tape, level and transit. Techniques in traverses, station surveying and construction surveys are explained.

Hirsch, S. Carl. *Mapmakers of America.* (Illus.) Viking, 1970. 176pp. $4.53. 70-102922. ISBN 0-670-45439-7; 0-670-45440-0 LB.

    JH-SH   The author discusses the need for maps and the effects that finished maps had on the development of the country. A considerable amount of technical information about surveying and cartography is introduced in telling the story of mapping explorers from the time of Coronado's search for Cibola to the present.

Thrower, Norman J.W. *Maps & Man: An Examination of Cartography in Relation to Culture and Civilization.* (Illus.) Prentice-Hall, 1972. vii + 184pp. $5.95; $2.95(p). 70-166141. ISBN 0-13-555961-8; 0-13-555953-7.

    C-P   A history of maps from the most primitive to the most recent which relates this form of communication to man's needs and his creative arts. A balanced discussion of essentially all map types is given.

527 CELESTIAL NAVIGATION

Gibbs, Tony. *Navigation: Finding Your Way on Sea and Land.* (Illus.) Watts, 1975. xviii + 88pp. $4.33. 75-12541. ISBN 0-531-00838-X. Gloss.;index;CIP.

    SH-C   This book provides elementary navigation information of considerable use to small boat owners. The author explains in some detail how to read nautical charts and maps and describes all the basic chart symbols.

Schlereth, Hewitt. *Commonsense Celestial Navigation.* (Illus.) Regnery, 1975. 231pp. $12.95. 74-27829. ISBN 0-8092-8279-8. Index;CIP.

    SH-C   The author delineates the logic of navigation, thereby reinforcing other concepts in science and mathematics, as well as providing a practical manual. Emergency situations are covered—navigation without a sextant or tables or chronometer, etc.

529 CHRONOLOGY (TIME)

Coleman, Lesley. *A Book of Time.* (Illus.) Nelson, 1971. 144pp. $5.95. 78-164969. ISBN 0-8407-6144-9. Bib.

    JH   A good introductory narrative which covers early calendars, calendar reforms, sundials, hour glasses, clocks, and the makers and repairers of clocks. There are chapters on the founding of the Royal Observatory at Greenwich, the development of the nautical almanac, and the perfection and periodic testing of chronometers for the Royal Navy.

Cousins, Frank W. *Sundials: The Art and Science of Gnomonics.* (Illus.) Pica Press (dist.: Universe Books), 1970. 247pp. $18.50. 72-105946. ISBN 0-87663-704-7.

    C-P   For those with a basic knowledge of astronomy and trigonometry, this treatise will be a delightful reading and learning experience. It provides a rich description of the history of sundials of many interesting types. Mathematical details concern zodiacal or declination lines, corrections for the equation of time, noon marks, dialing, scales, etc.; all with appropriate mathematical formulas and tables.

Dolan, Winthrop W. *A Choice of Sundials.* (Illus.) Stephen Greene, 1975. ix + 148pp. $10.00. 74-20525. ISBN 0-8289-0210-0. Index;CIP

    JH-SH-C   The history of sundials from early Egyptian, Greek and Chinese models is summarized and illustrated along with architectural uses to the present time. Sundial mottoes are also mentioned.

530 Physics

Acosta, Virgilio, Clyde L. Cowan and B.J. Graham. *Essentials of Modern Physics.* (Illus.) Harper&Row, 1973. xiii + 594pp. $12.95. 72-84326. SBN 06-040162-1.

C Includes quotations by pioneer physicists, pencil sketches and brief biographies. Chronologically develops the principles of physics through the period 1900-1969, but there is no mention of recent discoveries such as pulsars, quasars and tachyons. Diagrams, schematic summaries of selected topics, problems (including ones using BASIC programming), boxing of important formulas, use of standard notation, etc., make this book suitable for a college liberal arts course at the junior level.

Atkins, Kenneth R. *Physics, 3rd ed.* (Illus.) Wiley, 1976. xii + 818pp. $16.95. 75-11677. ISBN 0-471-03629-3. Index;CIP.

C For a 2-semester liberal arts course emphasizing analytical thinking. Atkin's approach is slanted toward modern physics: relativity, quantum mechanics, particle physics, nuclear processes, current searches for ultimate constituents and understanding of types of reactions, symmetry and conservation laws. No calculus and few equations; questions and problems are included.

Atkins, Kenneth R. *Physics—Once Over—Lightly.* (Illus.) Wiley, 1972. x + 370pp. $10.50. 76-177880. ISBN 0-471-03622-6.

C A short text for a 1-semester course for nonscientists, this book proceeds from classical mechanics through elementary particle physics, with samplings from special and general relativity. The writing is clear, and analogies are used whenever possible to illustrate fundamental principles. A mathematical appendix includes exponential notation, algebra, and limit processes.

Blackwood, Oswald H., William C. Kelly and Raymond M. Bell. *General Physics, 4th ed.* (Illus.) Wiley, 1973. x + 805pp. 72-6799. ISBN 0-471-07923-5. Index;CIP.

SH-C A simplified standard elementary physics text with an easily understood treatment of relativity and other basic theoretical and experimental material of 20th century physics. Requires only algebra.

Borowitz, Sidney, and Arthur Beiser. *Essentials of Physics: A Text for Students of Science and Engineering.* (Illus.) Addison-Wesley, 1971. viii + 568pp. $11.75. 70-131201.

C An important addition to the 1-year, calculus-based physics texts. The material is in the conventional order: mechanics, heat, electricity and magnetism, light, modern physics (quanta, molecules, atoms, and nuclei) and the solid state and elementary particles. Conservation principles are emphasized throughout. Discourages the cookbook approach and encourages the students to ask, "What is going on here?"

Flitter, Hessel Howard, and Harold R. Rowe. *An Introduction to Physics in Nursing, 6th ed.* (Illus.) Mosby, 1972. x + 273pp. $6.75(p). ISBN 0-8016-1596-8. Bib.

C Covers extremely well the nature of science, scientific inquiry, the history of the development of scientific theory, measurements, motion, energy, molecules and molecular phenomena, pressure, heat, sound and light, magnetism and electricity, atomic structure, nuclear physics, and quantum mechanics, with frequent reference to relevant nursing problems. Includes noncalculus mathematical discussions. Good chapter summaries, and study questions.

Flowers, B.H., and E. Mendoza. *Properties of Matter.* (Illus.) Wiley, 1970. xiii + 318pp. $15.50. 70-118151; ISBN 0-471-26497-0.

C Interprets the bulk properties of matter in terms of the microscopic properties
of molecules and their interactions in a manner which can be grasped by
beginning students. This book teaches what physics is all about. Excellent reference
or supplementary reading.

Halliday, David, and Robert Resnick. *Fundamentals of Physics.* (Illus.) Wiley, 1970.
xvii + 827pp. $13.95. 70-102867. ISBN 0-471-34430-3.

C This book is for a short, somewhat superficial course. Richly illustrated and
with excellent typography, it contains minimal calculus. At the end of most
chapters there is a list of challenging questions, and a good set of problems (and
answers to some) is provided for all chapters.

Hulsizer, Robert I., and David Lazarus. *The World of Physics.* (Illus.) Addison-
Wesley, 1972. x + 518pp. $7.96.

SH An outstanding attempt to integrate the usually compartmentalized topical
areas of high school physics. The pattern of the book is such that important
physical ideas are treated several times on ascending levels of sophistication. About
half of each page is devoted to unusually good drawings which illustrate the physical
principles in very clear and often "folksy" style. The approach is correct without
being rigorous.

Ivey, Donald G. *Physics in Two Volumes: Vol. 1, Classical Mechanics and Introductory
Statistical Mechanics.* (Illus.) Ronald, 1974. xxi + 818pp. $13.50. 73-93856. Index.
Hume, J.N. Patterson. *Vol. 2, Relativity, Electromagnetism and Quantum Physics.*
(Illus.) xvii + 500pp. $12.50. Index.

C These are unusually good undergraduate texts that emphasize the theoretical
and experimental bases of science. Terms and concepts are carefully defined,
and explanations are clear. In addition to problems, chapters include challenging
questions for discussion.

Lichten, William. *Ideas from Physics.* (Illus.) Addison-Wesley, 1973. iii + 284pp.
$3.36(p).

JH Encourages readers to find and ask the right questions. The cumulative nature
of physics is displayed through biographical anecdotes. Topics covered include atmospheric pressure, uniformly accelerated motion, momentum, electricity
and magnetism, nuclear energy, wave motion and light and more. Nonnumerical
problem sets.

Miller, Franklin, Jr., et al. *Concepts in Physics: A High School Physics Program, 2nd
ed.* (Illus.) Harcourt Brace Jovanovich, 1974. 500pp. $7.50. ISBN 0-15-362356-X.
Index.

SH This excellent text covers all of the usual topics and provides good illustrations, clear examples, thought-provoking questions and challenging problems. The only mathematical prerequisite is elementary algebra.

Reimann, Arnold. *Physics. Vol. 1: Mechanics and Heat.* (Illus.) Barnes&Noble, 1971.
xvii + 604pp. $6.50 ea. 73-137431. ISBN 0-389-00456-1. *Vol. 2: Electricity, Magnetism
and Optics.* xvii + 568pp. *Appendix A.* 126pp. ISBN 0-389-00459-6.

C First rate sources for science or engineering students who have some concur-
rent knowledge of calculus. Each subject is treated extensively enough to allow
extended reading beyond the scope of a particular course. The appendix includes
standard conversion factors, constants, tables of trigonometric functions, and common logarithms, a discussion of degrees of accuracy, and answers to selected problems which are often worked out in detail.

Romer, Robert H. *Energy: An Introduction to Physics.* (Illus.) Freeman, 1976. xvii + 628pp. $14.95. 75-35591. ISBN 0-7167-0357-2. Index;CIP.

C   For freshman liberal arts students. No calculus, but high school algebra and high school physics or chemistry needed. Romer focuses on energy in mechanics, heat, electricity and nuclear physics, but some of the usual topics of introductory physics are left out. Helps students think in a disciplined, quantitative and informed way about the technical aspects of societal energy problems. Includes lively examples and problems.

Theimer, Otto H. *A Gentleman's Guide to Modern Physics.* (Illus.) Wadsworth, 1973. xiv + 306pp. $9.95. 72-92159. ISBN 0-534-00281-1.

SH-C   A truly delightful guide to a world that seems to be moving too rapidly along to either progress or oblivion. Includes a chapter on "What is Science?" and concise excursions into the philosophy of science scattered throughout. Physics and astronomy are made understandable and enjoyable.

Tipler, Paul A. *Physics.* (Illus.) Worth, 1976. xxvi + 1026pp. $18.95. 74-82693. ISBN 0-87901-041-X. Index. *Instructor's Manual.* v + 189pp. *Study Guide* (with Granvil C. Kyker, Jr.) x + 345pp. $4.50(p). ISBN 0-87901-055-X.

C   For the standard 2- or 3-semester elementary physics course taken by engineering and science majors, covering the traditional range of topics and using predominantly SI units. The more difficult sections are clearly indicated and may be omitted. Includes review sections, simple exercises, and more difficult problems. The instructor's manual includes classroom demonstrations, additional applications and topics, suggested readings, and a film guide. The study guide contains chapter summaries, fundamental equations, review items, true-false and essay questions, and worked-out examples.

530.02 PHYSICS—MISCELLANY

Goldsmith, Donald, and Donald Levy. *From the Black Hole to the Infinite Universe.* (Illus.) Holden-Day, 1974. 330pp. $6.95(p). 73-86412. ISBN 0-8162-3323-3.

SH-C   The authors have attempted to show modern physics to nonscience students through a combination of fact and fiction. There are two parallel texts. One is a science fiction novelette, an episode of which opens each of the 15 chapters. The factual text describes the scientific subject matter quite clearly. Diagrams accompany the text.

Grossberg, Alan B. *Fortran for Engineering Physics: Electricity, Magnetism, and Light.* (Illus.) McGraw-Hill, 1972. viii + 246pp. $4.50(p). 72-2671. ISBN 0-07-024972-5.

C   Introduces computer programming techniques into the general physics laboratory course. There are too few sample programs, but it is still useful and informative. The development of mathematical problems makes conversion to computer programs relatively easy.

Kursunoglu, Behram, and Arnold Perlmutter (Eds.). *Impact of Basic Research on Technology.* (Illus.) Plenum, 1973. xv + 301pp. $16.50. 73-82141. ISBN 0-306-36901-X.

SH-C-P   This collection by experts in theoretical physics describes the historical development and ongoing work in various fields of physics without complicated mathematics. Topics include quantum physics, maser-laser development, collision physics, thermonuclear energy, superconductivity and television. For serious physics students, but also useful for students and teachers in history of technology.

Priest, Joseph. *Problems of Our Physical Environment: Energy, Transportation, Pollution.* (Illus.) Addison-Wesley, 1973. 389pp. $10.95. 72-9317. Gloss.;bibs.

C   This is a college physics text which covers energy and power; electric energy; thermodynamics; nuclear-fueled electric power plants; electric energy technology; motor vehicles; sound from motor vehicles; mass transportation; and energy and resources. Will be a useful reference for high school teachers.

530.03 DICTIONARIES

Thewlis, J. *Concise Dictionary of Physics and Related Subjects.* Pergamon, 1973. viii + 361pp. $16.50. 72-10122. ISBN 0-08-016900-7.

SH-C-P   In addition to coverage of physics proper, definitions of terms selected from related areas ranging from astronomy through crystallography, meteorology and physical metallurgy are included.

530.07 PHYSICS—STUDY AND TEACHING

McAlexander, John Aaron. *Experiments for Technical Physics.* (Illus.) Allyn&Bacon, 1973. vii + 286pp. $5.95(p). 72-89237.

C   This is a collection of traditional experiments, ranging from mechanics to nuclear physics, designed for 2-hour laboratory sessions. All graph paper and tables for recording data are provided.

Meiners, Harry F., (Ed.). *Physics Demonstration Experiments. Vol. 1: Mechanics and Wave Motion. Vol. 2: Heat, Electricity and Magnetism, Optics, Atomic and Nuclear Physics.* (Illus.) Ronald Press, 1970. 1395pp. $30.00set. 69-14674. Index.

SH-C   Provides aid and encouragement to practitioners at every level of sophistication. Ideas for demonstrations range from those for elementary school children through those for general science, physical science, and physics courses. The set should be available to all who teach courses involving physics concepts and in all school and college libraries.

Trigg, George L. *Landmark Experiments in Twentieth Century Physics.* (Illus.) Crane, Russak, 1975. x + 310pp. $18.50; $9.50(p). 74-21664. ISBN 0-8448-0602-1; 0-8448-0603-X. Index.

C-P   This book is suitable as collateral reading for upper college physics majors and for high school and lower-level college teachers. Each of the 16 "landmark" experiments selected by Trigg has been awarded one or more Nobel prizes. A typical chapter contains a description of the historical setting and of the method used, the experimental results and their subsequent impact on physics.

Walker, Jearl. *The Flying Circle of Physics.* (Illus.) Wiley, 1975. 224pp. $7.50(p). 75-5670. ISBN 0-471-91808-3. Index;CIP.

SH-C-P   There are 619 problems from the traditional areas of physics as well as from all areas of everyday activity. No physics teacher, library or physics laboratory should be without this book.

Whittle, R.M., and J. Yarwood. *Experimental Physics for Students.* (Illus.) Chapman&Hall (distr.: Halsted), 1974. 370pp. $16.25. SBN 412-09770-2. Index.

C-P   A physics lab manual with an innovative approach on a sophisticated level. Provides an almost complete discussion of the statistical treatment of random error. Excellent for "open-end" type physics laboratories.

530.09 PHYSICS—HISTORY AND BIOGRAPHY

Bernstein, Jeremy. *Einstein.* Viking, 1973. xii + 242pp. $6.95; $1.95(p). 72-76429. SBN 670-29077-7; 670-01959-3.

 SH-C   This marvelous book is crisply written, with an unyielding dedication to scientific spirit and preciseness and a sure grasp of the subject. It focuses on the genesis and critical reception of Einstein's major papers. Personal details add to the main narrative.

Clark, Donald W. *Einstein: The Life and Times.* (Illus.) World, 1971. xv + 718pp. $15.00. 71-149419.

 SH-C-P   Traces Einstein's life and provides a graphic demonstration of the interaction between science and politics. Clark illustrates Einstein's work with many analogies. A splendid volume for anyone interested in the history of ideas or the evolution of modern science.

Crawford, Deborah. *Lise Meitner, Atomic Pioneer.* Crown, 1969. 192pp. $3.95. 70-90997.

 JH   While Lise Meitner's life lacks the glamour of Madame Curie's—it was her life-long scientific associate, Otto Hahn, who was awarded a Nobel Prize—she fought against prejudice in becoming a physicist, suffered from exposure to both radiation and mercury vapor, met most of the leading physicists of her time as an intellectual equal, and finally fled Germany because of Hitler's purges.

Drake, Stillman, (Ed.). *Galileo Galilei: Two New Sciences: Including Centers of Gravity and Force of Percussion.* (Illus.;trans.) Univ. of Wisc. Press, 1974. xxxix + 323pp. $12.50; $4.50(p). 73-2043. ISBN 0-299-06400-X; 0-299-06404-2. Index;gloss.;CIP.

 SH-C-P   This book outlines Galileo's efforts to separate physics from metaphysical consideration and speculation and return it to basic, experimental studies. A glossary of English terms used for Galileo's mathematical expressions is included as is a history of the earlier translations of Galileo's manuscripts.

Hahn, Otto. *Otto Hahn: My Life.* (trans.) Herder and Herder, 1970. xii + 240pp. $6.50. 71-110791.

 SH-C   The discoverer of the fission of uranium writes the story of his life—his education (as a chemist), his early discovery of isotopes, his work during World War I with poison gas, and his efforts to reestablish German science after World War II. Technical terms are avoided.

Heilbron, J.L. *H.G.J. Moseley: The Life and Letters of an English Physicist, 1887-1915.* (Illus.) Univ. of Calif. Press, 1974. xiii + 312pp. $15.00. 72-93519. ISBN 0-520-02375-7. Index.

 C   A fully documented historical biography, this volume offers a discussion of Moseley's crucial experiments on the use of x-ray diffraction techniques to investigate the atomic structure of matter. All known letters to or from Moseley are included. A definitive work.

Heisenberg, Werner. *Physics and Beyond: Encounters and Conversations.* Harper &Row, 1971. xvi + 247pp. $7.95. 78-95963.

 SH   Heisenberg, formulator of the concept of indeterminancy, inventor of matrix mechanics and 5-matrix theory, has succeeded brilliantly in conveying the excitement and fervor of physics during the years 1920–1965. He reconstructs numerous discussions with Einstein, Rutherford, Planck, Bohr, Pauli, Fermi, Dirac, Schrödinger and von Weizacker, and the spirit of the times.

Hoffmann, Banesh, with Helen Dukas.   *Albert Einstein: Creator and Rebel.* (Illus.) Viking, 1972. xv + 272pp. $8.95. 70-186740. ISBN 670-11181-3.

   SH   Few books devoted exclusively to a popularized treatment of relativity convey the essential notions so clearly and accurately. Technical competence accompanies a good biography, and the important elements in Einstein's life are examined one at a time and followed through to their conclusions.

Kittel, Charles.   *Introduction to Solid State Physics, 4th ed.* (Illus.) Wiley, 1971. xv + 766pp. $14.95. 74-138912. ISBN 0-471-49021-0.

   C-P   Extensive rewriting, new illustrations and nine new topics, including flux quantization, the Kondo effect, and the applications of magnetic resonance make this a useful advanced reference. The material on superconductivity, magnetic resonance, van Hove singularities and flux quantization is noteworthy.

Lindsay, Robert Bruce.   *Men of Physics: Julius Robert Mayer, Prophet of Energy.* Pergamon, 1973. viii + 238pp. $8.25. 72-8045. ISBN 0-08-016985-6.

   C   Mayer first constructed the philosophico-empirical basis for the law of the conservation of energy. Lindsay, a professional physicist, provides his biography, a historico-critical analysis of his important contributions to science, and an English translation of his five most important papers on the general concept of energy.

Livingston, Dorothy Michelson.   *The Master of Light: A Biography of Albert A. Michelson.* (Illus.) Scribner's, 1973. xi + 376pp. $12.50. 72-1178. SBN 684-13443-8.

   C   Michelson's daughter gives a rare, intimate glimpse into the private life of one of America's greatest experimental physicists. Here we can find discussions of both his divorce and his professional struggles with colleagues and university administrators. Recommended without reservation for any college or university library.

Reid, Robert.   *Marie Curie.* (Illus.) Saturday Review Press/Dutton, 1974. 349pp. $8.95. 74-3469. ISBN 0-8415-0317-6. Index.

   SH-C   The author examines many phases of Madame Curie's life and shows her as an understandable person and a brilliant physicist. An excellently written biography.

Seeger, Raymond J.   *Men of Physics: Benjamin Franklin: New World Physicist.* (Illus.) Pergamon, 1974. xii + 190pp. $7.50. 73-7981. ISBN 0-08-017648-8.

   SH   A brief biography, with selected scientific letters and appropriate background material. Contains many revealing insights into Franklin's character, and treats his ideas and experiments on evaporation, thermal conductivity, surface tension and hydrodynamics, electrostatic winds, atmospheric electricity, lightning rods, etc. Instructive, entertaining and gives insight into physical phenomena as seen through the eyes of an intelligent and observant 18th century scientist.

Segre, Emilio.   *Enrico Fermi, Physicist.* (Illus.) Univ. of Chicago Press, 1970. x + 276pp. $6.95. 71-107424. ISBN 0-226-74472-8.

   SH-GA-C-P   Professor Segre's study of the life of Enrico Fermi (1901–1954) is a superb contribution to the history of immigrating European scholars. Fermi was one of the 20th century's greatest theoretical physicists and experimentalists, and this book should be read by all who wish to understand science in the first half of this century.

Silverstein, Alvin, and Virginia Silverstein.   *Harold Urey: The Man Who Explored from Earth to Moon.* (Illus.) John Day, 1971. 79pp. $3.49. 74-125573.

   JH   The primary theme is Urey's scientific career, but an adequate amount of personal history is included. Urey's discovery of deuterium and his subsequent work on isotope separation, the use of isotopic oxygen ratios to determine paleotemperatures, the atmospheric origin of organic compounds necessary for life, and the origin of the moon and planets are covered. Could be inspiration for a scientific career.

530.1 PHYSICS—THEORIES

Berkson, William. *Fields of Force: The Development of a World View from Faraday to Einstein.* (Illus.) Halsted/Wiley, 1974. xiii + 370pp. $19.75. 73-13458. ISBN 0-470-07029-3. Indexes;CIP.

   C Introduces various theories on force fields by describing and comparing the experimental work of Faraday, Maxwell, Hertz, Lorentz and Einstein. A college-level understanding of physics is essential to understand the argument that the true motivation for the theory of force fields is the desire to develop a world view.

Inglis, Stuart J. *Physics: An Ebb and Flow of Ideas.* (Illus.) Wiley, 1970. xv + 424pp. $9.95. 70-101973. Bibs.

   SH-C The approach is historical, from Aristotle through nuclear and particle physics. Two chapters are devoted to the failures of classical physics and special relativity, and three chapters cover the atom and its nucleus. Questions and problem sets end each chapter.

Nourse, Alan E. *Universe, Earth and Atom: The Story of Physics.* (Illus.) Harper & Row, 1969. xiii + 688pp. $10.00. 69-13493.

   SH-C A highly readable account of the ideas of physics for the layperson or beginning student. Emphasis is on the awe, surprise and grandeur of physics. Covers force, energy, gravitation, electricity, waves, light, Einsteinian relativity and quantum physics. The treatment of the micro-universe and the structure of the atom are extremely lucid, elementary accounts.

Rothman, Milton A. *Discovering the Natural Laws: The Experimental Basis of Physics.* (Illus.) Doubleday, 1972. xii + 227pp. $5.95; $1.45(p). 78-171318.

   SH-C Brilliantly yet simply presents the concepts, logic, and methodology leading to the contemporary view of the physical universe. Not a chronological account; rather it conveys the logical bases of laws through a progression of definitions, hypotheses, and, especially, experimental tests. Covers especially well the four fundamental interactions: gravitational, weak nuclear, electro-magnetic and strong nuclear attractions.

Sachs, Mendel. *Ideas of the Theory of Relativity: General Implications from Physics to Problems of Society.* Halsted/Wiley, 1974. xv + 190pp. $9.95. 74-1291. ISBN 0-470-74832-X. Index;CIP.

   C Sachs treats relativity from the physics point of view, but demands a rather sophisticated level of understanding in his attempt to clarify a rather abstruse mathematical subject. He addresses the curious humanities student rather than the scientist.

Solomon, Joan. *The Structure of Space: The Growth of Man's Ideas on the Nature of Forces, Fields and Waves.* (Illus.) Halsted/Wiley, 1974. 219pp. $10.95. 73-8543. ISBN 0-470-81221-4.

   SH-C This volume popularizes astronomy and astrophysics. Taking time and space as central ideas, it presents a nonmathematical survey. The reader is introduced to the theory of relativity, gravity waves, and the results of radio astronomy.

Struble, Mitch. *The Web of Space-Time: A Step-by-Step Exploration of Relativity.*

(Illus.) Westminster, 1973. 174pp. $5.95; $3.95(p). 72-12850. ISBN 0-664-32527-0; 0-664-34005-9.

SH-C An outline of the special and general theories of relativity, the physics that preceded those theories and some of the applications. A "broad brush" approach which will probably excite young persons already interested in physics to a further investigation.

Taylor, John G. *The New Physics.* (Illus.) Basic, 1972. xi + 224pp. $7.50. 78-174817. ISBN 0-465-050662.

SH-C The reader will find arresting ideas, challenging questions, and a growing excitement in learning how this accelerating narrative will turn out. A lucid, nonmathematical account of the tantalizing approach to unity of theory in understanding processes ranging from the nucleus to the universe. Brief explanations are given of important experiments which contribute to contemporary views.

531 MECHANICS OF SOLIDS

Barger, V., and M. Olsson. *Classical Mechanics: A Modern Perspective.* (Illus.) McGraw-Hill, 1973. xi + 305pp. $11.95. 72-5697. ISBN 0-07-003723-X.

C Presents numerous examples of mechanics in modern life. Includes such topics as the optimum design of drag racers, a grand tour of the outer planets, the flight of boomerangs, satellite stabilization and ocean tides, in addition to traditional topics. Requires little math. Numerical results are compared with experimental data to demonstrate the validity and relevance of the calculations.

Bowden, Frank P., and David Tabor. *Friction: An Introduction to Tribology.* (Illus.) Anchor/Doubleday, 1973. xi + 178pp. $5.95; $2.50(p). 72-84969. ISBN 0-385-05109-3; 0-385-05558-7.

C This exceptionally clear and logically planned book requires a knowledge of first-year physics. The authors discuss properties of materials, emphasizing the relationship between internal structures (molecular) and behavior, friction of metallic and nonmetallic materials, and lubrication and problems associated with friction. Includes electron micrographs.

Eisenstadt, Melvin M. *Introduction to Mechanical Properties of Materials.* (Illus.) Macmillan, 1971. xii + 444pp. $14.95. 77-115297.

C Describes macroscopic properties, atomic structures, the structure of solids, elastic behavior, temperature effects, instruments used for these studies, diffusion, phase transitions and strengthening mechanisms. The treatment is adequate for a first course in materials science. Assumes some background in calculus but none in thermodynamics.

Jennings, B.R., and V.J. Morris. *Atoms in Contact.* (Illus.) Oxford Univ. Press, 1974. viii + 95pp. $9.75; $4.95(p). ISBN 0-19-8518048; 0-19-8518099. Index.

C This brief book uses the basic properties of atomics and inter-atomic forces to examine some of the important characteristics of solids and liquids, including mechanical, electrical, and optical properties. A brisk writing style helps the reader gain a feeling for the subject without getting lost in details.

Kibble, T.W.B. *Classical Mechanics, 2nd ed.* (Illus.) Halsted/Wiley, 1973. xii + 254pp. $13.50. 73-8910. ISBN 0-470-47395-9.

C-P The author has directed this text toward those aspects of classical mechanics which are fundamental to an understanding of quantum mechanics and relativity. Suitable for teachers and the student with a good background in physics, calculus, and vector analysis.

Meriam, J.L.   *Dynamics, 2nd ed.* (Illus.) Wiley, 1971. xiii + 480pp. $12.95. 71-142138. ISBN 0-471-59601-9.

C   Clear and detailed illustrations, numerous sample problems, and a good selection of problems—many with answers—are well combined in this textbook. Theory is developed rigorously and concisely for a first course in engineering mechanics covering particles, rigid bodies, and nonrigid systems, each with its appropriate coverage of kinematics and kinetics.

Meriam, J.L.   *Statics, 2nd ed.* (Illus.) Wiley, 1971. xii + 378pp. $11.95. 71-136719. ISBN 0-471-59595-0.

C   Organized so that advanced material can be omitted or included. There are 680 problems of varying difficulty; each carefully detailed and remarkably realistic. About half of the book is devoted to the concepts of force and applications of equilibrium. The author acknowledges the role of vector analysis, but mainly relies on the intuitive and elegantly simple appeal of equilibrium. Outstanding!

Rosenberg, H.M.   *The Solid State: An Introduction to the Physics of Crystals for Students of Physics, Materials Science and Engineering.* Oxford Univ. Press, 1975. 235pp. $21.00; $7.25(p). ISBN 0-19-851832-3; 0-19-851833-1. Index;bib.

C · The basic elements of the physics of solids are described, with particular emphasis on the properties of waves in a periodic structure: basic structure and nomenclature, waves, defects and disorders, strength of materials under various conditions, detection of dislocations, thermal conductivities, metals, insulators, semiconductors, simple application, etc.

Shames, Irving H.   *Introduction to Solid Mechanics.* (Illus.) Prentice-Hall, 1975. xv + 490pp. $18.95. 75-4452. ISBN 0-13-497503-0. Index.

C   This textbook is designed specifically for a sophomore course on materials for engineering students. The book is well written, includes many excellent examples and problems, and emphasizes the tensor concept of stress and strain.

532 MECHANICS OF FLUIDS

National Committee For Fluid Mechanics Films.   *Illustrated Experiments in Fluid Mechanics: The NCFMF Book of Film Notes.* (Illus.) MIT Press, 1972. ix + 251pp. $3.00(p). 72-5264. ISBN 0-262-14014-4; 0-262-64012-0.

C-P   Based on the scripts from 21 lecture films which provide visual conceptions of various fluid phenomena. The book will be very valuable to the student as a study aid; it will also be of great assistance in visualizing fluid phenomena for the student who does not have access to the films. It is highly recommended, particularly for those students who do have access to the films.

Pierce, John.   *Almost All About Waves.* (Illus.) MIT Press, 1974. 213pp. $8.95. 72-11501. ISBN 0-262-16055-2. Index;CIP.

SH-C-P   Suitable for high school students or college freshmen who have had an introduction to calculus and physics. Pierce begins with elementary concepts and proceeds to develop the theory and the experimental aspects of waves in a variety of media—solid, liquid, gas and vacuum.

Vennard, John K., and Robert L. Street.   *Elementary Fluid Mechanics, 5th ed.* (Illus.) Wiley, 1975. xvii + 740pp. $14.50. 74-31232. ISBN 0-471-90587-9. Index;bibs.;CIP.

C   The emphasis here is on principles and physical concepts rather than mathematical manipulation alone, making this book valuable to a wide number of scientific disciplines. Each chapter is followed by an ample number of comprehensive problems.

534 SOUND

Chedd, Graham. *Sound: From Communication to Noise Pollution.* (Illus.) Doubleday, 1970. 187pp. $5.95. 111152.

SH-C   Gives the reader a reasonably broad and clear understanding of present knowledge and technology in the area of sound and provides more information about sound than is offered in the usual physics textbook. Major topics are speech and hearing, music, sonar, ultrasonics, noise, and noise pollution. Excellent explanation of the mechanics of human hearing.

Kock, Winston E. *Seeing Sound.* (Illus.) Wiley-Interscience, 1971. viii + 93pp. $7.95. 74-168644. ISBN 0-471-49710-X.

SH-C-P   This authoritative, compact book explains many techniques of sound analysis and their usefulness. The 88 illustrations are unusual and informative. Structural patterns of complex sounds are discussed with reference to the Potter sound spectrograph, the Dudley Vocoder, and the Bogert Vobanc. Fairly technical but nonmathematical.

Tannenbaum, Beulah, and Myra Stillman. *Understanding Sound.* (Illus.) McGraw-Hill, 1973. 176pp. $5.50. 72-9574. Index.

SH   The physics of sound is outlined, including the mechanical origins of sounds, behavior of sound waves, mechanical and biological reception of sound, the production of animal sounds (including the human voice), the mammalian ear, uses of sound waves, and ultra-1 and infrasonic waves. The text is clear, interesting and instructive and contains suggestions for demonstrations and experiments.

535 LIGHT

Adler, Irving. *The Story of Light.* (Illus.) Harvey House, 1971. 123pp. $4.50. 79-93519. ISBN 0-8178-4751-0.

JH   An understanding of light far beyond introductory optics, the entire electromagnetic spectrum is encompassed and the relation of light to atomic structures is well developed. Fundamentals of the periodic table of elements are brought in by derivation of their brightline spectra and Ritz terms. The relation of light to weather making and stored-energy sources such as wood and coal is also covered.

Froman, Robert. *Science, Art, and Visual Illusions.* (Illus.) Simon&Schuster, 1970. 127pp. $4.50. 77-86947. ISBN 0-671-65084-X.

JH-SH   Visual illusions are explained from the interrelated aspects of physiological perception, psychology, mathematics, physics and art. From simple linear and geometrical presentations the text moves on to a consideration of various modern works of art. While there are many books that deal with visual illusions, this is the first to attempt a complete multidisciplinary analysis.

Heavens, O.S. *Lasers.* (Illus.) Scribner's, 1973. 159pp. $9.95. 73-2053. SBN 684-13399-7.

SH-GA   The authors explain the basic physical phenomena required to understand laser action in qualitative terms, often drawing examples and similarities from everyday experience. An introductory work for the general reader.

Klein, H. Arthur. *Holography: With an Introduction to the Optics of Diffraction, Interference, and Phase Differences.* (Illus.) Lippincott, 1970. 192pp. $4.95. 77-117232.

SH   This technically accurate text is augmented with numerous illustrations and includes only the absolute minimum of mathematical expressions. The development of holography is discussed in historical order and in terms of the people who did the work—a good device with which to hold the attention of the young reader.

Stambler, Irwin. *Revolution in Light: Lasers and Holography.* (Illus.) Doubleday, 1972. 159pp. $4.95. 75-157428.

    SH-C Subjects covered are well described by the chapter titles: "Making Coherent Light," "3-D Miracle: Holography," "One Breakthrough After Another," "Lasers Come of Age," "New Communications Revolution," and "It's Only the Beginning." Many of the advances in lasers and holography are well treated from the historical standpoint.

Tolansky, S. *An Introduction to Interferometry, 2nd ed.* (Illus.) Halsted/Wiley, 1973. x + 253pp. $9.15(p). 72-10729.

    C-P An excellent review of the fundamentals of interferometry and its applications. The nature of interference is presented with sufficient mathematics to describe adequately the superposition of waves, but only a knowledge of basic trigonometry and elementary vectors is required for understanding. Includes diffraction, light sources, various practical interferometers and their applications, interference phenomena in crystals, interference microscopes, holography, and interference spectroscopes.

537 ELECTRICITY AND ELECTRONICS

Gregory, J.M. *Alternative Currents.* Methuen (dist.: Barnes&Noble.), 1971. $2.50(p). ISBN 0-423-84270-6.

    C-P A short introduction to the subject of alternating current electricity. Covers how alternating current is generated, waveform properties, phase and phasor diagrams, circuit analysis, meters, transformers, various types of motors, sinusoidal waveforms, and the practical employment of the concepts discussed. A familiarity with trigonometry, elementary calculus, and basic physics is assumed.

Kinariwala, Bharat, et al. *Linear Circuits and Computation.* (Illus.) Wiley, 1973. 598pp. $15.95. 72-13821. ISBN 0-471-47750-8.

    C This book stresses formal analytical techniques and numerical computational techniques. Each chapter contains a separate section that develops a major computational procedure. Recommended for use in the first one or two courses in electrical circuits.

Meyer, Herbert W. *A History of Electricity and Magnetism.* (Illus.) MIT Press, 1971. xvii + 325pp. $10.00. 70-137473. ISBN 0-262-13070-X. Index;bib.

    SH A good history of electricity and magnetism for the interested but nontechnical reader. Many inventions and discoveries are included and drawn together in a simple and understandable manner. Some fine illustrations are included from the collection of the Burndy Library.

Rojansky, Vladimir. *Electromagnetic Fields and Waves.* (Illus.) Prentice-Hall, 1971. xvi + 464pp. $11.50. 79-113606.

    C Starting with certain electric circuit notions, the author discusses current, charge, conductivity, and scalar fields. Then, after several chapters on necessary mathematical concepts, he returns to current and proceeds logically through electrostatics, magnetostatics, time-varying fields, Maxwell's equations, various wave phenomena and radiation. Easy to read and very informative.

Taylor, A.W.B. *Superconductivity.* (Illus.) Wykeham, 1970. xii + 95pp. $3.00. ISBN 0-85109-120-2.

    C-P An elementary knowledge of differential equations and electromagnetism is required to understand this phenomenological treatment of superconductivity. Taylor discusses the magnetic properties of the superconductive state, Types 1 and 2 superconductors; and the physics of superconductivity. Among the topics discussed are the Meissner effect, Londons' equation, Cooper pairs, and the Josephson junction. Applications illustrate the concepts introduced in this qualitative introduction.

539 MOLECULAR, NUCLEAR, AND ATOMIC PHYSICS

Arya, Atam P. *Elementary Modern Physics.* (Illus.) Addison-Wesley, 1974. xii + 548pp. $13.95. ISBN 0-201-00304-X. Index.

  C-P   The book contains chapters on relativity, waves and particles, introductory quantum mechanics, atomic and molecular structure and properties, the essential properties of solids, nuclear structure and reactions, and elementary particles. For students who have already had calculus and classical physics.

Ashby, Neil, and Stanley C. Miller. *Principles of Modern Physics.* (Illus.) Holden-Day, 1970. xvi + 513pp. $12.50. 71-113182.

  C   Includes probability, relativity, quantum mechanics, atomic physics, statistical mechanics, nuclear physics, and elementary particles. The computer-generated probability density plots for diffraction and solutions for the Schrödinger equation are notable.

Cohen, Bernard L. *Nuclear Science and Society.* (Illus.) Anchor/Doubleday, 1974. xii + 268pp. $2.95(p). 74-3555. ISBN 0-385-04427-5. Index;CIP.

  SH-C-P   Cohen's book continues the fine "Science Study Series" tradition of providing readable, up-to-date books by prominent scientists. This book provides a comprehensive, timely, and impeccably accurate treatment of subjects such as the physics of nuclear interaction, biological effects of radiation, and the environmental impacts of nuclear power.

Ellis, R. Hobart, Jr. *Knowing the Atomic Nucleus.* (Illus.) Lothrop, Lee&Shepard, 1973. 127pp. $4.50. 72-11985. ISBN 0-688-41295-5; 0-688-51295-X.

  JH-SH-C   A simple presentation of the structure of the atom, including its nucleus, and some effects produced by radioactive atoms. Includes biographical sketches.

Graham, Billie J., and William N. Thomas. *An Introduction to Physics for Radiologic Technologists.* (Illus.) Saunders, 1975. x + 331pp. $12.50. 74-12911. ISBN 0-7216-4200-4. Index;gloss.;CIP. *Instructors Manual.*

  C   Electricity and modern physics, x-rays and x-ray circuits, radioactivity and related topics (processing radiographs, image intensification, radiographic image in motion, ultrasonography and thermography) provide the necessary physics background for radiologic technology training programs.

Hurst, G.S., and J.E. Turner. *Elementary Radiation Physics.* (Illus.) Wiley, 1970. xi + 166pp. $7.95. 70-94921. ISBN 0-471-42472-2.

  C   All the concepts necessary for an understanding of this branch of physics are here. The book can be successfully used after a 1-year, algebra-based general physics course.

Stehle, Philip. *Physics: The Behavior of Particles.* (Illus.) Harper&Row, 1971. x + 434pp. $10.95. 78-141172. ISBN 0-06-046411-9.

  C   Starts with the geometry and laws of motion and the experiments that apply these ideas of particle motion to microscopic phenomena. This leads to consideration of the atom and subatomic particles, to systems of many particles (kinetic theory) and finally to electrons in solids. Encourages students to treat the mathematics somewhat casually, and to linger in the places where actual experiments are described.

Tayler, R.J.   *The Origin of the Chemical Elements.* (Illus.) Springer-Verlag, 1972. ix + 169pp. $6.50. 70-189454. ISBN 0-387-91100-6.

   C   This introduction to theories of the evolution of the chemical elements in the
        universe deals with concepts of nucleosynthesis of elements, their proliferation throughout the galaxies and the theoretical and experimental rationale for these concepts. Intended for physics and astronomy students, but could be useful for chemistry students.

Wilson, Jane, (Ed.).   *All in Our Time: The Reminiscences of Twelve Nuclear Pioneers.* (Illus.) Bulletin of the Atomic Scientists, 1975. 236pp. $3.45(p). 75-12223. Index.

   SH-C   This is a collection of first-person accounts dealing with the explosion of
          the first atomic bomb in 1945. Suitable for all who enjoy firsthand tales of adventure, whether scientists or not.

540 Chemistry

See also 660 Industrial Technologies.

Becker, Ralph S., and Wayne E. Wentworth.   *General Chemistry.* (Illus.) Houghton Mifflin, 1973. xiii + 779pp. $12.95. 72-5642. ISBN 0-395-160002-2.

   C   This is a well-written text designed for an introductory, full year course in
        general chemistry for science and engineering majors. There are chapters on organic compounds, thermodynamics, and chemical equilibria as applied to both inorganic and organic compounds. The appendices are valuable references.

Choppin, Gregory R., and Russell H. Johnsen.   *Introductory Chemistry.* (Illus.) Addison-Wesley, 1972. xiii + 499pp. $11.95. 77-140838.

   C   It would be difficult to find an introductory text that presents the principles of
        modern chemistry in a more lucid and comprehensive manner than this one. This book will admirably assist beginning college students who are seriously interested in pursuing the study of chemistry as a professional goal.

Day, R.A., Jr., and Ronald C. Johnson.   *General Chemistry.* (Illus.) Prentice-Hall, 1974. xiii + 609 pp. $12.95. 73-12382. ISBN 0-13-349340-7.

   C   This freshman chemistry text is based on a course at Emory University for
        chemistry majors and other preprofessional students. The mathematical level is through algebra only, and thermodynamics is developed only at the qualitative level. The book employs a spiral approach in which topics are treated first in an elementary way and then at a more advanced level later on.

Fernandez, Jack E.   *Modern Chemical Science.* (Illus.) Macmillan, 1971. xii + 288pp. $8.95. 70-122298.

   JH-SH   A careful balance is maintained among theories, facts and concepts by
           providing a partly historical treatment. All important fields are mentioned—including energy considerations and the statistical nature of chemistry and energy, chemical reactions, mechanisms, and an introduction to some important chemical processes. The mathematical considerations are very limited and the approach is largely descriptive.

Giddings, J. Calvin.   *Chemistry, Man, and Environmental Change: An Integrated Ap-*

proach. (Illus.) Canfield/Harper&Row, 1973. viii + 472pp. $10.95. 72-6264. ISBN 0-06-3827905.

C This book is for a college course in environmental chemistry for nonmajors. The first five chapters establish a background in chemistry and the last five are devoted to environmental topics, stressing their chemical origins.

Grillot, Gerald F. *A Chemical Background for the Paramedical Sciences, 2nd ed.* (Illus.) Harper&Row, 1974. x + 591pp. $10.95. 73-9288. ISBN 0-06-042511-3. Index;CIP.

C Modern theoretical approaches and the older historical styles are well balanced in this book. The author begins with elementary physical chemical principles, carries on through organic chemistry and biochemistry, and ends with physiological chemistry. Should be appealing to and popular with most students.

Gymer, Roger G. *Chemistry: An Ecological Approach.* (Illus.) Harper&Row, 1973. xxi + 801pp. $12.95. 72-87882. SBN 06-042565-2.

C This text establishes the human-environment relationship and retains a good balance of basic principles. The topics are presented in a complexity-related order, culminating in life as the most highly organized process. Includes aspects and sources of energy, their utilization, and environmental impact. Contains problems and chapter summaries.

Hankins, Warren, and Marie Hankins. *Introduction to Chemistry.* (Illus.) Mosby, 1974. xv + 470pp. $11.00. 73-8600. ISBN 0-8016-2041-1. Index;CIP;bib.

C Recommended for an introductory course to help nonscience majors see the role of science and technology in modern culture. Topics discussed in the major sections (general, physical and organic chemistry; biochemistry; nuclear chemistry and environmental problems) are excellent in selection and development.

Hill, John W. *Chemistry for Changing Times, 2nd ed.* (Illus.) Burgess, 1974. xviii + 456pp. $8.95. 74-75960. SBN 8087-0836-8. Indexes.

SH-C Half the book is devoted to the essentials of chemistry; the remainder considers various fields of application—food, air, water, drugs and others. Recommended both for self-instruction and for courses of the "Chemistry for the Citizen" type.

Kabbe, Fred, and Lois Kabbe. *Chemistry, Energy, and Human Ecology.* (Illus.) Houghton Mifflin, 1976. xii + 447pp. $11.95. 75-27126. ISBN 0-395-19833-X. Gloss.; index.

C The theme is that individuals must be informed as to the chemical basis of environmental problems so as to intelligently participate in the setting of public policy. Introduces the major conceptual divisions of chemistry, then following chapters deal with either a component of the environment or a social problem having environmental implications. Also discusses the development of social institutions.

Kieffer, William F. *Chemistry: A Cultural Approach.* (Illus.) Harper&Row, 1971. 461pp. $10.95. 71-137803. ISBN 0-06-043638-7.

SH Chemistry is thoroughly explained in very simple terms and content. The excitement of modern chemistry is brought in through the discussion of atomic chemistry, organic chemistry, molecular biology, and the environment. Should be required reading for high school seniors.

Longo, Frederick R. *General Chemistry: Interaction of Matter, Energy, and Man.* (Illus.) McGraw-Hill, 1974. xxi + 765pp. $13.95. 73-13814. ISBN 0-07-038685-4.

C Portraits and brief biographical sketches of chemists and a brief historical background of each topic enliven this unusually comprehensive general chemistry text. Theory, industrial applications, biochemistry and everday applications are lucidly and concisely discussed.

Medeiros, Robert W. *Chemistry: An Interdisciplinary Approach.* (Illus.) Van Nostrand Reinhold, 1971. xv + 555pp. $10.95. 79-147388.

   C   The book presents basic concepts and their relationship to other disciplines and in their historical context. Discusses stoichiometry, atomic structure, the periodic table, the earth's crust, extraterrestrial environments, chemistry of life processes, chemical manufacturing, and chemistry and the environment. A good solid text for a nonmajor chemistry course.

Muhler, Joseph C., et al. *Introduction to Chemistry.* (Illus.) Xerox, 1972. xiv + 533pp. $8.50(p). 72-157446. ISBN 0-536-00653-9.

   SH-C   Major emphasis is on the development and understanding of science through inductive and deductive reasoning. The text material is further reinforced by study questions at the end of each chapter. Well designed and well organized, it fills a need for the student who wishes to acquire a general background in chemistry.

Nordmann, Joseph. *What is Chemistry: A Chemical View of Nature.* (Illus.) Harper&Row, 1974. xiii + 706pp. $12.95. 73-8370. ISBN 0-06-044854-7. Index; CIP. Lab. Manual: *What Chemists Do.* (Illus.) xiv + 289pp. $5.95(p). ISBN 06-044855-5. Index.

   C   A freshman chemistry text for nonscience majors, the volume provides an overview of nature from the chemist's standpoint and a description of the efforts of scientists to understand nature's complexities. Nordmann deals with the character as well as the content of science, and concentrates equally on theory and application.

Ouellette, Robert J. *Introductory Chemistry, 2nd ed.* (Illus.) Harper&Row, 1975. xvi + 704pp. $12.95. 74-11650. ISBN 0-06-044962-4. Gloss.;index;CIP.

   C   This book is intended for a 1-year terminal course for students of nursing, agriculture, etc., and presumes a limited mathematical background. Four main sections cover the physical nature of matter, inorganic chemistry, organic molecules and physiological applications. The format, organization and sequential development of material are all well done.

Ouellette, Robert J. *Understanding Chemistry.* (Illus.) Harper&Row, 1976. xiv + 441pp. $11.95. 75-30718. ISBN 0-06-044968-3. Gloss.;index;CIP.

   C   This text for a 1-semester course in chemistry for nonmajors covers theoretical topics with almost no descriptive chemistry of the elements and their compounds. Includes learning objectives, summaries, numerous questions and problems, and solved examples. Key points are set off visually from the rest of the text.

Parry, Robert W., et al. *Chemistry: Experimental Foundations.* (Illus.) Prentice-Hall, 1970. x + 630pp. $7.12.

   SH   Development of principles from observation is stressed in this authorized revision of the CHEM Study textbook; experimentation is stressed. The revision clarifies and expands the previous coverage and adds information on transition elements, molecular biology of the gene and astrochemistry (Apollo 8).

Risen, William M., Jr. and George P. Flynn. *Problems for General and Environmental Chemistry.* (Illus.) Appleton-Century-Crofts, 1972. xii + 440pp. $5.95(p).

   C   Well written and to the point, with problems which are extensive and vary in difficulty. Complete solutions are provided for all problems in the last half of the book. The chapters on solids, organic chemistry and biochemistry should be particularly useful supplements to many general chemistry texts.

Rossotti, Hazel. *Introducing Chemistry.* (Illus. Penguin, 1975. 344pp. $3.95(p). ISBN 0-1402-1864-5. Index.

SH The book is written in a colloquial, nontechnical style with sparing use of chemical formulas and very little mathematical symbolism. The text is divided into five parts, beginning with an excellent introduction on the art and practice of the chemist. Recommended for the student with little background in science.

Sherwood, Martin. *The New Chemistry.* (Illus.) Basic, 1974. xi + 322pp. $9.50. 72-89183. SBN 465-05002-6. Index.

C-P Excellent review of chemistry and its likely future applications. Particularly recommended for advanced students, but some parts should be read by almost everyone.

Smith, Richard Furnald. *Chemistry for the Million.* (Illus.) Scribner's, 1972 x + 175pp. $7.95. 77-37219. ISBN 684-12771-7.

JH-SH-C-P A background text in chemistry which is not only of interest and comprehensible to the junior and senior high school student but also will hold the interest of the professional scientist as well. Chapters on alchemy, the chemical revolution, Lavoisier, atoms and molecules, water, and the anatomy of atoms are of particular interest.

Stoker, H. Stephen, and Spencer L. Seager. *Environmental Chemistry: Air and Water Pollution, 2nd ed.* (Illus.) Scott, Foresman, 1976. vi + 234pp. $3.95(p). 75-22104. ISBN 0-673-07978-3. Index;bib;CIP.

SH-C The authors provide examples of new data, interpretations of old, and a basis for accepting new ideas and conclusions based on research. The concept of pollution is explored along with the difficulty of explaining "unpolluted." The authors condense a large volume of information, supported by a generous list of references, that will stimulate discussion among persons from high school age upward.

White, Wilma L., Marilyn M. Erickson, and Sue C. Stevens. *Chemistry for Medical Technologists.* (Illus.) Mosby, 1970. xii + 710pp. $15.00 77-116590. ISBN 0-8016-5431-9.

C The contents are complete and excellent. There is some background provided for each subject, including normal and abnormal values, the various methods used, and the types of specific equipment available. Auto-analyzer and manual methods, chromatography and immunoelectrophoresis are included. Still an excellent reference for medical technology courses.

540.1 CHEMISTRY—THEORY

Aylesworth, Thomas G. *The Alchemists: Magic into Science.* (Illus.) Addison-Wesley, 1973. 128pp. $4.75. 72-7495. ISBN 0-201-00143-8.

JH The author presents brief biographies of notorious alchemists and describes situations involving them and their discoveries, including chemicals used in dyes, medicine, glass, water-proofing, sleeping potions and a pain-killing drug. Recommended supplementary science reading.

Dobbs, Betty Jo Teeter. *The Foundations of Newton's Alchemy or "The Hunting of the Greene Lyon."* (Illus.) Cambridge Univ. Press, 1976. xv + 300pp. $22.50. 74-31795. ISBN 0-521-20786-X. Index;CIP.

C-P Dobbs explains Newton's interest in alchemy in terms of the intellectual atmosphere of the 17th century. Gives a conceptual background and covers Newton's own experimentation. Thoroughly documented and referenced.

Faraday, Michael. *Chemical Manipulation.* (Illus.) Halsted/Wiley, 1974. viii + 656pp. $35.00. 73-22707. ISBN 0-470-25435-1. Index;CIP.

SH-C   First published in 1827, this book presents Faraday's methods and philoso-
     phy of chemical experimentation. Many experiments are timeless, although
much that he describes is only of historical interest. Every chemist or potential
chemist should read this book.

Guerlac, Henry. *Antoine-Laurent Lavoisier: Chemist and Revolutionary.* (Illus.) Scribner's, 1975. 174pp. $7.95; $2.95(p). 75-7596. SBN 0-684-14221-X; ISBN 0-684-14222-8. Index;bib.

C-P   Guerlac describes Lavoisier's intellectual contributions to the revolution in
     chemistry, the role Lavoisier played in clearing up the theory of acids and
establishing the constitution of water, and his contribution to the demise of phlogis-
ton. A valuable reference.

Hannaway, Owen. *The Chemists and the Word: The Didactic Origins of Chemistry.* (Illus.) Johns Hopkins Univ. Press, 1975. xiii + 165pp. $10.00. 74-24380. ISBN 0-8018-1666-1. Index;CIP.

C-P   This book traces the emergence of chemistry as an independent discipline at
     the turn of the 17th century. The author traces this development through the
writings of A. Libavius and O. Croll, who held contrasting views. Recommended to
serious students of the history of chemistry or science.

Jorpes, J. Erik. *Jac Berzelius, His Life and Work.* (Illus.;trans.) Univ. of Calif., 1971. 156pp. $8.00. 75-91801. ISBN 0-520-01628-9.

SH-C-P   Jac Berzelius helped transform alchemy into chemistry. He developed
     the blowpipe method of analysis, was the discoverer of cerium,
selenium, lithium, vanadium and thorium; developed the concepts of isomerism,
catalysis and proteins; and devised the present system of chemical symbols based on
the first letters of the Latin names of the elements. This is an important documentation of the early history of chemistry.

Knight, David M. *Classical Scientific Papers—Chemistry: 2nd Series.* American Elsevier, 1970. xiii + 441pp. $15.00. 74-12241. ISBN 0-444-19646-3.

SH-C-P   Reprints in facsimile a number of related, important papers in the devel-
     opment of chemistry, including those by Prout, Berzelius, Crookes,
Kopp, Ampere, De Chancourtois, Odling, Mendeleef, Berthelot, Rayleigh, Soddy
and Ramsey. The editor provides a considerable understanding of the history of
chemistry.

Vlassis, C.G. *Alchemy Revisited: Chemistry Experiments for Today.* (Illus.) Oxford Univ. Press, 1976. 122pp. $4.50(p).

C   Vlassis begins with an excellent section on safety and a description of labora-
     tory apparatus that will give the beginner an excellent foundation. Chemical
reactions, balancing equations, scientific notation, logarithms, electrolysis, organic
materials and spectroscopy and radiation (two topics not usually addressed in
elementary manuals) are covered. Only 15 experiments are included, but they are
well chosen and, unlike most beginning chemistry manuals, they graphically portray
principles of chemistry. An excellent job.

540.3 Dictionaries and Handbooks

Cahn, R.S. *An Introduction to Chemical Nomenclature, 4th ed.* (Illus.) Halsted/Wiley, 1974. 128pp. $6.95(p). 73-12479. ISBN 0-470-12931-X. Index;CIP.

C-P   Although this concise introduction to chemical nomenclature has a British
     emphasis, parallels are also drawn to American usage. Geared toward pro-

fessional chemists and graduate students, the book will also be useful from sophomore level on.

Chen, Philip S. *A New Handbook of Chemistry.* (Illus.) Chemical Elements, 1975.
212pp. $2.25. 75-2567. Index;gloss.

SH-C   The first section of this handbook for beginning students contains standard
five-place logarithms, and the second contains tables usually found in inorganic and some general chemistry texts. Latin and Greek roots for chemical terms
and a listing of chemical birthdays are also included.

Gordon, Arnold J., and Richard A. Ford. *The Chemist's Companion: A Practical
Handbook of Practical Data, Techniques, and References.* (Illus.) Wiley, 1973.
xii + 537pp. $14.95. 72-6660. ISBN 0-471-31590-7. Bib.

C-P   There are chapters on properties of molecular systems, properties of atoms
and bonds, kinetics and energetics, spectroscopy, photochemistry,
chromatography, experimental techniques, mathematical techniques, as well as
supplier lists and subject indexes. In addition to data presented in tables or figures,
there are brief reviews of some topics.

Grant, Julius. *Hackh's Chemical Dictionary, 4th ed.* (Illus.) McGraw-Hill, 1969.
xi + 738pp. $29.50. 61-18726.

SH-C   Hackh's defines familiar, unfamiliar, new, old, short, long, easy and hard
words, many with encyclopedic-type definitions. The numerous tables (historical table of elements, energy conversion factors, insecticides, common indicators
and their properties, properties of liquid fuels, fungicides, primary constants, colloi-
dal systems, amino acids, and more) and the definitions show a high degree of careful
scholarship.

Weast, Robert C., (Ed.). *CRC Handbook of Chemistry and Physics: A Ready Reference
Book of Chemical and Physical Data, 57th ed.* Chemical Rubber Co., 1976. 2344pp.
$29.95. 13-11056. ISBN 0-87819-456-8.

SH-C-P   Of primary interest to advanced students and professional scientists, this
is also a standard reference which should be owned by every senior high
school, college, public and special scientific library.

540.7 CHEMISTRY—STUDY AND TEACHING

Brescia, Frank, et al. *Chemistry: A Modern Introduction.* (Illus.) Saunders, 1974.
xvii + 644pp. $14.50. 73-80974. ISBN 0-7216-1983-5. Index. *Instructor's Guide.* 144pp.

C   Pruned to the essentials, this introduction is aimed at students with general
backgrounds. The book provides an ample foundation in chemistry with motivational guidelines for students later seeking to specialize.

Daniels, D.J., (Ed.). *New Movements in the Study and Teaching of Chemistry.* (Illus.)
Temple Smith (distr.: Transatlantic Arts), 1975. 272pp. $12.50. ISBN 0-85117-0773.

C   These 21 essays deal with the content of some of the new chemistry courses
and methods of teaching them. The current focus is on principles and their
derivation rather than facts. All but two contributors are from the United Kingdom
and include internationally known figures. Topics include behavioral objectives, the
Keller plan, chemistry for the less able and environmental chemistry.

Fine, Leonard W. *Chemistry Decoded.* (Illus.) Oxford Univ. Press, 1976.
xiii + 446pp. $11.50. 75-4213. Gloss.;index. *Instructor's Manual.* (Illus.) 131pp.

SH-C   An introduction to chemistry for the nonscience student. The historical
background of many topics is presented. Topics in physics are introduced

when they will aid in understanding chemistry, and the relationship of chemistry and biology is shown. The calculations involved are relatively simple.

Golden, Sidney. *General University Chemistry: A Developmental Approach, 2 vols.* (Illus.) Oxford Univ. Press, 1975. $6.95ea.(p). 75-4027. Indexes. *Vol. 1:* 704pp. ISBN 0-19-501881-8. *Vol. 2:* 694pp. ISBN 0-19-510882-6. *Responses to Queries, Problems, and Exercises.* 237pp. $2.50(p).

C-P   A nontraditional approach, with rigorous development of scientific princi-
        ples. Volume 1 covers thermodynamics, thermochemistry, laws of chemical change, theory of gases and chemical equations; phase equilibria, acids and bases, electrolytes, descriptive chemistry, the solid state and chemical equilibria. The second volume utilizes the concepts of the first to treat atomic and molecular structure and properties, nuclear chemistry, reaction kinetics, electrochemistry and some descriptive organic chemistry. The problems are exceptionally good, as is the narration of the intellectual history of chemistry.

Greenstone, Arthur W., et al. *Concepts in Chemistry.* (Illus.) Harcourt Brace, 1970. xi + 705pp. $5.25. Bib. *Concepts in Chemistry: Teachers Manual with Answer Key.* (Illus.) 324pp. $1.50.

SH   Very carefully written with an excellent format. Achieves a balance between
       theory and experiment, between the historical or classic and the wave mechanical or modern approaches, between basic and applied aspects of chemistry, and between description and illustration. The problems are simple but thought provoking.

Ledbetter, Elaine W., and Jay A. Young. *Keys to Chemistry.* (Illus.) Addison-Wesley, 1973. vii + 332pp. $8.64. Bib. *Laboratory Keys to Chemistry.* (Illus.) v + 139pp. $2.96(p).

SH   A text for students "who want to know what chemistry involves without
       becoming lost in theory and the memorization of facts." Each chapter starts with the performance objectives and a suggested order of study, with suggestions for creative work and creative writing. There are practice exercises, self-tests, and 13 appendixes, covering formula and equation writing, the gas laws, percentage composition, and weight/weight problems. Gives only cursory treatment of theories and quantitative ideas.

Loebel, Arnold B. *Chemical Problem-Solving by Dimensional Analysis: A Self-instructional Program.* (Illus.) Houghton Mifflin, 1974. xii + 367pp. $5.50(p). 73-8075. ISBN 0-395-16970-4.

SH-C   An easy-to-read, self-teaching text designed to aid students in their first
          experiences with chemistry.

Maas, Michael L. *Essentials of Chemistry.* (Illus.) Brown, 1971. xiv + 289pp. $6.50. 74-155159. ISBN 0-697-04905-1. Gloss. *A Laboratory Manual for Essentials of Chemistry.* (with George W. Slemmer) (Illus.) vii + 175pp. $3.95. ISBN 0-697-04906-X.

SH-C   Designed for a 1-semester course for nonscience majors. It contains the
          fundamentals of elementary chemistry and is readable. The manual provides traditional experiments with the names and illustrations of the common pieces of apparatus needed.

Merrill, Mary Alice. *Chemistry: Process and Prospect.* (Illus.) Merrill, 1973. xvii + 412pp. $9.95. 72-90998. ISBN 0-675-09023-7. *Lab. Manual.* vii + 248pp. $5.95(p). ISBN 0-675-09026-1. Gloss.;bib.

SH-C   Comprehensive and well written, both books are teaching aids for an intro-
          ductory or background course in general chemistry. The emphasis is on

preparation for reading chemistry intelligently. Explanations are precise and clear. Includes learning objectives and exercises.

O'Connor, Rod, and Charles Mickey. *Solving Problems in Chemistry: With Emphasis on Stoichiometry and Equilibrium.* (Illus.) Harper&Row, 1974. x + 402pp. $4.95(p). ISBN 06-044867-9.

SH-C   Intended to help students build mathematical skill in general chemistry problem-solving. Last section provides up-to-date problems in fields such as carbon monoxide poisoning, gas gangrene, and the economics of fertilizers. Also helpful for teachers desiring challenging and relevant examples and test questions.

Pierce, Conway, and R. Nelson Smith. *General Chemistry Workbook: How to Solve Chemistry Problems, 4th ed.* (Illus.) Freeman, 1971. viii + 369pp. $2.95. ISBN 0-7167-0157-X.

SH-C   Contains concise discussion and illustrations of the usual topics, such as dimensions, density and specific gravity, stoichiometry, the ideal gas law, chemical equilibrium in gases, electrochemistry, the ion-electron method of balancing equations, colligative properties of solutions, pH, acid-base equilibria, solubility product, complex ions, calorimetry, thermochemistry, and nuclear chemistry.

Pottenger, Francis Marion, III, and Edwin E. Bowes. *Fundamentals of Chemistry.* (Illus.) Scott, Foresman, 1976. 494pp. $13.95. 75-28449. ISBN 0-673-07876-0. Index;CIP. *Fundamentals of Chemistry in the Laboratory.* 208pp. $4.95(p). ISBN 0-673-07877-9.

C   This well-written and illustrated textbook covers measurement; matter, gases and change of state; atomic theory and the periodic table; stoichiometry; atomic structure; bonding; and solids, liquids and solutions. Chapters on equilibrium, acids and bases, and oxidation and reduction are optional. The directions for the laboratory experiments are explicit, and interpretation of data via graphing is needed for the majority of experiments.

Santiago, Paul. *An Audio-Tutorial Introduction to Chemistry.* (Illus.) Houghton Mifflin, 1975. xv + 192pp. $5.95. ISBN 0-395-17745-6. Gloss. Accompanying cassettes.

SH-C   This self-paced instructional program (cassette tapes and a workbook) consists of 22 lessons. The topics covered are equivalent to a 1-quarter course in introductory chemistry. Each lesson deals with a specific, fundamental topic. Answers to the questions and problems are appended.

Seese, William S. *In Preparation for College Chemistry.* (Illus.) Prentice-Hall, 1974. ix + 239pp. $4.95(p). 73-8519. ISBN 0-13-453662-2. Gloss.;index;CIP.

SH   Seese's conversation style readily adapts this book to individualized instruction for any person wishing to master the essentials of elementary chemistry in preparation for college-level course work. Simple, direct, and concise treatment of subject matter makes this volume a valuable source of much chemical information for any student.

Soltzberg, Leonard, et al. *BASIC and Chemistry: New Impression.* (Illus.) Houghton Mifflin, 1975. xi + 254pp. $5.95(p). 74-15587. ISBN 0-395-21720-2. Index.

C   BASIC is applied here to chemical problem-solving at the undergraduate level: stoichiometry, gases, solutions, equilibrium, radiochemistry, molecular structure determinations, atomic theory and chemical bonding, interconversion of units, linear least-square fitting and thermodynamics.

541 PHYSICAL CHEMISTRY

Andrews, Frank C. *Thermodynamics: Principles and Applications.* (Illus.) Wiley, 1971. xii + 288pp. $9.95. 77-150607. ISBN 0-471-03183-6.

 C-P   This slim volume on the fundamentals of macroscopic thermodynamics is meant to follow introductory courses in chemistry, physics, and calculus. Unlike others, this text sharply divides the development of underlying physical and mathematical principles from the specific applications of thermodynamics in chemistry, physics, biology, engineering, and earth sciences.

Bett, K.E., et al. *Thermodynamics for Chemical Engineers.* (Illus.) MIT Press, 1975. xii + 505pp. $24.95. 75-24573. ISBN 0-262-02119-6. Index.

 C-P   An unusually clear introduction to the basic principles of thermodynamics and the application of macroscopic concepts to chemical engineering, with a brief treatment of statistical thermodynamics and modern methods of predicting thermodynamic properties of gaseous systems at the molecular level. Requires a knowledge of differential and integral calculus and numerical analysis.

Campbell, J.A. *Chemical Systems: Energetics, Dynamics & Structure.* (Illus.) Freeman, 1970. xiv + 1095pp. $12.50. 75-75627. ISBN 0-7167-0145-6.

 C   A first-rate freshman chemistry text intended for the student majoring in science or engineering. The text is open-ended and expresses the spirit of science, showing its past and continuing evolution. Topics range from basic physics to chemical technology, the main emphasis being on structural and physical chemistry.

Dawson, B.E. *Kinetics and Mechanisms of Reactions.* (Illus.) Methuen (distr.: Barnes&Noble), 1973. 68pp. $3.50(p). ISBN 0-423-87510-8; 0-423-87520-5(p).

 C-P   Presents fundamental theoretical concepts of chemical kinetics and gives suggestions for open-ended kinetic experiments applying these concepts. Presents the value and use of models, reaction order, derivations of rate equations, half-lives, temperature and catalytic effects, and a cohesive theoretical interpretation of the models introduced.

Dillard, Clyde R., and David E. Goldberg. *Chemistry: Reactions, Structure, and Properties.* (Illus.) Macmillan, 1971. xvi + 654pp. $10.95. 78-121670.

 C   This college text is strongly oriented to the physical chemistry approach in first-year chemistry. The authors assume a strong preparation in high school chemistry, physics and algebra (some problems require calculus). The four parts, chemical reactions, atomic and molecular structure, properties of matter in bulk, and representative descriptive chemistry topics should provide adequate background for those planning to major in chemistry.

Howald, Reed A., and Walter A. Manch. *The Science of Chemistry: Periodic Properties and Chemical Behavior.* (Illus.) Macmillan, 1971. xiii + 689pp. $12.95. 70-121679.

 C   The authors treat periodicity, electronic structure, basic thermodynamics and bonding, then refer to these topics thereafter to emphasize that chemistry consists of related ideas. The treatment is far from traditional, with much more descriptive inorganic chemistry than is presently usual. The treatment is also rigorous and deserves to be looked at by all teachers of first-year college chemistry.

Lee, Garth L. *Principles of Chemistry: A Structural Approach.* (Illus.) International, 1970. xii + 713pp. $11.25. 70-117426. Index.

 SH-C   Introduces the elements and their inorganic compounds, atoms and molecules, the three states of matter, kinetic theory, bonding, atomic structure, rates of chemical reactions and chemical equilibrium. A very useful chemistry text.

Masterton, William L., and Emil J. Slowinski. *Chemical Principles, 3rd ed.* (Illus.) Saunders, 1973. xi + 707pp. $12.95. 72-82809. ISBN 0-7216-6172-6. *Instructor's Manual.*

C   This text presents general chemistry from the point of view of the physical chemist for both chemistry and nonchemistry majors. Contains appendixes, problems, detailed examples set off from the main text, and interesting comments in the margins. The style is smooth, lucid and informal. The lab manual covers all the principles plus applications to standardizations, group properties and identifications.

Morgan, Ralph A. *Collisions, Coalescence and Crystals: The States of Matter and Their Models.* (Illus.) Methuen Educational (distr.: Harper&Row), 1973. vi + 58pp. $2.75(p). SBN 423-84580-2; 423-86090-9.

SH-C   This brief survey, written at a slightly elevated level without being excessively complicated, employs the kinetic theory as a model for the properties of gases and compares ideal gases with real gases. Vibrational, rotational, and translational energies of gases are discussed, along with macroscopic properties of solids, the properties of a liquid, and the origin of forces in solids. A worthwhile reference since it deals with ionic, covalent, hydrogen, and metallic bonding.

Nash, Leonard K. *Chemthermo: A Statistical Approach to Classical Chemical Thermodynamics.* (Illus.) Addison-Wesley, 1972. xii + 207pp. $3.50(p). 72-183668.

C   For the reader with some knowledge of chemistry and of calculus, this brief text will be interesting and useful in a first course or as a review. The mathematical derivations are complete and descriptions of concepts are unusually simple and clear. Examples show how to solve the 70 problems included in an appendix.

Nash, Leonard K. *Elements of Statistical Thermodynamics, 2nd ed.* (Illus.) Addison-Wesley, 1974. 138pp. $4.95(p). 73-16553. ISBN 0-201-05229-6. Index.

C   Nash's text is a treatment of statistical mechanics for above average freshmen, and it should be an excellent supplementary book for use by any undergraduate chemistry student.

Pauling, Linus. *General Chemistry, 3rd ed.* (Illus.) Freeman, 1970. xiv + 959pp. $12.50. 78-75625. ISBN 0-7167-0148-0.

C   The subject matter leans rather heavily toward the physical and theoretical; theories of atomic and molecular structure and bond types are covered exceptionally well. Quantum mechanics, wave functions of the hydrogen electron, statistical mechanics, thermodynamics and chemical equilibrium are presented. There is correlation of descriptive chemistry with electronic structures of atoms, particularly with electronegativity. Logical, clear, and understandable.

Romer, Alfred, (Ed.). *Radiochemistry and the Discovery of Isotopes.* (Illus.) Dover, 1970. xii + 261pp. $3.50. 74-91273. ISBN 0-486-62507-9.

C   A fascinating collection of journal articles which form the basis of our present knowledge of radiochemistry. Considerable valuable historical comment. A treasury of the history of radioactivity, decay chains and radiochemistry, it should be in every college and university library.

White, John M. *Physical Chemistry Laboratory Experiments.* (Illus.) Prentice-Hall, 1975. xii + 563pp. $12.95. 74-11029. ISBN 0-13-665927-6. Index;bib;CIP.

C   Forty laboratory experiments cover experimental techniques, gases, thermodynamics, kinetics, spectroscopy, bulk electric and magnetic properties, electrochemistry, macromolecules, and molecular structure.

Williams, Virginia R., and Hulen B. Williams. *Basic Physical Chemistry for the Life Sciences.* (Illus.) Freeman, 1973. xviii + 524pp. $14.95. 73-7514. ISBN 0-7167-0171-5.

C   This text is suited for college juniors who intend to pursue professional careers as biochemists, biophysicists or chemical physicists. The topics are well chosen and provide a coordinated course in the fundamentals of physical chemistry. Useful for collateral reading for chemistry majors at all levels beyond the first year of college.

Wood, A. *Problems in Physical Chemistry.* Oxford Univ. Press, 1974. 167pp. $6.50(p). ISBN 0-19-855134-7.

C   This excellent book contains 312 problems, including thermodynamics (139 problems), phase equilibria (26), electrochemistry (48), kinetics (46), surface chemistry (9), crystallography (20), quantum chemistry and spectroscopy (21) and spectrophotometry (3). The book is intended to be used with a textbook in physical chemistry to provide the background necessary for solving the problems. SI units are used throughout.

542 LABORATORY TECHNIQUES AND APPARATUS

Bender, Gary T. *Chemical Instrumentation: A Laboratory Manual Based on Clinical Chemistry.* (Illus.) Saunders, 1972. xi + 291pp. $9.50(p). ISBN 0-7216-1694-1.

C-P   Students of analytical chemistry, particularly medical technology, will find very clear explanations of almost all of the instruments used in clinical chemistry labs, from basic visible absorption spectroscopy to gas and thin layer chromatographics. Includes a very simple account of the basic theory of analytical instruments. Numerous simple experiments and a "safety appendix" on the chemicals used in the experiments.

Diamond, P.S., and R.F. Denman. *Laboratory Techniques in Chemistry and Biochemistry.* (Illus.) Halsted/Wiley, 1973. 523pp. $21.50. 72-14168. ISBN 0-470-21255-1. Index;bib.

C   Contains sections on basic laboratory materials and methods, techniques of purification, electrometric methods, inorganic and organic analyses, absorptiometry, chromatography and electrophoresis, automation in the laboratory, special techniques (vacuum, polarography, radio chemistry, mass spectrometry), biochemical laboratory apparatus and methods, biochemical compounds and enzymes.

Slater, Carl D., et al. *Infrared Spectroscopy.* (Illus.) Willard Grant, 1974. viii + 56pp. $2.50(p). 74-976. ISBN 0-87150-706-2. Index;CIP.
Thompson, Clifton C. *Ultraviolet–Visible Absorption Spectroscopy.* (Illus.) Willard Grant, 1974. vi + 90pp. $2.95(p). 73-20135. ISBN 0-87150-705-5. Index;CIP.

C   These introductions to instrumentation both provide a lucid and concise explanation of the "hows" and "whys" of one instrumental method. Intended for instruction of undergraduates in chemistry, the texts each include a summary of factual information, sample problems and solutions and problems for student solution.

Steere, Norman V., (Ed.). *CRC Handbook of Laboratory Safety, 2nd ed.* Chemical Rubber Co., 1971. xv + 854pp. $26.00. 67-29478.

SH-C   This handbook, a compilation by some 40 contributors, is devoted to the prevention and control of accidents, injuries, fires, and losses wherever there are chemical hazards, particularly in laboratories. An essential addition to every chemistry laboratory, whether high school, college, clinical or industrial.

543 GENERAL ANALYSIS

Clarke, H.T., and B. Haynes. *A Handbook of Organic Analysis: Qualitative and Quantitative, 5th ed.* (Illus.) Edward Arnold (distr: Crane, Russak), 1975. x + 291pp. $35.00; $16.00(p). ISBN 0-7131-2460-1. Index.

   C   This is a complete revision of a text first published in 1911, with three new
       chapters: the preparation of derivatives, the use of spectroscopic methods, and
the quantitative determination of reactive groups. A fine reference text.

Coulson, E.H., A.E.J. Trinder, and Aaron E. Klein. *Test Tubes and Beakers: Chemistry for Young Experimenters.* (Illus.) Doubleday, 1971. 134pp. $4.95. 78-139011.

   JH-SH   An excellent instruction manual in chemistry for any interested student
       with access to basic laboratory hardware and chemicals. Chemical proce-
dures and analysis can be learned from the many open-ended experiments, which
range from the simple "limewater test" to the Tollins test. (Some require guidance.)
The book is well designed, with clear, instructive diagrams.

Harris, Walter E., and Byron Kratochvil. *Chemical Separations and Measurements: Background and Procedures for Modern Analysis.* (Illus.) Saunders, 1974. x + 284pp. $6.50(p). 73-88261. ISBN 0-7216-4535-6. Index. *Teaching Introductory Analytical Chemistry.* (Illus.) vii + 123pp.

   C   This laboratory manual for quantitative analysis emphasizes practical rather
       than theoretical aspects of the subject. Significantly more weight is given to
methods of analytical separation than is found in other texts. Clearly written, with a
useful teacher's guide to the experiments to be performed.

Kenner, C.T. *Analytical Separations and Determinations: A Textbook in Quantitative Analysis.* (Illus.) Macmillan, 1971. xx + 395pp. $9.95. 76-12389.

   C   This text does not require knowledge of physical chemistry or calculus. Gen-
       eral principles are emphasized throughout, and sections are devoted to the
reliability of measurements and to laboratory techniques. There is theoretical treatment of such topics as equilibrium, precipitation, neutralization, and oxidation-
reduction. Well organized and clearly written.

Laszlo, Pierre, and Peter J. Stang. *Organic Spectroscopy: Principles and Applications.* (Illus.) Harper&Row, 1971. xii + 275pp. $13.95. 75-148447. ISBN 0-06-043852-5.

   C-P   The judicious use of mass spectroscopy, nuclear magnetic resonance, in-
       frared, and ultraviolet spectroscpy for the determination of molecular struc-
ture, in addition to spin resonance, optical rotatory dispersion, circular dichroism,
and a discussion of the quantum mechanical basis for their application are all well
covered in this text.

Pietrzyk, Donald J., and Clyde W. Frank. *Analytical Chemistry: An Introduction.* (Illus.) Academic, 1974. xx + 667pp. $13.95. 73-18946. ISBN 0-12-555150-9. CIP.

   C   This excellent introduction to analytical chemistry will be easy for any student
       to use. Discussions of biological, clinical, and environmental applications
make appropriate reading for a 1-semester course on analytical chemistry for health-
science majors.

Slowinski, Emil J., et al. *Chemical Principles in the Laboratory: With Qualitative Analysis.* (Illus.) Saunders, 1974. x + 351pp. $6.25. ISBN 0-7261-8366-5. *Teacher's Guide.* iii + 127pp. $4.75.

   C   A comprehensive laboratory manual which also includes a set of advance
       assignments to insure student familiarity before engaging in laboratory work.
Interesting additions are a special computer program to aid teachers in grading, and a

teacher's guide which is one of the best and most useful available. Recommended for general chemistry students.

Stranks, D.R., et al. *Chemistry: A Structural View, 2nd ed.* (Illus.) Cambridge Univ. Press, 1970. x + 516pp. $12.50. 72-129936. ISBN 0-521-07994-2.

C   Stoichiometry, bonding, equilibria and descriptive chemistry are covered, but analytical chemistry is emphasized. Factual chemistry is treated through applications of the analytical principles described. Problems and questions are excellent.

546 INORGANIC CHEMISTRY

Chiswell, B., and D.W. James. *Fundamental Aspects of Inorganic Chemistry.* (Illus.) Wiley, 1969. xiii + 250pp. $6.95. 73-91783.

C   Either for an honors freshman course or for a second-year course in inorganic chemistry. Most of the fundamental concepts of bonding and stereochemistry are treated with emphasis on the rationalization and elucidation of principles which are then used to further the students' understanding of chemical relationships. No problems or questions, however.

Emeleus, H.J., and A.G. Sharpe. *Modern Aspects of Inorganic Chemistry, 4th ed.* (Illus.) Halsted/Wiley, 1973. xv + 677pp. $15.50. 73-2577. ISBN 0-470-23902-6.

C-P   The authors have presented an in-depth discussion of a number of topics, beginning with a brief review of physical methods. Also discussed are the structure and energetics of crystals and molecules and principles of aqueous and nonaqueous matter. Designed as a reference for undergraduate and graduate students.

Pass, Geoffrey, and Haydn Sutcliffe. *Practical Inorganic Chemistry: Preparations, Reactions and Instrumental Methods, 2nd ed.* (Illus.) Halsted/Wiley, 1974. xvi + 239pp. $7.95. 74-4163. ISBN 0-470-66896-2. Index;CIP.

C   This superb laboratory manual provides valuable collateral material for an inorganic chemistry course on the intermediate college level. Sections range from simple exercises to the complexities of coordination chemistry and homogeneous catalysis. Students are advised to exercise caution while performing some of the experiments.

547 ORGANIC CHEMISTRY

Adams, Roger, John R. Johnson, and Charles F. Wilcox. *Laboratory Experiments in Organic Chemistry, 5th ed.* (Illus.) Macmillan, 1970. xvi + 528pp. $8.95. 70-87890. Bib.

C   Modifications include more emphasis on spectroscopic, chromatographic and qualitative analytical techniques, advanced syntheses, and the principles on which techniques are based. Topics include separation and purification, qualitative and quantitative identification, syntheses and literature searches. Highly recommended for those whose mode of laboratory instruction is traditional.

Allinger, Norman L., et al. *Organic Chemistry, 2nd ed.* (Illus.) Worth, 1976. xxii + 1024pp. $19.95. 75-18431. ISBN 0-87901-050-9. Index. *Organic Nomenclature: A Programmed Study Guide.* viii + 130pp. $2.95(p).

C   A well-planned and designed introductory text for science majors, with a three-phase approach: the structure of organic molecules, their reactions and synthesis, natural products and special topics. Optional, more advanced material is set off throughout. The accompanying guide is useful and should save lecture time. Generally excellent.

Bates, Robert B., and John P. Schaefer. *Research Techniques in Organic Chemistry.* (Illus.) Prentice-Hall, 1971. xvii + 125pp. $7.95. 74-140411. ISBN 0-13-774489-7.

   C   Invaluable information for all undergraduate organic chemists engaged in research. The sections of the book dealing with an introduction to chemical literature, sources of supplies and equipment, and aids to writing journal articles provide excellent help. Includes many practical aspects of major reactions and isolation and structure determination techniques.

Billmeyer, Fred W., Jr. *Synthetic Polymers: Building the Giant Molecule.* (Illus.) Doubleday, 1972. ix + 182pp. $6.95; $1.95(p). 77-171279.

   SH-C   Provides an interesting and adequate understanding of this product of current technology. The author begins with a readily understandable explanation of oxygen molecules, hydrocarbon molecules and macromolecules, then proceeds to a much more sophisticated level. Recommended for chemistry students, science majors and intelligent laypersons.

Braun, Loren L. *Essentials of Organic and Biochemistry.* (Illus.) Merrill, 1972. v + 378pp. $9.95. 72-180761. ISBN 0-674-09166-7.

   SH-C   Covers present-day concepts of chemical bonding, a detailed, systematic (and somewhat pragmatic) discussion of organic compound classes; amines, amino acids/proteins; and biochemistry (enzymes, lipids, metabolism, etc.). Provides a sound introduction to organic chemistry and should also be useful in the training of students in other fields requiring some background in chemistry.

Brown, William H. *Introduction to Organic Chemistry.* (Illus.) Willard Grant, 1975. xi + 468pp. $13.00. 74-31129. ISBN 0-87150-709-0. Index;CIP.

   C   This text, for a 1-semester course in organic chemistry, assumes a background of general chemistry. The first two chapters are devoted to bonding and the dependence of physical and chemical properties on molecular structure. The remaining 13 chapters are concerned with designated classes of compounds, stereoisomerism and optical activity, and lipids and nucleic acids.

Butler, George B., and K. Darrell Berlin. *Fundamentals of Organic Chemistry: Theory and Application.* (Illus.) Ronald, 1972. x + 1113pp. 79-128351.

   C   The major aspects of organic chemistry are adequately presented. In the first chapters dealing with the syntheses and reactions of organic compounds, sufficient use is made of the concepts of chemical bonding and electronic effects, laying the groundwork for a more detailed discussion of mechanisms of organic reactions. There is enough material to challenge students at all levels of interest and motivation.

Clar, E. *The Aromatic Sextet.* (Illus.) Wiley, 1972. x + 128pp. $4.95(p). 72-616. ISBN 0-471-15840-2.

   SH-C-P   Clar, one of the great chemical scholars of our time, has devoted much of his life to the study of polycyclic hydrocarbons and aromaticity. Here he establishes the theoretical basis of bonding necessary for aromaticity and then constructs an entire system of chemistry, illustrating much of modern bonding theory and spectral interpretation. His writing style is a bit difficult; even so, the book is interesting, informative, and comprehensible even to nonexperts.

DePuy, Charles H., and Kenneth L. Rinehart, Jr. *Introduction to Organic Chemistry, 2nd ed.* (Illus.) Wiley, 1975. vii + 323pp. $12.50. 74-19442. ISBN 0-471-20350-5. Index;CIP.

   C   Early introduction of stereochemistry, extensive use of the R and S designations for the configurations of asymmetric carbon atoms and emphasis on the application of basic principles to the chemistry of familiar compounds characterize this text for a 1-semester chemistry course for nonmajors.

Gutsche, C. David, and Daniel J. Pasto. *Fundamentals of Organic Chemistry.* (Illus.) Prentice-Hall, 1975. viii + 1240pp. $18.95. 74-9714. ISBN 0-13-333443-0. Index;CIP.

   C   This broad introduction to the complexity and breadth of organic chemistry includes a rapid survey of organic structures, recent developments in biochemistry, and special topics that cover current areas of research in the field. A little too encyclopedic for use in a 1-year chemistry course, but it is still a good choice.

Jackson, Richard A. *Mechanism: An Introduction to the Study of Organic Reactions.* (Illus.) Oxford Univ. Press, 1972. xiii + 136pp. $2.95(p).

   C-P   Treats the study of the mechanism of an organic reaction as an intellectual exercise. Chapters deal with products, kinetics, intermediates, stereochemistry, and other evidence by means of which mechanisms may be deduced. There are frequent case-history examples of how mechanisms can be established, and the author shows both the pitfalls and the triumphs.

Krubsack, Arnold J. *Experimental Organic Chemistry.* (Illus.) Allyn&Bacon, 1973. xvi + 455pp. $10.95. 72-77622.

   C   This laboratory text attempts to guide the student toward independence in laboratory techniques. There is a detailed and comprehensive discussion preceding each procedure. An excellent reference work.

Monson, Richard S., and John C. Shelton. *Fundamentals of Organic Chemistry.* (Illus.) McGraw-Hill, 1974. viii + 438pp. $10.95. 73-13860. ISBN 0-07-042810-7. Index;CIP.

   C   Text for a short course in organic chemistry for students in biological or health fields. Theory and practice are carefully explained so that students with only one basic chemistry course will find this book readable and understandable.

Moore, James A. *Elementary Organic Chemistry.* (Illus.) Saunders, 1974. vi + 353pp. $11.50. 73-86382. ISBN 0-7216-6528-4.

   C   This brief text is designed for students in health-related sciences. The clear and uncluttered narrative includes a general discussion of biochemical process and basic organic chemistry.

Snyder, Carl H. *Introduction to Modern Organic Chemistry.* (Illus.) Harper&Row, 1973. xiv + 530pp. $11.95. 72-8260. ISBN 0-06-046342-2. CIP.

   C   A textbook designed for science majors and devoted to structural organic chemistry (spectrometry, stereochemistry and nomenclature, aliphatic reaction mechanisms, and descriptive chemistry of alcohols, aliphatic halides, ethers and hyrdocarbons) and to aromatics, carbonyls and carbohydrates, carboxylic acids and derivatives, amines, amino acids and proteins.

Stacy, Gardner W. *Organic Chemistry: A Background for the Life Sciences.* (Illus.) Harper&Row, 1975. xv + 412pp, $12.95. 74-26805. ISBN 0-06-046399-6. *Solutions to Problems and Study Aids.* (Illus.) 172pp. $3.95(p). SBN 60-046389-9.

   C-P   There is material dealing with organic synthesis and functional group reactions, structural and bonding fundamentals, nomenclature, stereochemistry, fats and oils, carbohydrates, nucleic acids, peptides and proteins, terpenes, vitamins, antibiotics and pesticides and methods of structural determination. Especially useful for students in the life sciences and related fields.

Sykes, Peter. *The Search for Organic Reaction Pathways.* (Illus.) Wiley, 1972. xii + 247pp. $6.25(p). 72-4192. ISBN 0-470-84130-3. Index;bib.

C-P Pathways of organic reactions are shown to be derived from simple kinetic measurements or other, more complex techniques involving the use of isotopes, reaction intermediates in multistep processes, stereochemical criteria and structure-activity correlations, singly or in combination. Well organized and readable.

Trahanovsky, Walter S. *Functional Groups in Organic Compounds.* (Illus.) Prentice-Hall, 1971. x + 149pp. $8.95; $4.95(p). 78-152087. ISBN 0-13-331967-9; 0-13-331959-8.

C Should find use as a text as well as for supplementary reading: The first third is a very well-done introduction to organic nomenclature, the middle section relates physical and spectral characteristics to molecular structures, and the final third discusses the chemistry of functional groups. It is a sound introduction which offers a good deal of material in easily digested form.

548 CRYSTALLOGRAPHY

Drummond, A.H., Jr. *Molecules in the Service of Man.* (Illus.) Lippincott, 1972. 173pp. $5.95. 72-3718. ISBN 0-397-31242-3.

SH-C Drummond provides a remarkably clear explanation of the relationship between molecular structure and observable properties of compound or crystal. He includes a brief résumé of atoms and how they combine, simple and complicated structures, clathrates, zeolites, chelates, liquid crystals, detergents, insecticides, polymers, and ideas on the future of chemical manufacturing.

Phillips, F.C. *An Introduction to Crystallography, 4th ed.* (Illus.) Wiley, 1971. ix + 351pp. $12.50. 77-127036. ISBN 0-582-44321-0.

C-P This fourth edition of a popular British text is clear, concise and well illustrated. It describes the basic symmetry and geometric patterns of crystalline materials both on the external morphological and on the atomic levels. Significant mathematical relationships are explained. Using an historical approach, the author derives the 7 crystal systems and 32 crystal classes. For each class the symmetry, the crystal forms, representative substances and pertinent diagrams are given.

Phillips, William Revel. *Mineral Optics: Principles and Techniques.* (Illus.) Freeman, 1971. ix + 249pp. $12.50. 78-134208. ISBN 0-7167-0251-7.

C-P Can be used by students in a 1-semester optical crystallography course and for individual study or review of the use of the petrographic microscope. Included are generalizations about relationships between atomic structure and optical properties, a section on refractometry, and some good illustrations of the analysis of optic orientation diagrams.

549 MINERALOGY

Arem, Joel. *Rocks and Minerals.* (Illus.) Bantam, 1973. 145pp. $1.45(p). 72-90792. Index.

JH-SH This book is a miniature course in geology, proceeding from basic information on minerals and rocks to crystals, properties and classification of minerals, gems and gem minerals and rocks and rock formations. Includes tables of minerals and of elements and excellent, almost 3-dimensional appearing pictures to help in identification of specimens.

Boegel, Hellmuth. *The Studio Handbook of Minerals: A Guide to Gem and Mineral*

Collecting. (Rev. and ed. by John Sinkankas) (Illus.) Viking, 1972. 304pp. $8.95. 74-117068. ISBN 670-68015-X.

JH-SH   This work compares favorably with other elementary treatments and consists of general mineralogy, descriptions of important and common minerals, and determinative tables. Illustrations of 154 mineral specimens comprise 48 color plates.

Boltin, Lee, and John S. White, Jr.   *Color Underground: The Mineral Picture Book.* Scribner's, 1971. 62pp. $6.95. 75-143948. ISBN 0-684-12384-3.

SH-C   Color photographs of 47 selected specimens in the mineral collections of the Smithsonian Institution reveal their beauty and characteristics. The text provides an introduction to minerals in general and explains the principles of crystallography and the seven crystal systems with the aid of diagrams. This is an enjoyable book to read and will be valuable for collateral study by any beginning student of mineralogy or crystallography.

Court, Arthur, and Ian Campbell.   *Minerals: Nature's Fabulous Jewels.* (Illus.) Abrams, 1975. 318pp. $35.00. 74-6267. ISBN 0-8109-0311-3. Index;CIP.

JH-SH-C   The book contains 183 illustrations in black-and-white and color, complete with very good captions. Some comments about minerals and geology in general are provided.

Frondel, Clifford.   *The Minerals of Franklin and Sterling Hill: A Check List.* (Illus.) Wiley, 1972. 94pp. $9.95. 72-8230. ISBN 0-471-28290-1.

C-P   Contains brief notations about more than 230 mineral species. Includes a brief history from before the Revolution through the latter half of this century of the mining operations which led to extensive studies of the geology and mineralogy of this area. The area's scientific importance resulted from its uncommon elements such as manganese, beryllium, boron, arsenic and antimony and the consequent assemblage of most unusual minerals. A basic knowledge of geology and mineralogy is required.

Hurlbut, Cornelius S., Jr.   *Dana's Manual of Mineralogy, 18th ed.* (Illus.) Wiley, 1971. ix + 579pp. $14.95. 72-114411. ISBN 0-471-42225-8.

SH-C   For over a century this book has been widely used by students, amateurs and professionals. Substantial revisions have been made to incorporate advances in the science. A classroom text and an authoritative reference work on mineralogy.

Roberts, Willard Lincoln, et al.   *Encyclopedia of Minerals.* (Illus.) Van Nostrand Reinhold, 1974. xiii + 693pp. $69.50. 74-1155. ISBN 0-442-26820-3. CIP.

C-P   Brief descriptions of about 2200 minerals provide information on classes, physical properties, localities and other features. Excellent color photographs of microscopic crystals are a notable addition to the book. Although most users would need more depth, this encyclopedia brings together a wide range of data.

Schnubel, Henri-Jean.   *Gems and Jewels: Uncut Stones and Objects d'Art.* (Illus.) Golden Press, 1972. 80pp. $5.50. 74-167718.

JH-SH   The introduction provides a very brief historical account of gemstones, and classification of the principal gems and gemstones. The 180 color photographs of objects selected from the treasures in museums and in private collections, are more magnificent than those found in some more advanced and more expensive works.

Sorrell, Charles A.   *Minerals of the World: A Field Guide and Introduction to the*

Geology and Chemistry of Minerals. (Illus.) Golden/Western, 1973. 280pp. $5.50; $3.95(p). 72-95509. Index;bib.

SH-C-P   The book is designed for the use of serious amateur mineralogists and beginning students of all ages. The author provides an introduction to mineralogy and crystallography as well as an in-depth documentation of the major minerals, with some description of their uses and availability.

Tindall, James R., and Roger Thornhill. *The Collector's Guide to Rocks and Minerals.* (Illus.) Van Nostrand, 1975. 256pp. $17.95. 74-5944. ISBN 0-442-28551-5. Gloss.;index;CIP;bib.

SH-C   In sophistication and content this outstanding book goes beyond most hobbyist's guides. After a synopsis of theories about the earth's origin and a physical and biological history of our planet, the authors describe formation, properties and tests for various minerals. Suggestions for collecting, cutting and polishing methods, and a source list of books and maps are valuable additions.

549.99 MINERALOGY OF THE MOON

Cooper, S.F., Jr. *Moon Rocks.* (Illus.) Dial Press, 1970. 197pp. $5.95. 72-111452.

SH   A lively account of the events and people at the Manned Spacecraft Center from June 1969 to January 1970. Reports on the preparations made for receiving moon rocks, how the rocks were handled, and some of the scientific results. The focus is on the statements of a score or more of the scientists at Houston during this period. Provides a fascinating picture of scientists at work in an uncharted area.

Levinson, Alfred A., and S. Ross Taylor. *Moon Rocks and Minerals: Scientific Results of the Apollo 11 Lunar Samples with Preliminary Data on Apollo 12 Samples.* (Illus.) Pergamon, 1971. xiv + 222pp. $11.50. 70-140580. ISBN 0-08-016669-5.

C   The authors cover all of the important categories of descriptive studies of lunar rocks and soil, mineralogical and chemical features of the lunar material, biological investigations, petrology and implications for the origin of lunar rocks, age determinations, various effects of cosmic ray bombardment, and seismic and magnetism experiments. An excellent introduction to the moon for the layperson.

550 Sciences of the Earth and Other Worlds

Calder, Nigel. *The Restless Earth: A Report on the New Geology.* (Illus.) Viking, 1972. 152pp. $10.00. 71-178178. ISBN 670-59530-6.

SH-C   Tells the fascinating story of the earth's evolution in a smoothly and clearly written narrative. There are maps, charts, diagrams and spectacular photographs. Excellent diagrams explain plate movement as the mechanism of continent building and the cause of earthquakes.

Cargo, David N., and Bob F. Mallory. *Man and His Geologic Environment.* (Illus.) Addison-Wesley, 1974. xx + 548pp. $12.95. 73-8236. ISBN 0-201-00892-0. Index.

C   The authors discuss population, water supply, soils, mineral resources, energy, surface geological processes, earthquakes and vulcanism, mining and construction, health and waste disposal. A well-done environmental approach to geology for those with a prior knowledge of introductory geology.

Cloud, Preston, *Adventures in Earth History.* (Illus.) Freeman, 1970. xv + 992pp. $17.50; $8.50(p). 79-94871. ISBN 0-7167-0426-0; 0-7167-0252-5.

C   A history of the earth made relevant to ecology, with 83 articles on historical geology, geochronology and climatology, supplemented with selections on the origin of the universe and of life, the rise of humans, and even family planning programs.

Creative, Editors of. *Geology of the Earth.* (Illus.) Creative Educational Society, 1971. 39pp. $4.95. 74-140639. ISBN 0-87191-072-1.

   JH-SH   The almost lyrical text touches upon soil and how it is made; springs, caves, geysers, hot springs and artesian wells; crystals of galena, quartz, and sulfur; and two gemstones, diamond and topaz. The pictures are lovely.

CRM, Inc. *Geology Today.* (Illus.) CRM, 1973. xi + 529pp. $13.95. 72-93822. Bib.

   C   People and events in the history of science are brought together by 23 authors, who together discuss the major trends in geology today. Exceptional artwork and photographs.

Donn, William L. *The Earth: Our Physical Environment.* (Illus.) Wiley, 1972. xi + 621pp. $13.95. 79-37431. ISBN 0-471-21785-9.

   C   One of the best introductory geology texts available. Theories are presented undogmatically and both the pros and the cons are considered carefully. This is an ideal book for academic advisers, and every high school science teacher should have it available for regular reference.

Durrenberger, Robert W., (Compiler). *Dictionary of the Environmental Sciences.* (Illus.) National Press, 1973. iv + 282pp. $7.95; $4.95(p). 78-142370. ISBN 0-87484-150-X.

   SH-C   This volume provides a list of terms and concise definitions of their use in environmental sciences. The illustrations aid greatly in defining many of the terms. Also includes the geologic time scale and a table of conversion factors.

Eardley, A.J. *Science of the Earth.* (Illus.) Harper&Row, 1972. xii + 468pp. $10.50. 70-168355. ISBN 06-041841-9.

   SH-C   Presents an integrated view of the physical nature of the earth and how it evolved. Discusses the solid earth, the oceans, the atmosphere, and environmental science. Theories are examined in terms of recent observations and theoretical advancements. Highly recommended as a text, especially for a first course in modern environmental science.

Gilluly, James, et al. *Principles of Geology, 4th ed.* (Illus.) Freeman, 1975. 527pp. $12.95. 74-23076. ISBN 0-7167-0269-X. Index;bibs.; CIP.

   SH-C   Outlines the major ideas of geology: a brief review of the science, the earth's size and shape, minerals, rocks, fossils, geologic maps, earthquakes, weathering, hydrology, glaciers, oceans, mineral resources and more. Appendixes cover geologic maps, rock and mineral identification, fossils and chemical data.

Laporte, Leo F. *Encounter with the Earth.* (Illus.) Canfield/Harper&Row, 1975. xiv + 538pp. $12.95. 74-18454. ISBN 0-06-384780-9. Index; glosses.; bibs.;CIP.

   C   This is an excellent introduction to geology and to human impacts on the earth's resources. Laporte attempts to establish the potentials and limitations of our resources by describing how they occur, to what degree they are renewable and what are the best estimates of their abundances. For nonmajors in the sciences.

Millard, Reed, and Science Book Associates. *Careers in the Earth Sciences.* (Illus.) Messner, 1975. 191pp. $6.95. 75-17845. ISBN 0-671-32752-6; 0-671-32753-4. Index;gloss.;CIP; bib.

   JH-SH   Explores the challenges involved in solving such problems as the search for new energy sources, geothermal power generation, mineral exploration, water resource evaluation and weather prediction and control. Outlines the careers available with and without college degrees in the earth sciences.

Press, Frank, and Raymond Siever. *Earth.* (Illus.) Freeman, 1974. xi + 945pp. $13.95. 73-21594. ISBN 0-7167-0261-4. Index;gloss.;bib;CIP.

   C   A perfect text and reference book for geology majors. Begins with the basic tools of geology, geologic history, methods of investigation and basic materials. Plate tectonics and experimental geology are used as examples of how geology works in this most complete coverage of modern views of geologic concepts.

Putnam, William C. *Geology, 2nd ed.* (Rev. by Ann Bradley Bassett.) (Illus.) Oxford, 1971. xv + 586pp. $14.95. 74-146947.

   C   Both physical and life science majors should find this text well worth reading. Each well-written and informative topic is linked to human affairs in the past and present. There are also appendixes on lunar geology, evolution, and historical geology.

Rhodes, Frank H.T. *Geology.* (Illus.) Golden, 1972. 160pp. $4.95. 72-150741. Bib.

   JH-SH-C   A traditional introductory text in content but not format. Subjects are discussed in a very condensed text and illustrated by abundant colored diagrams, figures, and photographs. A well-balanced treatment.

Strahler, Arthur N. *Principles of Earth Science.* (Illus.) Harper&Row, 1976. viii + 434pp. $14.95. 75-26638. ISBN 0-06-046451-8. Gloss.;index;CIP.

   C   An outstanding, comprehensive, current and beautifully illustrated text. Sections include astronomy, atmospheric and oceanic circulation, geology, geomorphology and hydrology, and topics range from plate tectonics to Apollo 11 samples. The approach is historical and good use is made of the diagrams and other illustrations.

Strahler, Arthur N. *Principles of Earth Science.* (Illus.) Harper&Row, 1976. viii + 434pp. $14.95. 75-26638. ISBN 0-06-046451-8. Gloss.;index;CIP.

   C   Clearly written chapters contain numerous fine photographs, maps, and some outstanding block diagrams; each chapter includes a human interest essay, a vocabulary list and review questions. Plate tectonics, while covered, is not integrated. A solid work, covering most of the traditional material.

Wyllie, Peter J. *The Dynamic Earth: Textbook in Geosciences.* (Illus.) Wiley, 1971. xiv + 416pp. $17.50. 73-155909. ISBN 0-471-96889-7. Index.

   C-P   A harvest of achievements and discoveries in global tectonics characterizes the current revolution in geosciences and are reflected in this judicious yet comprehensive report on geology, geophysics, and geochemistry. Ideal for advanced courses, and as a reference work for all geoscience students.

Young, Keith. *Geology: The Paradox of Earth and Man.* (Illus.) Houghton Mifflin, 1974. xi + 526pp. $13.50. 73-11947. ISBN 0-395-05561-X. Index;gloss.;bib.

   C   Emphasis is on humans' relation to the world, how we in our time on this earth have affected geology, and the desperate need to curtail our destructive effects on the earth. Sections on ecosystems, geological hazards and planning of population and evolution. An excellent textbook.

Zumberge, James H., and Clemens A. Nelson. *Elements of Geology, 3rd ed.* (Illus.) Wiley, 1972. xii + 431pp. $11.95. 79-180247. ISBN 0-471-09673-9.

   SH-C   Treats the earth's astronomical setting, the crust, the internal forces that produce changes in the crust and the agents of erosion that modify the earth's surface. Global tectonics and sea-floor spreading are discussed, as are all the traditional historical and evolutionary events.

550.9 GEOLOGY—HISTORICAL AND GEOGRAPHICAL TREATMENT

Baars, Donald L. *Red Rock Country: The Geological History of the Colorado Plateau.* (Illus.) Doubleday, 1972. 264pp. $9.95. 71-157573.

SH-C   Written for the non-geologist who would like to understand the significance and history of Colorado Plateau scenery. Arrangement of the topics is by decreasing age of the rock units. The story serves as a vehicle to explain fundamental geologic processes and principles of interpretation.

Briggs, Peter. *Laboratory at the Bottom of the World.* (Illus.) McKay, 1970. ix + 180pp. $5.95. 76-132161. Bib.

JH-SH-C   This excellent popular account of scientific activities in the Antarctic is the result of a National Science Foundation press tour, including flights to the South Pole station and to Byrd station, and discussions at the McMurdo Sound Headquarters. Ionospheric physics, oceanography, glaciology, biology, and especially geology are discussed. Emphasis is on the earth's history and the reinforcement of continental drift theories.

Dorr, John A., Jr. and Donald R. Eshman. *Geology of Michigan.* (Illus.) Univ. of Michigan Press, 1970. viii + 476pp. $15.00. 69-17351. ISBN 0-472-08280-0.

SH-C-P   Presents basic principles of physical and historical geology and specific aspects such as continental glaciation and hydrology. A remarkable introduction to and synthesis of history, minerals, fossils, etc., illustrated by magnificent drawings, diagrams, and photographs.

Energlyn, William D.E. *Through the Crust of the Earth.* (Illus.) McGraw-Hill, 1973. 127pp. $10.95. 73-4907. ISBN 0-07-019514-5.

JH-SH-C   Begins with the origin of man and mythical ideas about the underworld and progresses to a discussion of caves, mines and economic geology, and natural catastrophes. Evolution of the earth, plate tectonics, ancient life and fossils are discussed. The hypothetical nature of ideas about the interior of the earth and its development are stressed.

Harrington, John W. *To See A World.* (Illus.) Mosby, 1973. xii + 140pp. $4.75. 72-91625. ISBN 0-8016-2058-9.

JH-SH-C   An informal philosophy of geology without the systematic approach or the detail required of a text. Of value as a supplement to any beginning geology course or for independent study.

Hunt, Charles B. *Natural Regions of the United States and Canada.* (Illus.) Freeman, 1974. xii + 725pp. $14.95. 73-12030. ISBN 0-7167-0255-X. Indexes;CIP;bibs.

GA-C   Natural features and processes are discussed first, the distinctive features of regions due to the geologic characteristics and crustal history and the modifying processes of erosion, deposition and climate that have produced present-day landforms, soils and vegetation. A satisfactory text and a fine general reference.

Leveson, David. *A Sense of the Earth.* (Illus.) Doubleday, 1972. 176pp. $3.50(p).

SH-C   Not a text on geology, this collection of interesting essays presents some geological facts and present-day theories, enriched with striking photographs of the American West. Shows the meaning that a real understanding of the earth can have for everyone.

Marvin, Ursula B. *Continental Drift: The Evolution of a Concept.* (Illus.) Smithsonian Institution Press, 1973. 256pp. $12.50. 72-9575. ISBN 0-87474-129-7. Index;bib.

JH-SH-C   A fascinating, well-documented and comprehensive account of Alfred Wegener's 1912 hypothesis through its years of rejection to the remarkable conversion still taking place among earth scientists. Discusses geographical, geological, and geophysical speculations; alternative hypotheses; and the renaissance in earth science of the 1950s and 1960s.

Matthews, William H., III. *Invitation to Geology: The Earth Through Time and Space.* (Illus.) Natural History Press, 1971. x + 148pp. $5.95. 70-123701.

  SH-C The reader is introduced to historical and scientific concepts in such a way that he or she can readily see how present-day geological thought evolved. A good refresher for practicing geologists or educators.

Moore, Ruth. *The Earth We Live On, 2nd ed.* (Illus.) Knopf, 1971. xiv + 445pp. $8.95. 56-8924. ISBN 0-394-46968.

  SH-GA An outstanding biography of the earth which traces attempts to explain its origin and changes from ancient mythology to modern geophysics. The theories and accomplishments of Hutton, Cuvier, Agassiz, Lyell, Logan, Hall, Powell, Dutton, Urey, Goldschmidt and Mason, Kulp, Wegener, and others are presented along with biographical information.

Tabor, Rowland W. *Guide to the Geology of Olympic National Park.* (Illus.) Univ. of Wash. Press, 1975. xv + 144pp. $5.95(p). 74-32254. ISBN 0-295-95392-6; 0-295-95395-0(p). Index;gloss.;CIP.

  SH-C Tabor covers basic principles of geology and traces the geologic history of the mountains. Detailed geologic guides along 20 itineraries are supplied.

551 PHYSICAL AND DYNAMIC GEOLOGY

Bauer, Ernst. *Wonders of the Earth.* (Illus.) Watts, 1973. 128pp. $5.95. 73-3785. SBN 531-02116-5.

  JH-SH Attractively illustrated with many excellent color photographs and line drawings, this book deals with changes in the earth's appearance which have resulted from internal forces. There are chapters on vulcanism and geothermal activities. Background in geology is not required.

Cazeau, Charles, et al. *Physical Geology: Principles, Processes, and Problems.* (Illus.) Harper&Row, 1976. x + 518pp. $13.95. 75-25962. ISBN 0-06-041209-7. Gloss.;index;CIP.

  C A lucid and well-organized text for introductory geology courses or for the layperson. Includes student objectives and review questions.

Clark, Sydney P., Jr. *Structure of the Earth.* (Illus.) Prentice-Hall, 1971. xi + 131pp. $6.50; $2.75(p). 74-166213. ISBN 0-13-854653-3; 0-13-854646-0(p).

  C Although limited in scope, this text will be good supplementary reading for students taking traditional physical geology courses and who have a background in basic physics. Includes a brief introduction to the internal divisions of the earth and the distribution of volcanoes and earthquakes, geomagnetism, plate tectonics, the earth's gravity field, and the origin and movement of heat and determination of temperatures at different depths in the earth.

Corliss, William R. *Mysteries Beneath the Sea.* (Illus.) Crowell, 1970. vi + 170pp. $5.95. 71-127608. ISBN 0-690-57082-1.

  SH-C Covers particularly well the genesis of the earth and geological evolution. Arguments for and against various hypotheses are objectively given along with the applicable scientific evidence. Interesting and highly recommended as an introduction.

Davis, Alan. *Inside the Earth.* (Illus.) Grosset&Dunlap, 1973. 48pp. $1.95. 72-12134. ISBN 0-488-00729-0; 0-448-03532-4.

JH The organization of the book is basically that of an historical survey (myths
superceded by scientific investigations followed by projections for the future), with much space given to descriptions of interesting phenomena (tectonics,
earthquakes, tunneling, minerals and prospecting, fuels, ceramics, glasses, etc.).

Flint, Richard Foster, and Brian J. Skinner. *Physical Geology, rev. ed.* (Illus.) Wiley,
1974. vii + 497pp. $12.95. 73-11429. ISBN 0-471-26440-7.

C Incorporates the theory of plate tectonics with the basics of physical geology.
Includes information on the energy that drives Earth's dynamic processes, the
moon and planets, and the effects of humans. This edition contains much new material and is better arranged and easier to read.

Foster, Robert J. *Physical Geology, 2nd ed.* (Illus.) Merrill, 1975. ix + 421pp. 74-
77506. ISBN 0-675-08776-7. Index;gloss.;bibs.

SH-C-P There are discussions on environmental relationships and on emerging
new topics, such as the practical aspects of hazard mitigation or effects
of trace elements in the soil and on health. A well-written textbook for the beginning
geology student.

Gass, I.G., Peter J. Smith, and R.C.L. Wilson (Eds.). *Understanding the Earth: A
Reader in the Earth Sciences.* (Illus.) MIT Press, 1971. 355pp. $12.50; $7.50(p). 79-
151800. ISBN 0-262-07046-4; 0-262-57024-6.

C An interesting and readable collection of 25 articles on minerals and rocks,
measuring geologic time, the earth-moon system, sedimentary petrology, the
evolutionary record, the origin of life, oxygen and evolution, the primitive earth,
continental drift and plate tectonics, and recent hypotheses about geologic problems.
Recommended for both beginning and more advanced courses and for laypersons.

Hamblin, W. Kenneth. *The Earth's Dynamic Systems: A Textbook in Physical Geology.*
(Illus.) Burgess, 1975. xi + 578pp. $14.95. 74-80147. SBN 8087-0845-7. Gloss.;index.

C Spectacular ERTS color photographs make this a particularly appealing text.
Useful maps accompany what may be the best comprehensive, modern treatment of astrogeology. Separation of ideas from supportive data make each topic
particularly easy to follow.

Judson, Sheldon, et al. *Physical Geology.* (Illus.) Prentice-Hall, 1976. xiii + 560pp.
$14.50(p). 75-31853. ISBN 0-13-669655-4. Gloss.; index;CIP; bibs.

C-P This is a well-planned introductory geology text for majors and nonmajors.
Clear and up-to-date information on plate tectonics, earth crust deformation
and earthquakes; the geology of meteorites, the moon and Mars; earth ecology; and
useful earth materials and fuels.

Lauber, Patricia. *This Restless Earth.* (Illus.) Random House, 1970. 129pp. $3.50;
$1.50(p). 73-102387.

JH The fiery birth of the island Surtsey and its eventual habitation, earthquakes,
and the information provided by their resulting waves, famous volcanic eruptions, and mountain building are discussed. Convection currents, geosynclines and
theories of continental drift are introduced, and the shape of contintents, glaciation,
magnetic anomalies both on land and on the ocean floor, fossil evidence in a number
of instances, lithologic evidence, even faults traced from one continent to another,
are all discussed. Excellent!

Matthews, William H., III. *The Earth's Crust.* (Illus.) Franklin Watts, 1971. 92pp.
$3.75. 76-134367. ISBN 0-531-00724-3.

JH Gives a good sense of the layers composing the earth, the types of rocks
found in the crust, volcanoes and earthquakes, and the important metallic and
nonmetallic minerals in the crust. Excellent descriptions of seismic waves, the

theories of continental drift and the spread of oceanic crust. Should spark an interest in geology.

Matthews, William H., III. *Science Probes the Earth: New Frontiers of Geology.* (Illus.) Sterling, 1970. 176pp. $4.95. 77-90798. ISBN 0-8069-3068-8.

JH-SH Covers prediction of earthquakes and volcanoes, reasons for economic losses during times of earthquake activity, methods for reducing damage, especially in active fault zones, and uses of remote sensors to detect sea floor variations. Aerial, satellite, and infrared photography, astrogeology, theories of continental drift and exploiting mineral resources in the sea and on land are also discussed.

Page, Lou Williams. *Ideas from Geology.* (Illus.) Addison-Wesley, 1973. vi + 349pp. $3.36. ISBN 0-201-5653-4. Bib.

SH-C A brief, precise introduction to general aspects of physical geology, from rocks to structures to geologic processes, written in a fast-moving, informal style. Emphasizes the use of the present as a key to the past. Some portions are written in very simple terms and others in rather detailed technical form. Includes review questions.

Smith, Peter J. *Topics in Geophysics.* (Illus.) MIT Press, 1973. x + 246pp. $10.95. 73-382. ISBN 0-262-19115-6.Bib.

C An introduction to the earth's component parts and their interrelationships with position, heat, magnetism, stresses, tectonics, etc. This book is suitable for the reader with training in the physical sciences.

Stacey, Frank D. *Physics of the Earth.* (Illus.) Wiley, 1969. xi + 324pp. $11.95. 70-81330. ISBN 0-471-81955-7.

SH-C An astonishingly good book. Topics are covered briefly yet richly; the mathematics is simplified; and the entire range of solid-earth geophysics, beginning with cosmology and planetary considerations, and progressing through rotation, geodesy, the gravity field, seismology, earth structure, geomagnetism and paleomagnetism, is here.

Waters, John F. *The Continental Shelves.* (Illus.) Abelard-Schuman, 1975. 142pp. $5.95. 75-6697. ISBN 0-200-00157-4. Index;CIP;bib.

JH-SH-GA An interesting and well-illustrated volume on the nature and importance of the continental shelves.

551.07 GEOLOGY—STUDY AND TEACHING

Earth Science Curriculum Project. (Robert E. Boyer, series ed.) (Illus.) Houghton Mifflin, 1971. $14.00 set; $1.60 ea. Glosses.; bibs.
Field Guide to Rock Weathering by Robert E. Boyer. 38pp. ISBN 0-395-02615-6.
Field Guide to Soils by Henry Foth and Hyde S. Jacobs. 38pp. ISBN 0-395-02616-4.
Field Guide to Layered Rocks by Tom Freeman. 46pp. ISBN 0-395-02617-2.
Field Guide to Fossils by James R. Beerbower. 54pp. ISBN 0-395-02618-0.
Field Guide to Plutonic and Metamorphic Rocks by William D. Romey. 53pp. ISBN 0-395-02619-9.
Color of Minerals by George Rapp, Jr. 30pp. ISBN 0-395-02620-2.
Field Guide to Beaches by John H. Hoyt. 46pp. ISBN 0-395-02621-0.
Field Guide to Lakes by Jacob Verduin. 46pp. ISBN 0-395-02622-9.
Field Guide to Astronomy Without a Telescope by William A. Dexter. 54pp. ISBN 0-395-02623-7.
Meteorites by Carleton B. Moore. 46pp. ISBN 0-395-02624-5.

JH-SH-C A pamphlet series of field guides, for earth science students and teachers, each organized into sections covering development of the title subject, preparation for field trips, follow-up activities, glossary and a list of references. The illustrations are excellent and important. From the standpoint of topic development, practical field suggestions and reading level, these field guides should find extensive use by high school and college students as well as amateur earth scientists.

Hamblin, W.K., and J.D. Howard. *Exercises in Physical Geology, 4th ed.* (Illus.) Burgess, 1975. vi + 233pp. $6.96(p). ISBN 0-8087-0853-8.

  SH-C   This exercise book, clearly written and practically self-contained, would be a useful laboratory manual for an introductory geology course. The usual topics are covered: crystal growth, mineral identification, rock types, air photo and map interpretation, etc.

Marean, John H., Dudley G. Cate, and Edward C. Coppin. *Earth Science: Investigating the Earth, A Laboratory Approach.* (Illus.) Addison-Wesley, 1970. xi + 324pp. $6.40. *Teacher's Guide.* xi + 97pp. $2.60.

  JH-SH   The usual earth science topics are dealt with, including cartography, soil, water, the atmosphere, erosion, mountain-building processes, weather, the oceans, the solar system, time, and the history of the earth. The narrative includes specific instruction for open-ended laboratory investigations, simulations, and model-building exercises. The *Teacher's Guide* includes practical suggestions for teaching, detailed instructions for laboratory preparation, and answers to end-of-chapter questions.

Robertson, Forbes, and Frederick C. Marshall. *Historical Geology: Manual of Laboratory Exercises, 3rd ed.* (Illus.) Burgess, 1975. vi + 170pp. $6.95(p). ISBN 8087-1841-X.

  SH-C   Two attractive features are illustrations from published, professional sources and a colored strip map of the United States. Structural interpretation is emphasized; stratigraphy (including significant concepts not usually found in historical geology manuals) is stressed; and portions of maps are used in exercises to interpret regional geologic history. Paleontological concepts and exercises use fossils for analysis and interpretation.

Siegal, Barry S. *Geological Sciences Laboratory Manual, 2nd ed.* (Illus.) Burgess, 1975. vii + 195pp. $6.50(p). ISBN 0-8087-1978-5.

  C   This manual contains 17 exercises dealing with minerals and rocks, map and air photo interpretation (including geologic structures), geologic processes, energy considerations, city planning, etc.

551.1 GROSS STRUCTURE AND PROPERTIES

Anderson, Alan H., Jr. *The Drifting Continents.* (Illus.) Putnam's, 1971. 192pp. $4.29. 70-154788.

  JH   There are many new and forceful arguments in favor of drift: the paleomagnetic evidence in land rocks; magnetic striping on the sea floor; seismic analysis of transform faults; the discovery of the oceanic ridges and Pacific trenches. This book describes these discoveries in an interesting and informative fashion for young students.

Golden, Frederic. *The Moving Continents.* (Illus.) Scribner's, 1972. 124pp. $6.95. 75-162765. ISBN 0-684-12511-0. Bib.

  SH-C   Skillfully reviews the various theories of the origins of continents and goes on to demonstrate how recent discoveries tend to support the theories of continental drift.

Johnson, Helgi, and Bennett L. Smith (Eds.). *The Megatectonics of Continents and*

Oceans. (Illus.) Rutgers Univ. Press, 1970. xii + 282pp. $12.50. 69-13555. ISBN 0-8135-0625-5. Bibs.

   C-P   An important volume containing 12 outstanding, well-illustrated papers that deal with a variety of topics related to the exciting frontier of developments concerning continental drift. Skillfully edited and very readable.

Seyfert, Carl K., and Leslie A. Sirkin.   *Earth History and Plate Tectonics: An Introduction to Historical Geology.* (Illus.) Harper&Row, 1973. viii + 504pp. $12.95. 72-8262. SBN 06-045919-0. Bib.

   C   Geologic history is depicted by relating sedimentation and fossil fauna and flora to the drifting and tectonics of continental and oceanic plates throughout geologic time. The geologic history of each of the continents is summarized, with the emphasis on North America.

Sullivan, Walter.   *Continents in Motion: The New Earth Debate.* (Illus.) McGraw-Hill, 1974. xiv + 399pp. $17.50. 73-17315. ISBN 0-07-062412-7. Index; CIP.

   SH   Sullivan provides an excellent and extremely well-written historical survey and statement of the present status of plate tectonic theory.

Tarling, Don, and Maureen Tarling.   *Continental Drift: A Study of the Earth's Moving Surface, rev. ed.* (Illus.) Anchor/Doubleday, 1975. xiv + 142pp. $1.95(p). 74-12858. ISBN 0-385-06384-9. Index;CIP.

   SH-C   Covers all important phases of plate tectonics, starting with earliest concepts of continental drift. The geophysical evidence of seafloor spreading is reviewed as well as driving mechanisms and topics such as earthquake patterns and hot spots. Readable and enjoyable by the nonspecialist.

Young, Patrick.   *Drifting Continents, Shifting Seas: An Introduction to Plate Tectonics.* (Illus.) Watts, 1976. 89pp. $4.33. 75-40242. ISBN 0-531-00848-7. Index;CIP; gloss.;bib.

   JH-SH   An account of the ongoing revolution in basic earth science, detailing the evolution of the present theories. Young does an excellent job of explaining complex theories. Quite readable as a reference, but somewhat restrictive as a text.

551.2 PLUTONIC PHENOMENA

Brown, Billye Walker, and Walter R. Brown.   *Historical Catastrophes: Earthquakes.* (Illus.) Addison-Wesley, 1974. 191pp. $4.95. 73-15617. ISBN 0-201-00546-8.

   JH   Current knowledge about the causes of earthquakes is presented through a discussion of nine earthquakes, presented as eyewitness accounts. The authors show the different results of these earthquakes due to the differences in depth of focus and in types of buildings involved. A very readable book.

Lauber, Patricia.   *Earthquakes: New Scientific Ideas about How and Why the Earth Shakes.* (Illus.) Random House, 1972. 81pp. $3.95. 72-1808. ISBN 0-394-82373-1.

   JH-SH   Good coverage of a fast moving field. Young people will come away from this well-illustrated and easily read volume with a much improved appreciation of this ever-changing planet and an understanding of why earthquake-caused changes will always be with us.

MacDonald, Gordon A.   *Volcanoes.* (Illus.) Prentice-Hall, 1972. xii + 510pp. $18.00. 78-37404. ISBN 0-13-942219-6.

   C   A description of volcanoes and their activity for the beginning geology student and fascinating reading for the nonspecialist. MacDonald conveys the drama and excitement of a great volcanic eruption, gives definitions and presents the humanistic aspects. Explanations of volcanic rocks and magmas, types of lava flows, kinds of volcanic features, methods of predicting eruptions and suggestions for lessening destruction are given.

Marcus, Rebecca B. *The First Book of Volcanoes and Earthquakes, rev. ed.* (Illus.) Watts, 1972. 96pp. $3.75. 72-2301. ISBN 0-531-00661-1. Index;gloss.

    JH Discussions of the genesis of Paricutin Volcano and of the interrelationships between volcanoes, earthquakes and plate tectonic theory are presented. The text includes reports of a recent California earthquake; covers geysers and springs, touching on their potential power; and details the history of the seismometer.

Navarra, John Gabriel. *Nature Strikes Back.* (Illus.) Natural History, 1971. 224pp. $5.95. 70-140942.

    SH-C This is an interesting depiction of natural phenomena such as tornadoes, hurricanes, floods, avalanches, volcanoes, and other destructive systems. Presents an elementary physical explanation of the various phenomena and intriguing photographs.

Ollier, Cliff. *Volcanoes.* (Illus.) MIT Press, 1969. 177pp. $7.95. 79-103009. ISBN 0-262-15011-5.

    JH-SH-C A reference containing a useful description of the common volcanic rock-forming minerals, a development of the classification of volcanic eruptions and volcanoes, and a systematic analysis of volcanic features and their geomorphic relationships.

551.3 SURFACE AND EXOGENOUS PROCESSES AND THEIR AGENTS

Post, Austin, and Edward R. Lachapelle. *Glacier Ice.* (Illus.) Univ. of Wash., 1971. 110pp. $20.00. 75-152334.

    JH-SH-C-P A brief, explanatory text keyed to photographs that summarize glacier formation, mass balance, physical features, ice flow, fluctuations, and effects of glaciers on landscape. Geographical coverage is centered on the Olympics, Cascades, and Coastal Range in western North America and on the Alaskan icefields. Will supplement standard texts and provide pleasant reading and viewing.

Schultz, Gwen. *Ice Age Lost.* (Illus.) Anchor/Doubleday, 1974. xvii + 342pp. $10.00. 73-13280. ISBN 0-385-05759-8. Index.

    SH-GA-C Schultz provides well-documented references on animal and plant life and poses some of the possible causes for "ice climates," climactic change and repeated advances and retreat of the mountain and lowland glaciers and their effects on our basic resources.

Schultz, Gwen. *Icebergs and Their Voyages.* (Illus.) Morrow, 1975. 95pp. $5.95. 75-9958. ISBN 0-688-22047-9; 0-688-32047-3(p). Index;CIP.

    JH-SH A brief historical perspective leads into a discussion of the formation, movement, and vital statistics of both Northern and Antarctic icebergs. An excellent nontechnical introduction.

Tallcott, Emogene. *Glacier Tracks.* (Illus.) Lothrop, Lee&Shepard, 1970. 128pp. $4.95. 70-101479.

    JH Glacier features such as moraines, eskers, kettles, and till are described. The action of the ice in carving a cirque or horn is explained in a readable, easily understood manner. Different types of glaciers are discussed, along with the ice ages and human habitations in those times. The last chapters discuss Antarctica, history, and current research. A good introduction to glaciology.

551.4 GEOMORPHOLOGY

Adams, George F., and Jerome Wyckoff. *Landforms.* (Illus.) Golden, 1971. 160pp. $4.95. 77-141074.

 SH Information is presented on mineralogy, petrology, structural geology and surficial processes. The relationship between bedrock geology (lithology and structure) and landforms is beautifully presented and illustrated and easy to understand.

Blyth, F.G.H., and M.H. De Freitas. *A Geology for Engineers, 6th ed.* (Illus.) Edward Arnold (distr.: Crane, Russak), 1975. x + 557pp. $10.75(p). ISBN 0-7131-2440-7; 0-7131-2441-5(p). Index.

 C-P A wealth of basic, detailed information about the origin, occurrence and relationships between geological formations is contained in this completely revised edition. Recommended as a reference for structural engineers, an excellent text for graduate or undergraduate construction engineering courses and fascinating reading for a layperson.

Brunsden, Denys, and John C. Doornkamp (Eds.). *The Unquiet Landscape.* (Illus.; previously pub. in *British Geographical Magazine.*) Indiana Univ. Press, 1975. 171pp. $15.00. 74-8404. ISBN 0-253-36171-0. Index;CIP.

 JH-SH A basic introduction to geomorphology for the layperson, although few of the fascinating frontiers of geology are discussed. Cycles, climate and geology are covered, and there are brief biographies of scientists in this field.

Davies, G.L. *The Earth in Decay: A History of British Geomorphology 1578–1878.* (Illus.) Elsevier, 1970. xvi + 390pp. $16.00. 75-99798. Index;bib.

 C Describes 300 years of the history of geomorphology in Britain: the researchers, their theories, and the scientific and religious climates that influenced them. Carefully researched, with 620 references. Readable and very well organized; a knowledge of geomorphology is not necessary for understanding.

Dennis, John G. *Structural Geology.* (Illus.) Ronald, 1972. x + 532pp. $15.00. 74-181417.

 C A modern approach which incorporates the research results of the last decade. Continuous and discontinuous structures and introductory topics in geotectonics are covered. A teachable textbook also suitable for self-study.

Pitty, Alistair F. *Introduction to Geomorphology.* (Illus.) Methuen (distr: Barnes &Noble), 1971. xvi + 526pp. $16.00; $8.00(p). ISBN 0-416-11730-9; 0-416-29760-9.

 C Pitty focuses on the chemistry of rock weathering, the physics of mass transport on slopes and in streams, the hydraulics and evolution of rivers and drainage networks, the influence of climate on all of these, and the correlation of visible features of the present-day landscape with the climatic and tectonic events of the Pleistocene period. Provides an excellent, broad overview.

Price, R.J. *Glacial and Fluviological Landforms.* (Illus.) Scotland: Oliver&Boyd, 1973. viii + 242pp. £15.00. ISBN 0-05-002646-1.

 C Discusses the process of acceptance of certain theories, the bases of descriptive and generic classification systems, the limitations of evidence, and the terminology of the field. Stimulating, well documented and informative.

551.46 OCEANOGRAPHY

See also 574.92 Marine Biology.

Boyer, Robert E. *The Story of Oceanography.* (Illus.) Harvey House Pubs., 1975.

OCEANOGRAPHY   189

125pp. $5.89. 74-25425. ISBN 0-8178-5141-0; 0-8178-5142-9(p). Index;gloss.;bib.

SH This book deals with multiple aspects of the three-fourths of the earth's surface which is covered with water. Includes careers in oceanography, shorelines, marine plants and animals, continental drift, topography, waves and tides, etc.

Boyer, Robert E. *Oceanography.* (Illus.) Hubbard, 1974. 48pp. $4.95. 74-1649. ISBN 0-8331-1707-6. Index;CIP.

SH A remarkable 50-page condensation of a wealth of material on oceanography. Discusses seawater, circulation, the coastal zone, physiography, sea floor spreading, marine life, and ocean resources. An authoritative synopsis.

Briggs, Peter. *200,000,000 Years Beneath the Sea.* (Illus.) Holt, Rinehart&Winston, 1971. x + 228pp. $7.95. 78-138885. ISBN 0-03-085983-2.

SH-C A well-written narrative of the voyage of the *Glomar Challenger* that began August 11, 1968. It is the story of the ship, the personnel, and some of their scientific findings which confirmed a number of interesting theories on the oceans and continents.

Capurro, Luis R.A. *Oceanography for Practical Engineers.* (Illus.) Barnes&Noble, 1970. x + 175pp. $4.95. 71-126339.

SH-C-P The author concentrates on physical oceanography. Although written for practicing engineers, the book would be a valuable reference in college courses and parts could even be used as an introduction to oceanography in high schools.

Davis, Richard S., Jr. *Principles of Oceanography.* (Illus.) Addison-Wesley, 1972. xiv + 434pp. $10.95. 77-167992.

SH-C Begins with a brief history of oceanography, followed by a treatment of ocean basins, and of their physical, chemical, biological and geological characteristics.

Duxbury, Alyn C. *The Earth and Its Oceans.* (Illus.) Addison-Wesley, 1971. xv + 381pp. $10.95. 73-131202.

C The author presents an overview of the oceans as an environment, in a well-organized form. The main thrust is in physical science rather than environment. Basic information about the earth before the oceans; basins and borders; energy transfer; absorption of heat; and the physical and chemical properties of seawater. Some physics background helpful.

Horsfield, Brenda, and Peter Bennet Stone. *The Great Ocean Business.* (Illus.) Coward, McCann&Geoghegan, 1972. x + 360pp. $12.95. 74-172629.

SH-C Combines scientific, commercial, and historical developments in oceanography, using a dramatic approach. The varied topics include thinking on the splitting of Pangea and Panthalassa, continental drift theory, plate tectonics, the American aerospace companies' entrance into the field of oceanography, and undersea techniques for drilling for oil. Marine geology is emphasized.

King, Cuchlaine A.M. *Introduction to Physical and Biological Oceanography.* (Illus.) Arnold (distr.: Crane, Russak), 1975. 372pp. $29.00; $13.50(p). ISBN 0-7131-5735-6; 0-7131-5736-4. Index.

C-P Physical aspects of oceanography are emphasized here, with half the text on physical oceanography. Two chapters are concerned with biological oceanography, specifically productivity and exploitation of the oceans, and one chapter is on uses and problems of the ocean. Useful in any introductory course in oceanography.

Schlee, Susan. *The Edge of an Unfamiliar World: A History of Oceanography.* (Illus.) Dutton, 1973. 398pp. $10.95. 72-82705. SBN 0-525-09673-6.

SH-C-P   Covers oceanography in 19th century America, British oceanography before 1870, geological oceanography, the sea in motion, fisheries problems, American oceanography in the early 20th century, geophysical studies and the new theory of sea-floor spreading. An interesting book on the history of science and an excellent reference.

Scientific American. *Oceanography: Readings from* **Scientific American.** (Illus.) Freeman, 1971. vi + 417pp. $11.00; $5.75(p). 70-172644. ISBN 0-7167-0981-3; 0-7167-0980-5.

SH-C-P   The selections are readable, understandable, and well illustrated; they include history, geology, biology (particularly behavior), applied oceanography, resources, aquaculture and engineering. An excellent book for beginning courses in marine science.

Smith, F.G. Walton. *The Seas in Motion.* (Illus.) Crowell, 1973. vi + 248pp. $7.95. 72-83772. ISBN 0-690-72329-6.

SH-C   An overview of the movement of the seas, loaded with factual information and of interest to specialists in oceanography as well as to technologists and generalists. Provides a good appreciation of the nature and power of the seas and a particularly good account of the "tidal" wave that struck Kauai in 1946. Easy-to-read and nonmathematical.

Taber, Robert W., and Harold W. Dubach. *1001 Questions About the Oceans and Oceanography.* (Illus.) Dodd, Mead, 1972. xiv + 269pp. $7.50. 73-184136. ISBN 0-396-0649605. Bib.

JH-SH   The question-answer theme is used throughout to deal with such sections as marine geology, physical properties of sea water, sea ice, air-sea interaction, chemistry of sea water, marine biology, ecology, food from the sea, man and the sea, myths and legends, and pollution. An exciting and easily read book with well-selected photographs.

Thurman, Harold V. *Introductory Oceanography.* (Illus.) Merrill, 1975. vi + 441pp. $14.95. 74-33701. ISBN 0-675-08699-X. Index;gloss.

SH-C   The author discusses the history of oceanography, the chemical nature of the oceans, marine geology, physical oceanography and marine biology. Useful in a course for college nonscience majors, as well as in high school oceanography or marine science courses.

Voss, Gilbert L. *Oceanography.* (Illus.) Western, 1972. 160pp. $4.95. 72-84762. Index;bib.

SH   A comprehensive text, covering marine geology; physical, chemical, meteorological, biological, and fisheries oceanography; and ocean engineering and including a discussion of new techniques and industries derived from the parent-science. Many color photographs, diagrams and charts.

Weyl, Peter K. *Oceanography: An Introduction to the Marine Environment.* (Illus.) Wiley, 1970. xvii + 535pp. $12.50. 72-93300. ISBN 0-471-93774-4.

SH-C   One of the most complete and best organized introductions to the marine sciences available. Weyl introduces the system of measurements used; the history of life on earth; oceanography, cartography and mapping; and then discusses the earth as a heat engine, the sea floor, continental drift, marine flora and fauna, reefs, estuaries, seas, deep ocean circulation and changes in climate. Some basic procedures are described in detail, along with reviews of basic physical concepts and laws related to environmental topics. Technically accurate and entirely readable.

551.4607 DEEP SEA EXPLORATION

See also 627.7 Underwater Operations.

Barton, Robert. *Oceanology Today: Man Explores the Sea.* (Illus.) Doubleday, 1970. 192pp. $5.95. 78-111142.

    GA-C This well-written, knowledgeable overview of the marine sciences briefly covers subjects such as extraction of oil and gas from the ocean floor, fishing, mining, submersibles, and desalinization of the ocean water.

Heezen, Bruce C., and Charles D. Hollister. *The Face of the Deep.* (Illus.) Oxford Univ., 1971. vii + 659pp. $25.00. 77-83038. Bib.

    SH-C-P A superb account of the deep sea: animal life, geological processes that have shaped the ocean floor, currents that modify surface features, and the chemistry of the ocean. An excellent introduction to the few solved and many unsolved problems of the deep-sea floor. Outstanding illustrations and seven appendixes provide a wealth of information.

Herring, Peter, and Malcolm R. Clarke (Eds.). *Deep Oceans.* (Illus.) Praeger, 1971. 320pp. $18.50. 59-10356.

    C-P A remarkably complete historical account of oceanography, a basic review of the physics, chemistry, and biology of the sea, and some explanation of the ships and the hardware used by the oceanographer are but some of the choice components in this book for the lay reader or the serious undergraduate. Some of the photographs of marine life provide an opportunity to view forms of life hitherto known only to specialists.

Idyll, C.P., (Ed.). *Exploring the Ocean World: A History of Oceanography, rev. ed.* (Illus.) Crowell, 1972. vii + 296pp. $14.95. 72-78266. ISBN 0-690-28610-4.

    SH-GA-C The fundamentals of physical oceanography, biology, physics, chemistry, food resources, culture of marine fishes and invertebrates, ecology and pollution, marine archeology and undersea exploration are described.

McFall, Christie. *Underwater Continent: The Continental Shelves.* (Illus.) Dodd, Mead, 1975. 120pp. $5.50. 75-11851. ISBN 0-396-07175-9. Index;CIP.

    JH-SH This is a description of the continental shelves, the methods by which they are investigated and the factors that affect their character and use. The author discusses related activities such as fishing, mariculture, archeology and mining. An excellent reference.

Scott, Frances, and Walter Scott. *Exploring Ocean Frontiers.* (Illus.) Parent's Magazine Press, 1970. xii + 220pp. $4.95. 71-107231. ISBN 0-8193-0321-6; 0-8193-0322-4(p).

    C The authors examine people's use of the sea from the time of the Phoenicians to today's aquanauts and multi-nation, distant-water fishing fleets. Detailed sections on submersibles and underwater habitats. Will give some basic understanding of the international aspects of the oceans.

Soule, Gardner. *The Greatest Depths: Probing the Seas to 20,000 Feet and Below.* (Illus.) Macrae Smith, 1970. 194pp. $5.95. 70-87988. ISBN 0-8255-8350-0.

    JH-SH Exciting and interesting. Piccard and Walsh on the *Trieste* in the deepest ocean depth in the Marianas Trench, the drift of the *Ben Franklin* in the Gulf Stream, other deep-sea ventures of the past decade, Beebe and Barton in their novel bathysphere, earlier work by John and James Ross, the *Beagle,* the *Bull-Dog* and the *Challenger* are all here. Good scientific and technical accuracy; easy reading.

Wertenbaker, William. *The Floor of the Sea: Maurice Ewing and the Search to Under-*

stand the Earth. (Illus.) Little, Brown, 1974. xii + 275pp. $10.00. 74-13064. ISBN 0-316-93121-7. Index;CIP.

    SH-C-P   The author discusses exciting discoveries at sea, ranging from early explorations on the continental shelf to concepts of new global tectonics following World War II. New theories based on evidence collected under Maurice Ewing's leadership are presented.

551.48 LIMNOLOGY

Cole, Gerald A. *Textbook of Limnology.* (Illus.) Mosby, 1975. ix + 283pp. $12.00. 75-4850. ISBN 0-8016-1015-X. Index;CIP;bib.

    SH-C-P   This is a good introductory text, designed for students with a good background in physical and biological science. A useful reference.

Lind, Owen T. *Handbook of Common Methods in Limnology.* (Illus.) Mosby, 1974. vii + 154pp. $5.95(p). 74-8422. ISBN 0-8016-3017-7. CIP.

    SH-C   This pocket-sized, field-oriented handbook covers major aspects of classical limnology and includes some techniques suitable for beginning researchers.

551.5 METEOROLOGY

Berger, Melvin. *The National Weather Service.* (Illus.) Day, 1971. 124pp. $3.96. 72-135276.

    JH   In providing an account of the weather service's activities, the author explains the work of technicians and forecasters, and describes instruments and equipment used. The work of tornado, flood, and air pollution forecasters, climatologists, hurricane hunters, and spaceflight meteorologists is covered. Weather control research is touched upon.

Harris, Miles F. *Opportunities in Meteorology, rev. ed.* (Illus.) Universal, 1972. 184pp. $5.50. 77-184504.

    SH-C   A still useful compendium of information for individuals seeking an overview of the areas of possible employment as well as an insight into salaries, employment outlook, and education requirements.

Hughes, Patrick. *A Century of Weather Service: A History of the Birth and Growth of the National Weather Service.* (Illus.) Gordon&Greach, 1970. xii + 212pp. $10.00; $5.00(p). 78-107947.

    SH-C   An interesting and reasonably detailed account of the growth and scope of weather services which clearly delineates its growing pains. Deals with the initial development of a meteorological service, efforts at probing the atmosphere, the special problems arising from the military's need for weather information, and future possibilities.

Scorer, Richard. *Clouds of the World: A Complete Color Encyclopedia.* (Illus.) Stackpole, 1972. 176pp. $29.95. 72-115. ISBN 0-8117-1961-8. Index;bib.

    SH-C-P   This book transcends the role of cloud atlas and conveys the author's love for the beauty of our changeable skies. Every type of cloud and many optical phenomena are illustrated, many in full-color photographs. There are excellent discussions of the atmospheric hydrodynamics and thermodynamics associated with various classes of clouds.

Smith, Norman F. *The Atmosphere.* (Illus.) Steck-Vaughn, 1975. 42pp. $3.95. 74-23597. ISBN 0-8114-7770-3. Index;CIP.

    JH   Smith describes the various strata, their temperatures and pressures and the phenomena that are characteristic of each. A section on pollutants in the air and their effects on climate is included. Satisfactory supplementary reading at this level.

Tricker, R.A.R. *The Science of the Clouds.* (Illus.) American Elsevier, 1970. $6.95, 144pp. 74-116584. ISBN 0-444-19656-0.

JH-SH Tricker combines an excellent set of cloud pictures with 52 activities on weather to create an extremely worthwhile resource. Many of the activities are novel, but presented with explanations understandable to the layreader. The text is highly accurate yet nonmathematical.

Trowbridge, Leslie W. *Experiments in Meteorology: Investigations for the Amateur Scientist.* (Illus.) Doubleday, 1973. ix + 270pp. $6.95. 70-171324. ISBN 0-385-08238-X.

JH-SH-C Provides materials lists, procedures, and the "reasons for" 31 experiments. Leading questions guide the way to understanding the results. Experiments range from a simple compiling of weather observations, to building equipment to demonstrate the effect of the earth's rotation on circulation, to building an electroscope to measure electric potential in the atmosphere. Students may need assistance from their teacher; skills are required in mathematics, mechanics and chemistry, and interpretation which most junior high school students probably do not have.

Wachter, Heinz. *Meteorology: Forecasting the Weather.* (Illus.) Watts, 1973. 128pp. $5.95. 73-3786. SBN 531-02115-7.

JH-SH Atmospheric pressure, humidity, clouds, etc., are discussed lucidly, with scores of striking color photos. Weather forcasting is treated as data collection and map production; synoptic and numerical meteorology are ignored. Includes a good discussion of the causes of climate change and its economic consequences.

551.55–.56 ATMOSPHERIC DISTURBANCES

Jennings, Gary. *The Killer Storms: Hurricanes, Typhoons, and Tornadoes.* Lippincott, 1970. 207pp. $4.95. 78-101899.

JH Jennings describes the creation of these storms and many interesting phenomena associated with them. Includes a description of forecasting and tracking facilities.

Mason, B.J. *Clouds, Rain and Rainmaking, 2nd ed.* (Illus.) Cambridge Univ. Press, 1976. 189pp. $12.95. 74-16991. ISBN 0-521-20650-2. Index.

C-P The book covers advances since 1962 on the effects of nuclei on the formation of clouds, rain, snow and hail. There are experiments which encourage the readers to observe the detailed phenomena for themselves.

Ross, Frank, Jr. *Storms and Man.* (Illus.) Lothrop, Lee&Shepard, 1971. 192pp. $4.95. 70-148486.

SH-C This is an interesting and technically accurate account of weather disturbances. Ross describes climatology, meteorology, instrumentation, hurricanes, tornadoes, thunderstorms, winter storms, hurricane hunting with aircraft, hurricane tracking and warning, and attempts at storm control.

Tufty, Barbara. *1001 Questions Answered About Storms and Other Natural Air Disasters.* (Illus.) Dodd, Mead, 1970. xvi + 368pp. $7.50. 74-112901.

JH-SH Hurricanes, tornadoes, thunderstorms, hailstorms, and space storms are discussed. Questions concern a storm's characteristics, where, when and how it caused damage, and what research is underway. Observation techniques with radar, satellites, and lasers are discussed. A good anecdotal compendium.

Uman, Martin A. *Understanding Lightning.* (Illus.) Bek Technical, 1971. 166pp. $6.50. 70-150762.

SH   The format is question-and-answer, such as why is lightning zig-zag? What should I do if I am caught outdoors in a thunderstorm? How many thunderstorms are there in the world at one time? Excellent photographs and diagrams; a superior resource text.

551.6 CLIMATOLOGY AND WEATHER

Anderson, Bette Roda. *Weather in the West: From the Midcontinent to the Pacific.* (Illus.) American West, 1975. 223pp. $20.00. 73-90799. ISBN 0-910118-48-5. Gloss.;index;CIP.

JH-SH-C-P   How the geography of the West shapes the weather is beautifully presented. Topics range from prehistoric climate to numerical forecasting and from Indian rainmaking to modern weather modification. The discussion of weather is blended with history and folklore, and the excellent technical explanations aren't obtrusive. Recommended for scientists and laypersons alike.

Battan, Louis J. *Weather.* (Illus.) Prentice-Hall, 1974. viii + 136pp. $7.95. 73-17080. ISBN 0-13-947770-5; 0-13-947762-4(p). Index;bib;CIP.

SH   Battan begins with general features of the earth's atmosphere and proceeds through a discussion of air motion, fronts and cyclones, clouds, precipitation, the hydrologic cycle and severe storms.

Boesen, Victor. *Doing Something About the Weather.* (Illus.) Putnam's, 1975. 120pp. $6.95. 75-10440. SBN GB-399-60955-5; TR-399-20465-2. Index.

JH-SH   An overview of meteorology and climatology and of the recent advances in weather control and modification. Very easy reading.

Calder, Nigel. *The Weather Machine.* (Illus.) Viking, 1975. 143pp. $14.95. 75-1087. ISBN 0-670-75425-0. Index;CIP;bib.

SH-GA-C   A wide range of techniques, research and weather theory is discussed in this supplement to the BBC television program of the same name. The author traces the development of weather forecasting, including satellite and computer uses, and correlates human history with past climatic conditions. Current research on tree rings, sea temperatures and human effects on the long-term weather picture are also discussed.

Claiborne, Robert. *Climate, Man and History.* Norton, 1970. 444pp. $8.95. 68-20815. ISBN 0-393-06370-4.

SH-C   Claiborne relates climate to human history, beginning with the earlier ice ages. His well-documented but politically and culturally biased account covers the four major advances during the last ice age. A unique effort in relating climate to human history.

Fleagle, Robert G., et al. *Weather Modification in the Public Interest.* American Meteorological Society/Univ. of Wash. Press, 1974. ix + 88pp. $5.95. 74-590. ISBN 0-295-95321-7. Index;CIP.

C-P   The major problems of weather modification and their impacts are discussed. The decision-making process and the legal, economic, social, military and international implications of tampering with the weather are stressed. Intended mainly for policy-making officials, the book is also useful to teachers of meteorology and in secondary school earth science courses.

Gribbin, John. *Forecasts, Famines and Freezes: Climate and Man's Future.* (Illus.) Walker, 1976. 132pp. $8.95. 74-31920. ISBN 0-8027-0483-2. Index.

C The many complexities of climatic change, from the paleoclimatic range to short-term anomalies, are covered in this succinct, up-to-date and readable book, which presents complex theories in a simple, accurate way. Climate and its effect on human activities, especially future food production, is stressed.

Smith, Keith. *Principles of Applied Climatology.* (Illus.) Halsted/Wiley, 1975. 233pp. $17.50. 74-20976. ISBN 0-470-80169-7. Index;CIP.

C This is an anthology of facts and descriptive illustrations. A competently written textbook, but lacks problem-solving concepts and excitement.

Winkless, Nels, III, and Iben Browning. *Climate and the Affairs of Men.* (Illus.) Harper's Magazine Press/Harper&Row, 1975. 228pp. $8.95. 74-27305. ISBN 0-06-12950-7. CIP.

SH-C The authors explore the question, are the social and cultural affairs of humans controlled by natural phenomena? Voluminous data are presented and interesting correlations shown.

551.7 HISTORICAL GEOLOGY (STRATIGRAPHY)

Anstey, Robert L., and Terry L. Chase. *Environments Through Time: A Laboratory Manual in the Interpretation of Ancient Sediments and Organisms.* (Illus.) Burgess, 1974. vi + 136pp. $4.95(p). SBN 8087-0117-7. Bibs.

C The manual contains 17 exercises in interpreting the evidence found in sediment and sedimentary rocks so that geologic histories of local and regional areas can be worked out. Pre-supposes a basic preparation in geologic principles, including paleontology.

Dunbar, Carl O., and Karl M. Waage. *Historical Geology, 3rd ed.* (Illus.) Wiley, 1969. xi + 556pp. $9.95. 72-89681. SBN 471-22507-X.

SH-C The general principles of stratigraphy and paleontology, geologic time, earth dynamics, the pre-Paleozoic history of the earth, and the physical and biological history of the Paleozoic, Mesozoic, and Cenozoic eras are treated. Emphasis is on the evolution of the North American continent.

Eicher, Don L. *Geologic Time, 2nd ed.* (Illus.) Prentice-Hall, 1976. 150pp. $7.95; $4.25(p). 75-25824. ISBN 0-13-352492-2; 0-13-352484-1. Index;CIP.

C Eicher deals with principles and methods of stratigraphy, including a historical review, sedimentary environments and structures, time-stratigraphic record, correlation, mapping, paleogeography and biostatigraphy; and radiometric age determination. Changes are minor in this revision; plate tectonics is updated, but not its historical-geology applications. Precambrian ages are revised, and the section on age determination of the earth is expanded and updated. Organized, thorough, clearly presented and well illustrated. For a rigorous 1-semester course and a reference.

Flint, Richard Foster. *Glacial and Quaternary Geology.* (Illus.) Wiley, 1971. xii + 892pp. $24.95. 74-141198. ISBN 0-471-26435-0.

C-P An overview of late Cenozoic climate and glaciation and the development of glacial theory is followed by a discussion of modern glaciers. The text develops such topics as glaciation, stratigraphy, climatology, deep sea floors, Quaternary fossils and a consideration of causes. The diagrams, tables and illustrations are excellent. An appropriate reference for public and college libraries.

Selley, Richard C. *Ancient Sedimentary Environments.* (Illus.) Cornell Univ. Press, 1970. xi + 237pp. $6.50. 72-134437. ISBN 0-8014-0606-4.

C-P Descriptions and historical interpretations of specific sequences of strata, wide-ranging in space and time and selected to illustrate environments in which deposition of sediment has occurred since the Late Precambrian. Each case history is followed by an account of economic products commonly associated with a particular type of deposit. Compressed, lucid, informal and witty.

Silverberg, Robert. *Clocks for the Ages: How Scientists Date the Past.* (Illus.) Macmillan, 1971. 238pp. $5.95. 74-123134.

    JH-SH A well-written book which describes the historical development of ideas about the antiquity of the earth. The development of absolute dating methods is traced in detail, from genealogical analysis of the Bible through rates of sedimentation, increasing salinity of the oceans, cooling of the earth, organic evolution, varve and tree-ring chronology, radio-active disintegration, obsidian hydration, and fission tracks.

552 PETROLOGY

Garrels, Robert M., and Fred T. Mackenzie. *Evolution of Sedimentary Rocks.* (Illus.) Norton, 1971. xvi + 397pp. $11.50. 76-129500. ISBN 0-393-09959-8.

    C An overview of the history of the earth as seen through the eyes of a chemist. The motions of the crust and mantle, plate tectonics, continental drift, geosynclines, and the origin of mountain ranges are described by explaining where sediments come from. Very readable, with extensive charts, tables and diagrams.

Pettijohn, F.J. *Sedimentary Rocks, 3rd ed.* (Illus.) Harper&Row, 1975. xii + 628pp. $19.95. 74-12043. ISBN 0-06-045191-2. Indexes;CIP.

    C-P Emphasis is on describing and classifying the various families of sedimentary rocks and understanding their textures and structures. Will serve as a handy reference.

553 ECONOMIC GEOLOGY

See also 333.7 –.9 Conservation of Resources.

Adler, Irving. *Petroleum: Gas, Oil and Asphalt.* (Illus.) John Day, 1975. 48pp. $5.95. 75-2431. ISBN 0-381-99624-7. Index;CIP.

    JH Adler tells the story of oil and gas concisely, including definitions, formation, where to find oil and gas and how to extract it. Touches on the chemistry of petroleum compounds, with some structural diagrams included.

Cameron, Eugene N., (Ed.). *The Mineral Position of the United States, 1975 –2000.* (Illus.) Univ. of Wisc. Press, 1973. xvii + 159pp. $1.50(p). 72-7983. ISBN 0-299-06300-3; 0-299-06304-6(p).

    C-P A great deal of data concerning our country's mineral consumption habits is presented. The book is geared toward solutions to potential mineral problems based on maintaining as nearly as possible our present consumption patterns. It tends to slight the environmental implications of future policies.

Darden, Lloyd. *The Earth in the Looking Glass.* (Illus.) Anchor/Doubleday, 1974. ix + 324pp. $7.95. 73-9151. ISBN 0-385-02595-5. Index; CIP.

    JH-SH-C-P A superbly written and fascinating account of NASA's Earth Resources Technology Satellite (ERTS). A lively narrative and copious photographs relate specific applications of ERTS to problems of energy and mineral resources, the resource potential of the oceans, and the changing environment of wilderness areas.

Davies, Delwyn. *Fresh Water: The Precious Resource.* Natural History Press (distr.: Doubleday), 1969. 155pp. $5.95. 69-17356.

JH   After an introduction to the interesting properties of water, the volume is organized in independent units, including water uses and distribution, the roles of water in relation to human and to other life, and the supply and reservoirs of water. Attention is given to wetlands, irrigation, agriculture, pollution and the future.

Deming, H.G.   *Water: The Fountain of Opportunity.* (Illus.) Oxford Univ. Press, 1975. xix + 342pp. $12.50. 74-16657. ISBN 0-19-501841-9. Index.

SH-C   The author examines soils, rivers, climate, life in simple cells and complex organisms, water as a carrier of energy, pollution and desalination. He presents a wealth of data on water, its distribution and characteristics, in a non-mathematical form.

Landes, Kenneth K.   *Petroleum Geology of the United States.* (Illus.) Wiley, 1970. xi + 571pp. $24.95. 77-101975. ISBN 0-471-51335-0. Index;bib.

C-P   Excellent descriptions of pertinent types of petroleum accumulations for the U.S. and good maps and cross-sections explain the geology. Not a text, but a descriptive volume on oil and gas occurrences. Covers physiography and surface geology along with geologic history, structure and stratigraphy; discusses possible future discovery areas and secondary recovery potential of known fields.

McDivitt, James F., and Gerald Manner.   *Minerals and Men: An Exploration of the World of Minerals and Metals, Including Some of the Major Problems that Are Posed, rev. enlarged ed.* (Illus.) Resources for the Future/Johns Hopkins Univ. Press, 1974. xiii + 175pp. $7.50. 73-8138. ISBN 0-8018-1536-3. Index;CIP.

SH-C   The authors provide broad coverage of the essential interplay of geology, economics, geography and politics in the exploitation of mineral raw materials. Cycles of surplus and shortage in the mineral industries are chronicled.

Menard, H.W.   *Geology, Resources, and Society: An Introduction to Earth Science.* (Illus.) Freeman, 1974. xi + 621pp. $12.95. 73-17151. ISBN 0-7167-0260-6. Gloss.;index;CIP.

SH-C   Designed for students not majoring in earth science, this work provides a useful text for courses in environmental geology. Emphasis is given to the interactions between humankind and earth materials or processes.

Park, Charles F., Jr., with Margaret C. Freeman.   *Earthbound: Minerals, Energy, and Man's Future.* (Illus.) Freeman, Cooper, 1975. 279pp. $3.95(p). 73-87688. ISBN 0-87735-317-4; 0-87735-318-2(p). Index.

SH-C-P   The authors attempt to answer the questions, do we have too many people in the world, too few mineral resources, too little energy, and too little unpolluted space to maintain our present civilization? The conflicts between producers and environmentalists are presented, and the authors emphasize the need for educated and intelligent action on a worldwide basis to solve the problems.

Skinner, Brian J.   *Earth Resources.* (Illus.) Prentice-Hall, 1969. ix + 149pp. $5.95; $2.50(p). 76-89702. ISBN 0-13-222661-8; 0-13-222653-7.

SH-C   Nonrenewable mineral resources of the earth are discussed (abundant metals, scarce metals, chemical and fertilizer minerals, building materials, energy sources and water). Geologic features and processes of origin, geographic distribution, reserves, production, consumption and uses are treated.

Young, Bob, and Jan Young.   *The Search for Oil in America.* (Illus.) Messner, 1971. 190pp. $4.50. 73-139085. ISBN 0-671-32373-3. Index;gloss; bib.

JH-SH   Starting with a sketch of Edwin Drake and the drilling of the first oil well in Pennsylvania, the authors tell about the early oil discoveries and briefly describe some of the personalities, calamities, and political implications connected with the growing petroleum industry. Well written and interesting.

553.8 GEMS

Arem, Joel. *Gems and Jewelry.* (Illus.) Grosset&Dunlap, 1976. 159pp. $2.95. 75-14682. ISBN 0-448-12145-X; 0-448-13328-9. Index.

JH-SH-C   Discusses values of gemstones, their characteristics, rules for cutting, historical comments about famous gemstones and ancient uses and superstitions, minerals which are considered "collectors' items," man-made synthetics and imitations, and how to buy gemstones.

Desautels, Paul E. *The Gem Kingdom.* (Illus.) Random House, 1971. 252pp. $17.95. 70-158812.

SH-C-P   A fine coffee-table book. The color photographs of rough and cut precious and semi-precious stones and of their collection and preparation are magnificent. There is also scientific and technological content relating to geology, mineralogy, crystallography, etc.

Gubelin, Eduard. *The Color Treasury of Gemstones.* (Illus.; trans.) Crowell, 1975 (c. 1969). 138pp. $9.95. 75-15715. ISBN 0-690-00986-0. Gloss.;CIP.

MacFall, Russell P. *Minerals and Gems: A Color Treasury for Collectors and Guide to Hunting Locations.* (Illus.) xi + 242pp. $17.50. 74-28082. ISBN 0-690-00687-X. Index;CIP.

SH-C-P   The first book provides a comprehensive and easily accessible survey of the gemstone realm. *Minerals and Gems* is a how-to book for the serious collector. In combination, the two provide an in-depth reference set on the world of gems, appealing and useful to the novice and the professional.

MacFall, Russell P. *Gem Hunter's Guide: How to Find and Identify Gem Minerals, 5th rev. ed.* (Illus.) Crowell, 1975. 323pp. $7.95. 74-19469. ISBN 0-690-00656-X. Gloss.;bib.;CIP.

JH-SH-C   Besides finding and indentification, MacFall also deals with the characteristics of important gem materials, the use of black light and various testing methods and methods for judging quality and value.

Rutland, E.H. *An Introduction to the World's Gemstones.* (Illus.) Doubleday, 1974. 192pp. $14.95. 73-19294. ISBN 0-385-05191-3. Gloss.;index.

JH-SH-C   A magnificent work that includes excellent colored illustrations. Chapters cover origin, structure, collecting, lore and history of gemstones. A useful data table completes this outstanding volume.

559.91 GEOLOGY OF THE MOON

See also 549.99 Mineralogy of the Moon.

Mutch, Thomas A. *Geology of the Moon: A Stratigraphic View, rev. ed.* (Illus.) Princeton Univ. Press, 1972. ix + 391pp. $22.50. 70-38387. ISBN 0-691-08110-7.

SH-C-P   This interesting and readable book presents a sensible balance between old and new scientific information. The clear explanations, beautiful photographs, historical reviews and the discussions of analogous geological studies on the earth make this an excellent textbook. (Discusses only the Apollo moon landing results.)

560 Paleontology and Paleozoology

Cartner, William C. *How We Know What on Earth Happened Before Man Arrived.* (Illus.) Sterling, 1972. 96pp. $3.95. 76-180466. ISBN 0-8069-3046-2; 0-8069-3047-0.

JH   A good introduction to fossil study. Covers the geologic time table; paleozoic, mesozoic and cenozoic rock systems; plant and animal fossils; oil formation; and fossil evidence for the evolution of humans and other primates. Discusses field work, field sketching, specimen preparation, photography, museum study and fossil clubs. Includes a state-by-state listing of areas of interest.

Colbert, Edwin H.   *Wandering Lands and Animals.* (Illus.) Dutton, 1973. xxi + 323pp. $12.50. 76-158602. ISBN 0-525-22976-0.

SH-GA-C   Integrates patterns of vertebrate migration and distribution through geologic time with the continental drift postulated in modern plate tectonic theory. Written for a general audience, the book contains accounts of personal discovery of paleontologic evidence for the existence of Gondwanaland, Laurasia, and Pangaea, and the evolution and geography of the amphibians, reptiles and mammals from Paleozoic to modern time.

Desmond, Adrian J.   *The Hot-Blooded Dinosaurs: A Revolution in Palaeontology.* (Illus.) Dial/Wade, 1976. 238pp. $12.95. 76-190. ISBN 0-8037-3755-6. Gloss.; index; CIP.

SH-C   Outlines past ideas and presents the case for the new concept of the endothermic dinosaur. Presents the evidence in an understandable, nonmathematical way. For scholars and for laypersons with some knowledge of biology.

Flint, Richard Foster.   *The Earth and Its History: An Introduction to Physical and Historical Geology.* (Illus.) Norton, 1973. xii + 407pp. $9.95. 72-10879. ISBN 0-393-09377-8.

C   A well-written and readable supplementary text for liberal arts students. Historical geology is briefly covered, concepts of evolution and the nature of fossils and the glacial ages and man are discussed in more detail. There is no discussion of the prehistoric evolution of culture.

Glut, Donald F.   *The Dinosaur Dictionary.* (Illus.) Citadel, 1972. 218pp. $12.50. 70-147829.

SH-C   Presents a listing, with data, of virtually every known genus of dinosaurs. The author outlines some of the problems in the study of dinosaurs and presents a breakdown of the two orders into suborders, infraorders and families. Drawings and photographs contribute to making this an excellent book.

Howard, Robert West.   *The Dawnseekers: The First History of American Paleontology.* (Illus.) Harcourt Brace Jovanovich, 1975. xiii + 314pp. $8.95. 74-30407. ISBN 0-15-123973. Index;CIP.

SH-C   Biographical sketches of major natural historians and paleontologists who lived before and during the 19th century and who produced the evidence of the great antiquity of the earth are presented. The development of vertebrate paleontology, paleontologists, natural history museums, and the art of exhibiting fossil vertebrates are discussed. A factual and interesting report.

Jaworowska, Zofia Lielan.   *Hunting for Dinosaurs.* (Illus.; trans.) MIT Press, 1969. 177pp. $7.95. 73-87288. ISBN 0-262-11039-X; 0-262-61007-8(p).

SH-C   This book is the story of three expeditions to the Gobi Desert sponsored by the Warsaw Academy of Sciences. The author explains how her party prospected for, discovered, excavated and transported numerous dinosaurian bones and eggs discovered by the paleontologists.

Kurten, Björn.   *The Age of Mammals.* (Illus.) Columbia Univ. Press, 1974 (c. 1971). 250pp. $4.95(p). 79-177479. ISBN 0-231-03647-7. Indexes.

SH-C   An extraordinary book that provides animal history as readable evolutionary adventure relevant to the history of humans. The author describes the
appearance and evolutionary changes in mammalian orders during the seven major
periods in the Cenozoic. In addition, he deals with Australia, South America, the
Pleistocene Ice Age and the appearance of humans.

Lanham, Url. *The Bone Hunters.* (Illus.) Columbia Univ. Press, 1973. xii + 285pp.
$12.95. 73-5596. ISBN 0-231-03152-1. Index;bibs.

C   Lanham describes the history and development of vertebrate paleontology in
the United States during the 19th century, with emphasis on the last third of the
century. The major part of the work deals with the Cope-Marsh controversy. Recommended to all scientists and to nonscientists interested in paleontology and the
West.

Ley, Willy. *Worlds of the Past.* (Illus.) Golden, 1971. 143pp. $4.95. 72-142461. ISBN
0-307-16826-3.

JH-SH-GA   Describes the development of life as interpreted by fossils in 12 segments of geologic time and includes an excellent introduction that
defines fossils. The writing is anecdotal and lucid. The illustrations are either origi-
nals or are selected from leading museums and publications.

Raup, David M., and Steven M. Stanley. *Principles of Paleontology.* Freeman, 1971.
x + 388pp. $11.50. 79-120302. ISBN 0-7167-0247-9.

C-P   A stimulating introductory textbook on paleontology. Skillfully written in a
uniformly lucid style and well illustrated, the book covers fossil preservation
and identification, grouping of species, ontogenetic variation, biostratigraphy,
geochemistry, and the use of the computer in paleontology.

Silverberg, Robert. *Mammoths, Mastodons and Man.* (Illus.) McGraw-Hill, 1970.
223pp. $5.50. 73-107296.

JH-SH   Attempts to reconstruct the history of mammoths and mastodons are
traced. This story takes us to the Siberian tundra, to Europe and to the
New World. We also learn something about the human contemporaries of the extinct
elephants. The story is fascinating and is told in an easy, understandable fashion.

Spinar, Zdenek V. *Life Before Man.* (Illus.) American Heritage, 1972. 228pp. $8.95.
72-1866. ISBN 07-060240-9; 07-060241-7. Gloss.

SH-C   An introduction to the unfolding of life on earth from its earliest appearance
to the beginnings of recorded history. Most of the book consists of color
plates depicting the artist's and author's concepts of a myriad of extinct plants and
animals, along with short explanations. Brief classification outline and taxonomy.
Suitable for browsing.

Thenius, Erich. *Fossils and the Life of the Past.* (Illus.; trans.) Springer-Verlag, 1973.
x + 194pp. $5.90(p). 76-183484. ISBN 0-387-90039-X. Index;bib.

C-P   Indicates the importance, the working methods, and the aims of paleontological research. The approach is topical, covering processes of fossilization,
fossils in folklore, techniques and methods of research, the use of fossils in correlation of strata, fossils as evidence for organic evolution, trace fossils, paleoecology,
paleogeography, and "living fossils." A geological time scale is appended. The book
is outstanding in its field. A background in earth science is helpful.

Tidwell, William D. *Common Fossil Plants of Western North America.* (Illus.) Brigham
Young Univ. Press, 1975. 197pp. $9.95; $6.95(p). 75-4640. ISBN 0-8425-1298-5;
0-8425-1301-9. Gloss.;index;CIP;bib.

C   A unique guide for identification and study, useful to both an amateur collector
and to a paleobotanist. Contains an historical overview of the plant kingdom;

descriptions of fossils and their preservation, methods of study, collecting and processing; and petrified fossils. Includes an outline key of ferns and fern-like foliage. The excellent drawings and photographs make this perhaps the most beautiful and helpful book available on these often neglected plants.

Zappler, Lisbeth, and Georg Zappler. *The World After the Dinosaurs: The Evolution of Mammals.* (Illus.) Natural History Press, 1970. 183pp. $4.95. 72-116266.

JH-GA   An abundance of interesting material on mammalogy and vertebrate paleontology which can be understood and enjoyed by both teenagers and inquisitive adults. Not a textbook format; the material flows freely.

570 Life Sciences

Grobstein, Clifford. *The Strategy of Life, 2nd ed.* (Illus.) Freeman, 1974. 174pp. $4.95; $2.95(p). 73-18061. ISBN 0-7167-0591-5; 0-7167-0590-7. Index;CIP.

SH-C   Grobstein's thesis is that all of life, from the simplest to the highest order of living being, has a "strategy" or efficient plan of function. He covers phenomena from the nuclear level to that of the universe in 11 chapters.

Jacker, Corinne. *The Biological Revolution: A Background Book on Making a New World.* (Illus.) Parents', 1971. 226pp. $4.95. 71-158625. ISBN 0-8193-0524-3.

JH-SH-C   Admirably presents several complex topics, including molecular biology, human embryology, organ transplantation, chemical and biological weapons, neuroscience, and the origins of life. The current and potential roles these discoveries will play in changing the nature of human life and society are described. The final chapter, "Controlling the Biological Revolution," is a thoughtful essay on the ethical emptiness of science.

Science Foundation for Physics. *Molecules to Man.* (Illus.) Heinemann, 1972 (distr.: Crane, Russak, 1973). 410pp. $8.75. ISBN 0-435-60235-7.

C-P   A collection of essays which describes the continuity which exists between molecules, primitive organisms and man, beginning with the structures, activities and evolution of biologically important molecules; the biochemistry of immunity; mammalian adaptation; population resources and environment (by P.R. Ehrlich); and the interaction of science with society. Useful to teachers of high school and college biology; requires a good understanding of chemistry and biology.

Thurber, Walter A., et al. *Exploring Life Science.* (Illus.) Allyn&Bacon, 1975. viii + 438pp. $10.00. 74-12748. Index;bib.

JH   The 29 chapters develop concepts of cells, the human body, behavior and the environment. Each theme has three parts: exposition of the known facts and statement of limits of knowledge; research projects which the student may carry out; and extension and utilization of knowledge based on investigative work.

Tortora, Gerard J., and Joseph F. Becker. *Life Science.* (Illus.) Macmillan, 1972. xi + 770pp. $11.95. 74-153765. Index;gloss.;bib.

C   This is a good textbook for an introductory course in the life sciences, well designed, with many excellent drawings and photographs. The review questions before each chapter guide students to the major points made in that chapter.

Wong, Harry K., et al. *Ideas and Investigations in Science: Life Science.* (Illus.) Prentice-Hall, 1973. 410pp. $7.96. ISBN 0-13-449744-9.

JH-SH   Presents life science concepts in a stimulating, comprehensible, and meaningful way for the student who has not previously understood or been interested in science. Much of the content is in cartoon form. Contains mainly illustrated instructions for laboratory work, with a minimum of written instructions. All writing is in the vernacular of today's students. Each lesson is a structured, auto-instructional activity. Deemphasizes terminology and detail; laboratory material requirements are neither extensive nor expensive.

572 HUMAN RACES

See also 301.2 Culture and Cultural Processes.

Bova, Ben. *The Weather Changes Man.* (Illus.) Addison-Wesley, 1974. 140pp. $5.50. 73-16480. ISBN 0-201-00555-7. Index;CIP.

JH-SH Demonstrates the impact of climate on human evolution, including aspects of climate-dictated changes in the human body, cultural organization, and human behavior.

Buettner-Janusch, John. *Physical Anthropology: A Perspective.* (Illus.) Wiley, 1973. xiii + 572pp. $12.95. 72-14093. ISBN 0-471-11785-4. Gloss.;index.

C A Perspective is a somewhat limited and personalized view of physical anthropology. The coverage is about the same as the author's *Origin of Man,* but with less emphasis on living primates and biochemical aspects of heredity and more on human anatomy, hominid fossils and the controversial aspects of racial categories.

Goldsby, Richard A. *Race and Races.* (Illus.) Macmillan, 1971. xi + 132pp. $5.95. 77-133082.

SH-C-P Goldsby provides a fascinating, lucidly written and clearly argued exposition on races, ethnocentrism, and the concept and meaning of race. Excellent illustrations.

Goodall, Vanne, (Ed.). *The Quest for Man.* (Illus.) Praeger, 1975. 240pp. $17.50. 72-79550.

C This is a collection of essays which "attempt to draw together some of the skeins of modern scientific and metaphysical thought concerning human evolution." Provides increased understanding of our physical origins, material and intellectual development, and future potentialities.

Klass, Morton, and Hal Hellman. *The Kinds of Mankind: An Introduction to Race and Racism.* Lippincott, 1971. 219pp. $5.95. 73-117231.

JH-SH-C A good introduction to the concept of race, the science of physical anthropology, and, indirectly, to the broader problem of racism generated by ethnocentrism. Taxonomy and humans leads into a discussion of the history of attempts to classify humankind and the use of language, genetic traits and characteristics of physical appearance to enumerate the kinds of people. The useful concluding chapter is "Why Study Race?"

Napier, John. *The Roots of Mankind.* (Illus.) Smithsonian Institution Press, 1970. 240pp. $6.95. 72-112772. ISBN 0-87474-103-3.

SH-GA-C This well-organized and highly readable volume views human origins against a tapestry of primate behavior and adaptability and directs the reader's attention to the interrelation of dietary habits, growth, locomotion and ecology in primate history.

Osbourne, Richard H., (Ed.). *The Biological and Social Meaning of Race.* (Illus.) Freeman, 1971. viii + 182pp. $6.00; $2.95(p). 75-150652. ISBN 0-7167-0935-X; 0-7167-0934-1.

SH-C Contains 10 articles from *Eugenics Quarterly,* mostly by anthropologists. The articles explicitly dispel biological and social assumptions that underlie prejudicial attitudes toward different racial groups. Each article summarizes research data in nontechnical language and concludes with a demonstration that the evidence supports the concept of one unitary human race.

573.2 ORGANIC EVOLUTION AND GENETICS OF HUMANS

See also 575 Organic Evolution and Genetics.

Dobzhansky, Theodosius. *Genetic Diversity and Human Equality.* Basic, 1973. xii + 128pp. $5.95. 73-76262. SBN 465-02671-0.

SH-C   A distinguished scientist considers the role of genetics in the quest for social equality, discussing genetic diversity in humans and other species, the complexities of racial biology, our place in the universe, and our self-views. A biology course is necessary background.

Fraser, F. Clarke, and James J. Nora. *Genetics of Man.* (Illus.) Lea&Febiger, 1975. xii + 270pp. $14.50. 75-4743. ISBN 0-8121-0484-6. Index;CIP.

C-P   The authors present basic concepts of genetics and familiarize the reader with the scope of human genetics from analysis to counseling. Suitable for students interested in a premedical curriculum.

Gallant, Roy A. *How Life Began: Creation Versus Evolution.* (Illus.) Four Winds, 1975. vix + 214pp. $7.95. 75-12996. ISBN 0-590-17363-4. Index;CIP;bib.

JH-SH-C   A clear presentation of the merits and demerits of both sides of the creation versus evolution controversy. Discusses the historical development of the problem and the gradual acceptance of evolutionary theory.

Greenblatt, Augusta. *Heredity and You: How You Can Protect Your Family's Future.* Coward, McCann&Geoghegan, 1974. 256pp. $7.95. 73-93773. SBN 698-10588-5. Index.

SH-GA   An optimistic account of the current status of medical genetic treatment and counseling is presented. Excellent fare for the lay-person.

Halacy, D.S., Jr. *Genetic Revolution: Shaping Life for Tomorrow.* Harper&Row, 1974. x + 207pp. $6.95. 73-4085. ISBN 0-06-011713-3.

C   Halacy reviews developments in genetics from the theories of Hippocrates and Aristotle through the current research in genetic engineering and its possible effects on the development of societies of the future. Analogies are repeatedly drawn to Aldous Huxley's *Brave New World.* Recommended for collateral reading for classes in the biological disciplines.

Hamiltion, Michael, (Ed.). *The New Genetics and the Future of Man.* Eerdmans, 1972. 342pp. $6.95. 70-188248. ISBN 0-8028-3416-7.

C-P   Presents papers on three major topics of concern: new beginnings of life, genetic therapy and pollution and health. From their differing points of view, lawyers, theologians and philosophers discuss each subject after it has been presented by a scientist. Provides timely discussions for thoughtful people.

Ipsen, D.C. *Eye of the Whirlwind: The Story of John Scopes.* (Illus.) Addison-Wesley, 1973. v + 159pp. $4.25. 72-4777. ISBN 0-201-03172-8.

JH-SH   This biography is based primarily on the memoirs of John Thomas Scopes himself. The major focus of the biography is the "monkey trial," an event so fascinating that no embellishment is necessary to hold the reader's interest.

Jacob, Francois. *The Logic of Life: A History of Heredity.* (Trans.) Pantheon, 1974. viii + 348pp. $8.95. 73-18010. ISBN 0-394-47246-2.

C-P   An excellent review of the development of genetics as a science. Jacob considers how various discoveries led to intellectual revolutions, new notions of man's place in the universe, and the larger problem of how science develops and scientific breakthroughs occur. Offers a novel explanation of evolution and of the potential for further evolution of mankind.

McKusick, Victor A. *Human Genetics, 2nd ed.* (Illus.) Prentice-Hall, 1969. xv + 221pp. $6.95. 69-19671.

C-P   A stimulating detailed account of modern human genetics emphasizing the behavior of genes in kindreds, in individuals, and in populations. Explains analytical techniques used by geneticists, including linkage analysis, cytogenetic correlations and determination of protein structure and gene action. Also presents some genetic implications of social forces and a few social implications of human genetics.

Moody, Paul Amos. *Genetics of Man, 2nd ed.* (Illus.) Norton, 1975. ix + 507pp. $9.95. 75-2259. ISBN 0-393-09228-3. Gloss.;index;CIP.

SH-C   This completely rewritten edition covers a broad range of human genetics principles and problems. The author emphasizes scientific approaches used by researchers to solve genetics problems and gives enough background information for concept comprehension. Excellent reference for high school students and a good text for college human genetics courses.

Nagle, James J. *Heredity and Human Affairs.* (Illus.) Mosby, 1974. xii + 337pp. $10.00. 73-14547. ISBN 0-8016-3620-5. Index. CIP.

C   Nagle attempts to explain the possible origins of life and provides discussions of evolution, basic human reproduction and artificial birth control. An excellent biology text for the nonmajor.

Porter, Ian H., and Richard G. Skalko (Eds.). *Heredity and Society: Proceedings of a Symposium on Heredity and Society.* Academic, 1973. xi + 324pp. $11.00. 72-77366. ISBN 0-12-562850-1.

SH-C-P   One of the best books on the topic, providing a background in the art, history and ethics of medical genetics. Clear and accurate; good scope.

Scheinfeld, Amram. *Heredity in Humans.* (Illus.) Lippincott, 1971. xiv + 303pp. $6.95. 70-159730.

SH-C   The text is factual, well organized, easy to read but not entertaining. In systematic fashion Scheinfeld proceeds from explanations of the inheritance of such traits as sex, baldness, and basic eye color, to the more complex metabolic disorders and diseases, and concludes with discussions of intelligence, behavior and longevity. Genetic-counseling procedures are outlined.

Smith, Anthony. *The Human Pedigree.* Lippincott, 1975. 308pp. $10.00. 74-34175. ISBN 0-397-00876-7. Index;CIP.

C-P   The author discusses human evolution and the future of humanity from the standpoint of heredity. Population control, contraception, abortion, the role of eugenics, intelligence, and general human genetics are covered, as are probability, breeding of animals and sex ratios. Useful for biologists, anthropologists and students.

Williams, R.J. *Evolution and Human Origins: An Introduction to Physical Anthropology.* (Illus.) Harper&Row, 1973. viii + 277pp. $9.95. 70-178116. SBN 06-047117-4. Bibs.

SH-C   The first chapter places current theories of evolution in historical perspective. Then there are six chapters on human genetics and seven on the evolution of man. A useful introduction.

573.3 Prehistoric Humans

Clark, J. Desmond. *The Prehistory of Africa.* (Illus.) Praeger, 1970. 302pp. $8.50. 77-108243.

SH-C-P   The author, one of the foremost authorities on African prehistory, provides the only complete one-volume synthesis of this enormous segment of the cultural and biological heritage not only of Negroes, but of all humankind. The main methods of prehistorians are explained in this lively yet succinct work.

Cole, Sonia. *Leakey's Luck: The Life of Louis Seymour Bazett Leakey, 1903–1972.* (Illus.) Harcourt Brace Jovanovich, 1975. 448pp. $14.95. ISBN 0-15-149456-8. Gloss.; index.

JH-SH-C   The book provides the reader with an intimate view of Leakey, man and scientist. Text and photographs are excellent; treatment is thorough.

Goode, Ruth. *People of the Ice Age.* (Illus.) Macmillan, 1973. viii + 151pp. $5.95. 72-85191. ISBN 0-02-736420-8.

JH-SH   Traces human physical and cultural development. Hunting techniques, toolmaking, artwork and ways of life are described; intriguing hypotheses concerning reasons, methods or accidents behind many early accomplishments are examined; and evolving human relationships are explored. Physical evolution emerges less sharply, but a clear explanation of the role of mutation and natural selection is offered.

Kennedy, Kenneth A.R. *Neanderthal Man.* (Illus.) Burgess, 1975. v + 106pp. $2.95(p). 75-12186. ISBN 0-8087-11164.

C   This is as much an essay on historical changes in scientific reasoning as it is an analysis of the significance of the Neanderthal remains. Discussion of Neanderthal demography, ability to speak, belief system, and other matters make this text a fascinating reader and teaching tool.

Kurten, Björn. *Not From the Apes.* (Illus.) Pantheon, 1972. vii + 183pp. $5.95. 72-154018. ISBN 0-394-47123-7.

C-P   A major departure from current views on human evolution that surveys the fossil evidence for human evolution by examining the skeletal characteristics of fossil forms of man and man's ancestors. The author argues that men and apes separated some 35 million years ago, dismissing as inaccurate data that are accepted by most students of human evolution. An important book, should be widely read and discussed.

McKern, Sharon S., and Thomas W. McKern. *Tracking Fossil Man: An Adventure in Evolution.* (Illus.) Praeger, 1970. xiv + 174pp. $6.95. 75-125489. Gloss.;bib.

JH-SH-C-GA   Probably the best book on the subject for students in introductory anthropology or physical anthropology courses. It is equally suitable for high school, college undergraduate, and educated adult readers. The book describes the course of human evolution by discussing both the fossils and the ways in which they were found and interpreted.

Prideaux, Tom, et al. *Cro-Magnon Man.* (Illus.) Time-Life, 1973. 160pp. $7.95. 73-79435.

JH-SH-C   This is a good, readable summary of the climax of hunting and gathering. The authors incorporate recent archeological work which has greatly expanded our understanding of human adaptation to the changing environment of the last glacial period. High school and undergraduate instructors as well as students will find the book useful.

Shapiro, Harry L. *Peking Man.* (Illus.) Simon&Schuster, 1975. $7.95. 190pp. 74-19396. ISBN 0-671-21899-9. Index;CIP.

   SH-C Shapiro discusses the original discoveries, discoverers, and their methods and the conditions surrounding the disappearance of the fossils, and then describes his own involvement in the search. Data on the physical appearance of Peking Man are presented, and logical behavioral reconstructions are made. Excellent collateral reading for any introductory anthropology course.

574 BIOLOGY

Baer, Adela S., William E. Hazen, et al. *Central Concepts of Biology.* (Illus.) Macmillan, 1971. vii + 385pp. $8.95. 70-113932.

   C Descriptions of cellular and subcellular structure are followed by discussions of cellular metabolism, genetics and reproduction, ecology and evolution. The constant emphasis on the interrelationships of these areas is a major strength of the book.

Bailey, Paul C., and Kenneth A. Wagner. *An Introduction to Modern Biology, 2nd ed.* (Illus.) Intext, 1972. xvi + 560pp. $10.50. 72-183720. ISBN 0-7002-2361-4.

   C The discussions of genetics, evolution and ecology have sufficient background material preceding them to enable the student to understand the concepts discussed. The initial material presented is cell chemistry, organization, differentiation and reproduction. Concluding chapters discuss biological functions and diversity of animal and plant populations.

Beaver, William C., and George B. Noland. *General Biology.* (Illus.) Mosby, 1970. xiv + 546pp. $10.00. 74-99910. Bib.

   SH-C A truly fine work which follows the previous edition to a great extent. Discusses biology in everyday life; cells, cell division and organization; characteristics, chemical and physical properties and kinds of life; viruses, monera, and plants; animal and human biology; the continuity of life, genes and gene action; organisms and environment, with emphasis on behavior. Well-chosen review questions and topics, and an overall quality which will surely motivate the reader.

Becker, George C. *Introductory Concepts of Biology.* (Illus.) Macmillan, 1972. xvi + 335pp. $8.50. 72-151170.

   SH-C A unique approach to the teaching of biological concepts, this excellent book should be attractive to the nonscience major. It covers all the basic and relevant concepts of biology. The clarity of the writing and the straightforward diagrams give the book an attractive format, and it would be suitable for use as a beginning major's text.

Bernard, Claude. *Lectures on the Phenomena of Life Common to Animals and Plants: Vol. 1.* (Illus.;trans.) Charles C. Thomas, 1974. xxxv + 288pp. $12.95. 73-4297. ISBN 0-398-02857-5. CIP.

   C This is a collection of Bernard's lectures on the problems of physiology and a summary of his own work showing the applicability of experimentation to discover the commonalities which unite all living things. Valuable both to students of the history of science and as ancillary reading for advanced science courses.

Biological Science Curriculum Study. *Biological Science: Interaction of Experiments and Ideas, 2nd ed.* (Illus.) Prentice-Hall, 1970. 434pp. $7.72. ISBN 0-13-077008-6. *Teacher's Guide.* xvi + 125pp. $2.00. ISBN 0-13-077016-7.

SH-C   This BSCS volume is aimed at bright high school students or college
freshmen capable of independent study. Exercises are generally well written and include an unusually fine balance of open-ended plant and animal laboratory
exercises. Especially good experiments on the nutritional requirements of plants and
of animal behavior are clearly detailed.

Brandwein, Paul F., et al. *Life: A Biological Science.* (Illus.) Harcourt Brace
Jovanovich, 1975. xii + 500pp. $6.75. ISBN 0-15-365900-9. Index;gloss. *Teacher's
Manual.* xx + 203pp. $2.52(p). ISBN 0-15-365915-7.

JH   This is a sequenced life sciences textbook with seven conceptual schemes:
the interdependence of living things and their environment, the cell as the
basic unit of structure for living things, the interchange of matter between living
things and their environment, plant evolution, animal evolution, variation in living
things, and the consequences of modifying the environment.

Brandwein, Paul F., et al. *Life: Its Forms and Changes, 2nd ed.* (Illus.) Harcourt
Brace, 1972. xii + 563pp. $5.40. ISBN 0-15-366330-8. *Teacher's Manual, 2nd ed.* ISBN
0-15-366331-6.

SH   A novel approach to the teaching of biology, each unit develops one concept,
and each is concise and crammed full of interesting information. The color
photographs are excellent. Particularly interesting are the units on biochemistry, frog
dissection, photosynthesis, geology, the continuity of life and invertebrates. Each
chapter contains suggestions for investigations.

Clark, Mary E. *Contemporary Biology: Concepts and Implications.* (Illus.) Saunders,
1973. xiv + 707pp. $11.50. 72-86447. ISBN 0-7216-2597-5.

SH-C   This is a comprehensive biology book designed for senior high students or
those in the first year of college. The effect of biology on human affairs and
ecology is emphasized. Highly recommended.

Curtis, Helena. *Biology, 2nd ed.* (Illus.) Worth, 1975. xxv + 1065pp. $14.95. 74-27183.
ISBN 0-87901-040-1. Gloss.;index. *Study Guide* (by Vivian Manns Null). ix + 311pp.
$4.95(p). ISBN 0-87901-044-4.

C   This second edition is better looking, better illustrated and more interesting,
although the explanations are lengthy. Evolution is a unifying principle for
organizing information on cells, organisms and populations; there is heavy emphasis
on angiosperms and primates. A high quality text for beginning biology courses.

Curtis, Helena. *Invitation to Biology.* (Illus.) Worth, 1972. xii + 587pp. $9.75. 70-
181459. ISBN 0-87901-020-7. Bib.

SH-C   Covers all the areas essential for the nonmajor and does it well. The characteristics of life, cell structure and function, genetics, a brief survey of plants
and animals, plant and animal physiology, human physiology (reproductive, nervous,
endocrine, respiratory, circulatory, digestive systems), ecology and evolution are
discussed. Questions are thought provoking.

Eisman, Louis, and Charles Tanzer. *Biology and Human Progress, 4th ed.* (Illus.)
Prentice-Hall, 1971. xiii + 530pp. $8.40. ISBN 0-13-076943-6. Index;gloss.

JH   Designed for students who are not science-oriented. The book is divided into
nine units: exploring life, how the body works, food, behavior, scientific
control of disease, reproduction, heredity, life in past ages, and ecology.

Gerking, Shelby D. *Biological Systems.* (Illus.) Saunders, 1969. xiv + 480pp. $8.50.
69-10566. ISBN 0-7216-4100-8.

SH-C   Topics provide a cultural overview of biology, leading the reader from the
cell as the fundamental unit of life through molecular biology, physiology,

biochemistry, ecology and evolution. The historical development of several scientific concepts shows the evolution of science and the rapid proliferation of scientific knowledge. There is appropriate emphasis on the relation of certain topics to humans, and function is emphasized more than morphology.

Goldsby, Richard A., et al. *Biology.* (Illus.) Harper&Row, 1976. 862pp. $14.95. 75-24812. ISBN 0-06-042646-2. Gloss.;index.

SH   Covers phenomena at the cellular level, including molecular biology, cell replication and genetics; plant and animal classification; development of the organism; physiology; evolution; and ecology. Peripheral subjects include contraceptive methods, pharmacology with a special reference to drug dependence, extraterrestrial life and exobiology. The illustrations, including electron micrographs, are of special value. Includes a chemistry appendix.

Hammen, Carl Schlee. *Elementary Quantitative Biology.* (Illus.) Wiley, 1972. 144pp. $3.95(p). 72-000026. ISBN 0-471-34721-3.

C   Should be required of *every* beginning biology student. Biological problems are selected from a broad range of subdisciplines and are lucidly analyzed. Problems are presented for student solution, with answers given in an appendix. Other appendixes include log tables, a periodic table and t-distributions. The orientation and clarity make it most attractive.

Jessop, N.M. *Biosphere: A Study of Life.* (Illus.) Prentice-Hall, 1970. xiii + 954pp. $11.50. 70-79114.

C   A general biology text presenting ecological relationship as the selective factor operating to shape the evolution of life at all its levels of organization. A smoothly written, coherent, well-illustrated volume. The author has included far more molecular biology and taxonomy than is found in other such texts.

Koob, Derry D., and William E. Boggs. *The Nature of Life.* (Illus.) Addison-Wesley, 1972. xvii + 494pp. $9.95. 73-140837.

SH-C   Designed for use by liberal arts students, it differs from other biology books in several aspects: less emphasis on plants and animals than is usual; "life" is viewed as a fourth state of matter characterized by a remarkably high level of organization; chapters end not with problems but with readings. Includes two chapters on the environment.

Korn, Robert W., and Ellen J. Korn. *Contemporary Perspectives in Biology.* (Illus.) Wiley, 1971. x + 606pp. $12.50. 79-140178. ISBN 0-471-50376-2.

SH-C   Four major themes—energy, the cell and cellular control, evolution, and population-environmental interaction—provide an introduction to modern biology. In a clear and logical progression, each chapter builds on the foundation laid by preceding ones and emphasizes the common origin and interdependence of all forms of life. Especially useful for interdisciplinary courses.

McElroy, William D., et al. *Biology and Man.* (Illus.) Prentice-Hall, 1975. xiii + 656pp. $12.95. 74-29048. ISBN 0-13-076695-X. Index;gloss.;CIP.

SH-C   Deals with modern cell biology, human physiology, and human relationships with the environment. An excellent introductory biology text.

Marshall, P.T. *The Development of Modern Biology.* (Illus.) Pergamon, 1969. xi + 129pp. $4.75; $2.75(p). 70-92112.

SH-C   The author devotes one chapter to the development and origins of the scientific method and then discusses plant and animal classification, evolution, mechanisms of heredity, some aspects of the biology of medicine, and the development of agriculture.

Morholt, Evelyn L., and Paul F. Brandwein. *Biology: Patterns in Living Things.* (Illus.) Harcourt Brace Jovanovich, 1976. vii + 376pp. $7.95. ISBN 0-15-362432-9. Index.

    JH-SH This biology textbook uses the inquiry method. Includes activities and questions. The reading level is fifth grade, but the content is suitable for the average junior high school student and is highly recommended for high school students with certain learning disabilities.

Noland, George B., and William C. Beaver. *General Biology, 9th ed.* (Illus.) Mosby, 1975. xiv + 579pp. $12.95. 75-1498. ISBN 0-8016-3686-8. Gloss.;index;bib;CIP. *Instructor's Manual.* 42pp.

    C The whole organism concept is stressed in this introductory text for beginning college students. General biological principles are followed by sections on various groups of organisms from viruses to humans. This revision has made the text more readable and improved the format.

Painter, John H. *Biology Today.* (Illus.) CRM Books, 1972. xxxi + 1020pp. $14.95. 72-176334.

    C Suitable as a major text and invaluable as a supplement, particularly in introducing biological concepts to nonbiology majors. Both thorough in its coverage of problems in contemporary biology and refreshing in its approach to classical themes. The human organism, particularly reproductive and drug-related behavior, is emphasized. Remarkable in its use of graphics to inform.

Phillips, Edwin A. *Basic Ideas in Biology.* (Illus.) Macmillan, 1971. xiii + 716pp. $10.95. 77-80308.

    C Forsakes the usual approach in which the physical and chemical fundamentals are presented first. Instead, a section on organismal biology discusses scientifically much of what already lies within the student's experiences. Material is presented as it was experimentally developed. Discussions of specific experiments provide exceedingly broad coverage which is clear and accurate.

Ross, Herbert H. *Biological Systematics.* (Illus.) Addison-Wesley, 1974. 345pp. $12.95. 73-2141.

    C-P This volume incorporates recent information from a variety of disciplines which bear upon biological systematics. Ross presents an orderly discussion of the nature of systematics, its methodology and interpretative aspects, and provides many fine examples from botany and zoology to illustrate various points of his discussion. A good treatment and a useful reference for teachers.

Sherman, Irwin W., and Vilia G. Sherman. *Biology: A Human Approach.* (Illus.) Oxford Univ. Press, 1975. 553pp. $13.95. 74-12669. Index. *Teacher's Manual.* ix + 142pp.

    C This introductory text for nonmajors adds relevant social and political applications to basic human biology. The style of writing is clear and concepts such as cell structure, genetics and organism functions are explored in depth. Students will be challenged and interested.

VanNorman, Richard W. *Experimental Biology, 2nd ed.* (Illus.) Prentice-Hall, 1970. viii + 304pp. $9.95. 74-105444.

    SH-C Extremely well organized and useful as a text in high schools for students well prepared in physical sciences and mathematics. Useful also to the college sophomore preparing for independent research. An excellent, comprehensive review of literatures and their availability and use to the researcher and clinician is included.

Villee, Claude A. *Biology, 7th ed.* (Illus.) Saunders, 1977. xviii + 980pp. $15.95. 76-014694. ISBN 0-7216-9023-8. Index;gloss.

C   A classically oriented introductory text. Human physiology and reproduction are emphasized. The book contains an excellent description of a science of facts, but short-changes the dynamic experimental science of the phenomena of living things.

Wallace, Robert A. *Biology: The World of Life.* (Illus.) Goodyear, 1975. xvii + 511pp. $12.95. 74-31026. ISBN 0-87620-114-1. Gloss.;index;CIP.

C   The author begins this nonmajor's introductory biology textbook with Darwin's theory of evolution and the origin of life, then, after some background chemistry, goes into energetics and DNA. Special interest sections (e.g., evolutionary implications of war) and an appendix on classification are included.

Wilson, Edward O., et al. *Life on Earth.* (Illus.) Sinauer, 1973. xi + 1053pp. $11.95. 73-78919. ISBN 0-87893-934-2.

C   A collaboration by eight prestigious biologists who provide a well-organized view of the broad field of biology, from molecules through cells and organisms to populations and societies. Photographs are outstanding. Sections of papers and sentences or paragraphs are summarized and highlighted for special attention. There is no single theme, thus permitting flexible use of the book as a text or reference.

574.03 BIOLOGY—DICTIONARIES AND ENCYCLOPEDIAS

Gray, Peter. *The Encyclopedia of the Biological Sciences, 2nd ed.* (Illus.) Van Nostrand Reinhold, 1970. 1027pp. $24.95. 77-81348.

SH-C   One of the finest one-volume encyclopedias of the biological sciences; intended to provide succinct and accurate information for biologists in those fields in which they are not themselves experts. Also useful for librarians, for teachers in high schools and their students, and for nonprofessionals. Contains scholarly articles by scientists covering all of the biological sciences and their interrelations with other disciplines. Many brief biographical sketches.

Lapedes, Daniel N., (Ed.). *McGraw-Hill Dictionary of the Life Sciences.* (Illus.) McGraw-Hill, 1976. xiv + 946pp. 76-17817. ISBN 0-07-045262-8.

SH-C   Contains some 20,000 terms and definitions, identified by fields which include zoology, microbiology, genetics, anatomy, ecology, chemistry, physics and thermodynamics. The various appendixes include clinical chemistry and cytology values and animal, plant and bacterial taxonomy through class, subclass and order. Much of the material was previously published in the *McGraw-Hill Dictionary of Scientific and Technical Terms (1974).* Definitely should be in high school and public libraries of whatever size.

Steen, Edwin B. *Dictionary of Biology, 2nd ed.* Barnes&Noble, 1975. vii + 630pp. $12.50(p). 70-156104. ISBN 0-06-480827-0.

C   A dictionary of biological terms which will serve as an excellent and convenient reference source for both undergraduate and graduate students of biology. This handy paperback includes most of the terminology currently in use. The format is very clear.

574.07 BIOLOGY—STUDY AND TEACHING

Berger, Melvin. *Tools of Modern Biology.* (Illus.) Crowell, 1970. 215pp. $4.50. 73-94788.

JH-SH   The role of observation, the use of classification, the importance of discussion and bibliographic research, the use of statistics in biological research and scientific instruments and their uses (simple, compound, and electron microscopes; centrifuges; chromatography; x-rays; radioactive tracers and radiation; and computers) are discussed. Provides a good exposition of the evolution of scientific method.

Biological Sciences Curriculum Study. *Biology Teacher's Handbook, 2nd ed.* Wiley, 1970. 656pp. $8.95. 78-93488. ISBN 0-471-07308-3.

C-P   Explains the BSCS approach to the teaching of biology as a process of scientific inquiry; compares and explains the content of the three versions of BSCS; and provides a summary of the principles and concepts of the physical sciences, of statistical methods, and of the principles of biochemistry that high school teachers should know as background. Also explains the many types of supplementary materials that have been produced by BSCS.

Biological Sciences Curriculum Study. *Research Problems in Biology: Investigations for Students, 2nd ed.* (Illus.) Oxford Univ. Press, 1976. $10.00ea.; $4.00ea.(p). 75-39505. *Series 1:* xv + 221pp. ISBN 0-19-502063-4. *Series 2:* xv + 215pp. ISBN 0-19-50264-2. *Series 3:* xiv + 225pp. ISBN 0-19-502065-0. Bibs.

SH-C   Designed to acquaint students with the process of scientific investigation. Each volume contains over 40 problems, each by an experienced researcher, covering behavior, physiology, ecology, microbiology, genetics and growth, form and development. The problems vary from those with detailed experimental procedures to questions without suggestions as to methodology. The time commitment, technical capability and level of sophistication needed vary. Most useful as a reference for Science Fair projects, etc.

Burnet, Macfarlane. *Genes, Dreams and Realities.* (Illus.) Basic, 1971. 232pp. $7.95. 72-177260. ISBN 0-465-02672-9.

JH-SH-GA   A brilliant, lucid, and enthusiastic overview of much of what has been and still is important in human biology. At the same time, it is a fine work of synthesis, of things such as genes, chromosomes, cancer, immunology, evolution, and much else. The writer spices his work with certain of his favorite controversial views.

Chiscon, J. Alfred, et al. *The Laboratory Experience: A Principles of Biology Manual, 2nd ed.* (Illus.) Burgess, 1976. x + 295pp. $6.95. ISBN 0-8087-0362-5.

C   For a 2-semester general biology course, stressing the unity of biological principles. The content ranges from microscopy to the kinetics of an immune response. If an instructor judiciously selects material, this manual could be useful.

Crum, Lawrence E. *Classroom Activities and Experiments for Life Science.* (Illus.) Parker, 1974. 224pp. $8.95. 74-621. ISBN 0-13-136226-7. Index;CIP.

SH-P   This lively manual invites students to participate actively in experimentation, and suggests activities which are alive to current interests. Written for both students and teachers.

Korn, Robert W., and Ellen J. Korn. *Investigations into Biology, 2nd ed.* (Illus.) Wiley, 1971. x + 502pp. $5.95(p). 72-126228. ISBN 0-471-50380-0.

C   One of the finest laboratory manuals in introductory biology. The variety and depth of the exercises are particularly impressive, and their conceptual approach will help develop a research attitude in the student. It can be adapted to either a 1- or 2-semester course and used with any modern biological text.

Smith, Roger C., and W. Malcolm Reid. *Guide to the Literature of the Life Sciences, 8th ed.* Burgess, 1972. vi + 166pp. $6.50(p). 74-181748. ISBN 0-8087-1964-5.

SH-C   This eighth edition represents a continuing effort to survey the sources of biological information. It has been expanded to include all the biological sciences and has been updated, with new sources being added and less useful references deleted. The chapter on the preparation of a scientific paper is concise and easy to follow.

Worth, C. Brooke. *Of Mosquitoes, Moths and Mice.* (Illus.) Norton, 1972. 258pp. $8.95. 75-39017. ISBN 0-393-06390-9.

SH-C   Skillfully leads the reader through the frustrations, failures and accomplishments inherent in scientific research. There are chapters to delight the entomologist, the ecologist, the ornithologist and birdbander, the geneticist, herpetologist and mammalogist. Can also be enjoyed by professional biologists.

Young, Stephen. *Electronics in the Life Sciences.* (Illus.) Halsted/Wiley, 1973. iv + 198pp. $11.50. 73-8083. ISBN 0-470-97943-7.

SH-C-P   This book is for research workers or students in the life sciences who may need to measure or observe small signals or responses to stimuli in various organisms. Young covers the usual circuits and instruments encountered in regular work, and his descriptions are clear and concise.

574.1 PHYSIOLOGY

Dagley, S., and Donald E. Nicholson. *An Introduction to Metabolic Pathways.* (Illus.) Wiley, 1970. xi + 343pp. $11.95. 76-99975. ISBN 0-471-63706-8. Bib.

C-P   Strong features of the book are the accurate line drawings of the pathways with the compression of massive amounts of information into a relatively small space and the pairing of catabolic pathways with syntheses of the common metabolites. This work is a classic, providing easy access to the field of metabolism for beginning and advanced students and for the professional educator or researcher.

Grinnell, Alan, and Albert A. Barber. *Laboratory Experiments in Physiology, 9th ed.* (Illus.) Mosby, 1976. xiv + 199pp. $6.50(p). ISBN 0-8016-2978-0. Index.

SH-C   A very flexible manual containing many uncomplicated experiments, ranging from physiochemical and nerve-muscle experiments to radioisotope applications. Includes rationales for each experiment.

Hall, Thomas S. *History of General Physiology 600 B.C. to A.D. 1900. Vol. 1: From Pre-Socratic Times to the Enlightenment.* Univ. of Chicago Press, 1975 (c.1969). xii + 419pp. $28.50 set; $6.50ea.(p). 69-16999. ISBN 0-226-31360-3; 0-226-31353-0. Index;bib. *Vol. 2: From the Enlightenment to the End of the Nineteenth Century.* vii + 399pp. 69-16999. ISBN 0-226-31360-3; 0-226-31354-9. Index;bib.

C-P   The author is concerned with what makes "the difference between living and nonliving things." He traces the history of ideas on this subject and related "classic questions" in physiology by excerpting from the writings of various philosophers, biologists and physicians.

Halstead, Beverly, and Jennifer Middleton. *Bare Bones: An Exploration in Art and Science.* (Illus.) Univ. of Toronto Press, 1973. 119pp. $7.95. ISBN 0-8020-1971-4; 0-8020-0314-1 (microfiche).

C   Covers the art and science of bones. Treats the morphological, physiological, embryological, phylogenic and pathological aspects, and also discusses the technology, culture and aesthetics of bones. Illustrated with quality drawings and photographs, it makes a memorable gift.

Marean, John H., Odell Johnson, and Bernadette R. Menhusen. *Life Science: Inquiring into Life.* (Illus.) Addison-Wesley, 1972. xi + 318pp. $6.64. *Teacher's Guide.* xiii + 128pp. $2.80.

JH-SH   Life processes, including response to stimuli, use of water, transport, motion, food getting, food production, respiration, reproduction, and learning constitute the main content. Each process is developed with a variety of organisms, including human beings, as examples. The *Teacher's Guide* offers practical teaching suggestions, laboratory-preparation instructions, suggested time schedules, and answers to the end-of-chapter questions. Could form the basis for a good introduction to biology.

Rothstein, Howard. *General Physiology: The Cellular and Molecular Basis.* (Illus.) Xerox, 1971. xx + 602pp. $13.50. 78-119788. Bib.

C   A broad text concerned with the cellular and molecular aspects of those functional features shared by living organisms, including physiochemical background, cell membranes and transport of materials across them, cellular metabolism (including photosynthesis and internal structure), cell genetics, bioelectricity, and contractility. Rich in historical and experimental background. Methodology is an important factor in the development of each topic.

574.191 BIOPHYSICS

Arena, Victor. *Ionizing Radiation and Life: An Introduction to Radiation Biology and Biological Radiotracer Methods.* (Illus.) Mosby, 1971. xi + 542pp. 77-150126. ISBN 0-8016-0278-5. Index;bib.

SH-C   A good text to introduce radiation biology and radiotracer methods to undergraduates. Areas covered are nuclear physics, the use of radioisotopes in medicine and research, radiation accidents, and radiation biology. The text includes useful appendixes.

Arena, Victor. *Ionizing Radiation and Life: Laboratory Experiences.* (Illus.) Mosby, 1971. 184pp. $5.25.

SH-C   The manual could be used as part of a course in radiation biology, in practical on-the-job training for professionals and technicians who wish to work with radioisotopes, or in an advanced biochemistry laboratory to show the application of radioisotopes to biological research. Each experiment is presented in a logical, easy-to-understand manner, and is accompanied by a removable report form.

Bowen, T.J., and A.J. Rowe. *An Introduction to Ultracentrifugation.* Wiley, 1970. xviii + 171pp. $6.95. 79-129158. ISBN 0-471-09215-0.

C-P   A useful supplement to the manuals which are supplied by instrument manufacturers. It contains a helpful chapter on preparative ultracentrifugation as well as descriptions and comparisons of British and German instruments and of analytical attachments.

Christensen, Halvor N., and Richard A. Cellarius. *Introduction to Bioenergetics: Thermodynamics for the Biologist.* (Illus.) Saunders, 1972. xii + 224pp. $5.50(p). 70-168596. ISBN 7216-2588-6.

C   Will give the biology student an exciting and interesting introduction to thermodynamic principles, but some knowledge of metabolism and the chemistry of biological molecules is essential. Quantitative problems are sprinkled throughout, and a useful summary is provided, as is a final test covering all essential points.

Simon, Hilda. *Living Lanterns: Luminescence in Animals.* (Illus.) Viking, 1971. 128pp. $4.53. 73-136828. ISBN 0-670-43536-8; LB 0-670-43537-6.

JH   It is highly unusual to find so complex and specialized a subject covered so clearly at such an elementary level. One excellent additional factor is encouragement for the reader to consider a future in natural history research. The book's uniqueness in subject matter recommends it for all junior high school natural history libraries.

574.192 BIOCHEMISTRY

Berger, Melvin. *Enzymes in Action.* (Illus.) Crowell, 1971. 151pp. $4.50. 76-132291. ISBN 0-690-26735-5.

JH-SH   This short review of enzymology and its practical application in industry covers such topics as the role of enzymes in digestion, food production, and preparation of beverages; enzymes as drugs; in diseases; and how enzymes work. Describes historic and salient discoveries including some recent achievements.

Chedd, Graham. *The New Biology.* (Illus.) Basic, 1972. xiv + 306pp. $7.50. 74-174816. ISBN 0-465-04998-2.

GA-C   Very skillfully traces the historical studies and summarizes research on the functions of nucleic acids and proteins. After discussing nucleic acids, the author traces their role in protein production, then continues the discussion of proteins noting their unique structure and how they are involved in life processes.

Dickerson, Richard E., and Irving Geis. *The Structure and Action of Proteins.* (Illus.) Harper&Row, 1969. viii + 120pp. $9.95; $4.50(p). $6.50 (stereo suppl.). 69-11112.

C-P   Topics include makeup of living organisms; structural proteins; limitations on folding; molecular carriers; enzymes; the serum complement system and others. Finally, the supplemental stereo drawings will make the book invaluable to teachers and research workers in the field. The illustrations would nearly stand alone in presenting the story of the structure and action of the proteins.

Farago, Peter, and John Lagnado. *Life in Action: Biochemistry Explained.* (Illus.) Knopf, 1972. xv + 258pp. $6.95. 69-10705. ISBN 0-394-43320-3.

SH-GA-C   With their warm and friendly style, the authors make biochemistry appeal to the lay reader. Emphasis is placed on proteins and protein synthesis, metabolism, biochemistry of genetics, and biological control and regulation mechanisms. The book is not a textbook, but it is an introduction to a fascinating field for a nonscientist.

Hill, Leonard, Denis Bellamy, and Ian Chester Jones. *Integrated Biology.* (Illus.) Harper&Row, 1973. xiii + 354pp. $6.95. 72-8254. SBN 06-042829-5.

C   This unique introductory biochemistry text successfully integrates a structural, molecular and systematic approach. The principles of biochemistry, physiology, ecology and botany are combined in discussions of the nature and origin of life, morphology and biochemistry of cell organelles, intermediary metabolism, environmental stresses in evolution and organism-environment relationships, the physiology and biochemistry of the various systems (nervous, reproductive, etc.) and genetics. Good except for some oversimplified models.

Holmes, Frederic Lawrence. *Claude Bernard and Animal Chemistry: The Emergence of a Scientist.* Harvard Univ. Press, 1974. 541pp. $18.00. 73-88497. SBN 0-674-13485-0. Index.

C   Realistic biography focusing on the trials and errors of the "father of modern physiology." Explores the relationship between the development of Bernard's career and the thoughts and events which shaped the ultimate outcome of his work. Required reading for a thorough understanding of the foundations upon which modern experimental science is constructed.

Holum, John R. *Elements of General and Biological Chemistry: An Introduction to the*

Molecular Basis of Life, 3rd ed. (Illus.) Wiley, 1972. xi + 578pp. $10.95. 79-168648. ISBN 0-471-40846-8.

C   Well suited to a 1-semester introductory course which emphasizes biological chemistry for nonmajors. General topics, fundamentals of organic chemistry and biochemical topics are discussed. The format is attractive, with effective use of color in the illustrations and equations.

Johnston, David O., et al. *Chemistry and the Environment.* (Illus.) Saunders, 1973. xi + 452pp. $10.95. 72-90723. ISBN 0-7216-5185-2. *Instructor's Guide.*

SH-C   Written for nonscience majors; the breadth and depth of the chemical principles discussed are more than sufficient for them. But to understand some of the environmental problems, more chemistry is needed than is included. Emphasis is on biochemistry.

Lehninger, Albert L. *Biochemistry: The Molecular Basis of Cell Structure and Function, 2nd ed.* (Illus.) Worth, 1975. xxiii + 1104pp. $22.95. 75-11082. ISBN 0-87901-047-9. Index.

C-P   The book is organized around four themes: structures of biomolecules, catabolic and energy generating processes, anabolic and energy consuming processes, and macromolecular information transfer. Concise chapter summaries are provided. It is a valuable reference for biology and biochemistry courses.

Mahler, Henry R., and Eugene H. Cordes. *Biological Chemistry, 2nd ed.* (Illus.) Harper&Row, 1971. xiv + 1009pp. $19.00. 76-141169. ISBN 0-06-044172-0.

C-P   After a brief historical sketch, a chapter on equilibria and thermodynamics leads into consideration of the structures and functions of proteins, nucleic acids, carbohydrates and lipids. Molecular biology and the gene, biosynthesis of nucleic acids, and proteins are treated with a depth rarely seen in a textbook.

Mehler, Alan H., et al. *Biochemical Problems and Calculations.* (Illus.) Burgess, 1975. v + 137pp. $3.75. ISBN 0-8087-1376-0.

C   Systematic thought and construction in problem solution is emphasized. The problems are organized into 15 categories of general biochemistry covering most major topics, and basic concepts are well developed. Addressed to the student in introductory biochemistry.

Needham, Joseph, (Ed.). *The Chemistry of Life: Lectures on the History of Biochemistry.* Cambridge Univ., 1970. $9.50. 78-85733. ISBN 0-521-07379-0.

SH-C   Eight lectures by eminent Cambridge biochemists are presented as an introductory history of biochemistry. Covers significant developments since about 1800 in photosynthesis, enzymes and biological oxidations, microbiology, neurology, hormones, vitamins, foundations of modern biochemistry, and outstanding 19th century pioneers of biochemistry.

Patton, A. Rae. *The Chemistry of Life.* (Illus.) Random House, 1970. 129pp. $3.50; $1.50(p). 68-9798. ISBN 0-394-81910-1.

JH-SH-C   Written for the nonspecialist, this brief yet informative biochemistry book should interest readers in a wide range of ages. Good styling, handsome illustrations; an attractive and still useful book.

Riedman, Sarah R. *Hormones: How They Work, rev. ed.* (Illus.) Abelard-Schuman, 1973. 222pp. $5.95. 72-12072. ISBN 0-200-00005-5. Index.

JH-SH   In this revision of a book first published in 1956, the author describes the discovery of hormones, their discoverers, and hormonal activity in an interesting manner. A useful supplement for high school biology classes.

Scientific American. *Chemistry in the Environment: Readings from* Scientific American. (Illus.) Freeman, 1973. 361pp. $12.00; $5.95(p). 73-3172. ISBN 0-7167-0878-7; 0-7167-0877-9. Bibs.

C This volume deals with the chemical interactions of our industrial society with the environment. Covers the processes in the biosphere which influence life's chemistry and the perturbations caused by industrial societies (impact on land, energy use, atmospheric pollutants and biological effects of various pollutants).

Scientific American. *Organic Chemistry of Life: Readings from* Scientific American. (Illus.) Freeman, 1973. xii + 452pp. $12.00; $6.95(p). 73-12457. ISBN 0-7167-0884-1; 0-7167-0883-3.

SH-C These selections illustrate the chemistry of living cell processes, such as the induction of interferon, photosynthesis, the genetic code, fat metabolism and energy transformations. Also covered are the complex structural chemistry of insulin, ribonuclease, prostaglandins, hallucinogenic drugs, pheromones and the molecular evolution of cytochrome C. Will be useful to both the high school and college chemistry teacher for course enrichment.

Strong, F.M., and Gilbert H. Koch. *Biochemistry Laboratory Manual, 2nd ed.* Brown, 1974. xii + 260pp. $5.95. ISBN 0-697-04692-3. Index.

C The experimental work covered by this extensive manual takes students from the advanced topics of general chemistry through quantitative analysis, organic, and biological chemistry. Well-written, instructive experiments also involve descriptive biochemistry.

Stryer, Lubert. *Biochemistry.* (Illus.) Freeman, 1975. xii + 877pp. $19.95. 74-23269. ISBN 0-7167-0174-X. Index;CIP.

C-P A clearly written, comprehensive presentation of biochemistry with superb illustrations. Molecular biology is exceptionally well done while conformation, metabolism and molecular physiology are major themes. A highly recommended text for undergraduate or graduate courses, and also useful as a reference for high school teachers or students.

Yudkin, Michael, and Robin Offord. *Biochemistry.* (Illus.) Houghton Mifflin, 1974. x + 528pp. $14.95. 73-14495. ISBN 0-395-17199-7. Index.

C This text deserves high praise because it clearly establishes the unity of biochemistry in terms of structure and function. Emphasis is on the fields of molecular biology, and teachers will find this a useful text even for advanced high school biology or biochemistry. Valuable also for pre-med and nursing students and graduate students needing a guide to the periodical literature.

574.2 PATHOLOGY

Mattingly, P.F. *The Biology of Mosquito-Borne Disease.* (Illus.) American Elsevier, 1970. viii + 184pp. $6.50. 70-93924. ISBN 0-444-19723-0.

SH-C-P The 11 chapters are arranged into introductory sections on evolution and disease as an ecological system, followed by more specialized reviews of genetics, speciation, research, etc. The book closes with a brief history of mosquito-borne disease, eradiction methods and present and future prospects.

Steinhaus, Edward A. *Disease in a Minor Chord.* Ohio State Univ. Press, 1975. xviii + 488pp. $20.00. 75-4527. ISBN 0-8142-0218-7. Indexes;CIP.

C-P Fleas and other small insects are the subjects of this lively account of the history of the study of infectious disease and pathology in invertebrates. For anyone interested in nucleic acid biosynthesis, hormonal control and regulation of life processes.

Wilson, David. *Body and Antibody: A Report on the New Immunology.* (Illus.) Knopf, 1971. xii + 345pp. $8.95. 70-154943. ISBN 0-394-46157-6.

SH-C   An interesting account of the modern science of immunology and its impact on our understanding of human reaction to foreign substances. Includes a discussion of our rapidly developing understanding of the mechanism of immune systems and the theory of self and the recognition of self.

574.3 DEVELOPMENT AND MATURATION

Balinsky, B.I. *An Introduction to Embryology, 3rd ed.* (Illus.) Saunders, 1970. xviii + 725pp. $10.00. 71-103563.

C-P   Excellent presentations are given on the storage and transmission of genetic information in germ cell maturation, in fertilization, in the synthesis of DNA, RNA, and proteins in cleavage, and in the selective action of genes in differentiation and growth. Placental form and physiology in various mammalian orders and hormones in ovulation and pregnancy are discussed as well in this excellent introduction to animal embryology from the genetic, experimental, and physiological points of view.

Bonner, John Tyler. *On Development: The Biology of Form.* (Illus.) Harvard Univ. Press, 1974. 282pp. $10.00. 73-88053. SBN 674-63410-1. Index.

C-P   A general treatise on developmental biology which advances the author's view that pattern control is a function of the synthesis of substances, their timing, and localization in space. Bonner shows how evolution occurs and that biological order emerges from essentially random mutation.

Cousteau, Jacques-Yves. *The Act of Life.* (Illus.) World, 1973. 144pp. $7.95. 72-87710. ISBN 0-529-04937-6.

SH-C   Describes nontechnically and objectively the multitude of factors in the sea which affect animal reproduction; plant life is mentioned only cursorily. Topics include some general principles of evolution, the nature of DNA, asexual and sexual reproduction, types of embryonic development and the various kinds of post-natal care provided by sea-dwelling organisms. Many unusually beautiful color photographs.

Kohn, Robert R. *Principles of Mammalian Aging.* (Illus.) Prentice-Hall, 1971. xiii + 171pp. $7.95. 78-152446. ISBN 0-13-709444-2; 0-13-709436-1(p).

GA-C   Numerous characteristics contribute to the process of mammalian aging. Those discussed include chemical, extracellular, and intracellular aging; aging of cells and aging of animals. Each characteristic is discussed thoroughly and results of research on the subject are reported in a style suitable for both lay readers and specialists.

574.5 ECOLOGY

See also 301.3 Ecology and Community and 333.7 – .9 Conservation and Utilization of Natural Resources.

Bendick, Jeanne. *Adaptation.* (Illus.) Watts, 1971. 72pp. $3.95. 71-155406. ISBN 0-531-01437-1.

JH-SH   A fascinating book which explains how animals and plants have adapted to their environment. The text is well written and facts about animals and plants are correlated to show how each has adapted to particular environmental changes.

Boughey, Arthur S. *Fundamental Ecology.* (Illus.) Intext, 1971. x + 222pp. $3.95(p). 77-151647. ISBN 0-7002-2363-0. Index;gloss.;bib.

SH-C One book in a series intended to survey at the introductory level, and without overlap, the major fields of current environmental confrontation. With selections from the bibliography, this work can be used as a course text. It covers ecosystems; environments; population dynamics, evolution, interactions and behavior; and the nature and structure of communities.

Colinvaux, Paul A. *Introduction to Ecology.* (Illus.) Wiley, 1973. ix + 621pp. $12.50. 72-3788. ISBN 0-471-16498-4. Gloss.;bib.

SH-C In a discussion of the major problem areas which have engaged the attention of ecologists through the years, this author treats the errors and misconceptions as well as the successes of ecological research.

Darnell, Rezneat M. *Ecology and Man.* (Illus.) Brown, 1973. xii + 149pp. $2.50(p). 72-94579. ISBN 0-697-04521-8. Bibs.

SH-C The author shows that ecology can serve as a bridge between science and humanism. Two principal sections address "the natural systems" (where man is not a controlling influence) and "man and nature" (human ecology). Ecological relationships are treated at the individual, population, community and ecosystem levels.

Dorfman, Leon. *The Student Biologist Explores Ecology.* (Illus.) Richards Rosen, 1975. ix + 116pp. $4.80. 75-14158. ISBN 0-8239-0327-3. Gloss.;CIP.

SH Dorfman discusses the quality of the environment and our effect on it. Ecology, energy flow, cycling of nutrients, limits of the environment, interactions of populations, ecological successions, world biomes, and finally humans and the environment are presented. Useful to supplement a tenth-grade biology course.

Godfrey, Michael A. *A Closer Look.* (Illus.) Sierra Club, 1975. x + 148pp. $14.95. 75-8961. ISBN 0-87156-143-3. Index;CIP.

JH-SH-GA-C The reader is given a close, very personal view of both nature and the interrelationships in the biological community. Excellent photographs accompany the text.

Hungerford, Harold R. *Ecology: The Circle of Life.* (Illus.) Childrens, 1971. 92pp. $3.95. 79-153087.

JH Fills a large gap between natural history books and academic ecology texts. It is essentially a primer of ecology in which the author illustrates each point with fine black-and-white photographs and line drawings. The final chapter deals with humans' role in nature.

Kelly, Mahlon G., and John C. McGrath. *Biology: Evolution and Adaptation to the Environment.* (Illus.) Houghton Mifflin, 1975. xii + 558pp. $12.95. Gloss.;index.

SH-C The authors have tried to avoid the emotional crisis approach in this excellent text in environmental biology. Geared to nonmajors, topics range from metabolism to competition among populations. An emminently readable, useful text.

Klein, Stanley. *A World of Difference: Living Things of the World.* (Illus.) Doubleday, 1971. 63pp. $4.95. 70-131086.

JH A review of differences and similarities in animals and plants and an introduction to adaptations in ecosystems. Offers a great deal of information, logically presented, with down-to-earth examples which should appeal to students.

Kormondy, Edward J. *Concepts of Ecology, 2nd ed.* (Illus.) Prentice-Hall, 1976.

xiv + 238pp. $9.95; $5.95(p). 75-30848. ISBN 0-13-166470-0; 0-13-166462-X. Index;CIP.

C   An excellent, basic, nonmathematical text for post-general biology students.
Discusses the nature of ecosystems, their energy flow and biogeochemical cycles, ecology of populations, and the organization and dynamics of ecological communities. New figures have been added, others clarified, and discussion of predation/prey, stability/diversity, and population behavior/stability are updated.

Kucera, Clair L., and John R. Rochow. *The Challenge of Ecology.* (Illus.) Mosby, 1973. xiv + 226pp. $5.95(p). 72-87642. ISBN 0-8016-2803-2. Index;bib.

SH-GA-C   Covers interdependence in ecology, the historic point of view on the ecosystem, energy relationships in photosynthesis, the energy pyramid, biotic succession as part of the ecosystem and its effect on environment, the soil environment, terrestrial biomes, and the world ocean. Includes an appendix on environmental issues, discussion questions, a list of endangered species, and chapter summaries.

McGraw-Hill. *McGraw-Hill Encyclopedia of Environmental Science.* (Illus.) McGraw-Hill, 1974. 754pp. $24.50. 74-13065. ISBN 0-07-045260-1. Index;CIP.

SH-C   An authoritative, timely, tightly written compendium of data and factual discussions drawn from the publisher's *Encyclopedia of Science and Technology* and subsequent yearbooks. Definitions are precisely stated and there is minimal use of mathematical equations. Reference librarians will find this a shortcut to hitherto scattered data.

Meeker, Joseph W. *The Spheres of Life: An Introduction to World Ecology.* (Illus.) Scribner's, 1975. 123pp. $6.95. 74-11217. ISBN 0-684-13937-5. Gloss.;index;CIP.

SH-C   A concise summary of important problems in world ecology. Energy, population, the lithosphere and the biosphere are some of the complex, interrelated systems discussed. This should be in the library of any individual or institution truly concerned with the human role in world ecology.

Milne, Lorus, and Margery Milne. *The Arena of Life: The Dynamics of Ecology.* (Illus.) Doubleday/Natural History, 1971. 351pp. $15.00. 71-159519.

SH-C   An excellent, nontechnical introduction to ecology. The authors emphasize the broader aspects of the subject and present a lively discussion of principles, historical information, the work of various scientists, and the development of ideas. The chapter on the human environment is an excellent portrayal of the reasons why ecologists are alarmed about humanity's survival potential.

Milne, Lorus, and Margery Milne. *The Nature of Life: Earth, Plants, Animals, Man and Their Effect on Each Other.* (Illus.) Crown, 1971. 320pp. $17.50. 72-130316.

SH-C   The book is like a motion picture of the still evolving, ever changing earth and its many landforms, waters, and plant and animal populations. The final chapter is an interesting consideration of the transformation of humans from hunters and gatherers, to sowers, reapers, and animal herders, and then to exploiters of all forms of life.

Nason, Alvin, and Robert L. Dehaan. *The Biological World.* (Illus.) Wiley, 1973. xii + 736pp. $12.95. 72-8573. ISBN 0-471-63045-4.

C   Provides a basic background for understanding growth and reproduction in the biological world, and emphasizes interrelationships among organisms and between organisms and their environment. Each new term or concept is identified by bold-face type followed by a brief definition. Useful as a classroom text.

Odum, Eugene P. *Fundamentals of Ecology, 3rd ed.* (Illus.) Saunders, 1971. xiv + 574pp. $11.75. 76-81826. ISBN 0-7216-6941-7.

SH-C-P   This version has excellent additional topical material on applications and technology. No other general textbook on ecology is as clearly written as this uniquely designed text which not only intelligent citizens but also undergraduate and graduate students in a variety of fields will find tremendously useful as either a text or a reference.

Pringle, Laurence. *Ecology: Science of Survival.* (Illus.) Macmillan, 1971. 152pp. $4.95. 72-158171. Bib.

JH   Fundamental principles of ecology such as interaction of organisms and abiotic environment, energy flow, biogeochemical cycles, succession, and population changes are topics for discussion in this book. It could be put to good use as an ecology primer for early teens.

Rasmussen, Frederick, et al. *Man and the Environment: Life Science Investigations (Teacher's Edition).* (Illus.) Houghton Mifflin, 1971. xii + 417pp. $6.60. 71-131290.

JH   Four units comprise this textbook: investigating living things, the environment affects living things, living things affect each other, and people affect the environment. Questions are posed for students; the teacher's role is that of a guide. The emphasis is on environmental management and control of pollution.

Russell, Helen Ross. *Earth: The Great Recycler.* (Illus.) Nelson, 1973. 160pp. $5.95. 73-7823. ISBN 0-8407-6268-2.

JH   Russell surveys the earth as part of the solar system, elements, properties of matter, molecular building blocks, water, green plants as energy traps, soils, energy, food chains, people and the ecosystem, all in relation to our environmental problems.

Segerberg, Osborn, Jr. *Where Have All the Flowers, Fishes, Birds, Trees, Water, and Air Gone? What Ecology is All About.* (Illus.) McKay, 1971. ix + 303pp. $6.95. 75-142068.

SH-C   A wealth of information in a single readable text. If books such as this one could become mandatory reading for the whole student body in secondary schools, we might see an increase in understanding of ecological problems and a modification of current trends.

Sutton, David B., and N. Paul Harmon. *Ecology: Selected Concepts.* (Illus.) Wiley, 1973. xv + 287pp. $3.95(p). 73-8715. ISBN 0-471-83830-6.

SH-C   This programmed reader has cross-references to 10 ecology textbooks. Topics include the nature of ecosystems, energy in ecological cycles, and ecology of populations. Behavioral objectives are listed at the beginning of each chapter, the explanatory material is very readable, and key terms are underlined. A useful supplement.

Tribe, Michael A., et al. *Ecological Principles.* (Illus.) Cambridge Univ. Press, 1975. viii + 160pp. $19.95; $6.95(p). 75-6285. ISBN 0-521-20658-8; 0-521-20638-3. Index.

C   Prepared for a basic undergraduate biology course, this programmed-instruction handbook attempts to develop an understanding of the structure of ecosystems. Most of the emphasis is on dynamics, with few comments on genetic or evolutionary principles.

Wesley, James Paul. *Ecophysics: The Application of Physics to Ecology.* (Illus.) Charles C. Thomas, 1974. xxvi + 340pp. $19.75; $13.95(p). 73-11066. ISBN 0-398-02959-8. Index;CIP.

C-P   With a stimulating freshness of approach, this book answers basic questions
about the mechanical structure of life. The author predicts that humans may
become extinct in 200 years by their own devices—machines. Good source material
for introductory lectures on the physical and life sciences.

574.524 ECOLOGY OF COMMUNITIES—INTERSPECIES RELATIONSHIPS

Flader, Susan L.   *Thinking Like a Mountain: Aldo Leopold and the Evolution of an
Ecological Attitude Toward Deer, Wolves, and Forests.* (Illus.) Univ. of Missouri Press,
1974. xxv + 284pp. $12.50. 74-80390. ISBN 0-8262-0167-9. Index.

SH-C-P   Flader documents both Aldo Leopold's emergence as an influential
ecologist and the development of concepts of deer management from the
time deer were thought to be a disappearing species to the time of overabundance and
habitat destruction. Recommended for every ecologist and conservationist; the
scholarship is outstanding.

Lockley, Ronald.   *The Island.* (Illus.) Regnery, 1971 (c.1969). 227pp. $6.95. 76-
143845.

SH-C   Close observations of birds and European rabbits as they interacted on a
240-acre island off the coast of Wales are framed by a deep appreciation for
the land, its plants, its weather changes, and its quasi-isolation.

MacArthur, Robert H.   *Geographical Ecology: Patterns in the Distribution of Species.*
(Illus.) Harper&Row, 1972. xviii + 269pp. $12.95. 71-188200.

C-P   A major contribution to the field of ecology. The first section reviews the
major questions of population ecology, with a discussion of climates, competition and predation, foraging strategies and evolution of species. The remainder of
the book develops the pattern of species, with a consideration of species diversity,
comparison of regions, and the role of history in geographical ecology.

Patent, Dorothy Hinshaw.   *Plants and Insects Together.* (Illus.) Holiday House, 1976.
128pp. $5.95. 75-34205. ISBN 0-8234-0274-6. Index;CIP.

JH-SH   Valuable as collateral reading in biology, evolution or ecology. The in-
tricacies of the relations between insects and plants as brought about by
co-evolution are presented in a lucid and easy style without the flamboyance or
mystical superlatives of some "nature" books.

Read, Clark P.   *Parasitism and Symbiology.* (Illus.) Ronald, 1970. 316pp. $10.00.
75-110390.

C-P   Interaction between organisms living in association and among the organisms themselves are described. Reviewed are classification of symbiotes,
their relationships with and without disease; symbiotic adaptation, nutrition and
metabolism; genetics and genetic markers associated with various organisms;
evolutionary patterns which have led to modifications in symbiosis; and how infectious diseases have influenced our cultures and history. An indispensable reference
book.

Simon, Hilda.   *Partners, Guests and Parasites: Coexistence in Nature.* (Illus.) Viking,
1970. 127pp. $4.53. 71-106924. ISBN 0-670-54087-0.

JH-SH   Symbiosis, parasitism, and commensalism are illustrated in talented drawings as scientifically accurate as photographs. The book is an accurate but
simplified exposition of the complex interdependent relationships among various
species which constitute a vital aspect of the balance of nature and of biological
cycles. Worthwhile collateral reading for biology students.

574.526 ECOLOGY OF SPECIFIC ENVIRONMENTS

See also 574.9 Biology of Areas and Regions.

Adams, Richard. *Nature Through the Seasons.* (Illus.) Simon&Schuster, 1975. 108pp. $8.95. 75-7667. SBN 671-22107-8. Index.

> **SH**   The progression of seasons provides a background for discussion of various topics such as photoperiodism, food and energy, pollination, seed dispersal, fungi, bacteria and cycling of elements; food chains webs; overwintering and hibernation; and temperature control of animals. The landscape illustrations are excellent and very detailed.

Batten, Mary. *The Tropical Forest: Ants, Ants, Animals and Plants.* (Illus.) Crowell, 1973. 130pp. $4.95. 73-4196. ISBN 0-690-00138-X; 0-690-00139-8.

> **JH-SH**   Ecology is the theme in this discussion of plant relationships, layers of life, survival techniques, and animals. Well-developed ideas and comparisons; recent research is described. Excellent!

Bennett, Isobel. *The Great Barrier Reef.* (Illus.) Scribner's, 1974 (c.1971). 183pp. $17.50. 73-9491. SBN 684-13620-1. Index.

> **SH-C**   A pictorial atlas of the coral reefs off Australia's eastern coast. Numerous illustrations accompany an informative narrative on the history, geology, and biologic features of the reef system and the islands. General interest reading which is also good for reference.

Burton, Jane. *Animals of the African Year: The Ecology of East Africa.* (Illus.) Holt, Rinehart&Winston, 1972. 173pp. $10.95. 72-79639.

> **SH-C**   In this description of East Africa and its biota, one sees clearly the interrelationships among the region's geology, climate, plants and animals; with man included but perhaps with insufficient emphasis. The book is a natural for both classroom use and collateral reading in ecology, biology, earth sciences, botany and zoology courses.

Costello, David F. *The Desert World.* (Illus.) Crowell, 1972. 264pp. $7.95. 77-184973. ISBN 0-690-23513-5.

> **SH-C**   This sensitive narrative deals with the main aspects and principal life forms (plants and animals) of the five major deserts in North America. The very readable and vivid presentation is from the ecologist's point of view. Unusually clear pictures. Though not a textbook, it could serve as one unit of a series covering the biotic regions of North America.

Faulkner, Douglas. *This Living Reef.* (Illus.) Quadrangle/N.Y. Times, 1974. 179pp. $27.50. 73-92293. ISBN 0-8129-0455-9.

> **SH-GA-C**   Faulkner presents a reliable picture of a coral reef and its flora and fauna and offers an easy introduction to some of the principles of marine ecology and to the physiographic nature and history of the islands. The writing and photographs are excellent. Recommended for anyone who has a feeling for natural beauty, plants, animals and faraway places.

Geist, Valerius. *Mountain Sheep and Man in the Northern Wilds.* (Illus.) Cornell Univ. Press, 1975. 248pp. $10.00. 75-5481. ISBN 0-8014-0943-8.

> **SH-C**   This is a personal account of both Geist's adventures and his scientific work in the Canadian wilderness. He provides a close look at human society in the wilderness and at animal life. Those interested in ecology, ethology and human evolution will find this a valuable reference.

Johnston, Verna R. *Sierra Nevada.* (Illus.) Houghton Mifflin, 1970. xiii + 281pp. $7.95. 79-96064. Bib.

SH-GA Johnston vividly and accurately paints the natural history of the Sierra Nevada, drawing on rich historical accounts and recent research. The altitudinal zonation of vegetation and animal life in this great mountain range provides the framework for interesting observations of nature's interactions, relevant anecdotes and outstanding photographs.

Kavaler, Lucy. *Freezing Point: Cold as a Matter of Life and Death.* (Illus.) John Day, 1970. 416pp. $8.95. 78-107206.

SH-C Interaction of cold with biological materials forms the subject matter. The temperatures considered run from those that are chilly to humans down to the near-absolute-zero temperatures of the laboratory or outer space. Topics are the adaptations man, animals and plants have made to allow them to survive in cold temperatures; medical uses of cold; preservation by freezing of cells or tissue; and a speculative section dealing with the beginnings of life on a cold planet and the possibilities of human hibernation for long space travel.

Kirk, Ruth. *Desert: The American Southwest.* (Illus.) Houghton Mifflin, 1973. xv + 361pp. $10.00. 73-9902. ISBN 0-395-17209-8. Index;bib.

SH-C There is a wealth of material presented in this book about deserts. In addition, it contains a selective list of scientific names, maps on the inside covers, and an abundance of excellent photographs.

McCombs, Lawrence G., and Nicholas Rosa. *What's Ecology?* (Illus.) Addison-Wesley, 1973. v + 248pp. $4.00. ISBN 0-201-4513-3.

SH-C A "book of knowledge" type of presentation, providing general coverage of fundamental ecological issues concerning forest, fresh water, oceanic and other habitats. A reference for students as well as a particularly useful book for elementary and junior high school teachers.

McGregor, Craig, et al. *The Great Barrier Reef.* (Illus.) Time-Life, 1974. 184pp. $7.95. Index.

SH-C The authors describe the history and geography of the Great Barrier Reef, the natural history of its component organisms and devote some attention to the biology of the numerous islands that are scattered along the Reef. The numerous photographs are excellent.

MacKinnon, J., and K. MacKinnon. *Animals of Asia: The Ecology of the Oriental Region.* (Illus.) Holt, Rinehart&Winston, 1974. 172pp. $11.95. 74-19372. ISBN 0-03-014116-8. Gloss.;index;CIP.

SH Following a general introduction to the Asian area, the authors describe the different habitats, from desert wastes to tropical rain forests, and the various animals peculiar to each. Emphasis is on ecological relationships, and the book is very attractively done. A good book that will serve as reference material for term papers or just browsing.

Perry, Richard. *Life in Forest and Jungle.* (Illus.) Taplinger, 1976. 254pp. $8.95. 74-21573. ISBN 0-8008-4799-7. Index;bib.

SH-C As a brief synthesis of world literature on forest wildlife ecology, this book will be useful as supplementary reading for forestry students. Adaptations are described for more than 200 birds, mammals and social insects. Chapter organization is partly by biomes. Extensive bibliography.

Quilici, Folco. *The Great Deserts.* (Illus.; adapted by Margaret O. Hyde) McGraw-Hill, 1969. 128pp. $4.50. 70-82745.

JH-SH   Well-written histories of the human, physical and biological ecologies of the great desert regions are provided in this adaptation. Factors that differentiate the life struggles in these regions, and consequently the type of fauna and flora, are emphasized.

Russell, Franklin. *At the Pond, Vol. 1: Corvus the Crow.* (Illus.) Four Winds, 1972. 116pp. $5.50. 72-182120.

JH-GA   The balance among plant and animal life, maintained through disease, natural disaster and predation among species, is dramatically described. Events are seen primarily through the eyes of an old crow forced by injuries to live year-round at the pond. "Corvus" takes note of each small change in his territory: populations arriving and leaving, breeding and dying; food supplies fluctuating. Masterfully evokes a sense of the wilderness.

Russell, Franklin. *At the Pond, Vol. 2: Lotor the Raccoon.* (Illus.) Four Winds, 1972. 92pp. $5.50. 73-189140.

JH-GA   A sense of the timelessness of the natural world is the strongest impression the reader will carry away from this truly excellent book about life at the pond as seen through the eyes of "Lotor" the raccoon. While readers will not learn much about the technical details of raccoon anatomy or physiology, they will learn much about the general balance and the superimposed food/breeding cycles of the natural world.

Schwartz, George I., and Bernice S. Schwartz. *Life in a Log.* (Illus.) Doubleday, 1972. 129pp. $5.95. 79-177294.

JH   Brings into view a wide variety of biological observations and relationships while maintaining the unity of the subject matter. The material included reinforces many of the points made in elementary biology texts, and the section on projects and study suggestions increases the value of the book for school use.

Shepherd, Elizabeth. *Arms of the Sea: Our Vital Estuaries.* (Illus.) Lothrop, Lee&Shephard, 1973. 160pp. $4.95. 73-4951. ISBN 0-688-41558-X; 0-688-51558-4.

JH-SH   Shepherd introduces the reader to the basic concepts of biological production in water and to the natural and man-made factors that act upon the production processes. She describes the life cycles of birds, fishes, mammals and various invertebrates and also conveys a clear picture of the requirements and means of estuarine protection.

Stephenson, T.A., and Anne Stephenson. *Life Between Tidemarks on Rocky Shores.* (Illus.) Freeman, 1972. xii + 425pp. $15.00; $6.95(p). 79-152055. ISBN 0-7167-0687-3; 0-7167-0698-9. Bib.

SH-GA-C   A wealth of information on shore ecology, with emphasis on the distribution of animals and plants in the intertidal zones. Good, solid data are reported from North America, Africa, Australia, the British Isles, the Arctic, Chile, Mauritius and Bermuda. Well organized and clearly written; nomenclature and bibliography are dated but still useful. Suitable for supplementary reading or reference.

Stonehouse, Bernard. *Animals of the Antarctic: The Ecology of the Far South.* (Illus.) Holt, Rinehart&Winston, 1972. 171pp. $10.95. 72-76576. ISBN 0-03-091962-2.

SH-C   For the past several million years Antarctica has been a simple unbuffered ecosystem populated by specialized animals and plants. This book is about these organisms, their habitats and the teeming, biologically productive seas around the continent. But the book also warns us about the newest species on the scene— man.

Stonehouse, Bernard. *Animals of the Arctic: The Ecology of the Far North.* (Illus.) Holt, Rinehart&Winston, 1971. 172pp. $10.95. 74-162304. ISBN 0-03-086699-5.

JH-SH   Clear, short, informative sentences make this book outstanding for this age group. A multitude of color photographs with informative captions could almost make up an independent book. The biota of the Arctic are described and short life histories of many birds and mammals are given. Highly recommended.

Ursin, Michael J. *Life In and Around Freshwater Wetlands: A Handbook of Plant and Animal Life In and Around Marshes, Bogs, and Swamps of Temperate North America East of the Mississippi.* (Illus.) Crowell, 1975. 116pp. $6.95. 74-13632. ISBN 0-690-00673-X; 0-8152-0378-0(p). Index;CIP.

JH-SH   The book provides a basic, easily understood taxonomic framework for the common plants and animals of marshes, swamps and bogs. Each organism is illustrated, and there are descriptive and ecological notes. The only drawback is the rather brief introduction to wetlands.

Warner, William W. *Beautiful Swimmers: Watermen, Crabs and the Chesapeake Bay.* (Illus.) Atlantic/Little, Brown, 1976. xiii + 304pp. $10.00. 75-29289. ISBN 0-316-92326-5. CIP.

SH-C   This book is a must for every high school library, and a fine gift for both amateur and professional biologists. In an anecdotal, narrative style, Warner describes the wonderful complexity of marine life in Chesapeake Bay and the dependence of the fishermen on their environment. He captures especially well the intricate give-and-take and the critically timed changes in the life cycles of the myriad of organisms. Includes lovely sketches.

Wellman, Alice. *Africa's Animals: Creatures of a Struggling Land.* (Illus.) Putnam's, 1974. 191pp. $4.97. 72-94946. SBN GB-399-60838-9; TR-399-20349-4.

JH-SH   Wellman's book is based on six African ecosystems—thornbush, rain forest, grassed plains, swamps and bodies of water, highlands, and open bush and woodlands. The location, topography, weather, seasonal changes and typical vegetation of each ecosystem are described. For each system, one or two animals (species) are described in some depth.

574.65 ORGANISMS DELETERIOUS TO HUMAN INTERESTS

See also 574.524 Ecology of Communities—Interspecies Relationships.

Busch, Phyllis S. *Living Things that Poison, Itch, and Sting.* (Illus.) Walker, 1976. 128pp. $6.95. 74-31905. ISBN 0-8027-6218-2; 0-8027-6215-8. Index.

JH   Provides very elementary descriptive material on potentially harmful shellfish, jellyfish, snails, sea urchins, fish, insects and arachnids, and on poisonous plants. Includes scientific names for further reference.

Hopf, Alice L. *Misplaced Animals and Other Living Creatures.* (Illus.) McGraw-Hill, 1976. 136pp. $5.72. 75-10952. Index;bib.

JH   The theme of Hopf's book is the introduction of organisms into new areas by humans. Those discussed are pigeons, rats, gypsy moths, lampreys, "Brazilian" bees, water hyacinths, English sparrows, prickly pear, rabbits, pheasants, striped bass, brown trout, reindeer and others.

Silverstein, Alvin, and Virginia Silverstein. *Animal Invaders: The Story of Imported Wildlife.* (Illus.) Atheneum, 1974. 122pp. $5.95. 73-84835. ISBN 0-689-30146-4. Index.

JH-SH   This timely report on the ecological impacts of accidental or intentional transfers of wildlife focuses on the unexpected and sometimes severe effects animals may have on an alien environment and stresses the need for humans to maintain that delicate ecological balance necessary for harmony among wildlife.

574.87 CELL BIOLOGY

Ambrose, E.J., and Dorothy M. Easty. *Cell Biology.* (Illus.) Addison-Wesley, 1970. x + 500pp. $12.50. 73-137431.

   C   The authors begin with a brief review of the properties of cells and basic chemical and physical concepts for students deficient in these areas. Subsequent chapters emphasize the way in which cells interact in the development and functioning of whole organisms. The interaction of the part with the whole theme is well presented, and an outstanding feature is the incorporation of material from all biological areas—animal and plant, higher organism and microorganism.

Borek, Ernest. *The Sculpture of Life.* (Illus.) Columbia Univ. Press, 1973. xii + 181pp. $10.00; $2.95(p). 73-6831. ISBN 0-231-03425-3.

   SH-C   An account of cellular biology and growth, this book covers such topics as cell structure, metabolism, aggregations, sex, cytogenetics, growth and its regulation, and many other subjects. Useful collateral reading.

Brown, Walter V., and Eldridge M. Bertke. *Textbook of Cytology, 2nd ed.* (Illus.) Mosby, 1974. vii + 528pp. $16.00. 73-14625. ISBN 0-8016-0831-7. Index;CIP.

   C-P   The book is a general textbook of cytology, designed for the biology major, the pre-med student, or the biology minor. The authors present a fine balance of literature, reviews, and classical concepts. This is a rich and elegant collection of writing about cells.

Dowben, Robert M. *Cell Biology.* (Illus.) Harper&Row, 1971. xiv + 570pp. $12.95. 78-137802. ISBN 0-06-041698-X.

   C   Compared with others on this subject, this text has excellent style and organization, very accurate information and excellent illustrations. (But knowledge of biochemistry, biophysics, and 2 years of college mathematics is essential.) Presentations of and illustrative materials on biochemical, biophysical, and genetic concepts are derived from the work of leading scientists in their respective fields.

Gunning, Brian E.S., and Martin W. Steer. *Ultrastructure and the Biology of Plant Cells.* (Illus.) Arnold (distr.: Crane, Russak), 1975. vii + 312pp. $70.00. ISBN 0-7131-2494-6. Indexes.
Plant Cell Biology: An Ultrastructural Approach. (Illus.) Crane, Russak, 1975. 101pp. $8.95(p). 75-13749. ISBN 0-8448-0669-2.

   SH-C-P   In these books the many illustrations of the plant cell and its components exhibit an outstanding brilliance of detail and reproduction. The first book is a hardcover edition which includes extensive text; the second is paperback with only the illustrations and accompanying legends. Excellent!

Hall, D.O., and Shirley E. Hawkins (Eds.). *Laboratory Manual of Cell Biology.* (Illus.) Crane, Russak, 1975. 281pp. $9.75(p). 74-21529. ISBN 0-8448-0601-3.

   C-P   This comprehensive laboratory manual includes techniques of cell and tissue culture, tracers, phase microscopy, cytology, membrane surfaces, cell motility, physiological chemistry, growth, viruses, enzyme induction, differentiation and immunology.

Harris, Patricia J., (Ed.). *Biological Ultrastructure: The Origin of Cell Organelles.* (Illus.) Oregon State Univ., 1971. 128pp. $5.00. 52-19235. ISBN 0-87071-169-5.

   C-P   Reports advances in viewing and interpreting the structure of some of the organelles of the cell, and examines the idea that mitochondria and chloroplasts may have once been free-living invaders. The collection is outstanding in explaining methods, problems and interpretation, including some limitations of the data. An advanced but needed reference for college biology.

Kruse, Paul F., Jr., and M.K. Patterson, Jr. (Eds.). *Tissue Culture: Methods and Applications.* (Illus.) Academic, 1973. xxvii + 868pp. $22.00. ISBN 0-12-427150-2.

SH-C-P   One of the great students of tissue culture has put together the best, most complete and timeliest book on the subject yet written. Contributors are major tissue culture experts who developed a special technique or process and have written a short article on that technique alone. Even people who know nothing about tissue culture can understand what the process is and how it is done.

McElroy, William D., and Carl P. Swanson. *Modern Cell Biology, 2nd ed.* (Illus.) Prentice-Hall, 1975. xvi + 398pp. $9.25(p). 75-22218. ISBN 0-13-589614-2. Index;CIP.

SH-C   The book deals with the biochemistry of the cell and provides descriptions of the general properties of cells. The development of diploid organisms is covered and a segment on human evolution is included. Suitable for a one-semester introductory biology course.

Puck, Theodore T. *The Mammalian Cell as a Microorganism: Genetic and Biochemical Studies in Vitro.* (Illus.) Holden-Day, 1972. xi + 219pp. $14.95. 73-188127. ISBN 0-8162-6980-7.

C   Discusses the growth of single mammalian cells into distinct colonies, the characteristics of mammalian chromosomes, selective mutation of single genes, gene mapping, effect of radiation, and molecular biology of the reproductive cycle of cells. An excellent discussion of *in vitro* techniques contributes to a better understanding of the mechanisms of gene expression and cell differentiation in mammals. The human implications of the scientific developments are also discussed.

Simkiss, K. *Bone and Biomineralization.* (Illus.) Arnold (distr.: Crane, Russak), 1975. 60pp. $3.00(p). ISBN 0-7131-2492-X; 0-7131-2493-8(p).

C-P   Describes the fundamental physiological and biochemical processes associated with biomineralization, with the emphasis on bone. Exceptionally well organized, clear, concise and comprehensive. Presents several critical experiments.

White, M.J.D. *The Chromosomes, 6th ed.* (Illus.) Chapman&Hall (distr.: Halsted), 1973. 214pp. $11.75; $6.50(p). 73-7337. SBN 412-11930-7.

C-P   An introductory book which deals with the morphology of chromosomes, their behavior in resting and dividing cells, and the growth of applications of this knowledge in science and human affairs. A suitable reference, lucid and current.

574.88 MOLECULAR BIOLOGY

Ayala, Francisco J., (Ed.). *Molecular Evolution.* (Illus.) Sinauer, 1976. x + 277pp. $10.00(p). 75-36113. ISBN 0-87893-044-2. Indexes.

C-P   Underscores the utility of standard biochemical and molecular biological techniques in areas as diverse as population variability and the evolution of protein structure. The book would serve well as the nucleus of an interdisciplinary course, or a supplement for a biochemistry or molecular biology course. An excellent resource.

Borek, Ernest. *The Code of Life.* (Illus.) Columbia Univ., 1969 (c.1965). xi + 226pp. $2.25(p). 65-10944. ISBN 0-231-08630-X.

SH-C   An excellent and still useful introduction to the nucleic acids and to the fundamentals of genetics and molecular biology.

Olby, Robert. *The Path to the Double Helix.* (Illus.) Univ. of Wash. Press, 1975. xxiii + 510pp. $23.50. 74-10676. ISBN 0-295-95359-4. Index;CIP.

C-P   Robert Olby traces the evolution of a concept through the several generations of ideas that made the final one possible. The idea is the "Central
Dogma" of Watson and Crick, placing the genetic message on the DNA molecule and
suggesting a mechanism by which the information may be replicated.

Parker, Gary E., and Thomas R. Mertens. *Life's Basis: Biomolecules.* (Illus.) Wiley, 1973. xvii + 158pp. $2.95(p). 72-10744. ISBN 0-471-65919-3.

C   Students interested in molecular biology may obtain a minimum knowledge of
chemistry through this programmed guide for associating the appropriate
chemical terminology with the relevent biological phenomenology. Not a substitute
for a biochemistry course but adequate to introduce beginners, laypersons and nonscience students to the subject.

Raacke, Ilsa Dorothea. *Molecular Biology of DNA and RNA: An Analysis of Research Papers.* (Illus.) Mosby, 1971. xiii + 291pp. $6.90. ISBN 0-8016-4068-7.

C   Assists students in acquiring the skill of reading scientific journals through a
concise dialogue on logical procedures for approaching the literature. Presents
information about the molecular biology of DNA and RNA in a well-chosen compilation of papers on the subject, an analysis of each chapter, and questions and answers
following each chapter. Valuable collateral reading for the beginning student in genetics and molecular biology.

Sayre, Anne. *Rosalind Franklin and DNA.* (Illus.) Norton, 1975. 221pp. $8.95. 75-11737. ISBN 0-393-07493-5. CIP.

SH-C   Franklin's scientific beginnings are described and her attitudes and emotions discussed in detail. The book is intended to acknowledge Franklin's
work through which, using x-ray diffraction equipment, she had found the answers to
most questions about the DNA structure before Watson and Crick. There are 21
pages of notes and references. A unique chronicle of a scientist.

574.9 BIOLOGY OF AREAS AND REGIONS

Bakker, Elna S. *An Island Called California: An Ecological Introduction to Its Natural Communities.* (Illus.) Univ. of Calif., 1971. xvii + 357pp. $10.00. 70-107657. ISBN 0-520-01682-3.

C   This nature tour of California reads like a travelogue, yet it has sufficient
scholarly detail to make valuable collateral reading for students of natural
history. Organized by types of biomes, each chapter is a story in itself, which is
accompanied by photographs and diagrams that explain the peculiar characteristics
of the locality.

Burton, Maurice, (Ed.). *The Shell Natural History of Britain.* (Illus.) London House &Maxwell, 1970. 481pp. $12.95. ISBN 0-8277-0152-7.

SH-C   An interesting and attractive natural history survey of the British Isles;
includes a resumé of the geological evolution of Great Britain and miscellaneous facts, opinions, and personal recollections. Each division—plants, invertebrates, the seashore, fishes, amphibians and reptiles, birds and mammals—is written
by a well-known British specialist.

Carlquist, Sherwin. *Hawaii, a Natural History: Geology, Climate, Native Flora and Fauna Above the Shoreline.* (Illus.) Natural History Press, 1970. x + 463pp. $19.95. 72-101431.

SH-C   The geology, climate, volcanic origins and other salient features of the
geographic history are described, followed by a discussion of biological

phenomena, and the dispersal of life. Adaptive radiation, unique adaptations, arborescence, flightlessness in insects and birds, and loss of dispersibility of plants are some interesting sections of the book. There are biological discussions of the various unique ecological regions of the main islands.

Guggisberg, C.A.W. *Man and Wildlife.* (Illus.) Arco, 1970. 224pp. $12.50. 76-110614. ISBN 0-668-02282-5.

    SH-C Packed with detailed information on the interaction of man and wildlife from the times of our immediate prehuman ancestors to the present. It adds a guide to the world's national parks and wildlife sanctuaries, with data on extinct and endangered species.

Hutchins, Ross E. *Hidden Valley of the Smokies: With a Naturalist in the Great Smoky Mountains.* (Illus.) Dodd, Mead, 1971. viii + 214pp. $6.50. 74-160863. ISBN 0-396-06382-9.

    SH-C A treasurehouse of minutiae, mostly about plants, but with considerable attention paid to animals, their protection, care and study. Ranges from albino squirrels to poisonous plants, with lucid text and remarkably clear photographs.

Johnson, James Ralph. *The Southern Swamps of America.* (Illus.) McKay, 1970. 152pp. $4.25. 75-120817.

    JH Central to this adequate account of the major Southern swamps of the United States is the author's sense of proportion regarding the plants and animals and the swamps in which they live. The subject of conservation is prevalent as the simple narrative guides the reader through Okefenokee, Everglades and Corkscrew swamps before terminating at Reelfoot Lake.

Ketchum, Richard M. *The Secret Life of the Forest.* (Illus.) American Heritage Press, 1970. 108pp. $7.95. 73-117353.

    JH-SH The physiology of forest trees, ecological succession in forest communities, diversity of form in trees, forest management principles, and the uses of forest products are all described. The text is profusely illustrated with excellent color drawings. Not only should this book be in every library, but it also could serve easily as required reading in environmentally oriented biology classes.

Kinkead, Eugene. *A Concrete Look at Nature: Central Park (and Other) Glimpses.* (Illus.) Quadrangle/N.Y. Times, 1974. xii + 242pp. $8.95. 74-77936. ISBN 0-8129-0471-0. Index;CIP.

    SH Kinkead's 13 essays provide a new and personal look at Central Park and other areas for recreation in New York. Excellent coverage of the ecology, nature, and wildlife of this somewhat misunderstood section of New York.

Ross-MacDonald, Malcolm, (Ed.). *The World Wildlife Guide.* (Illus.) Viking, 1971. 416pp. $8.95. 77-163977. ISBN 0-670-79018-4.

    C This unique volume contains brief descriptions of the flora and fauna of the major geographical regions followed by detailed descriptions of over 650 national and state parks, reserves and sanctuaries throughout the world. Not only are typical plants and wildlife listed, but information is given on how to get to the areas, the optimum season for visiting, and much other useful information.

Sutton, George Miksch. *High Arctic: An Expedition to the Unspoiled North.* (Illus.) Eriksson, 1971. viii + 119pp. $12.95. 72-151434. ISBN 0-8397-3300-3.

    SH-GA The author—explorer, artist, ornithologist—reports what he did and saw during a summer expedition in the Canadian Arctic. The photographs and paintings are beautiful. Much attention is given to musk-oxen, wolves, hares, lemmings, gulls, snowy owls, jaegers, kittiwakes, sandpipers, sanderlings, ptarmigan, and people.

Thornton, Ian   *Darwin's Islands: A Natural History of the Galápagos*. Natural History, 1971. xiv + 322pp. $7.95. 69-20061.

> **SH-C**   A well-written and informative text accompanied by readable and accurate maps and diagrams plus a good assortment of black-and-white photographs details the processes of natural colonization of the islands by plants and animals, including the often adverse effects of animals accidentally introduced by man. Highly recommended for collateral or pleasure reading.

Zwinger, Ann H., and Beatrice E. Willard.   *Land Above the Trees: A Guide to American Alpine Tundra*. (Illus.) Harper&Row, 1972. xviii + 489pp. $15.00. 72-79702. SBN 06-014823-3. Gloss.;bib.

> **SH-C**   Describes the boulder fields and some of their plant and animal inhabitants; the talus slopes; the "cushioned or matted" plants; and the meadows, turfs, marshes and lakes; alpine areas in the western U.S.; the Mount Washington area of New England; and human use of the tundra.

574.92 MARINE BIOLOGY

See also 551.46 Oceanography.

Cousteau, Jacques-Yves.   *Oasis in Space*. (Illus.) World, 1972. 144pp. $7.95. 72-87710. ISBN 0-529-04936-8.

> **JH**   Concerned principally with ecology and with the close relationships between the oceans and other bodies of water. Discusses the marine environment, the nature and behavior of marine organisms and the impact of human activity. Contains superb color photographs, used as much to set the theme as to illustrate specific statements. Stimulating reading.

Creative, Editors of.   *Life in the Sea*. (Illus.) Creative Educational Society, 1971. 37pp. $4.95. 70-140638. ISBN 0-87191-073-X.

> **JH**   Physical, chemical and biological oceanography comprise this slim volume. Each chapter consists of a page of text and an excellent color photograph.

Fell, Barry.   *Introduction to Marine Biology*. (Illus.) Harper&Row, 1975. x + 356pp. $10.95. 74-31119. ISBN 0-06-042034-0. Index;CIP.

> **C**   Fell clearly demonstrates the interrelationships among the marine sciences. The organisms most likely to be encountered in oceanographic collections along the seashore are described. Suggestions for laboratory exercises abound, and no additional laboratory text should be needed.

Laurie, Alec.   *The Living Oceans*. (Illus.) Doubleday, 1973. 187pp. $6.95. 70-178897.

> **JH-SH-C**   A first-rate book, with eye-pleasing typography and excellently reproduced illustrations, most in magnificent color. Well-organized, lucid, and accurate sections on physical oceanography, planktonic animals, vertical migrations, the deep sea, the continental shelf and the littoral.

Lucas, Joseph, and Pamela Critch.   *Life in the Oceans*. (Illus.) Dutton, 1974. 216pp. $7.95. 74-3635. ISBN 0-525-14545-1. Gloss.;index.

> **JH-SH**   This capsule survey of marine life is designed to bring the reader in closer touch with biological oceanography. Authors treat the ocean as an environment where physical, chemical, and geological factors interact with biology.

McConnaughey, Bayard H.   *Introduction to Marine Biology, 2nd ed.* (Illus.) Mosby, 1974. x + 544pp. $13.95. 73-17290. ISBN 0-8016-3257-9. Index;gloss.;bib.;CIP.

C   This introductory text for students entering the field of marine biology deals
with marine environment and biota. The author looks at the oceans throughout
time and considers not only the progression of life forms over the ages but also such
topics as continental drift, world climates and the impact of man on the oceans.

Paysan, Klaus. *Creatures of Pond and Pool.* (Illus.) Lerner, 1971. 106pp. $6.95. 77-
102893. ISBN 0-8225-0562-2.

JH   Excellent descriptions of selected species of amphibians, reptiles and insects
that inhabit freshwater ponds. The textual material very clearly presents
information concerning structural characteristics, life history and breeding habits.
The description of each species is accompanied by a color photograph, and the
close-up photography is superb.

Perry, Richard. *The Unknown Ocean.* (Illus.) Taplinger, 1972. 288pp. $7.95. 71-
148828. ISBN 0-8008-7938-4.

JH-SH-C   In a clear and interesting fashion, Perry describes the natural history of
the sea, from plankton to giant whales, and discusses the balance of
interrelated activities of prey, predator and symbiont, their patterns of migrating,
breeding, feeding, camouflage and defense. The vivid and exciting descriptions cover
aspects of coastal, offshore and deep water life forms and provide penetrating insights into life histories, ecological relationships and conservation problems.

Russell-Hunter, W.D. *Aquatic Productivity.* (Illus.) Macmillan, 1970. xiii + 306pp.
$5.95. 77-113935.

SH-C   This introduction to some basic aspects of biological oceanography and
limnology covers fundamental concepts of ecology as they relate to marine
and freshwater ecosystems. Global productivity, the population problem, and future
exploitation of the sea are also dealt with in this thought-provoking book.

Schlieper, Carl, (Ed.). *Research Methods in Marine Biology.* (Illus.; trans.) Univ. of
Wash. Press, 1973. xiii + 356pp. $12.50. 72-6089. ISBN 0-295-95234-2.

SH-C   These easy-to-follow, cookbook-type instructions cover physical and
chemical oceanographic methods, collection and preservation of various
plants and animals, and laboratory cultures. Presentation is clear and well organized.

Schwartz, George I. *Life in a Drop of Water.* (Illus.) Doubleday, 1970. 174pp. $4.50.
71-92177.

JH   A fascinating introduction to the myriad components of plankton. Clear verbal descriptions and plentiful but carefully selected illustrations teach a good
deal about the essence of life, including metabolism, reproduction, and development,
and painlessly introduce taxonomy. Suggestions for the collection of aquatic organisms and their maintenance in a micropond are also given.

Silverberg, Robert. *The World Within the Ocean Wave.* (Illus.) Weybright&Talley,
1972. 114pp. $6.95. 78-186561.

JH-SH   For a nonspecialist reader, this book presents an excellent discussion of
oceanic organisms and their general biology, and a review of features of
the oceans, such as physical properties and biological life. The biology and ecology of
the phytoplankton and algae are evaluated, as is their role in primary production. The
natural history of the major zooplankton is described. The last chapter is devoted to
food from the sea and how humans have polluted the oceans.

Silverberg, Robert. *The World Within the Tide Pool.* (Illus.) Weybright&Talley, 1972.
119pp. $6.95. 78-186560.

JH-SH   Plants and animals found within the tide pools of the Atlantic and Pacific
are examined. Topics include the three major ecological zones of the tide

pool, the important biota in these zones, and the tidal forces which influence zonation in the shoreline environment, the biology and ecology of the animals and plants found within the tide pools, and how human pollutants are damaging and, in some instances, destroying tide pools.

Sterling, Philip. *Sea and Earth: The Life of Rachel Carson.* (Illus.) Crowell, 1970. 213pp. $4.50. 70-87157.

JH-SH   A well-researched, charmingly written biography of a gifted biologist and writer who has conveyed to thousands her love of nature, of the sea and its creatures, of life itself and her dedication to conservation through her books.

Teal, John, and Mildred Teal.   *The Sargasso Sea.* (Illus.) Atlantic/Little, Brown, 1975. xxxi + 216pp. $10.00. 74-34176. ISBN 0-316-84351-3. Index;CIP.

SH-C   This is a good biological description of the Sargasso Sea. There is an excerpt from John Teal's journal describing an oceanographic cruise to the sea, and descriptions of the *Sargassum,* the sea, and its migrants.

574.999 ASTROBIOLOGY

Aylesworth, Thomas G.   *Who's Out There? The Search for Extraterrestrial Life.* (Illus.) McGraw-Hill, 1975. 119pp. $5.72. 74-31236. ISBN 0-07-002637-8. Index;CIP.

SH-C   Aylesworth gives a comprehensive review of the arguments for and against life elsewhere. Numerous scientists and scientific findings concerning extraterrestrial life are quoted. The illustrations are excellent.

Bracewell, Ronald N.   *The Galactic Club: Intelligent Life in Outer Space.* (Illus.) Freeman, 1975. xii + 141pp. $6.96; $3.95(p). 74-23056. ISBN 0-7167-0353-X; 0-7167-0352-1. Index;CIP;bib.

C   Bracewell traces the development of current ideas about extraterrestrial life, giving both sides of each controversy. A wide range of scientific facts, theories and speculation is covered.

Engdahl, Sylvia Louise.   *The Planet-Girded Suns: Man's View of Other Solar Systems.* (Illus.) Atheneum, 1974. 201pp. $7.50. 73-84825. ISBN 0-689-30135-9. Index;CIP.

SH-C   This is a historical and religious look at prospects for intelligent life elsewhere in the universe. Reasonably accurate scientific discussions occur as frequently as philosophical and religious treatments of interstellar communication.

575 ORGANIC EVOLUTION AND GENETICS

See also 573.2 Organic Evolution and Genetics of Humans.

Campbell, Bernard, (Ed.).   *Sexual Selection and the Descent of Man: 1871–1971.* (Illus.) Aldine, 1972. x + 378pp. $14.75. 70-169510. ISBN 0-202-02005-3.

C-P   These essays, written mainly by well-known figures in the fields of biology, anthropology and genetics (Eiseley and Simpson, Dobzhansky, Mayr, Fox and Caspari, etc.), examine the perceptions of Charles Darwin in the light of present-day knowledge. The extent to which Darwin's observations and conjectures were valid is strikingly shown in discussions of natural selection, sexual selection in lower organisms, and human evolution.

Corwin, Harry O., and John B. Jenkins (Eds.). *Conceptual Foundations of Genetics: Selected Readings.* (Illus.) Houghton Mifflin, 1976. xii + 471pp. $7.95(p). 75-26092. ISBN 0-395-24064-6.

C-P   This reader for a 1-term genetics course includes the definitive papers detailing the discovery of DNA, the mode of its inheritance, the source of genetic variation, and the mechanisms of gene expression. The combination of classical and recent papers, with prefaces to indicate each paper's significance, is a great aid for the novice reader of scientific literature.

Crow, James F. *Genetics Notes, 7th ed.* (Illus.) Burgess, 1976. vi + 278pp. $5.95(p). 75-16798. ISBN 0-8087-0360-9. Gloss.;index.

    SH-C-P   An outstanding introduction to genetics by a world renowned geneticist. The balance between human, animal, plant, bacterial and viral genetics is exquisite; the organization is logical and the writing is excellent.

Darwin, Charles, and Thomas Henry Huxley. *Autobiographies.* (Illus.; edited by Gavin de Beer) Oxford Univ. Press, 1974. xxvi + 123pp. $10.75. ISBN 0-19-255410-7. Index;bib.

    SH-C   Darwin's autobiography forms the bulk of the book, while Huxley's is very short. Complete chronologies of each scientist's work are included.

Forrest, D.W. *Francis Galton: The Life and Work of a Victorian Genius.* (Illus.) Taplinger, 1974. x + 340pp. $14.95. 74-5819. ISBN 0-8008-2682-5. Index.

    C   Galton's life and professional accomplishments in genetics, teaching and other fields are detailed. An overuse of direct quotes sometimes causes tedious reading, but this is a well-documented and useful reference.

Gardner, Eldon J., and Thomas R. Mertens. *Genetics Laboratory Investigations, 6th ed.* (Illus.) Burgess, 1975. vi + 180pp. $5.50(p). ISBN 0-8087-0742-6.

    C   These 20 investigations cover basic genetics. Procedures demonstrate fundamental chromosome phenomena and quantitative aspects of genes; chromatography, phage recombination and microbial enzyme genetics are included.

Hanson, Earl D. *Animal Diversity, 3rd ed.* (Illus.) Prentice-Hall, 1972. 164pp. $7.95. 72-000015. ISBN 0-13-037168-8; 0-13-037150-5(p).

    C   Emphasizes the concepts and principles on which phylogeny is based and the problems yet to be solved. The author's interpretations are controversial, but the book's conclusions are sufficiently open-ended so that further investigations are encouraged. Exceptionally valuable as a brief, authoritative introduction to the principles of animal phylogeny.

Head, J.J., (Ed.). *Readings in Genetics and Evolution: A Collection of Oxford Biology Readers.* (Illus.) Oxford Univ. Press, 1973. 260pp. $11.25. ISBN 0-19-914149-5.

    SH-C   The articles in this composite volume are on mitosis; meiosis; genetic recombination in fungi; gene expression during cell differentiation; the nucleolus; and various aspects of evolution, beginning with the origin of life and continuing through discussions of homology, adaptation and origin and development of plant and animal groups. Excellent diagrams, drawings and photographs; lively text.

Hull, David L. *Darwin and His Critics: The Reception of Darwin's Theory of Evolution by the Scientific Community.* Harvard Univ. Press, 1973. xii + 473pp. $18.50. 72-81274. SBN 674-19275-3.

    C-P   Anthologizes reviews of the *Origin of Species* and kindred subjects, and discusses the contemporary views of the philosophy and logic of science that seemed to color the opinions of the reviewers. Darwin's comments on each review show his evaluation of the review or reviewer. Chapters on the inductive method, occultism, teleology, and essentialism complete the book.

Kuspira, John, and G.W. Walker. *Genetics: Questions and Problems.* (Illus.) McGraw-Hill, 1973. viii + 776pp. $8.50(p). 72-6855. ISBN 07-035672-6.

    C   Covers classical, molecular, and population genetics, with theoretical and practical questions about a wide range of organisms. Not a text in the usual sense, but rather a very extensive collection of questions and problems designed to complement a current genetics text and to assist students in their study and review of lecture topics. Very stimulating.

Lewis, K.R., and B. John. *The Matter of Mendelian Heredity, 2nd ed.* (Illus.) Wiley, 1973. 273pp. $8.75(p). 72-7450. ISBN 0-470-53199-1. Bib.

C   Emphasizes principles rather than facts so that a limited quantity of information stimulates critical evaluation of data. Covers Mendelian inheritance and segregation, reproduction, variation, selection, evolution, mutation, and the physical and chemical basis of heredity. This clear, accurate, well-illustrated presentation will supplement standard textbooks.

Ludmerer, Kenneth M. *Genetics and American Society: A Historical Appraisal.* Johns Hopkins Univ. Press, 1972. xi + 222pp. $10.00. 72-4227. ISBN 0-8018-1357-3.

SH-C   Centers principally on the up-and-down history of eugenics in the U.S., and the often reluctant courtship between eugenicists and geneticists. Well written and very carefully researched, often relying on direct interviews with geneticists. This book ably details what happens when apparently well-intentioned people use the public interest to pervert the course of science.

Luria, S.E. *Life: The Unfinished Experiment.* (Illus.) Scribner's, 1973. 167pp. $7.95. 72-1179. SBN 684-13309-1.

SH-C   Luria discusses the ordered process of life: evolution, how protein synthesis is regulated, some mysteries in immunology, speculations on the functioning of the mind, and much more.

Luria, S.E. *36 Lectures in Biology.* (Illus.) MIT Press, 1975. xvii + 439pp. $8.95(p). 74-19136. ISBN 0-262-12068-2; 0-262-62029-4. Index;CIP.

C   The lectures presented here are centered around the unifying concept that all living organisms have a program—a set of genetic information items underlying all vital functions—that evolves by mutation, genetic recombination and natural selection. A summary of basic facts of organic and physical chemistry is included. A useful reference for teachers and advanced students.

McKinney, H. Lewis. *Wallace and Natural Selection.* (Illus.) Yale Univ. Press, 1972. xix + 193pp. $12.50. 72-75203. ISBN 0-300-01556-9.

C-P   Does an excellent job of depicting the role of Alfred Russel Wallace in the development of the theory of natural selection. The development of Wallace's work is carefully annotated, as is the influence which Wallace's pre-1858 work had on both Lyell and Darwin. In addition, professional historians will find the references and precision of writing very useful.

Mayr, Ernst. *Populations, Species, and Evolution.* (Illus.) Harvard Univ. Press, 1970. xv + 453pp. $10.00. 79-111486. ISBN 0-674-69010-9.

C   A fluid, lucid account of the attributes and evolutionary significance of species and the history and origin of the modern synthetic theory of evolution. There are thorough discussions of the meaning of biological properties of species, of their population structure, and of the origin of new species. Careful definitions, alternative hypotheses, relevant references and findings from comparative anatomy, embryology, ethology, genetics, and paleontology are suitably utilized. Some thoughts about the evolutionary future of man provide a stimulating ending.

Merrell, David J. *An Introduction to Genetics.* (Illus.) Norton, 1975. xxi + 822pp. $14.95. 75-1076. ISBN 0-393-09247-X. Gloss.;index;CIP;bib.

C-P   A thorough introduction to genetic principles and applications. This book covers classic genetics plus units on the nature of the gene, concepts of gene action and recombination, population genetics including statistical tools, and the implications of genetics for human behavior and for ecological studies.

Pringle, J.W.S., (Ed.)   *Biology and the Human Sciences: The Herbert Spencer Lectures, 1970.* Oxford Univ. Press, 1972. viii + 140pp. $8.50; $3.50(p). ISBN 0-19-847122-4.

    SH-C   Intended to demonstrate the inseparability of the sciences from the liberal arts and natural history, using the particular examples of heredity and evolution. Topics included are language as a biological process; race, class and culture; overpopulation; and aggression. Not always convincing, but filled with interesting ideas which deserve careful consideration.

Rhodes, F.H.T.   *Evolution.* (Illus.) Golden/Western, 1974. 160pp. $4.95. 74-76432. ISBN 0-307-64360-3. Index.

    SH   A good introduction to the overall diversity of life and an overview of the origin and development of the concept of evolution. Evolutionary evidence is lucidly presented along with discussions of processes and future implications.

Rhodes, F.H.T.   *The Evolution of Life, 2nd ed.* (Illus.) Penguin, 1976. 330pp. $3.95(p). ISBN 0-14-020512-8. Gloss.;index.

    SH-C   This text for a first course in evolution is particularly well suited to students with no prior exposure to geology or paleontology. By placing emphasis on the fossil record, Rhodes provides the invaluable dimension of time. Clearly written, with few technical terms. Revisions in this edition are minor, mainly a new discussion of causes of changes in the atmosphere and its consequences to life and expansion of the chapter on the development of life.

Scientific American.   *Ecology, Evolution, and Population Biology: Readings from* **Scientific American.** (Illus.) Freeman, 1974. viii + 319pp. $12.00; $5.95(p). 73-17448. ISBN 0-7167-088804-1; 0-7167-0887-6. Index;CIP.

    C-P   This fine collection of articles covers the evolutionary process, the multiplication and dispersal of species, the growth and interaction of populations, and ecosystems. Ideal as a complementary text and as a resource for teachers.

Stebbins, G. Ledyard.   *Processes of Organic Evolution, 2nd ed.* (Illus.) Prentice-Hall, 1971. xiii + 193pp. $7.50. 74-152445. ISBN 0-13-723395-7; 0-13-723387-6(p).

    C-P   An excellent synthesis which presents historical background and discussions of known sources of variability, such as mutation and recombination; the effects of this variability in populations; hybridization; long-range phenomena; the evolution of man; and the respective roles of organic and cultural evolution. The presentation is very good, but because it is so concise, it is best used as supplemental reading.

Wilson, Edward O., and William H. Bossert.   *A Primer of Population Biology.* (Illus.) Sinauer, 1971. 192pp. $3.95. 73-155365. ISBN 0-97893-926-1.

    C-P   A relatively painless refresher for both students and professionals; the necessary mathematical operations and explanations are presented clearly and with adequate illustrations. A fine presentation of concise and mathematically illustrated expressions of the basic framework and principles of evolution.

575.1 GENETICS—HEREDITY AND VARIATION

Burns, George W.   *The Science of Genetics: An Introduction to Heredity, 2nd ed.* (Illus.) Macmillan, 1972. vii + 470pp. $9.95. 70-158173.

    C   Since the elucidation of the structure of DNA, of the genetic code, and of the operon model, introductory genetics texts have given so much attention to molecular genetics that frequently such important concepts as quantitative and population genetics, epistasis, and human genetics are given less than adequate coverage. This exception to the trend is a nicely balanced text with a lot of intriguing information and speculation on human karyotypes.

Cove, D.J. *Genetics.* (Illus.) Cambridge Univ. Press, 1971. vi + 215pp. $10.00; $3.95(p). 75-160089. ISBN 0-521-08255-2; 0-521-09663-4.

    SH-C   The book is small, yet deals extremely well with all phases of the genetic mechanisms. The concepts of life cycle, segregation, dominance, linkage, and finally, the chromosomal basis of heredity are developed.

Dobzhansky, Theodosius. *Genetics of the Evolution Process.* (Illus.) Columbia Univ. Press, 1970. 505pp. $10.95. 72-127363. ISBN 0-231-02837-7. Indexes;bib.

    C-P   This synthesis of current understanding of biological evolution is authoritative and well written. Included are discussions of the hereditary basis for continuity and change in organisms, natural selection as both a stabilizing and innovating force, effect of population size on evolutionary trends, evolution of the biological hierarchy, and reproductive isolation and speciation.

Fried, John J. *The Mystery of Heredity.* (Illus.) John Day, 1971. x + 180pp. $6.95. 75-107208.

    SH-C   An historical development of our knowledge of heredity and the mechanism involved in transmission of characters from generation to generation and a fairly up-to-date account of how this mechanism works. Fried also provides a very good, simple summary of major research projects which have elucidated the genetic process.

Gardner, Eldon J. *Principles of Genetics, 4th ed.* (Illus.) Wiley, 1972. xi + 527pp. $12.50. 71-168646. ISBN 0-47129130-7.

    C   Basic genetic concepts are presented clearly and concisely with an appreciation of their historical origins. The concept of the gene is presented in terms of classical Mendelian genetics, then in terms of modern molecular biology, then in terms of cytogenetics, and finally in terms of recombination and mutation. At the end of each chapter there are interesting questions.

Ghiselin, Michael T. *The Economy of Nature and the Evolution of Sex.* Univ. of Calif. Press, 1974. xii + 346pp. $12.95. 73-78554. ISBN 0-520-02474-5. Index.

    C-P   Ghiselin replaces old paradigms regarding nature and approaches its study from the standpoint of economy—that processes of nature are nonwasteful and based on a specialized division of labor between the two sexes. The bulk of the book is devoted to the progress of sexual reproductive mechanisms from primitive organisms to higher forms. Contains many references.

Land, Barbara. *Evolution of a Scientist: The Two Worlds of Theodosius Dobzhansky.* (Illus.) Crowell, 1973. 262pp. $4.50. 72-83788. ISBN 0-690-27214-6.

    JH-SH   This lively account of the life and scientific career of a distinguished geneticist traces his early years in Russia, student life during World War I and the revolution and subsequent study in the United States. His contributions to the biological theory of evolution and subsequent research on speciation and genetic load are well treated.

Levine, Louis. *Biology of the Gene, 2nd ed.* (Illus.) Mobsy, 1973. xiv + 358pp. $10.50. 72-89495. ISBN 0-8016-2987-X. *Laboratory Exercises in Genetics, 2nd ed.* (with Norman M. Schwartz). vii + 175pp. $4.95. ISBN 0-8016-2981-0.

    C   A balanced introduction to classical and molecular genetics. Departing from conventional treatment, the author begins with modern concepts of the structure and function of nucleic acids and the genetic code. Included are useful summaries, excellent review questions and problems and selected references. Applied aspects of genetics are omitted. The useful laboratory manual contains 10 experiments with Drosophila and 9 with microorganisms.

Levine, Louis. *Papers on Genetics: A Book of Readings.* (Illus.) Mosby, 1971. xii + 497pp. $6.75. ISBN 0-8016-2983-7.

C   A presentation of classical papers in genetics organized to provide the subject matter of the usual beginning text. Organization is logical rather than chronological, which makes for an ordered development of much original resource material.

Lewis, Kenneth R., and Bernard John. *The Organization of Heredity.* (Illus.) American Elsevier, 1970. ix + 241pp. $11.75; $5.95(p). 70-130964. ISBN 0-444-19637-4; 0-444-19638-2.

C   A thorough but succinct treatment of modern molecular and microbial genetics in three parts: the chemical basis of heredity, organization and transmission of genetic information, and functional organization of the genotype. The references are to the key papers in each of the areas discussed.

Moore, John A. *Heredity and Development, 2nd ed.* (Illus.) Oxford Univ. Press, 1972. 292pp. $2.95(p). 76-161890. ISBN 0-19-501492-8.

Moore, John A., (Ed.). *Readings in Heredity and Development.* (Illus.) Oxford Univ. Press, 1972. xi + 329pp. $3.95(p). 70-170264. ISBN 0-19-501492.

SH-C   Guides readers through the events and evolution of thought which have led to modern knowledge of genetics and developmental biology, beginning with Darwin. Problems are clearly stated, as are experiments designed to solve them. Includes chapters on human genetics and molecular genetics. The *Readings* presents original papers and expositions, beginning with Hippocrates and concluding with a 1970 presentation by Gunther Stent on molecular genetics. Together, these books provide an uncommonly good introduction.

Paterson, David. *Applied Genetics: The Technology of Inheritance.* (Illus.) Doubleday, 1969. 192pp. $5.95. 68-22672.

SH   An exceptionally well written book recounting the major triumphs of the science of genetics and some of their applications. Fairly complex genetic experiments are explained, and the range of subjects includes molecular genetics, plant and animal breeding, the development of microbes for their commerical value and medical genetics.

Rothwell, Norman V. *Understanding Genetics.* (Illus.) Williams&Wilkins, 1976. xiv + 486pp. $14.95. 74-12151. ISBN 0-683-07389-3. Index;CIP.

SH-C   One of the best texts for an introductory course, written by an experienced teacher who knows the areas which are difficult for students with only a general background in biology. Presents basic classical genetics and current molecular aspects. The historical development and the relationships to classical genetics are clearly pointed out. Good basic information.

Scientific American. *Facets of Genetics.* (Illus.) Freeman, 1970. 354pp. 78-99047. ISBN 0-7167-0949-X(p).

SH-C   A collection of articles from *Scientific American* (1945–1970) which covers the elements and processes of inheritance, the nature of the gene and how it works in cells, the development of individual organisms, the genetic basis of evolutionary diversity, and the present and potential impact of genetics on humans. Four authors are Nobel Laureates; others have been the leaders of the research in genetics in the past 25 years. An excellent supplement to any general genetics text.

Sheppard, P.M., (Ed.). *Practical Genetics.* (Illus.) Halsted/Wiley, 1974. xii + 337pp. $27.50. 73-9709. ISBN 0-470-78360-5. Index;CIP.

   C-P   Each chapter of this extremely useful reference book is written by a different specialist who describes challenging and interesting experiments at various levels of expertise. Clearly written, the guide supplies valuable technical information needed to conduct sophisticated experiments. Recommended for college teachers and students working on special problems.

Silverstein, Alvin, and Virginia Silverstein. *The Code of Life.* (Illus.) Atheneum, 1972. 89pp. $4.50. 77-175558.

   JH-SH   Explores the concept that all living things are very similar since all have cells which function in basically similar ways. The structure of DNA is described, and the authors show how DNA affects reproduction, inheritance of traits and mutation. Many of the definitive experiments are described.

Stubbe, Hans. *History of Genetics: From Prehistoric Times to the Rediscovery of Mendel's Laws.* (Illus.; trans.) MIT Press, 1972. ix + 356pp. $14.95. 73-148973. ISBN 0-262-19085-0.

   C-P   A unique book on the evolution of ideas concerning the generational continuities and changes of living organisms. Blends historical details with clarity and readability. Includes a 50-page bibliography.

576 MICROBIOLOGY

Alexander, Martin. *Microbial Ecology.* (Illus.) Wiley, 1971. x + 511pp. $12.95. 71-137105. ISBN 0-471-02054-0.

   SH-C   This survey of the ecology of microorganisms—principally bacteria and fungi and with less emphasis on protozoa—deals with the community and its development, interspecific relationships, and the effect of microorganisms on their surroundings. Appropriate as collateral reading for microbiology at the introductory or intermediate level.

Anderson, Dean A. *Introduction to Microbiology.* (Illus.) Mosby, 1973. x + 391pp. $10.75. 72-81116. ISBN 0-8016-0205-X. *Teacher's Guide.*

   C   Covers fundamental and practical aspects of microbiology, and microorganisms and the production of disease. Uses semantics to trace scientific terms. Includes practical applications, chemical properties of antimicrobial substances and their physiological effects, techniques used in examining serum, and references. No background in the physical sciences needed.

Aylesworth, Thomas G. *The World of Microbes.* (Illus.) Collins/Watts, 1975. 128pp. $6.90. 74-9391. ISBN 0-531-02120-3. Index.

   JH-SH   Excellent historical and fundamental aspects of microbiology are presented here in a simple and well-illustrated manner. Chapters on disease, ecology and viruses are included.

Bailey, W. Robert, and Elvyn G. Scott. *Diagnostic Microbiology, 3rd ed.* (Illus.) Mosby, 1970. x + 385pp. $8.95. 70-102126.

   C-P   Includes a brief but lucid description of general bacteriologic methods, handling of specimens and normal bacterial flora from various anatomical sites, preparation of media, and the identification of pathogenic organisms by morphological, biochemical, and serologic characteristics. Rickettsiae, viruses and fungi are also covered in limited but very clear fashion. The main attraction is the clarity of presentation and excellent organization.

Blazevic, Donna J., and Grace Mary Ederer. *Principles of Biochemical Tests in Diagnostic Microbiology.* (Illus.) Wiley, 1975. xiv + 136pp. $12.50. 75-17591. ISBN 0-471-08040-3. Index;CIP.

   SH-C-P   This useful book is a collection of much widely scattered information, together with citations to the original sources. The 27 biochemical procedures commonly used in the identification of enteric and related bacteria are clearly described and their chemical bases explained, both the enzymatic activities of the bacteria and the chemical reactions of the test reagents.

Brock, Thomas D. *Biology of Microorganisms.* (Illus.) Prentice-Hall, 1970. xii + 737pp. $12.95. 76-79113.

   C   Brock captures the excitement of modern microbiology as he reveals the basic biological mechanisms of microbes. The applied aspects of soil, water and medical microbiology, the recent breakthroughs in microbial structure and function, immunologic phenomena and genetics are interwoven with the basic discussions of the cell, taxonomy and classification, ecology and the more prosaic aspect of microorganisms.

Crabtree, Koby T., and Ronald D. Hinsdill. *Fundamental Experiments in Microbiology.* (Illus.) Saunders, 1974. vi + 349pp. $6.95(p). 73-89175. ISBN 0-7216-2733-1.

   C   This well-organized laboratory guide contains 35 exercises divided into seven major categories of life sciences. Sections on ecology and applied microbiology lead the student to interesting and challenging experiments. Useful, introductory manual.

Frobisher, Martin, et al. *Fundamentals of Microbiology, 9th ed.* (Illus.) Saunders, 1974. xvii + 850pp. $16.50. 73-88259. ISBN 0-7216-3922-4. Gloss.;index.

   C-P   A general presentation of microbiology fundamentals, including use of the microscope, chemical bases for microbial life, systematic classifications, and the relationship of microorganisms to their environment, as well as their role in waste disposal and food preservation. Suitable for reference collections.

Gebhardt, Louis P. *Microbiology, 4th ed.* (Illus.) Mosby, 1970. viii + 354pp. $9.25. 71-102121.

   SH-C   A lightweight survey for students with scant training in the sciences. About a third of the book is devoted to general principles, a third to sanitary and industrial microbiology, and a third to disease-producing microorganisms. The author consciously oversimplifies but writes with clarity. A good book for high school libraries.

Lechevalier, Hubert A., and Morris Solotorovsky. *Three Centuries of Microbiology.* Dover, 1974. 536pp. $5.00(p). 73-91785. ISBN 0-486-23035-X. Index.

   SH-C-P   This is a fascinating account of the evolution of various facets of microbiology. Each chapter may be read independently of the others. A vivid picture of both major and minor scientists influencing the development of microbiology is presented.

Levy, Julia, et al. *Introductory Microbiology.* (Illus.) Wiley, 1973. vi + 684pp. $14.95. 72-8781. ISBN 0-471-53155-3.

   C   This is a fine introductory text which begins with a short history of the field followed by cellular structure, growth and nutrition. There are also chapters on ecological relationships, viruses, immune and allergic responses, host-parasite relationships, disease in human populations and epidemiology. An excellent textbook for a general course in microbiology for those with a background in biology and chemistry.

Locke, David M. *Viruses: The Smallest Enemy.* (Illus.) Crown, 1974. ix + 502pp. $10.95. 71-185103. ISBN 0-517-514850. Index.

C-P This gifted writer and knowledgeable scientist has produced a very readable narrative of the history and development of virology. He also discusses our present knowledge of viruses and virus diseases of animals, plants and bacteria, especially those causing cancer. A work equally suited to classroom use or pleasure reading.

Patent, Dorothy Hinshaw. *Microscopic Animals and Plants.* (Illus.) Holiday House, 1974. 160pp. $5.95. 74-7575. ISBN 0-8234-0247-9. Index;CIP.

JH-SH Informative introduction to microscopic biology with instructions on microscope use and guides for studying the structure and morphology of various organisms. Ample, well-matched illustrations accompany a text in which there is minimal use of technical jargon. Geared toward the beginning student of biology.

Payne, William J., and Dean R. Brown. *Microbiology: A Programmed Presentation, 2nd ed.* (Illus.) Mosby, 1972. x + 283pp. $6.75(p). ISBN 0-8016-3770-8.

C-P This self-instruction program provides basically instructive or supplemental material required for introductory courses, foundation material for laboratory workers, and refresher material for professionals preparing for examinations necessary to advance in rank or professional status. The contents include "Introducing Microorganisms," "Growth and Physiology of Microorganisms," "Descriptive Microbiology," antimicrobial methods, pathogenicity and resistance, microbiology of water and air; food and milk, and useful chemical activities of microorganisms. Includes references. Useful even though it does not cover the entire field.

Poindexter, J.S. *Microbiology: An Introduction to Protists.* (Illus.) Macmillan, 1971. viii + 582pp. $10.95. 74-113933.

C The book is designed for students with a background in general chemistry and biology. Historical discussion is well outlined; cell structure well described in concise modern terms; and the use of electron microscopy for the study of ultimate cellular structure is indicated. Suitable as a reference.

Reid, Robert. *Microbes and Men.* (Illus.) Saturday Review Press, 1975. 170pp. $8.95. 74-24326. ISBN 0-8415-0348-6.

SH-C A fascinating introduction to the history of microbiology starting with the first vaccinations against smallpox. Reid also presents the work of the leaders in the field, such as Pasteur, from the human perspectives of jealousy, conceit and frustration. The style of the book is excellent.

Rosenberg, Nancy, and Louis Z. Cooper. *Vaccines and Viruses.* (Illus.) Grosset&Dunlap, 1971. 159pp. $4.50. 71-130853. ISBN 0-448-21414-8; LB 0-448-26184-7.

JH-SH Presents a review of the nature of viruses, scientists' attempt to conquer them with vaccines, details concerning natural and acquired defenses against these pathogens, and accounts of the success against a variety of viral diseases. These topics are lucidly and interestingly explained.

Silverstein, Alvin, and Virginia B. Silverstein. *Germfree Life.* (Illus.) Lothrop, Lee&Shepard, 1970. 96pp. $4.50. 76-103709.

JH-SH This brief book introduces the science of gnotobiology. Explained are basic microbiological techniques; various microorganisms and their helpful and harmful effect on animal hosts; the contributions of noted researchers; and the problems of devising tools and techniques, establishing controls, demonstrating cause and effect, and doing extensive testing of current or potential applications.

Smith, Paul F. *The Biology of Mycoplasmas.* (Illus.) Academic, 1971. x + 257pp. $14.50. 71-154397. ISBN 0-12-652050-X.

C-P   Lucid, systematic, detailed and sequentially arranged, this book reviews research conducted by scientists, including the author, on mycoplasmas and explains basic and biochemical comparisons of mycoplasmas to bacteria and bacterial L forms. Gross, colonial, and ultrastructural morphology of mycoplasmas; fermentative and nonfermentative energy production; and the relation of cellular structures to known cellular functions are covered. A highly recommended reference.

Stanier, Roger Y., et al. *The Microbial World, 3rd ed.* (Illus.) Prentice-Hall, 1970. vii + 873pp. $15.75. 70-110090.

C   Extensive introduction to microbiology emphasizing all aspects of bacteriology—such as cytology, physiology, taxonomy, ecology, economic importance and much more.

Wyss, Orville, and Curtis Eklund. *Microorganisms and Man.* (Illus.) Wiley, 1971. ix + 389pp. $9.50. 70-146674. ISBN 0-471-96900-1.

SH-C   This text is designed to stimulate students who are not biology majors to develop a deeper understanding of the diversity and scope of biological science. Includes microbiological techniques, bacterial anatomy, bacterial cell function, classification, genetics, fungi, viruses and mycoplasma. For collateral reading in high school biology and as a text for introductory college biology.

577 GENERAL NATURE OF LIFE

Cohen, Daniel. *How Did Life Get There?* (Illus.) Messner, 1973. 96pp. $5.29. 72-13149. ISBN 0-671-32573-6; 0-671-32574-4.

JH   Presents the mystery of dispersal of life forms through very elementary discussions of centripetal speciation, ecological isolation, hybridization, convergent evolution and geographic variation. Includes descriptions of the volcanic Galápagos, with accounts of Darwin's trip; Krakatoa, with its violent demise and subsequent rebirth; and Surtsey, the island created in 1963 off the coast of Iceland.

Margulis, Lynn. *Origins of Life: Proceedings of the Third Conference on Planetary Astronomy.* Springer Verlag, 1973. xi + 268pp. $14.80. 72-91514. ISBN 0-387-06065; 3-540-06065-0(p). Index;bib.

C-P   A well-edited discussion of the implications for determining the origins of life from data gathered on and about the Moon and Mars. The Conference participants present their views in an effective dialogue form.

Orgel, L.E. *The Origins of Life: Molecules and Natural Selection.* (Illus.) Wiley, 1973. vi + 237pp. $7.50; $4.25(p). 72-10534. ISBN 0-471-65692-5; 0-471-65693-3.

C   Traces the *in vitro* formation of the chemical building blocks of life, as deduced from laboratory experiments, current theories on the origin of the earth and the fossil record. Simple inorganic molecules easily form polymers of amino acids, nucleic acids and sugars. Evolution of biological organization is hypothesized from such beginnings through competition and natural selection.

Ponnamperuma, Cyril. *The Origins of Life.* (Illus.) Dutton, 1972. $7.95. 72-183747. SBN 0-525-17195-9; 0-525-04125-7(p).

SH-C   A very ambitious high school senior who is willing to do some collateral reading in chemistry could profit from this book, but much of the text would baffle the general reader. It is excellent collateral reading for chemistry majors, however, and is recommended as a general source of information on elementary organic geochemistry, the origin of life and exobiology.

Szent-Gyorgyi, Albert. *The Living State: With Observations on Cancer.* (Illus.) Academic, 1972. x + 113pp. $6.55. 72-82637.

> **SH-C** Develops a chain of reasoning which begins with a thermodynamic concept of the nature of life and culminates in a theory of carcinogenesis based on the loss of normal mechanisms for the control of cellular division. The argument takes the reader through the biologic properties of water, the mechanism of muscular contraction, and selected aspects of evolution. It continues with a discussion of the central role of the hydrogen ion within the cell and, following an elaboration of plant and animal defense mechanisms, arrives at a regulatory theory of cancer.

Thompson, Paul D. *Abiogenesis: From Molecules to Cells.* (Illus.) Lippincott, 1969. 192pp. $4.95. 78-82399.

> **SH-C** Spontaneous generation—the formation of living organisms de novo from lifeless matter—is treated in this well-organized, up-to-date account. There is a brief discussion of what life is, the biochemical structure of protoplasm, the chemistry of the cell during reproduction, abiogenesis through history, a brief account of the available geological data on the age of life on earth and some information from the early space program.

578 MICROSCOPY IN BIOLOGY

Galigher, Albert E., and Eugene N. Kozloff. *Essentials of Practical Microtechnique, 2nd ed.* (Illus.) Lea&Febiger, 1971. x + 531pp. $12.00. 70-152023. ISBN 0-8121-0356-4.

> **C** The book is accurate, authoritative, comprehensive and quite readable. The most notable addition is a chapter in which fixing, embedding, and sectioning of specimens for electron microscopy is discussed. This chapter considerably widens the potential audience of the text. It is highly recommended for use as a reference or textbook to all life scientists concerned with the microstructure of biological materials.

Goldstein, Philip, and Jerome Metzner. *Experiments With Microscopic Animals: Investigations for the Amateur Scientist.* (Illus.) Doubleday, 1971. 245pp. $7.95. 72-116207. Index;bib.

> **JH-SH** Two experienced biology teachers have developed an outstanding guide for the amateur who has the motivation to undertake serious investigations and thereby learn the basic skills of microscopy. The book details protozoa and other microscopic animals which can be easily raised and maintained by young experimenters. The appendixes cover simple lab techniques and equipment.

Tribe, Michael A., et al. *Light Microscopy.* (Illus.) Cambridge Univ. Press, 1975. 108pp. $14.95; $5.95(p). ISBN 0-521-20656-1; 0-521-20556-5. Index.

> **C** This self-instruction manual in the use of light microscopy covers several types of microscopes used in biology. Carefully written directions and explanations accomplish the self-instruction intent. An excellent book.

Vishniac, Roman. *Building Blocks of Life: Proteins, Vitamins, and Hormones Seen Through the Microscope.* (Illus.) Scribner's, 1971. 62pp. $6.95. 76-143959. ISBN 0-684-12381-9.

> **SH-C** The text explains briefly the discovery, structure, composition and physiological roles of proteins, enzymes, vitamins and hormones. The 31 outstanding color photographs aid greatly in understanding the text explanations. Professional or student microscopists, who are always seeking new techniques to aid in

enhancing the visibility and interpretation of their specimens or preparations, will value this book.

578.45 ELECTRON MICROSCOPY

Meek, Geoffrey A. *Practical Electron Microscopy for Biologists.* (Illus.) Wiley-Interscience, 1970. xviii + 498pp. $22.00. 74-116655. ISBN 0-471-59030-4.

    C   This is one of the best books on this subject. The first section deals with the electron microscope in clear language, presuming no real mathematical knowledge. The second part deals with the use of the electron microscope, and the third is on specimen preparation. This book can be unreservedly recommended as a textbook or reference for all biologists.

Swift, J.A. *Electron Microscopes.* (Illus.) Barnes&Noble, 1970. 88pp. $6.50. ISBN 0-389-04067-3.

    C-P   A brief, informative look at electron microscopes which presents an excellent, if cursory, survey of types of instruments, how they work, their limitations and how they may develop in the future.

Tribe, Michael A., et al. *Electron Microscopy and Cell Structure.* (Illus.) Cambridge Univ. Press, 1975. viii + 117pp. $14.95; $5.95(p). ISBN 0-521-20657-X; 0-521-20557-3. Gloss.;index.

    C   Emphasis is on the examination of electron micrographs and explanations of structure and function using the greater resolving power of electron microscopes. Sample preparation is discussed, but the techniques of operating an electron microscope are treated lightly. There are numerous excellent reproductions of micrographs. The question-and-answer format is effective for independent study. For advanced biology students; includes lists of films and tapes keyed to the book.

Wischnitzer, Saul. *Introduction to Electron Microscopy, 2nd ed.* (Illus.) Pergamon, 1970. viii + 292pp. $12.50. 77-93757.

    C   A significant, brief synopsis of the technology and physical theory of electron microscopy applied to biological materials. Comprehensive in scope, it ranges from basic physical and electrical theory to historic review of electron microscopy and an indexed survey of fine structure contributions to understanding of specific organ systems, tissues, and cell parts. An extensive section covers microscope development and applications such as freeze-etching and scanning electron microscopy.

579.4 TAXIDERMY

Grantz, Gerald J. *Home Book of Taxidermy and Tanning.* (Illus.) Stackpole, 1970. 160pp. $7.95. 77-85651. ISBN 0-8117-0805-5.

    SH-C   Far more than a series of "insert tab A into slot B" instructions, this is a well-rounded treatment of taxidermy methods, with tips and tricks which will help the beginner gain the confidence and skill necessary for a good job. Covers tools; preparation of fish, birds, small and large mammals; skins and tanning; supply sources and tanning services.

580 Botanical Sciences

580.74 EXHIBITS

Hyams, Edward. *Great Botanical Gardens of the World.* (Illus.) Macmillan, 1969. 288pp. $35.00. 73-87880. Indexes.

SH-C   Provides historical and background data on the gardens, their purpose, specialities and principal flora. An outline map keys the location of the world's 525 botanical gardens and lists their names and addresses. These famous gardens represent a great diversity of horticultural and architectural features; the illustrations are very good.

Whittle, Tyler. *The Plant Hunters.* (Illus.) Chilton, 1970. xii + 281pp. $8.95. 77-104717. ISBN 0-8019-5472-X.

SH-C   This entire book is devoted to those individuals who introduced both herbarium and horticultural specimens to the Western world. In addition to the perils of specimen-collecting journeys, Whittle outlines the philosophies behind the expeditions themselves and some of the difficulties encountered in transporting plants and establishing herbaria. An adventure story in the same spirit as the popular works of C.W. Beebe and Roy Chapman Andrews.

581 BOTANY

See also 635 Horticulture.

Alexander, Taylor R., Will R. Burnett, and Herbert S. Zim. *Botany: A Golden Science Guide.* (Illus.) Golden Press, 1970. 160pp. $4.95. 77-85477.

SH-C   An exceptionally good study guide and reference for advanced high school botany classes and beginning college courses. It will help students understand the basic concepts of botany, including plant classification, plant nutrition, anatomy, reproduction, inheritance, growth regulators and evolution.

Arnett, Ross H., and Dale C. Braungart. *An Introduction to Plant Biology.* (Illus.) Mosby, 1970. xiii + 497pp. $9.75. 75-108299. Gloss.;indexes.

C   An excellent text for a beginning course in botany, agronomy or horticulture. Easy-to-read, logically organized, yet the narrative flows easily from one topic to the next. Many concise examples; particularly good treatment of genetics, evolution and life cycles.

Blunt, Wilfrid, with William T. Stearn. *The Compleat Naturalist: A Life of Linnaeus.* (Illus.) Viking, 1971. 256pp. $14.95. 78-147393. ISBN 0-670-23396-X.

C   Although not as well known, Linnaeus was probably as important to 18th century biology as Darwin was to the 19th century. This book does an excellent job of making Linnaeus come alive. Good illustrations, most by Linnaeus's contemporaries, many in color.

Cronquist, Arthur. *Introductory Botany, 2nd ed.* (Illus.) Harper&Row, 1971. ix + 885pp. $14.95. 75-132658. ISBN 0-06-041431-6.

C   A taxonomic framework for coverage of cell and organelle biology, with numerous electron micrographs and an excellent discussion of molecular genetics as well as the traditional morphology and organismal physiology. Designed for either a 1-year evolutionary survey of the plant kingdom or for a 1-semester course on the flowering plants alone. The book is eminently readable and attractively produced.

Greulach, Victor A., and J. Edison Adams. *Plants: An Introduction to Modern Botany, 3rd ed.* (Illus.) Wiley, 1976. xii + 586pp. $13.95. 75-16134. ISBN 0-471-32769-7. Gloss.; index;CIP.

C   Ranging from simple chemistry to the complex physiology of plants, and from unicellular plants to the large seed plants, *Plants* would serve as a good text for quarter or 1-semester courses. Includes discussions of ecosystems and vegetational formations. Well organized and clear.

Jensen, William A., and Frank B. Salisbury. *Botany: An Ecological Approach.* (Illus.) Wadsworth, 1972. xi + 748pp. $14.95. 70-183575. ISBN 0-534-00092-4.

C   An ecological approach to botanical study that does not neglect molecular, cellular and evolutionary botany. Includes classical concepts of plant structure and function as well as a survey of the plant kingdom. Photoessays are a unique attraction. A delightful instructive text or reference.

Milne, Lorus, and Margery Milne. *The Nature of Plants.* (Illus.) Lippincott, 1971. 206pp. $5.95. 78-151473.

SH-GA-C   Quite readable and scientifically accurate; most suitable for collateral reading by students of biology or by interested adults, but not so suitable as a text. The plant kingdom is surveyed, including brief chapters on plant physiology and genetics; taxonomy, anatomy, and morphology; and plant evolution and economic botany.

Northern, Henry, and Rebecca Northern. *Ingenious Kingdom: The Remarkable World of Plants.* (Illus.) Prentice-Hall, 1970. xii + 274pp. $7.59. 76-110413. ISBN 0-13-464859-5.

JH-SH-C   The authors have written about the diversity of the plant kingdom and about complex botanical phenomena in a most understandable manner. Although the simple plants, algae, fungi, mosses, and ferns are discussed, the book focuses on the structure and behavior of flowering plants. Strong emphasis is placed on the timely subject of the role of plants in our environment.

Raven, Peter, and Helena Curtis. *Biology of Plants.* (Illus.) Worth, 1970. xi + 706pp. $11.95. 70-110761.

SH-C   An interesting narrative built around carefully selected concepts, techniques, and other important facets of plants. Refreshingly, plant life cycles have been simplified, interesting essays have been inserted, and there are outstanding pictures and diagrams. A good, understandable reference and text book.

Stocks, Dayna L., W.M. Hess and Darrell J. Weber. *Introductory Plant Biology Manual, rev. ed.* (Illus.) Burgess, 1973. iv + 140pp. $4.50. SBN 8087-1973-4.

SH-C   This good general laboratory manual for a 1-semester introductory botany course covers the great diversity of plant life and the necessity for classification. Exercises present levels of plant life, life cycles, seed dissemination, genetics and evolution, fundamental physiological processes, and an introduction to ecology. Relevant theory, a statement of objectives, needed materials and methods, pertinent questions, and adequate exercises are included; but the format may discourage independent or unusual observations.

Tortora, Gerard J., Donald R. Cicero and Howard E. Parish. *Plant Form and Function: An Introduction to Plant Science.* (Illus.) MacMillan, 1970. x + 563pp. $10.95; $3.95(p). 76-77510. Bib.

SH-C   Traditional format and content with emphasis on the structure and function of the flowering plant and on a survey of the plant kingdom. Molecular aspects of modern botany are treated in depth.

Tribe, Ian. *The Plant Kingdom.* (Illus.) Grosset&Dunlap, 1970. 155pp. $3.95. 70-120438.

JH-SH-GA   This general, popularized reference is a worldwide compendium of plants, their morphology, natural history and uses. Individual entries are encyclopedic. The narrative is pleasant to read and only eight pages lack realistic color drawings.

Weier, T. Elliot, C. Ralph Stocking and Michael G. Barbour. *Botany: An Introduction to Plant Biology, 4th ed.* (Illus.) Wiley, 1970. ix + 708pp. $12.50. 79-105385. ISBN 0-471-92467-9.

C   One of the most detailed and complete beginning botany books available. The plant world is explained in a straightforward, clear, and simple manner; ecology, genetics, plant breeding and all the usual topics are covered. The drawings, photographs and tables are extremely well done. There is a lab manual available, well correlated with the text.

581.1–.2 PLANT PHYSIOLOGY AND PATHOLOGY

Devlin, Robert M., and Allen V. Barker. *Photosynthesis.* (Illus.) Van Nostrand Reinhold, 1971. xiv + 304pp. $8.95. 76-156746. Index.

C-P   History of photosynthesis, nature of light, pigments involved, the structure and ultrastructure of the chloroplast, the reduction of carbon dioxide, photophosphorylation, photorespiration and the factors affecting the rate of photosynthesis are considered. Not a short or simple compilation, it presumes a strong background in biochemistry and biophysics because the essential biochemical mechanisms are emphasized. A chapter is also devoted to the bacterial photosynthesis.

Epstein, Emanuel. *Mineral Nutrition of Plants: Principles and Perspectives.* (Illus.) Wiley, 1972. ix + 412pp. $10.95. 75-165018. ISBN 0-471-24340-X.

C-P   Extensive references to classic and recent articles in journals, textbooks, and monographs are appended to each chapter of this survey of one aspect of plant physiology. This book would be of great value as a text for a course in mineral nutrition or as collateral reading for students of plant physiology.

Fogg, G.E. *Photosynthesis, 2nd ed.* (Illus.) American Elsevier, 1972. xii + 166pp. $3.95(p). 72-87526. ISBN 444-19-571-8.

C-P   Discusses the supply of light, carbon dioxide and water; absorption and conversion of light; biochemistry of carbon assimilation; relationship of photosynthesis to other biochemical processes; and the evolution and value of photosynthesis. Some 17 plates, including electron micrographs of the photosynthetic apparatus, show specific light responses due to photosynthesis and indicate the methods used for large scale algae production.

Hutchins, Ross E. *Galls and Gall Insects.* (Illus.) Dodd, Mead, 1969. 128pp. $3.46. 70-81626.

JH   A good account of the biology of gall and gall insects; covers gall-forming insects; types of galls, particularly of oaks and hickories; history and use of galls; leaf minors; and the gall midge family.

San Pietro, Anthony, (Ed.). *Experimental Plant Physiology.* (Illus.) Mosby, 1974. xii + 176pp. $6.75(p). 73-20407. ISBN 0-8016-4307-4. Bib.

C   This is a collection of 23 laboratory experiments which cover all facets of modern plant physiology. At the beginning of each chapter, there is an assessment of the problem and the current status of the specific research in that area.

Stevens, Russell B. *Plant Disease.* (Illus.) Ronald, 1974. vii + 459pp. $11.95. 73-77863. Index.

C-P   A general principles approach to plant pathology which includes a comprehensive treatment of pathogens, epidemics, and disease control. Useful to students with a background of general biology and as collateral reading for plant disease majors.

Tribe, Michael A., et al. *Photosynthesis.* (Illus.) Cambridge Univ. Press, 1975. viii + 77pp. $13.95; $4.95(p). ISBN 0-521-20820-3; 0-521-20821-1. Gloss.

C   This book introduces the fundamental principles of photosynthesis in a succinct manner. Subjects discussed are energy storage, absorption spectra and redox potentials. Recommended as a valuable aid in teaching photosynthesis at the beginning college level.

581.3−.4 PLANT GROWTH AND STRUCTURE

Bierhorst, David W. *Morphology of Vascular Plants.* (Illus.) Macmillan, 1971. xii + 560pp. $14.95. 70-112853.

C-P   The organization that gives rise to the form of organs and of plants themselves is developed against the background of continuing evolution as displayed through a taxonomic arrangement of orders of plants. Clear line drawings, photographs, and photomicrographs illustrate the various points under discussion. What constitutes a particular degree of evolutionary advance, what constitutes a relationship between organisms, what the significance is of given expressions of ontogeny and phylogeny are given in a highly scholarly manner.

Duddington, C.L. *Evolution and Design in the Plant Kingdom.* (Illus.) Crowell, 1970. 259pp. $7.95. 75-96481.

SH-C   A folksy, easy-reading book which covers the field adequately except for its lack of discussion of the biochemical aspects of plant science. Includes such topics as the evolution and function of specialized orchid flowers, various mechanisms of pollination, adaptation of plants to their environment through different life forms. British orientation.

Gentile, Arthur C. *Plant Growth.* (Illus.) Natural History Press, 1971. 144pp. $5.95. 70-132505.

SH   Gentile has combined topics on plant form and structure with those concerning the physiological processes of growth and development to produce a useful volume for the beginner. Short chapters on seed germination and dormancy, primary and secondary growth, environmental factors, growth regulators, abscission, fruit development, tropisms, and vegetative propagation are included.

Grant, Verne. *Genetics of Flowering Plants.* (Illus.) Columbia Univ. Press, 1975. xiv + 514pp. $20.00. 74-13555. ISBN 0-231-03694-9.

C-P   This first attempt to survey the field of higher plant genetics since the mid-1950s succeeds admirably. Four sections cover genes, gene linkage and genetic systems. Readily adaptable for use as an undergraduate or graduate text but most useful as a source of lecture material for instructors.

Grant, Verne. *Plant Speciation.* (Illus.) Columbia Univ. Press, 1971. 435pp. $15.00. Indexes;bib.

C-P   Reviews the phenomena of evolution not treated in the author's earlier book, *Origin of Adaptations.* Covers the nature and divergence of species, refusion and its consequences, derived genetic systems, and evolution of hybrid complexes. The book is clearly written, contains an enormous amount of information, and can even be recommended to the interested layreader.

Poling, James. *Leaves: Their Amazing Lives and Strange Behavior.* (Illus.) Holt, Rinehart&Winston, 1971. 114pp. $4.95. 77-80324. ISBN 0-03-081614-9.

JH   A compelling book about leaves. Although the approach emphasizes the strange and amazing, basic information about leaves is given. The topics include structure, physiology, ecology, insectivorous plant leaves, plant growth, economic and animal uses of leaves, and autumn coloration.

Rahn, Joan Elma. *How Plants Are Pollinated.* (Illus.) Atheneum, 1975. v + 135pp. $5.95. 75-9526. ISBN 0-689-30482-X. Gloss.;CIP.

JH   The how and why of plant pollination is covered in adequate detail, with the emphasis on mechanisms such as wind, animals, bees, birds, bats or water. Outstanding illustrations. Provides a good lesson on the value of observation.

Salisbury, Frank B. *The Biology of Flowering.* (Illus.) Natural History, 1971. 175pp. $5.95.

SH-C   Treats the responses of plants to temperature, to conditions of illumination, and to time, and introduces the reader to some of the biochemical and biophysical events that ensure the flowering process. A thoughtful and accurate introduction to botany, it would be worthwhile collateral reading for all students of biology.

Steward, F.C., and A.D. Krikorian. *Plants, Chemicals and Growth.* (Illus.) Academic, 1971. xii + 232pp. $4.50(p). 76-167780. Bib.

C   In a philosophical discourse rather than a typical review, the authors communicate the importance of understanding chemical growth as more extensive attempts are made to control regulatory substances in plants. Included are discussions of chemical regulation and bioresponse and of historical development of concepts of regulation of plant growth, postulations of growth initiation, mechanisms of growth regulation, and a wide-ranging classification of compounds according to their type of plant-growth effects. Excellent pictures and good diagrammatic representations.

Villiers, Trevor A. *Dormancy and the Survival of Plants.* (Illus.) Arnold (distr.: Crane, Russak), 1975. 67pp. $3.76(p). ISBN 0-7131-2516-0; 0-7131-2517-9.

SH-C   Research on dormancy in plants is effectively linked into a logical pattern as Villiers traces the importance of dormancy from broad ecological relationships to the cellular mechanisms affecting dormancy. Considers global climatic patterns, morphological and physiological aspects, and the phenomena associated with emergence from dormancy. Integrates many aspects of botany into a cohesive picture that will undoubtedly arouse the interest of students.

581.6 ECONOMIC BOTANY

See also 630 Agriculture and Related Technologies.

Angier, Bradford. *Field Guide to Edible Wild Plants.* (Illus.) Stackpole, 1974. 255pp. $4.95(p). 73-23042. ISBN 0-8117-0616-8; 0-8117-2018-7. CIP.

JH-SH   This guide is a must for anyone hoping to live off the land. Beginning with a directory of wild foods, the author describes over 100 edible plants and gives recipes for their use. Especially useful for students preparing projects.

Dodge, Bertha S. *Potatoes and People: the Story of a Plant.* (Illus.) Little, Brown, 1970. 190pp. $4.95. 77-97145.

JH   A most readable treatise on our potato's exciting history and the imprint it has made on mankind, interwoven with a description of various biological principles, particularly from the field of plant pathology.

Hall, Alan. *The Wild Food Trailguide.* (Illus.) Holt, Rinehart&Winston, 1973. xi + 195pp. $3.45(p). 72-91562. ISBN 0-03-007701-X. Index;gloss.

JH-SH-GA   Written for the general public as a guide to survival plants which can be found over broad sections of the U.S. and southern Canada. There is also a section on poisonous plants. Included are a good illustration of each plant, a range map, size, habitat, common names and the season when the plant can be found.

Hardin, James W., and Jay M. Arena. *Human Poisoning from Native and Cultivated Plants, 2nd ed.* (Illus.) Duke Univ. Press, 1974. xii + 194pp. $6.75. 73-76174. ISBN 8223-0303-5. Gloss.;index.

SH-C   A layperson's guide to poisonous plants of America. Narrations accompany nontechnical descriptions to assist in plant identification.

Healey, B.J. *The Plant Hunters.* (Illus.) Scribner's, 1975. 214pp. $8.95. 74-32295.

C   This discussion of the history of plant hunting begins with the early domestication of showy plants. Interesting illustrations and text; individual plant hunters and their often peculiar patrons are discussed in detail.

Joyce, C.R.B., and S.H. Curry (Eds.). *The Botany and Chemistry of Cannabis.* (Illus.) Churchill (distr.: Williams&Wilkins), 1970. xii + 217pp. $12.00. ISBN 0-7000-1479-9.

C-P   The proceedings of a conference organized by the Institute for the Study of Drug Dependence contains contributions on botany, chemistry and pharmacological aspects and makes an honest effort to give solid information. An excellent, although somewhat technical, reference.

Knap, Alyson Hart. *Wild Harvest: An Outdoorsman's Guide to Edible Wild Plants in North America.* (Illus.) Pagurian (distr.: Arco), 1975. xiv + 190pp. $8.95. ISBN 0-919364-97-7. Gloss.;indexes.

SH-C   This book attempts to combine the features of a survival manual with those of a cookbook. The book is organized by use: salad plants, potherbs, starchy staples, fruits and nuts. Many recipes are provided, some original.

Martin, Alexander C. *Weeds.* (Illus.) Western, 1972. 160pp. $4.95. 72-78574.

SH-C   Explains what weeds are, the harm they cause, their benefits and major habitats, and individual weeds in family sequence. Related species are kept together. Poisonous nature or economic importance is indicated. Generalized U.S. range map for each weed is included with each description.

Masefield, G.B., et al. *The Oxford Book of Food Plants.* (Illus.) Oxford Univ. Press, 1969. viii + 206pp. $10.00. Gloss.

JH-SH-C   Provides information on the origin, domestication, geographical distribution, botany and nutritional contribution of more than 400 plant varieties nourishing to humans. Some 19 classifications even include seaweeds, mushrooms, truffles, exotic water plants and wild British food plants.

Schery, Robert W. *Plants for Man, 2nd ed.* (Illus.) Prentice-Hall, 1972. viii + 657pp. $15.95. 72-140. ISBN 0-13-681254-6.

SH   A nonspecialist would do well to consult this encyclopedic volume which is a compilation of information from an immense and scattered field related to useful plants. Discusses the sources and uses of cell walls, plants yielding exudates and extractives, and plants used primarily for food and beverage. Discussions of forests are extensive and detailed, but overall the content is somewhat unbalanced.

Weiner, Michael A. *Man's Useful Plants.* (Illus.) Macmillan, 1976. 146pp. $6.95. 74-18469. ISBN 0-02-792600-1. Index;CIP;bib.

JH   This simply written book will be most useful, especially as background material for teachers introducing a botany course. Weiner discusses foods and beverages, medicines, fabrics, wood products and other materials made from plants. Includes anecdotes about plant life.

Wilson, Charles Morrow. *Green Treasures: Adventures in the Discovery of Edible Plants.* Macrae Smith, 1974. 184pp. $5.95. 74-4400. ISBN 0-8255-9104-X. Index;CIP.

JH-SH   An interesting introduction into the need for and joys of plant hunting. The book traces the search for and development of plants useful to man. Wilson traces the discovery and utilization of common plants, such as the papaya, avocado, cranberry and blueberry.

581.9 Plants—Geographical Distribution

Abbey, Edward, and Editors of Time-Life Books. *Cactus Country.* (Illus.) Time-Life, 1973. 184pp. $5.95. 72-91599.

JH-SH-C   Abbey describes the flora—especially the cactus plants—of the 69,000 square miles of southern Arizona and northern New Mexico. The photographs are outstanding. An excellent reference work.

Craighead, Frank C., Sr. *The Trees of South Florida, Vol 1: The Natural Environments and Their Succession.* (Illus.) Univ. of Miami Press, 1971. xvii + 212pp. $5.95. 75-107362. ISBN 0-87024-146-X.

SH-C   Much of the area discussed comprises Everglades National Park, and the book should be consulted by anyone interested in the park's ecology. The title is misleading, since herbaceous and marine communities are discussed in detail. Within each "province," geology, soil and hydrology, vegetation, and plant associations are discussed. Almost every community is illustrated by an adequate photograph, and areas where each may be found are identified.

Ewan, Joseph, (Ed.). *A Short History of Botany in the United States.* (Illus.) Hafner, 1969. ix + 174pp. $7.50. 74-75143.

SH-C-P   Prepared for the 11th International Botanical Congress held in Seattle in 1970, the book provides an overview of the development of various aspects of botanical science in the United States. Aside from the general matter, some twelve specific fields, ranging from paleobotany to genetics, phycology to horticulture, are treated.

Eyre, S.R., (Ed.). *World Vegetation Types.* (Illus.) Columbia Univ. Press, 1971. 264pp. $12.50. 78-147779. ISBN 0-231-03503-9.

C   A collection of authoritative works; includes six articles on mid-latitude grasslands, one on coniferous forests, and others on convergent evolution. Provides information on the original vegetation cover necessary for a full understanding of animal ecology and conservation. Although it could serve as a text in certain plant science courses, it is more appropriate as a reference.

Fitter, Richard, and Alastair Fitter. *The Wild Flowers of Britain and Northern Europe.* (Illus.) Scribner's, 1975. 336pp. $10.00. 74-3756. ISBN 0-684-13880-8. Gloss.;indexes.

SH-C   This book provides descriptions of trees, shrubs, and many of the wild plants of the United Kingdom and northern and western continental Europe. The descriptions are brief but adequate, and keys with color guides are included.

Hutchinson, G. Evelyn. *A Treatise on Limnology, Vol. 3: Limnological Botany.* (Illus.) Wiley, 1975. x + 660pp. $30.00. 57-8888. ISBN 0-471-42574-5. Indexes;CIP.

C-P   This volume on lake-dwelling plant life introduces the plants of limnological importance and their ecological classification; describes their life histories, structure and physiology; details many chemical factors that regulate their occurrence; and shows their world distribution. Of interest to chemists, geneticists, embryologists, and environmentalists, as well as ecologists.

Mohlenbrock, Robert H. *Flowering Plants: Flowering Rush to Rushes.* (Illus.) Southern Ill. Univ. Press, 1970. xiii + 272pp. $10.00. 69-16117. Gloss.;index;bib.
Flowering Plants: Lilies to Orchids. (Illus.) xii + 288pp. $10.00. 69-16118. Gloss.;index.

SH-C-P Two segments of a multivolume publication on the Illinois flora which describe families and genera in general but adequate terms. A state map indicating distribution by counties and beautiful black-and-white illustrations are provided for each species. Very good reference books for high school and college botany classes in and around Illinois, and the individual descriptions will be useful wherever the particular species are found.

Morley, Brian D. *Wild Flowers of the World.* (Illus.) Putnam's, 1970. 432pp. $15.00. 79-116143.

JH-SH-C Reproductions of the paintings are superb, and many species are illustrated here that are pictured nowhere else. Gives a concise introduction to plant nomenclature, morphology, classification, ecology and geography, plus an historical essay emphasizing botanical exploration.

Niehaus, Theodore F. *Sierra Wildflowers: Mt. Lassen to Kern Canyon.* (Illus.) Univ. of Calif. Press, 1974. 223pp. $3.95(p). 73-80828. ISBN 0-520-02742-6; 0-520-02506-7. Index.

SH-C-P Begins with a discussion of the geographical area, the reasons for leaving wildflowers intact and for using scientific names. The rest of the book is devoted to equipment, procedures and keys for identifying over 500 species. Very useful to amateurs and beginning botany students working in the region indicated.

Ornduff, Robert. *An Introduction to California Plant Life.* (Illus.) Univ. of Calif. Press, 1974. 152pp. $3.95(p). 73-85793. ISBN 0-520-02583-0; 0-520-02735-3. Indexes;bib.

SH-C-P A concise, well-written statement about the 5000 species of California flora. The book provides brief but adequate information on nomenclature, topography, climate and adaptation, plant communities and evolution. Maps of plant community distribution are good, but color reproductions and line drawings are only fair.

Ownbey, Gerald B. *Common Wildflowers of Minnesota.* (Illus.) Univ. of Minn. Press, 1971. 331pp. $9.75. 72-161439. ISBN 0-8166-0609-9. Index;gloss.

SH-C About 300 of the approximately 1800 species of wildflowers in Minnesota are beautifully illustrated by black-and-white drawings. Woody plants, grasses, and grass-like plants are excluded. The legend for each illustration is accompanied by a brief description of the species and a note on its occurrence in the state.

Page, Nancy M., and Richard E. Weaver, Jr. *Wild Plants in the City.* (Illus.) Quadrangle/N.Y. Times, 1975. x + 117pp. $3.95(p). 74-29034. ISBN 0-8129-0557-1. Index.

SH-C The most common and conspicuous plants found in many cities are listed; scientifically correct plant descriptions are provided in this interesting and useful product of a community project for disadvantaged children.

Wood, Carroll E., Jr., et al. *Student's Atlas of Flowering Plants: Some Dicotyledons of Eastern North America.* (Illus.) Harper&Row, 1974. 122pp. $2.95(p). 74-8069. ISBN 06-047207-3. Index.

JH-SH-C Accurate detail and the beauty of superb pen-and-ink drawings aid in identification of the floral components of 156 genera of eastern North

American flora. First-quality illustrations, accurate terminology, and a flexible format aid in the book's usefulness for biology laboratory and field exercises.

582.16 TREES

Agriculture, U.S. Department of. *A Tree Hurts, Too.* (Illus.; research by Alex. L. Shigo.) Scribner's, 1975. 28pp. $6.95. 74-14079. ISBN 0-684-14117-5.

> JH   This book presents a simplified overview of how trees protect themselves from diseases resulting from the entry of microorganisms into wounds or exposed woody tissue.

Holmes, Sandra. *Trees of the World.* (Illus.) Bantam, 1975. 159pp. $1.95(p). 74-83804. Gloss.;index.

> SH   A brief introduction explains what trees are, how they grow and where they can be found. Each tree genus is shown on one page, along with a brief description of the characteristics, distribution and economic uses of prominent species. Recommended to nature lovers of all ages.

Li, Hui-Lin. *Trees of Pennsylvania, the Atlantic States and the Lake States.* (Illus.) Univ. of Penn. Press, 1972. viii + 276pp. $17.50. 72-80376. ISBN 0-8122-7665-5. Gloss.

> SH-C   A useful guide to the identification of trees native to Pennsylvania and the Atlantic coastline. Sixty-four genera and 118 species are identified, including a few introduced species and most of the "large shrubs." Photographs and drawings of leaves, fruits, flowers and branches; winter keys.

Mirov, Nicholas T., and Jean Hasbrouck. *The Story of Pines.* (Illus.) Indiana Univ. Press, 1976. xi + 148pp. $6.95. 74-30899. ISBN 0-253-35462-5. Index;CIP;bib.

> SH-C   A survey of the natural history of pines by a world authority. He discusses the pine life cycle; its folklore, economic uses, ecology and distribution; and pine plantations, hybridization, and ornamental uses. Includes a species list, their general distributions and common names. Some of the text presumes advanced biological or chemical knowledge, but in general no background required.

Montgomery, F.H. *Trees of the Northern United States and Canada.* (Illus.) Warne, 1971. 144pp. $4.95. 75-131362. ISBN 0-7700-0324-9. Index;gloss.

> SH-C-P   Keys to genera and species, 136 numbered figures and numerous others, and 24 color photographs comprise this fine taxonomic field book. Montgomery treats about 170 species belonging to approximately four dozen genera in his keys. Includes a section summarizing the leaf and fruit characteristics.

Petrides, George A. *A Field Guide to Trees and Shrubs, 2nd ed.* (Illus.) Houghton Mifflin, 1972. xxxii + 428pp. $5.95. 76-157132. ISBN 0-395-13651-2. Gloss.

> SH-C-P   This edition of a valuable field guide to the woody flora of northeastern and central North America includes 646 species. An excellent and most useful book in which the use of technical terms is limited. The subject is covered exhaustively, and characteristics that will assist identification year-round are stressed. The great emphasis is on utility, with genera and species arranged by easily observed similarities rather than by traditional evolutionary specialization.

587–588 VASCULAR CRYPTOGAMS AND BRYOPHYTES

Davis, Bette J. *The World of Mosses.* (Illus.) Lothrop, Lee&Shepard, 1975. 64pp. $4.75. 74-10606. ISBN 0-688-41667-5; 0-688-51667-X. Gloss.;index;CIP.

JH While sharing her appreciation for the beauty, function and survivability of
mosses, the author also provides simplified coverage of their life cycles,
growth, identification and importance. Instructions for making a terrarium are included, and the illustrations are useful and attractive.

Grimm, William C., and Jean Craig. *The Wondrous World of Seedless Plants.* (Illus.)
Bobbs-Merrill, 1973. 128pp. $5.95. 73-1757. ISBN 0-672-51709-4. Bib.

JH-SH The seedless plants are the algae, fungi, lichens, liverworts, mosses, and
the ferns and their relatives. Their life cycles, habits, ecology, distribution
and uses are discussed; names and illustrations of common representatives of each
group are given.

589.2 FUNGI

Ashworth, J.M., and Jennifer Dee. *The Biology of Slime Moulds.* (Illus.) Arnold
(distr.: Crane, Russak), 1975. 67pp. $3.75(p). ISBN 0-7131-2511-X; 0-7131-2512-8(p).

SH-C Considers two species of slime moulds—one cellular and the other
acellular—chiefly from the perspective of developmental biology. Traces
the processes of differentiation and integrates them with genetics in a modern conceptual mode. For introductory students with an interest in developmental biology.

Bigelow, Howard E. *Mushroom Pocket Field Guide.* (Illus.) Macmillan, 1974. 117pp.
$3.50. 73-7682. ISBN 0-02-510650-3. Gloss.;index;CIP.

SH-C-P A handy pocket-sized book for the professional as well as amateur
botanist. The 61 species are beautifully illustrated in their natural
habitats by color photographs, and each photo is accompanied by a description of
habitat structure and plant color. Poisonous species are also designated.

Major, Alan. *Collecting and Studying Mushrooms, Toadstools and Fungi.* (Illus.)
Arco, 1975. 268pp. $12.00. 74-19896. ISBN 0-688-03725-3. Index.

SH-C-P The author focuses on the conspicuous mushrooms and toadstools, discussing how and where to collect them and their form. Emphasis is on
species of the British Isles, but North American species are also included. Color
plates are included.

Moore-Landecker, Elizabeth. *Fundamentals of the Fungi.* (Illus.) Prentice-Hall,
1972. xi + 482pp. $16.00. 75-160527. ISBN 0-13-339267-8. Indexes;gloss.

C Suitable for an undergraduate introductory course in mycology for students of
plant pathology, general botany, biology, or even as an elective for students of
other sciences. The text is well organized and carefully written. Topics treated are
morphology and taxonomy, physiology and reproduction, and ecology and utilization
by humans.

Perry, Phyllis J. *Let's Learn About Mushrooms.* (Illus.) Harvey House, 1974. 47pp.
$4.29. 73-92673. ISBN 0-8178-5172-0. Index.

JH This is a useful book for school children. There is a minimum of terminology,
some good illustrations and a commendable emphasis on the common edible
mushroom, Agaricus campestris, as a teaching model.

Smith, Alexander H. *A Field Guide to Western Mushrooms.* (Illus.) Univ. of Mich.
Press, 1975. 280pp. $16.50. 74-25949. ISBN 0-472-85599-9. Gloss.;index.

C Smith's mushroom guide provides relatively nontechnical descriptions of some
of the conspicuous fungi of an undefined region extending eastward for about a
thousand miles from the Pacific Ocean. Some 200 species are described and illustrated in color. Useful for both college and public libraries.

Smith, Alexander H., and Harry D. Thiers. *The Boletes of Michigan.* (Illus.) Univ. of Mich. Press, 1971. 428pp. $20.00. 77-107979. ISBN 0-472-85590-5.

C-P   Highly recommended to all mycologists interested in fleshy fungi. Because of the richness of this flora, this monograph should have a much wider audience than its title suggests. The introduction includes a taxonomically useful description of various important morphological features of the boletes, as well as a relevant discussion of habitat distribution and relationships of these fungi. The 300 illustrations are excellent.

Webster, John. *Introduction to Fungi.* (Illus.) Cambridge Univ. Press, 1970. 424pp. $10.50. 77-9314. ISBN 0-521-07640-4.

SH-C-P   The author takes an interesting and relevant approach to the study of fungi by emphasizing those that are readily available and encouraging students to seek out fungi on their own. Very good photographs and drawings by the author. Useful as an additional reference or collateral reading.

589.3 ALGAE

Chapman, V.J. *Seaweeds and Their Uses, 2nd ed.* (Illus.) Methuen (distr.: Barnes &Noble), 1970. xiii + 304pp. $14.50.

C-P   A brief discussion of the occurrence and distribution of seaweeds leads into a history of their use. The book is clearly written and to the point; numerous revisions have been made and new material added. Charts, graphs and drawings are excellent. A survey of seaweeds and an estimate of their quantities is presented.

Duddington, C.L. *Beginner's Guide to Seaweeds.* (Illus.) Drake, 1971. 187pp. $6.95. ISBN 0-87749-101-1.

JH-SH-C-P   A very interesting introduction, with some discussion of most areas of seaweed study. An extensive discussion on reproductive cycles is presented, as well as other functional, structural and ecological information. Not uncluttered nor in the most readable style, yet interesting and reasonably complete.

Prescott, George W. *How to Know the Freshwater Algae, 2nd ed.* (Illus.) Brown, 1970. 248pp. $7.95. 79-128177. ISBN 0-697-04859-4; 0-697-04858-6(p).

SH-C   A very useful guide to 530 genera of freshwater algae, revised and updated including improved typography and illustrations.

590 Zoological Sciences

590.74 EXHIBITS

See also 639.9 Conservation of Biological Resources and 500.9 Natural History.

Bracegirdle, Cyril. *Zoos Are News: Conservation or Extinction?* (Illus.) Abelard-Schuman, 1973. 176pp. $5.95. 72-000072. ISBN 0-200-71877-0. Index.

JH-SH   Describes the relationship of zoos to conservation of animal species and the study of animal behavior, the evolution of zoos from the first modern zoo of the 19th century to today's zoos which offer natural habitats, and presents stories about various species and a list of zoos.

Bridges, William. *Zoo Careers.* (Illus.) Morrow, 1971. 160pp. $4.95. 71-155989.

JH   Students thinking of zoo work, teachers preparing field trips for their classes, parents, and zoo buffs will all want to read this.

Elgin, Robert. *Man In a Cage.* (Illus.) Iowa State Univ. Press, 1972. xii + 122pp. $5.95; $2.95(p). 70-37950. ISBN 0-8138-1020-5; 0-8138-1050-7.

JH-SH   A fascinating account of Elgin's problems as manager of a children's zoo.

Hancocks, David. *Animals and Architecture.* (Illus.) Praeger, 1971. 200pp. $13.95. 70-101251.

C   Provides a history of shelter for animals. Architectural features of nests and
     dens are described. The transition from menageries, from merely caging animals to expose them to public view, to the development of modern zoological gardens provides the major message.

Perry, John. *The World's a Zoo.* Dodd, Mead, 1969. viii + 303pp. $6.95. 70-82618.

JH-SH   Problems facing zoo directors (acquisition of animals, adaptation to captivity, breeding; protection of rare animals in the wild, pets) are discussed. The "game preserve" zoo is covered, and the author also discusses, "Why bother about the animals?" Delightful and informative.

Shannon, Terry, and Charles Payzant. *New at the Zoo: Animal Offspring from Aardvark to Zebra.* (Illus.) Golden Gate, 1972. 80pp. $4.95. 72-78395. ISBN 0-87464-196-9.

JH-SH-C   Of the 57 species of animals briefly described and illustrated here, some
               are rare or in danger of extinction, and all are interesting. Notes concerning care, feeding and reproductive habits are of great interest, and the message is that zoos have become true centers of information and education.

Shannon, Terry, and Charles Payzant. *Zoo Safari: The New Look in Zoos.* (Illus.) Golden Gate, 1971. 80pp. $4.95. 70-157850. ISBN 0-87464-180-5.

JH-SH   Describes the extensive efforts being made by zoo personnel in the U.S.
             and Canada to prevent many animals from becoming extinct. Four of these efforts are described in some detail: the Zoo and Wild Animal Park at San Diego, the Alberta Game Preserve, the African Lion Safari and Game Farm near Toronto, and the Lion Country Safari of California.

Solandt, Barbara M. *Children of the Ark.* (Illus.) Univ. of Toronto Press, 1973. ix + 95pp. $7.95. 73-85093. ISBN 0-8010-2101-8.

JH-SH-GA   Brings into focus the increasingly important role of zoos in maintaining and breeding animals. Excellent full-page, black-and-white photographs of animals—mostly babies—are included.

591 ZOOLOGY

Hickman, Cleveland P. *Integrated Principles of Zoology, 5th ed.* (Illus.) Mosby, 1974. 1025pp. $14.95. 73-14546. ISBN 0-8016-2184-4. Bibs.

SH-C   Covers the diversity of animal life, its activity, continuity, and evolution,
        and the animal and its environment.

Villee, Claude A., et al. *General Zoology, 4th ed.* (Illus.) Saunders, 1973. xviii + 912pp. $12.95. 72-82815. ISBN 0-7216-9038-6.

C   This 4th edition retains its quality, style and clarity. There has been considerable change and improvement in the organization—the section on vertebrate form and function emphasizes mammals, particularly humans. Comprehensive.

591.03 ENCYCLOPEDIAS AND DICTIONARIES

Clement, Roland C. *Hammond Nature Atlas of America.* (Illus.) Ridge/Hammond, 1973. 255pp. $17.95. 73-7524. ISBN 0-8436-3511-2.

JH-SH-C   Provides general information for common trees, wild flowers, mammals, birds, reptiles and amphibians, fishes, and insects. Contains dis-

cussions of ecological relationships. Content selection, style, photography, reproductions, and color maps are superb.

Grzimek, Bernhard, (Ed.).   *Grzimek's Animal Life Encyclopedia, 13 Vols.* (Illus.) Van
Nostrand-Reinhold, 1972-1974. $29.95ea; $325.00set. 79-18378.

    SH-C   The 13 volumes consist of four on mammals, three on birds, two on fishes
          and amphibians, and one each on reptiles, insects, mollusks, and lower
animals. The appendixes of each volume contain the systematic classification of the
groups covered in the text and a four-part dictionary of animal names (English,
German, French and Russian). The excellent colored illustrations contribute greatly
to the text, which covers life history, behavior, ecology and conservation of the
species discussed. One of the most complete, interesting, and useful multivolume
sets ever compiled.

Leftwich, A.W.   *A Dictionary of Zoology, 3rd ed.* Constable (distr.: Crane, Russak),
1973. ix + 478pp. $13.50. ISBN 0-09-454972-9.

    SH-C   A comprehensive dictionary of zoological terms, this volume contains over
          7500 definitions. A valuable reference for all libraries, high school, public
or college.

Nayman, Jacqueline.   *Atlas of Wildlife.* (Illus.) John Day, 1972. 124pp. $10.00. 78-
38034.

    JH-SH-C   The earth's six principal zoogeographical regions are first specified,
          then splendidly related to the newer data on continental drift and earth
history. Highly interesting charts show the main groups of animals which are found in
each of the regions and the extent to which these characteristic forms are found in
other regions. The animal "portraits" are outstanding.

Rand McNally.   *The Rand McNally Atlas of World Wildlife.* (Illus.) Rand McNally,
1973. 208pp. $25.00. 73-3724. SBN 528-830147. Index;bib.

    JH-SH-C   Each zoogeographical region is discussed separately and completely
          with beautiful photographs and discussions of interrelationships.
Evolution is explained, and sections on the oceans, humans and wildlife, and national
parks are included. Contains maps. A magnificent volume.

591.042 RARE AND EXTINCT ANIMALS

See also 639.9 Conservation of Biological Resources.

Curry-Lindahl, Kai.   *Let Them Live: A Worldwide Survey of Animals Threatened with
Extinction.* Morrow, 1972. xix + 394pp. $9.95. 79-188750. Index;bib.

    SH-C   An indictment against the human is presented in this excellent account of
          rare and endangered species. Presents zoogeographical, ecological and
ethological material concerning animal species by continents, islands, other land
masses and the ocean. Recommended highly for all students in introductory biology.

Harris, John, and Aleta Pahl.   *Endangered Predators.* (Illus.) Doubleday, 1976. 84pp.
$5.95. 75-34051. ISBN 0-385-08038-7; 0-385-08012-3.

    JH   Fictional biographical sketches briefly describe how five predators (wolf, fox,
          coyote, cougar and bobcat) live and have evolved as part of natural ecosys-
tems. Most of the factual information is accurate and current, and the many pencil
drawings are beautiful. An interesting and painless way to learn about predators.

Laycock, George.   *The World's Endangered Wildlife.* (Illus.) Grosset&Dunlap, 1973.
152pp. $4.99. 72-92928. ISBN 0-448-26237-1. Index;bib.

    SH-C   Imaginative, informative and delightful; this book clearly defines the ad-
          verse influences responsible for decimation of some 100 species. Reasons

for the secure future of several other rare species are cited. Useful for conservation or environmental courses at any level; list of organizations concerned with endangered wildlife is included.

Littlewood, Cyril. *The World's Vanishing Animals: The Mammals.* (Illus.) Arco, 1970. 63pp. $4.50. 75-102638. ISBN 0-668-02201-9.

JH-SH   An attractive, brief factual source of information on 69 mammals that are becoming extinct. The physical characteristics, geographic environment, and habits of each are described, and maps locate the animals' places of origin and key the areas where they presently survive.

Littlewood, Cyril. *The World's Vanishing Birds.* (Illus.) Arco, 1973. 63pp. $5.95. 72-89859. ISBN 0-668-02889-0.

JH-SH   With only 60 species treated, the book obviously is not comprehensive. It is, however, representative of the world-wide situation of peril to wildlife from habitat destruction, displacement by man, and pollution. Half of each text page is a flat projection of a world map, with symbols for location. Nonemotional and matter-of-fact.

Milne, Lorus J., and Margery Milne. *The Cougar Doesn't Live Here Any More.* (Illus.) Prentice-Hall, 1971. x + 258pp. $9.95. 76-131870. ISBN 0-13-181149-5.

SH-C   Considers how animals in general and mammals in particular have evolved and how they survive or face extinction in a world of ever increasing human populations. Present-day problems are evaluated in terms of facts about earlier organisms and their environments as observed in the fossil record. Prehistoric domestication of animals and special evolutionary change factors are also treated.

Ricciuti, Edward R. *To the Brink of Extinction.* (Illus.) Harper&Row, 1974. 177pp. $4.95. 73-14314. ISBN 06-024982-X; 06-024983-8. Index.

JH-SH-C   Provides a comprehensive review of seven species of endangered wildlife (Puerto Rican parrot, Pere David's deer, European bison, bog turtle, osprey, Pine Barron's tree frog, and gray whale) ranging from their origins to their situation today.

Stewart, Darryl. *Canadian Endangered Species.* (Illus.) Gage (distr.: Vanguard), 1975. xx + 172pp. $12.95. Index.

SH-C   Stewart describes the effects of pesticides, air and water pollution, automobiles and highways, and the increasing urbanization of Canada on the survival of unique Canadian species. A total of 92 endangered mammals, birds, reptiles and fish are discussed.

591.072 STUDY, TEACHING AND RESEARCH

Clark, Eugenie. *The Lady and the Sharks.* (Illus.) Harper&Row, 1969. 269pp. $6.95. 74-83588.

JH-SH   Eugenie Clark first established that sharks have some intelligence and are trainable. This book tells of her work in setting up the Cape Haze marine lab and many of the investigations carried on there. Clark also discovered the secret of the grouper *Serranus,* true hermaphrodites containing both male and female organs and capable of behaving as either sex. Excellent reading.

Hickman, Frances M., and Cleveland P. Hickman, Sr. *Laboratory Studies in Integrated Zoology, 4th ed.* (Illus.) Mosby, 1974. ix + 463pp. $6.50(p).

C   This new edition of a highly successful manual helps students gain an understanding of how to make observations and perform scientific experiments. Incorporates the fundamentals of life sciences, microscope exercises, and a variety of laboratory aids, all of which are supplemented by accurate and detailed drawings. Recommended for general zoology courses.

Lauber, Patricia. *Of Man and Mouse: How House Mice Became Laboratory Mice.* (Illus.) Viking, 1971. 126pp. $4.31. 70-162671. ISBN 0-670-52058-6. Index;bib.

JH-SH   Although it is written simply, there are so few books about mice and the research at Jackson Laboratory that this one may have some value even for adults. Classical physiological genetics is well covered. A history of man's relation to mice, with emphasis on their scientific usefulness, is the main theme.

Mahoney, Roy. *Laboratory Techniques in Zoology, 2nd ed.* (Illus.) Halsted/Wiley, 1973. 518pp. $21.50. 72-13645. ISBN 0-470-56375-3.

C-P   Provides collection and culture methods as well as preservation, fixative, staining and/or other preparation techniques for all the phyla and most of the classes of the animal kingdom. A brief description of the taxa is also supplied. A useful reference for any biologist, teacher, researcher or student.

Murie, Olaus J. *Field Guide to Animal Tracks, 2nd ed.* (Illus.) Houghton Mifflin, 1975. xxi + 375pp. $6.95; $4.95(p). 74-6294. ISBN 0-395-19978-6; 0-395-18323-5. Index;CIP.

JH-SH-GA-C   Murie's study covers every mammal for which tracks have been obtained in North America, Mexico and Central America. It includes over 30 birds and some snakes also. Recommended for all wildlife lovers.

Pringle, Laurence, (Ed.). *Discovering Nature Indoors: A Nature and Science Guide to Investigations with Small Animals.* (Illus.) Natural History Press (distr.: Doubleday), 1970. 128pp. $4.50. 70-103134.

JH-SH   This handsomely designed and illustrated book is a goldmine for the young biologist; it will serve as an excellent reference and text for a sequence of studies. The well-written chapters are by science teachers, college professors, museum curators, researchers, and science writers. The authenticity and professional "surveillance" of this guide are above reproach.

Webster, David. *Track Watching.* (Illus.) Watts, 1972. 90pp. $4.50. 75-180167. ISBN 531-02030-4.

JH   Going beyond mere identification, Webster guides the reader into discovering what particular animals were doing and how different animals walk or run. Sketches and easy instructions explain how to collect tracks by plaster, paraffin, and other methods.

591.1 PHYSIOLOGY OF ANIMALS

See also 574.1 Physiology.

Cousteau, Jacques. *The Art of Motion.* (Illus.) World, 1973. 144pp. $7.95. 72-87710. ISBN 0-529-05073-0.
Window in the Sea. (Illus.) $7.95. 72-87710. ISBN 0-529-05049-8.

JH-SH-GA   These two books combine spectacular photography with easy text. In *Window in the Sea*, there are chapters on eyes and light which demonstrate eye positions, varieties of eyes and their relationship to camouflage and survival. *The Art of Motion* is a cohesive exploration of movement, obstacles to movement in water, and how these obstacles are overcome.

Eakin, Richard M. *The Third Eye.* (Illus.) Univ. of Calif. Press, 1974. xi + 157pp. $7.50. 72-97740. ISBN 0-520-02413-3.

C-P   In this entertaining account, most of the research is concerned with frogs, lampreys or lizards, but the author comments on the general significance of vertebrate pareital bodies (i.e., pineal bodies or "third eyes") as well as a number of other topics. Major emphasis is on the anatomical, developmental and physiological aspects of these primitive sense organs. Recommended to anyone with a basic knowledge of zoology.

Gordon, Malcolm S.   *Animal Physiology: Principles and Adaptations, 2nd ed.* (Illus.) Macmillan, 1972. xvi + 592pp. $13.75. 72-146941.

C   The usual physiological subjects are considered, but the emphasis is on the variations of function (particularly among vertebrates) and their evolutionary development and ecological significance. Major topics in environmental physiology are especially well covered. There are still only a few textbooks on comparative vertebrate physiology, and this is one of the best.

Ipsen, D.C.   *What Does a Bee See?* (Illus.) Addison-Wesley, 1971. 94pp. $4.75. 77-127785. ISBN 0-201-03165-5. Index.

JH   A worthwhile overview which includes detailed discussion of the experiments of Karl von Frisch and Karl Daumer, in particular, illustrating the nature of research and the necessity for painstaking experiments in order to find the facts of animal senses and behavior.

Mason, George F.   *Animal Feet.* (Illus.) Morrow, 1970. 95pp. $3.95. 70-118270.

JH   Discusses the uses of feet (both vertebrate and invertebrate) in defense, predation, burrowing, swimming, climbing, etc. Sections on evolution, the mutational process, and human feet. For sampling more than continuous reading.

Prince, J.H.   *Animals in the Night: Senses in Action After Dark.* (Illus.) Nelson, 1971. 143pp. $4.95. 71-145917. ISBN 0-8407-6142-2.

JH-SH-C   Covers the wide range of sense organs with which animals are able to orient themselves in the dark. The account is succinct and interesting. The book is well illustrated; descriptions are brief and undocumented.

Schmidt-Nielsen, Knut.   *Animal Physiology, 3rd ed.* (Illus.) Prentice-Hall, 1970. x + 145pp. $6.95. 71-110093. ISBN 0-13-037390-7.

SH-C   Not a comprehensive book on animal physiology, the main theme is rather a consideration of the physiological adaptations of animals (with more emphasis upon vertebrates) to their environments. The presentation is interesting and easily understood.

Schmidt-Nielsen, Knut.   *Animal Physiology: Adaptation and Environment.* (Illus.) Cambridge Univ. Press, 1975. xvi + 699pp. $15.95. 74-12983. ISBN 0-521-20551-4. Index.

C-P   The author follows an environmental approach to comparative animal physiology. The five parts of the book cover adaptations related to oxygen; food and energy; temperature; water; and movement, information and integration.

Silverstein, Alvin, and Virginia B. Silverstein.   *The Excretory System: How Living Creatures Get Rid of Wastes.* (Illus.) Prentice-Hall, 1972. 74pp. $4.95. 72-4190. ISBN 0-13-293654-2.

JH   Although the scientific vocabulary appears difficult, phonetic pronunciations and an adequate explanation for each term are given. The excretory processes of humans, water and land animals, and plants are compared, and some simple experiments are outlined. The book discusses recent research and provides insight into the problems of patients on kidney machines.

Silverstein, Alvin, and Virginia B. Silverstein. *The Muscular System: How Living Creatures Move.* (Illus.) Prentice-Hall, 1972. 76pp. $4.75. 70-37600. ISBN 0-13-606947-9.

JH   An interesting, well-illustrated and very readable account of movement in various species. Explores the ways the human muscular system is similar to that of other animals as well as the ways that the human muscular system is especially suited to human activities.

Simon, Hilda. *The Splendor of Iridescence: Structural Colors in the Animal World.* (Illus.) Dodd, Mead, 1971. 263pp. $25.00. 72-126295. ISBN 0-396-06208-3.

C-P   The author has produced superior results through the preparation of her own color separations for her outstanding illustrations. The book has two major sections, "The Anatomy of Structural Color," and "The Beauty of Iridescence." The electron microscope was used to study the structural iridescence of feathers, butterflies, beetles, fishes and mollusks. Biographies of numerous scientists who have contributed to the study of color are included. A beautiful book, recommended for all college and public libraries.

Smythe, R.H. *Vision in the Animal World.* (Illus.) St. Martin's, 1975. x + 165pp. $18.95. 75-13590. Index.

JH-SH   The structure and function of the eye are discussed succinctly in this fascinating volume. Smythe discusses in depth the mammalian and human eye and the visual apparatus of dogs, cats, horses, birds, fishes, reptiles and insects, and speculates on what animals actually see. The material is clearly presented, although there is some oversimplification of evolutionary principles and a tendency to utilize teleological explanations. A simple reference for comparative biology students.

Street, Philip. *Animal Reproduction.* (Illus.) Taplinger, 1974. 263pp. $8.95. 72-75665. ISBN 0-8008-0257-8. Index.

SH-C-P   This collection of typical examples of the life cycles and reproductive patterns of major groups of animals will be a helpful reference for biology teachers and serious students in related fields of animal biology.

Turner, C. Donnell, and Joseph T. Bagnara. *General Endocrinology, 5th ed.* (Illus.) Saunders, 1971. x + 659pp. $12.00. 79-135338. ISBN 0-7216-8932-9.

C   The strength of the approach here is that it allows phylogenetic comparison of endocrine function, rather than restricting the student to narrower concepts of mammalian endocrinology. The inclusion of neurosecretion in invertebrates and the discussion, although brief, of plant hormones contribute to the understanding of chemical mediators. Will serve as reference in general biology and zoology courses and as a teaching text for a course in endocrinology.

591.3 DEVELOPMENT AND MATURATION OF ANIMALS

Freedman, Russell. *Growing Up Wild: How Young Animals Survive.* (Illus.) Holiday House, 1975. 63pp. $5.95. 75-10795. ISBN 0-8234-0265-7. CIP.

JH   This is an account of the natural history of the first year of life of a leopard frog, a rattlesnake, a bald eagle, a beaver, a lion and a gorilla. Recommended for any future zoologist.

Jenkins, Marie M. *Embryos and How They Develop.* (Illus.) Holiday House, 1975. 194pp. $6.95. 74-23547. ISBN 0-8234-0254-1. Gloss.;index;CIP.

JH-SH   How reproduction occurs and life develops from a simple, one-celled organism through the more complex mammals, including humans, is treated in this well-illustrated book. (Metric units and English equivalents are used throughout.)

Saunders, John W., Jr. *Patterns and Principles of Animal Development.* (Illus.) Macmillan, 1970. ix + 282pp. $8.95. 75-79031. Gloss.

SH-C   Simple, clear style, didactic organization, and well-selected illustrations make a singularly useful supplementary reader. Origin and fertilization of the egg, germ-layer formation, body form, developmental principles, metamorphosis and regeneration, genetic control of development, mitosis-meiosis, and genetic control of protein synthesis are detailed.

Stonehouse, Bernard. *Young Animals: The Search for Independent Life.* (Illus.) Viking, 1974. 172pp. $10.95. 73-5153. ISBN 0-670-79423-6. Gloss.;index;CIP.

JH-SH-C   High-quality color photographs complement this natural history of all types of multicellular young animals. Demonstrates various species' behavioral strategies for their perpetuation through the relationships of the young to the adults and to the environment. Excellent for both collateral reading or reference.

591.5 ECOLOGY OF ANIMALS

Fogden, Michael, and Patricia Fogden. *Animals and Their Colors: Camouflage, Warning, Coloration, Courtship, and Territorial Display.* (Illus.) Crown, 1974. 172pp. $9.95. ISBN 0-517-514893. Index;gloss.

SH-C   The Fogdens describe the role of coloration and the diverse relationships among animals' coloration, habitats and behavior. The color photographs are beautiful. Suitable for casual reading and for supplementary reference in classes on general zoology, behavior or evolution.

Hancocks, David. *Master Builders of the Animal World.* (Illus.) Harper&Row, 1973. x + 144pp. $8.95. 73-5456. SBN 06-011757-5.

JH-SH-C   Written with exquisite concern for the complexities of modern evolutionary biology; 175 illustrations and a thoroughly interdisciplinary and exciting text cover all manner of animal-constructed dwellings.

Holden, Raymond P. *Wildlife Mysteries.* (Illus.) Dodd, Mead, 1972. 111pp. $4.25. 72-720. ISBN 0-396-06511-2.

JH-SH   Nine stories, all about animals, are told clearly and accurately and make fascinating reading. Although there is no common denominator beyond scientific discovery, they demonstrate nature's infinite variety and ingenuity.

Johnson, James Ralph. *Animals and Their Food.* David McKay, 1972. ix + 114pp. $4.50. 75-186432. Bib.

JH-SH   Categorizes animals according to the patterns ooff ttheof their fofooonsumption: the herbivores, the carnivores and the omnivores. Points out how the eternal search for food affects the land on every continent and how humans have upset the ecological food chain.

Milne, Lorus, et al. *The Secret Life of Animals: Pioneering Discoveries in Animal Behavior.* (Illus.) Dutton, 1975. 214pp. $29.95. 75-10073. ISBN 0-525-19932-2. Index.

JH-SH   The 10 main chapters cover the senses, programs for living, migration, nonmigrants, mating, parenthood, groupings, parasitism, commensalism and predation. Each chapter begins with a general introductory statement, followed by excellent photographs with extended legends.

Moss, Cynthia. *Portraits in the Wild: Behavior Studies of East African Mammals.*

(Illus.) Houghton Mifflin, 1975. xv + 363pp. $12.50. 75-20284. ISBN 0-395-20722-3. Index;CIP.

SH-C This fine book is more a progress report on field research than an encyclopedia, but the data are scientifically sound. Moss discusses the elephant, giraffe, rhinoceros, zebra, antelope, baboon, cat and hyena.

Naumov, N.P. *The Ecology of Animals.* (Illus.;trans.;ed. by Norman D. Levine.) Univ. of Ill. Press, 1972. x + 650pp. $17.50. 71-170965. ISBN 0-252-00219-9.

C-P This translation of a standard Russian text illustrates the state of ecology in Russia and allows it to be compared with the same science in the western world. Covers ecology of individuals, ecology of populations and ecology of associations. Energy dynamics is omitted, as is well-known contemporary work by some western authors.

Rood, Ronald. *Animals Nobody Loves.* (Illus.) Stephen Green, 1971. 215pp. $6.95. 70-118222. ISBN 0-8290-0128-7.

JH-SH-C An interesting study of spiders, wolves, rats, fleas, mosquitoes, bats and snakes. Shows these and other animals to be uniquely successful examples of adaptation, sources of food, pollinators, contributors to our economic welfare, and as links in the ecological relationship of all things. Along with little-known physiological and behavioral facts about these animals, Rood recounts many interesting myths.

Roth, Charles E. *Walking Catfish and Other Aliens.* (Illus.) Addison-Wesley, 1973. 176pp. $4.95. 72-7436. ISBN 0-201-06528-2. Index.

JH The author discusses mammals, birds, insects and fishes alien to the Western Hemisphere. An adequate and interesting historical background is given in each case. The photographs are excellent.

Shuttlesworth, Dorothy E. *Animals that Frighten People: Fact Versus Myth.* (Illus.) Dutton, 1973. 122pp. $4.95. 73-77458. ISBN 0-525-25745-4.

JH Instructive information on animals, including some current field research, is blended with a story-like style of writing. Organisms discussed include the wolves, great cats, bears, gorillas, bats, snakes, alligators, crocodiles, sharks, octopi, squids, birds of prey, spiders, and scorpions.

Simon, Hilda. *Chameleons and Other Quick-Change Artists.* (Illus.) Dodd, Mead, 1973. 157pp. $7.95. 73-1656. ISBN 0-396-06801-4.

JH-SH Contains anecdotes about the author's own experiences. Examples of animals from terrestrial, aquatic and marine environments illustrate color adaptation, while drawings demonstrate the dramatic color changes which can occur. A theory of color changes for both camouflage and communication purposes is cogently argued.

Von Frisch, Otto. *Animal Camouflage.* (Illus.) Collins/Watts, 1973. 128pp. $5.95. 72-11226. SBN 531-02113-0.

JH-SH-C Von Frisch shows camouflage in the animal kingdom to be basically a function of optical phenomena. Most of the text is an explanation of the photographs and illustrations. This is a beautifully done work and a recommended addition to any library.

Wilson, Edward O. *Sociobiology: The New Synthesis.* (Illus.) Belknap/Harvard University Press, 1975. ix + 697pp.$20.00. 74-83910. ISBN 0-674-81621-8. Gloss.;index.

C-P After many years, the biological bases of social interactions are now emerging as a major field of study. Drawing from manydisciplines, Wilson explores specific social phenomena such as communication using comparative examples from different species. This book will undoubtedly stand as a classic, and students and scholars in the biological or social sciences will enjoy the lucid, provocative ideas presented.

591.52 SPECIFIC ENVIRONMENTS, ADAPTATIONS, AND BEHAVIOR PATTERNS

See also 156 Comparative Psychology.

Edmunds, M. *Defence in Animals: A Survey of Antipredator Defences.* Longman, 1974. xvii + 357pp. $14.50. 73-92246. ISBN 0-582-44132-3. Gloss.:index.

C-P This is an analysis of the means by which animals escape their predators. Physical weapons, color, behavior, and a great many other defenses are described. The evolutionary implications of defensive structures and behaviors are stressed. A valuable reference.

Freedman, Russell, and James E. Morriss. *Animal Instincts.* (Illus.) Holiday House, 1970. 159pp. $4.95. 77-102432.

JH Deals with those functions, actions or activities an animal does instinctively, sometimes even automatically, in response to an appropriate "signal" or stimulus. Explores the nature of instinct, migration, home boundaries, animal court-ship and raising progeny, and mentions some fundamentals of genetics, natural selec-tion, and behavior. Documented discussions, well-chosen photographs.

Jarman, Cathy. *Atlas of Animal Migration.* (Illus.) John Day, 1972. 124pp. $10.00. 72-1748.

JH-SH-C Designed as a companion volume to Jacqueline Nayman's *Atlas of Wildlife* (see 591.03). The theme of migration by birds, mammals, fishes, reptiles and amphibians, and insects is strikingly developed and illustrated with color maps and other graphic aids, including animal "portraits" of the very highest quality.

Laycock, George. *Wild Travelers: The Story of Animal Migration.* (Illus.) Four Winds, 1974. 110pp. $5.95. 74-7397. ISBN 0-590-07312-5. Index;CIP.

JH-SH Laycock presents a series of essays concerning animal migration phenomena as they vary with the changing seasons. An informative de-scription of bird-banding and other wildlife research techniques is included.

McCoy, J.J. *Wild Enemies.* (Illus.) Hawthorn, 1974. xii + 210pp. $6.95. 73-9307. Index.

JH-SH Includes chapters devoted to the coyote, wolf, mountain lion, bear, bob-cat and lynx, and fox species and hawks, owls and eagles. The natural history is traced as well as the history of human conflict with these predators. McCoy argues for their ecological significance and presents a balanced view of predator control. For supplementary reading in ecology courses.

Marchant, R.A. *Where Animals Live.* (Illus.) Macmillan, 1970. 170pp. $4.95. 69-18240.

JH-SH Animal survival and reproductive behavior in single, family, and societal living arrangements of various vertebrates is the theme here. Especially interesting chapters are on the control of the populations, cave dwellers, and organ-ized societies.

Orr, Robert T. *Animals in Migration.* (Illus.) Macmillan, 1970. xv + 303pp. $12.50. 72-81553.

SH-C The migrations of insects, fishes, amphibians, reptiles, bats, cetaceans, pinnipeds, herbivores and birds are reviewed. Environmental factors and physiological changes, orientation, hazards of migration, and methods of studying migration are discussed.

Street, Philip. *Animal Migration and Navigation.* (Illus.) Scribner's, 1976. 144pp. $8.95. 75-30276. ISBN 0-684-14516-2. Index.

JH-SH-C   A review of problems and questions in the field, many of which still have only partial answers. Street reports on progress made toward their understanding, drawing on primary sources.

Street, Philip. *Animal Partners and Parasites.* (Illus.) Taplinger, 1975. x + 209pp. $10.95. 72-6630. ISBN 0-8008-0255-1. Index.

C   The many and diverse parasitic relationships that occur among animals are well covered. Fine for supplemental reading at the college level and also recommended as a reference for high school biology teachers and their classes.

591.59 ANIMAL COMMUNICATION

Cohen, Daniel. *Talking with the Animals.* (Illus.) Dodd, Mead, 1971. 135pp. $3.95. 75-134321. ISBN 0-396-06280-6.

JH-SH   Recounts historical fascination with the notion of human-animal communication and captures the reader's imagination with explanations of scientific advances in this field. Cohen covers historical cases of "talking" animals and then proceeds to a discussion of modern theories and work. The photographs are charming and enhance this clear but not simplistic overview.

Cosgrove, Margaret. *Messages and Voices: The Communication of Animals.* (Illus.) Dodd, Mead, 1974. 144pp. $4.95. 73-2134. ISBN 0-396-06809-X. Index;gloss.

JH-SH   The author discusses the languages used among members of nonhuman animal species. Begining with an introduction to various communication modes, she then describes ways in which postures, sounds, smells and movements are used by various species to communicate. An excellent, nontechnical introduction.

Prince, J.H. *Languages of the Animal World.* (Illus.) Nelson, 1976. 160pp. $6.95. 75-19464. ISBN 0-8407-6470-7. Gloss.;index;CIP.

JH-SH   A survey of communication between animals for the nonspecialist, from an animal behavior and anatomical viewpoint, with simplified physiological explanations. Vertebrates, insects, and crustaceans are considered. The simplified explanations are often extremely clear and useful, but some are inaccurate or teleological. Still, very useful as collateral reading or for reference. No special background is assumed, but general biology would be desirable.

Tinbergen, Niko, and Hugh Falkus. *Signals for Survival.* (Illus.) Oxford Univ. Press, 1971. 80pp. $8.00.

JH-SH   A very attractive book, useable by a wide age range and equal in value to many more extensive treatments of animal behavior. Although it deals specifically only with one species of sea gull, it has general application.

591.9 GEOGRAPHICAL TREATMENT OF ANIMALS

Ayer, Margaret. *Animals of Southeast Asia.* (Illus.) St. Martin's, 1970. $4.95. 149pp. 76-103149.

JH-SH   The interplay between animals and people in the realms of religion, legend and custom are explored. The final chapter about methods of collecting animals for zoos is especially informative.

Burton, Maurice. *Animals of Australia.* (Illus.) Abelard-Schuman, 1969. 134pp. $3.75. 70-92293. Index.

JH   A good introduction to some of the unusual and interesting animals of Australia deriving from the continent's long isolation (geologically speaking) from the rest of the world's land masses. Covered are egg-laying and pouched mammals, and birds, reptiles, frogs, toads, termites, and honeypot ants. Simple illustrations provide for general identification.

Cott, Hugh B.   *Looking at Animals: A Zoologist in Africa.* (Illus.) Scribner's, 1975. 221pp. $14.95. 74-29192. ISBN 0-684-14249-X. Index.

SH-C   The book is concerned with the natural history of animals indigenous to Tanzania and Kenya. The author also comments on many animals and birds and their habits—peculiar and normal. The photography is excellent.

Grzimek, Bernhard.   *Among Animals of Africa.* (Illus.) Stein&Day, 1970. 368pp. $12.50. 78-122418. ISBN 0-8128-1324-3.

SH-C   Beautiful color photographs and large amounts of data concerning African animals both in their native habitat and in the zoo are the major attractions of this volume, but the lack of focus and failure to draw conclusions are somewhat frustrating.

Hough, Richard.   *Galápagos—The Enchanted Islands.* (Illus.) Addison-Wesley, 1975. 48pp. $5.50. 74-30436. ISBN 0-210-09190-9. Index;CIP.

JH   The author discusses the major animals found on the islands, their biology, evolution, and current status. Animals discussed include tortoises, frigate birds, albatrosses, pelicans, penguins, seals and flamingoes. Each is treated from an ecological perspective.

Long, Tony.   *Mountain Animals.* (Illus.) Harper&Row, 1972. 152pp. $5.95. 70-185892. ISBN 06-012666-3; 06-012667-1(p). Indexes.

SH-C   Describes mammals (including humans), birds, insects, fish, amphibians and reptiles that live, have lived, or have tried to live in the mountains of the world. Organization is by class and habitat of animal, permitting comparisons of niche appropriateness, parallelism, evolutionary trends, etc. Of value in zoology, biology, and ecology classes and as a source of excellent photographs.

Thompson, Ralph.   *An Artist's Safari.* (Illus.) Dutton, 1970. 148pp. $19.50. 77-124306.

C   The delightful combination of natural history notes and observations and hundreds of action sketches and drawings made on a safari in Kenya and Tanzania add up to solid natural history in an unusual form.

591.92 MARINE ZOOLOGY

Chester, Michael.   *Water Monsters.* (Illus.) Grosett&Dunlap, 1973. 114pp. $4.99. 72-92925. ISBN 0-448-26232-0.

JH   An enjoyable, descriptive introduction to the strangers and the strangeness of the sea. The book is well written and employs a minimal amount of scientific terminology. The only disappointing characteristic is the reliance on drawings instead of photographs of the various species.

Cousteau, Jacques-Yves.   *Quest for Food.* (Illus.) World, 1973. 144pp. $7.95. 72-87710. ISBN 0-529-05010-2.

JH-SH   An account of the ceaseless search for food by marine animals. Covers the physiological needs, including normal maintenance of body functions, growth, reproduction and transportation; the problems involved in obtaining food; and methods available for human harvesting of marine resources. Distinctions between observed fact and secondary assumption are carefully maintained.

Dozier, Thomas A. *Dangerous Sea Creatures.* (Illus.) Time-Life Films (distr.: Little, Brown), 1976. 128pp. $7.95. 75-45283. ISBN 0-913948-04-7. Index.

JH-SH   Excellent photography and composition make this volume a thing of true beauty; literature excerpts from *Jaws, Kon-Tiki, The Edge of the Sea*, the *Odyssey*, etc., provide a well-rounded and interesting reader. Useful as an introduction and a reference.

Faulkner, Douglas, and C. Lavett Smith. *The Hidden Sea.* (Illus.) Viking, 1970. 148pp. $14.95. 72-117062. ISBN 0-670-37067-3.

SH-C   The photographs reveal some extraordinary details of representative marine species. The text is a series of essays which describe the marine environment and sponges, corals, mollusks, crustaceans, echinoderms, fishes, cleaning symbiosis, and various dangerous marine animals.

Greenberg, Jerry, and Idaz Greenberg. *The Living Reef: Corals and Fishes of Florida, the Bahamas, Bermuda and the Caribbean.* (Illus.) Seahawk, 1972. 110pp. $5.95(p). 70-187354. ISBN 0-913008-00-1.

JH-SH-C   Life in and about a coral reef exhibits a gradual shift of dominant forms in a discernable pattern of ecological succession. A sampler of coral reef communities is captured in excellent color photos. The marvelous adaptations in color, body form and habits of these animals and nonflowering algal plants are well outlined.

Walker, Braz. *Oddball Fishes and Other Strange Creatures of the Deep.* (Illus.) Sterling, 1975. 192pp. $5.95. 75-14510. ISBN 0-8069-3726-2; 0-8069-3727-0. Index.

JH-SH   Provides good synopses of a variety of unusual forms of marine organisms (unusual in the sense of phylogenetic development). Walker's style is journalistic, but he covers his scientific material well and introduces scientific names of animals. (Also useful to science teachers as an interesting review of biological phenomena.)

Wise, William. *Monsters of the Deep.* (Illus.) Putnam's, 1975. 64pp. $4.49. 72-97316. SBN GB-399-60844-3; TR-399-20355-9. Gloss.

JH   There are descriptions and assorted facts about giant squid, whale-sharks, blue whales, skates and rays, eels, poisonous zebra and stone fishes, giant clams, dolphins, deep-living gulpers and the coelacanth. Well written and lavishly illustrated.

592 INVERTEBRATES

Hickman, Cleveland P. *Biology of the Invertebrates, 2nd ed.* (Illus.) Mosby, 1973. x + 757pp. $13.50. 72-83970. ISBN 0-8016-2170-4.

C   Deals with organisms at the generic level. Structural-functional relationships within phylal subdivisions are well explained. Parasitic metazoans, insects and annelids are given cursory treatment, but sections on coelenterates and mollusks are well done. An accurate, clearly written and useful text.

Jacobson, Morris K., and Rosemary K. Pang. *Wonders of Sponges.* (Illus.) Dodd, Mead, 1976. 79pp. $4.95. 75-38363. ISBN 0-396-07300-X. Gloss.;index;CIP.

JH   Classes of freshwater and marine sponges (including some unusual varieties), biocycling in reefs and the ways of sponge dwellers are described. Early accounts of sponges (back to Aristotle), their early and current uses, collection and preservation are included.

Jenkins, Marie M. *Animals Without Parents.* (Illus.) Holiday House, 1970. 192pp. $4.95. 79-119800.

SH-GA-C   The author presents accurate and excellent descriptions of the biology
of invertebrate animals. The section on sponges is very good, and the
detailed presentations on asexual reproduction are clearly written, vivid descriptions. Especially suitable as collateral reading by general biology students.

Nichols, David, and John A.L. Cooke. *The Oxford Book of Invertebrates: Protozoa,*
Sponges, Coelenterates, Worms, Molluscs, Echinoderms, and Arthropods (Other than
Insects). (Illus.) Oxford, 1971. vii + 218pp. $11.00. ISBN 0-19-910008-X.

JH-SH   With emphasis primarily on intertidal marine organisms, the very wide
range of invertebrate phyla are covered through representative and familiar specimens from each group. The simple natural history format is illustrated with a
profusion of water-color plates of type specimens. The scientific name of each organism and some details on its life habits, peculiarities, and natural habitat are given.
A must for high school science libraries.

Smith, J.E., (Ed.). *The Invertebrate Panorama.* (Illus.) Universe, 1971. vii + 406pp.
$12.50. 79-103104. ISBN 0-87663-154-5.

C-P   Describes in a nontaxonomic way invertebrate animals (excluding insects
and protists). The chapters by various authors are well integrated and are
profitable reading for someone unfamiliar with invertebrate taxonomy.

Zeiller, Warren. *Tropical Marine Invertebrates of Southern Florida and the Bahama*
Islands. (Illus.) Wiley, 1974. 132pp. $19.95. 74-2467. ISBN 0-471-98153-2. Index-
es;CIP.

JH-SH-C   This color-illustrated book aids in identifying 183 invertebrate species
inhabiting the waters off Florida and the Bahama Islands. Useful for the
marine naturalist as well as the hobbyist.

594 MOLLUSCA

Cook, Joseph J. *The Changeable World of the Oyster.* (Illus.) Dodd, Mead, 1974.
80pp. $4.50. 73-7094. ISBN 9-396-06847-2. Index.

SH-C   An informative guide to the unique anatomy and environment of the oyster
which identifies natural hazards, including marine predators and human
threats to the oyster population. Cook also dispels some of the major myths about
oysters. Suitable for students familiar with animal biology.

Cousteau, Jacques-Yves, and Philippe Diole. *Octopus and Squid: The Soft Intelligence.*
(Illus.;trans.) Doubleday, 1973. 304pp. $9.94. 72-76141. ISBN 0-385-06896-4. Gloss.;
index.

JH-SH-C   The text is somewhat rambling and loosely organized, and the superb
photographs are not referred to in the text. A classification of the
Cephalopoda, discussion of principal genera, an article on the octopus' relationship
to man, and a short account of physiological research on the giant nerve fiber are
appended.

Humfrey, Michael. *Sea Shells of the West Indies: A Guide to the Marine Molluscs of the*
Caribbean. (Illus.) Taplinger, 1975. 351pp. $19.95. 74-20213. ISBN 0-8008-7014-X.
Indexes.

SH-C   Humfrey briefly describes and illustrates 497 species of seashells, generally
those which exceed 12 millimeters in length and which are accessible in
depths of less than 63 meters. Includes superb illustrations, information on collecting, preserving and storing shells, and brief accounts of their uses and value.

Jacobson, Morris K., and William K. Emerson. *Wonders of the World of Shells: Sea,*

Land, and Fresh-Water. (Illus.) Dodd, Mead, 1971. 80pp. $3.95. 73-143287. ISBN 0-396-06326-8.

JH   Gives a brief account of the various molluscs, the shells they form, where they are found, methods of shell collection, and their importance as food and as vectors of disease organisms. Clearly and simply written.

Jenkins, Marie M.   *The Curious Mollusks.* (Illus.) Holiday House, 1972. 224pp. $5.95. 72-75597. ISBN 0-8234-0209-6.

JH-SH   This book is not a field guide to shells; rather, it presents the living animal, with emphasis on the soft parts. It is popular in approach and includes examples of experiments.

Morris, Percy A.   *A Field Guide to Shells of the Atlantic and Gulf Coasts and the West Indies, 3rd ed.* (Illus.) Houghton Mifflin, 1973. xxviii + 330pp. $7.95; $4.95(p). 72-75612. ISBN 0-395-16809-0.

SH-C-P   There are 1035 species and subspecies here, each illustrated by at least one photograph. Descriptions and information about range and habitat are provided. Will be of great value to experienced collectors and as a classroom reference.

Oliver, A.P.H.   *Guide to Shells.* (Illus.) Quadrangle/N.Y. Times (distr.: Harper &Row), 1975. 320pp. $9.95. 75-4307. ISBN 0-8129-0565-2. Index;bib.

SH-C   This identification manual for the average shell collector is oriented toward gastropods, rather than bivalves and other molluscs. Each shell is briefly described in nontechnical terms.

595.1-.4 WORMS, ARTHROPODA, CRUSTACEA, ARACHNIDA

Bird, Alan F.   *The Structure of Nematodes.* (Illus.) Academic, 1971. xi + 318pp. $16.50. 79-163440.

C-P   This very detailed description includes an excellent presentation of the collection and handling of nematodes for observation by standard gross, microscopic and special techniques. Discusses nematode anatomy in different stages of development, and compares the different nematode species. There is good correlation of anatomy and function.

Burt, D.R.R.   *Platyhelminthes and Parasitism: An Introduction to Parasitology.* (Illus.) American Elsevier, 1970. viii + 150pp. $8.75. 77-104784. ISBN 0-444-19697-8.

C   An introduction to the basic aspects of parasitology with emphasis on flatworms. Historical accounts related to parasitic diseases, the various classes and orders of Platyhelminthes, animal associations, the morphology and replication stages, host-parasite relations, and host specificity and phylogeny are discussed.

Cook, Joseph J.   *The Curious World of the Crab.* (Illus.) Dodd, Mead, 1970. 96pp. $3.23. LB 71-105288.

JH   Sixteen "Various Crabs of the World" are discussed most informally. Horseshoe crabs are included; crab harvesting and its economic importance are mentioned. Most new or difficult words are explained; biology background is not needed to enjoy the book.

Cook, Joseph J.   *The Nocturnal World of the Lobster.* (Illus.) Dodd, Mead, 1972. 80pp. LB $3.95. 70-143289. ISBN 0-396-06420-5.

JH   This small, well-illustrated book describes the lobster's sex life, habits and haunts. History and poetry are tied into this volume in a very clever way.

Croll, Neil A.   *The Behaviour of Nematodes: Their Activities, Senses and Responses.* (Illus.) St. Martin's, 1971. ix + 117pp. $8.25. 75-124954.

SH-C-P   Croll examines the astonishing variety of activities that the nematodes display, with emphasis on mechanisms to explain the observed behavior. He discusses responses to specific stimuli and concludes with the mechanisms of orientation.

Fabre, J. Henri.   *The Life of the Spider.* (Trans.) Horizon, 1971. xi + 404pp. $7.50. 79-151015. ISBN 0-8180-1705-8.

SH-C   This book has long been out of print, and this recent edition should find enthusiastic new readers as well as old friends. It is not for the impatient reader who is looking for pictures and short descriptive notes; rather, it is a book to read carefully, enjoying the mental images which develop.

Huxley, T.H.   *The Crayfish: An Introduction to the Study of Zoology.* (Illus.) MIT Press, 1974 (c.1880). xxviii + 371pp. $12.50. 73-12751. ISBN 0-262-08069-9. Index;CIP.

C-P   Huxley's compelling prose still maintains its relevance for today's biological researcher in this reprint of an 1880 book examining morphology, development and distribution of the crayfish. Huxley demonstrates that exacting observation and careful inference can supply what often is missing in modern experimental biology.

Ross, Arnold, and William K. Emerson.   *Wonders of Barnacles.* (Illus.) Dodd, Mead, 1974. 78pp. $4.95. 74-3783. ISBN 0-396-06971-1. Index.

JH-SH   The authors impress upon the reader the abundance and significance of barnacles by discussing barnacle diversity in habitat, life style (free-living, parasitic, commensal) and morphology. Some details of reproduction, and development to settling, growth, feeding and defense, are given. Fine for supplementary reading.

Savory, Theodore.   *Introduction to Arachnology.* (Illus.) Crane, Russak, 1974. xiv + 111pp. $7.50. ISBN 0-584-10144-9. Gloss.;index.

C-P   This British reference book is mainly of interest to those doing research work on spiders, but it is also suitable as a general reference for anyone wishing to learn more about past research on arachnids.

Snow, Keith R.   *The Arachnida: An Introduction.* (Illus.) Columbia Univ. Press, 1970. 84pp. $5.00. 70-109151. ISBN 0-231-03419-9. Gloss.;index.

SH-C   Snow does a reasonably good job of providing a simplified introduction to the phylum Arthropoda and class Arachnida. External features, anatomy, physiology, and life histories of mites, ticks, spiders, scorpions, and other arachnids provide the beginning student with basic information.

Taylor, Herb.   *The Lobster: Its Life Cycle.* (Illus.) Sterling, 1975. 80pp. $5.95. 74-31699. ISBN 0-8069-3480-8; 0-8069-3481-6. Index.

JH-SH   Taylor describes the life cycle of the Maine lobster in considerable detail. Anatomy of adults, production of eggs, hatching, larval stages, molting, growth and habits of young and adult are also discussed. Should be compulsory reading for biology students and others concerned with the drastic depletion of lobster stocks due to overfishing.

595.7 INSECTS

Barbosa, Pedro, and T. Michael Peters (Eds.).   *Readings in Entomology.* (Illus.) Saunders, 1972. ix + 450pp. $6.50(p). 73-183443. ISBN 7216-1541-4.

C-P   Entomology majors should find this set of readings stimulating. The articles were selected to introduce the student to the more dynamic aspects of entomology and are keyed to various textbooks. Sections are on the status and future of entomology; morphology, taxonomy and physiology; insect ecology; insect behavior; and insect control and medical entomology.

Borror, Donald J., and Dwight M. DeLong. *An Introduction to the Study of Insects, 3rd ed.* (Illus.) Holt, Rinehart&Winston, 1971. xiii + 812pp. 73-96139. ISBN 0-03-082861-9.

SH-C   More than a guide to insect classification, the book would be an excellent text for an introductory college course in entomology which involves both field and laboratory work. The volume is also an excellent advanced reference for secondary school, college and public libraries.

Callahan, Philip S. *The Evolution of Insects.* (Illus.) Holiday House, 1972. 192pp. $5.95. 72-80654. ISBN 0-8234-0213-4.

SH-C-P   A fine discussion of not only evolution of insects but also evolution in general. Many biology teachers as well as professional entomologists will acquire much interesting information from this book which belongs in every school library.

Callahan, Philip S. *Insects and How They Function.* (Illus.) Holiday House, 1971. 191pp. $4.95. 75-141403. ISBN 0-8234-0181-2.

SH   Insects are presented to the reader in diagrams, photographs, and stereoscan electron micrographs. Callahan describes insects' variability and significance to man; their maintenance systems; their communications systems, both internal and external; and the mating, fertilization and deposition of eggs.

Chapman, R.F. *The Insects: Structure and Function.* (Illus.) American Elsevier, 1969. xiii + 819pp. $13.75. 71-75216. ISBN 0-444-19758-3.

C   A complete readable presentation of the life of the insects, combining descriptions of structure, function, and behavior. Topics include head parts; food getting, preferences, ingestion and utilization; feeding habits; symbiotic relationships; the thorax and movement; the abdomen and reproduction and development; the cuticle and respiration and excretion; the nervous system and sensory mechanism; and the blood with its hormones and pheromones.

Clark, J.T. *Stick and Leaf Insects.* (Illus.) Shurlock (distr.: Transatlantic Arts), 1974. viii + 65pp. $6.95. ISBN 0-903330-10-5. Index.

SH-C-P   The general biology of these insects is discussed, including life history, structure, feeding and defense mechanisms. Caging of specimens and simple experiments with living stick insects are described.

Hutchins, Ross E. *Insects and Their Young.* (Illus.) Dodd, Mead, 1975. 125pp. $5.50. 74-20862. ISBN 0-396-07062-0. Index;CIP.

JH-SH   Using personal anecdotes from his own research, the author develops a logical, interesting description of the biological classification schema. All insect forms are described based on four types of life histories. Useful collateral reading and an excellent reference book.

Leftwich, A.W. *A Dictionary of Entomology.* Crane, Russak, 1976. 360pp. $17.50. 75-27143. ISBN 0-8448-0820-2. Bib.

SH-C   This British dictionary contains 3000 descriptions of insect species entered by generic or common names, 700 supergeneric taxa, 700 anatomical and physiological terms, and an insect classification outline. Supplements but does not substitute for Toree-Bueno's *A Glossary of Entomology.*

Linsley, E.G., and J.L. Gressitt (Eds.). *Robert Leslie Usinger: Autobiography of an Entomologist.* (Illus.) Pacific Coast Entomological Society, 1972. xiii + 330pp. $15.00.

JH-SH-C   The reader is taken to many places around the world as this naturalist scientist reviews his life activities. Of special interest are the reasoned approaches to gathering, collating and publishing research data. A particularly interesting section on the *Cimicid* (bedbug) is devoted to the trials of taxonomy.

Little, V.A.   *General and Applied Entomology*. (Illus.) Harper&Row, 1972. xi + 527pp. $13.95. 79-181540. ISBN 06-044023-6.

C-P   Introductory chapters include insects and related animals, anatomy and physiology, and classification of insects. The bulk of the text consists of a phylogenetic treatment of the orders, with recognition characteristics. End chapters include information on ecology, insect control, collecting, mounting and preservation of specimens.

Mallis, Arnold.   *American Entomologists*. (Illus.) Rutgers Univ. Press, 1971. xvii + 549pp. $15.00. 78-152316. ISBN 0-8135-0686-7.

SH-C   Provides biographical accounts of 203 entomologists who lived between 1749 and 1966—their publications, accomplishments, and personalities. Delightful reading from which the reader can learn much about the history of entomology in the U.S. and Canada.

Rhine, Richard.   *Life in a Bucket of Soil*. (Illus.) Lothrop, Lee&Shepard, 1972. 96pp. $4.25. 72-155756.

JH   Suggests an approach to beetles and other insects similar to the pegging-out of a square foot of soil and studying it. Worms, snails, wood lice, spiders, ants, springtails and the many-legged crawlers are other forms described as part of this fascinating world.

595.705 INSECT BEHAVIOR

Askew, R.R.   *Parasitic Insects*. (Illus.) American Elsevier, 1971. xvii + 316pp. $11.50. 73-137683. ISBN 0-444-19629-3.

GA-C-P   Hundreds of intriguing facts and information on parasitic insects are found in this volume. Although written for college biology students, it can be enjoyed by the interested lay reader.

Callahan, Philip S.   *Insect Behavior*. (Illus.) Four Winds Press, 1970. 155pp. $4.95. 79-105342. Index;gloss.;bib.

JH-SH   A brief introduction, with chapters on environment, morphology, reproduction, feeding, migration, camouflage, habitats, flight, communication, and economic importance. Presents various projects for the beginner.

Dalton, Stephen.   *Borne on the Wind: The Extraordinary World of Insects in Flight*. (Illus.) Reader's Digest Press (distr.: Dutton), 1975. 160pp. $18.95. 75-6517. ISBN 0-88349-052-8. CIP.

JH-SH-C   Dalton offers an informative account of insect flight, introductory comments regarding the eight insect orders represented and various details about the 55 insects shown in the 74 unbelievable color photographs.

Hickin, Norman E.   *Household Insect Pests: An Outline of the Identification, Biology and Control of the Common Insect Pests Found in the Home*. (Illus.) St. Martin's, 1974. 176pp. $10.00. 72-95358. Index.

SH   A very informative and well-illustrated book, covering basic morphology, metamorphosis and classification; the appearance, economic importance and life cycle of about 75 domiciliary insects; and the importance of identifying insects before attempting pest control. Written for the British homeowner, but most are also found in the United States.

Kaufmann, John. *Insect Travelers.* (Illus.) Morrow, 1972. 126pp. $4.95. 72-1546. ISBN 0-688-20036-2; 0-688-30036-7. Bib.

JH   Helps readers become aware of behavior patterns in insects and includes many varied examples of seasonal and daily behavioral cycles. Unfortunately, many of the diagrams are complex, but it could serve as a reference or supplementary reader at this level.

Nachtigall, Werner. *Insects in Flight: A Glimpse Behind the Scenes in Biophysical Research.* (Illus.;trans.) McGraw-Hill, 1974. 150pp. $12.95. 70-172030. ISBN 0-07-045736-0. Index;CIP.

C   Ample illustrations accompany this lucid and informative introduction to insect flight. Includes explanations of insect physiology and the dynamics of flight. Suitable for entomology or general biology courses.

Norsgaard, Ernestine J. *Insect Communities.* (Illus.) Grosset&Dunlap, 1973. 120pp. $4.99. 73-4462. ISBN 0-448-26247-9.

JH   The author explains how ants care for their young, how and where they live, and how they get a meal without becoming a meal. Termites and honeybees are also discussed. The author comments on possible human parallels in insect organization.

Oldroyd, Harold. *Elements of Entomology: An Introduction to the Study of Insects.* (Illus.) Universe Books, 1970. x + 311pp. $7.95. 77-116574. ISBN 0-87663-127-8. Bib.

SH-C   Fills the gap between textbooks of entomology and introductory zoology or biology textbooks. Deals with diversity of insects and their relatives; vision, sound, flight, adaptation, feeding, behavior; and relationships between humans and insects. This book is easily readable and has definitive statements on some controversial phenomena, with excellent illustrations and photographs.

Patent, Dorothy Hinshaw. *How Insects Communicate.* (Illus.) Holiday House, 1975. 127pp. $5.95. 75-6699. ISBN 0-8234-0263-0. Index;CIP.

JH-SH-GA-C   The author explains various complicated ways of communication in representative groups of common social and nonsocial insects. Interesting general reading.

Schneirla, T.C. *Army Ants: A Study in Social Organization.* (Ed. by Howard R. Topoff.) (Illus.) Freeman, 1971. xx + 349pp. $12.00. 70-149408. ISBN 0-7167-0933-3.

C-P   This is the primary reference on army ant behavior. The behavior of the colony at rest and on the move, of the queen, the males and young queens, and of the individual in relation to the colony is described. In addition, there are attempts to explain the development of unique behavior.

Varley, G.C., et al. *Insect Population Ecology: An Analytical Approach.* (Illus.) Univ. of Calif. Press, 1974. x + 212pp. $7.95(p). 73-89367. ISBN 0-520-02667-5.

SH-C-P   Nine self-contained chapters, including "Parasites and Predators," "Climate and Weather," "Life Tables," and "Biological Control," are provided along with experiments and exercises. Liberally supplied with figures, tables and references, this book is an excellent reference.

595.709 INSECTS—GEOGRAPHICAL TREATMENT

Borror, Donald J., and Richard E. White. *A Field Guide to the Insects of America North of Mexico.* (Illus.) Houghton Mifflin, 1970. 404pp. $5.95. 70-80420.

JH-SH-C   Provides simplified identification of the more common of the 88,000 insect species that have been identified in North America; covers collecting, preserving, and observing insects in the field. Line drawings emphasizing the key characters by which the order is identified and 16 color plates are included.

Healy, Anthony, and Courtenay Smithers. *Australian Insects in Color.* (Illus.) A.W. Reed, 1971. 112pp. $5.95. ISBN 0-589-07105-X.

    JH-SH   Emphasis is on characteristic inhabitants of various ecological areas including deserts, mountains, rivers, and the Australian bush. Gives brief physical descriptions, life histories, behavioral adaptations, and relationships with other animals. Magnificent color photographs.

Linsenmaier, Walter. *Insects of the World.* (Illus.;trans.) McGraw-Hill, 1972. 392pp. $25.00. 78-178047. ISBN 07-037953-X. Indexes.

    SH-C   This very fine book on insect life includes self-contained chapters on insect morphology, metamorphosis, insect orders, social behavior, insects as "living works of art," and insect mimicry and camouflage. The reader is referred constantly to superb illustrations which show exactly what the author is discussing. Includes index which gives the scientific names and world locations of all insects illustrated in the book.

Swan, Lester A., and Charles S. Papp. *The Common Insects of North America.* (Illus.) Harper&Row, 1972. xiii + 750pp. $15.00. 75-138765. ISBN 06-014181-6.

    JH-SH   Brings order and relative simplicity to the vast array of insect nomenclature. It will enable the amateur to find the insect in question, its name, and much additional information. Beautifully illustrated; useful to amateur entomologists, agriculturists and ecologists.

Tweedie, Michael. *Atlas of Insects.* (Illus.) John Day, 1974. 128pp. $10.00. 73-12022. ISBN 0-381-98258-0. Index;CIP.

    SH-C-P   Insects are discussed according to zoogeographic region inhabited— Palearctic, Nearctic, Oriental, Ethiopian, Neotropical, Australian, etc. The text is well researched and written and will provide excellent recreational reading for both amateur and serious student.

595.72–.79 ORTHOPTERA, THYSANOPTERA, HEMIPTERA, DIPTERA, LEPIDOPTERA, HYMENOPTERA

Andrews, Christopher. *The Lives of Wasps and Bees.* (Illus.) American Elsevier, 1970. 204pp. $5.75. 70-91272. ISBN 0-444-19733-8.

    SH-C   The biology, behavior, and general ecology of bees and wasps is presented in enough detail for the professional. Strongly recommended as collateral reading or reference for general entomology courses and advanced high school biology classes.

Baron, Stanley. *The Desert Locust.* (Illus.) Scribner's, 1972. xiv + 228pp. $7.95. 73-180626. ISBN 684-12702-4.

    C-P   This review of locust depredations, both historical and contemporary, and the explanation of the significance of entomological research are skillfully written. Interesting as well as informative reading, this book is an on-the-scene account of the battle of the 1967–69 locust plague.

Emmel, Thomas C. *Butterflies: Their World, Their Life Cycle, Their Behavior.* (Illus.) Chanticleer/Knopf, 1975. 260pp. $35.00. 75-10407. ISBN 0-394-49958-1. Gloss.;index.

    SH-C   A detailed life history of the butterfly is provided, accompanied by excellent diagrams. Butterfly behavior is covered briefly with fascinating discussions on flight and mimicry. A substantial portion of the text is devoted to families, their representative species and their world distribution.

Evans, Howard E., and Mary Jane West Eberhard. *The Wasps.* (Illus.) Univ. of Mich. Press, 1970. vi + 265pp. $3.45. 71-124448. ISBN 0-472-00118-3; 0-472-05018-4(p).

    SH-C In a book packed with information gained through years of painstaking observation of one of the most advanced groups of the insect world, Evans and Eberhard examine both social and solitary wasps. These are compared with ants, bees, and other insects in a great book for the naturalist and ecologist.

Hutchins, Ross E. *The Bug Clan.* (Illus.) Dodd, Mead, 1973. 127pp. $4.25. 72-11254. ISBN 0-396-06771-9.

    JH-SH Well-written, technically accurate descriptions of two orders of insects: Hemiptera and Homoptera. Excellent, close-up photographs will interest entomologists, and the text could provide an evening of refreshing reading. The only weakness is lack of sufficient information on how the bugs and other insects relate to other arthropods. A reliable reference.

Hutchins, Ross E. *Grasshoppers and Their Kin.* (Illus.) Dodd, Mead, 1972. 144pp. $4.50. 70-184186. ISBN 0-396-06503-1.

    JH-SH An authoritative and highly readable discourse, with excellent photography. Separate chapters deal with the grasshoppers, the katydids, the crickets, the mantids and the phasmids, as well as with the territorial extent of each group.

Lavine, Sigmund A. *Wonders of the Fly World.* (Illus.) Dodd, Mead, 1970. 64pp. $3.50. 75-121979. ISBN 0-396-06213-X.

    JH-SH A readable, brief book about flies, with a discussion of myths, superstitions, and literature references; information on morphology, the life cycle, courtship, behavior, flies beneficial and detrimental to man; and specific examples of insects (by common or scientific names) with notes on their biology.

Riley, Norman D. *A Field Guide to the Butterflies of the West Indies.* (Illus.) Quadrangle/N.Y. Times (distr.: Harper&Row), 1976. 224pp. $12.50. 74-25436. ISBN 0-8129-0554-7. Index;gloss.;bib.

    SH-C-P One of a series of valuable area guides, this is by an experienced lepidopterist for the collector visiting the West Indies. Provides background information, detailed descriptions of species, and information on where and when they can be found. Includes common and scientific names, check list, distribution table, etc.

Shuttlesworth, Dorothy. *The Story of Flies.* (Illus.) Doubleday, 1970. 63pp. $3.95. 68-17817.

    JH The distinguishing characteristics of Diptera, "true flies," are described, followed by general comments on the physiology, morphology and reproduction. Chapters are devoted to the house fly; the tsetse, horse and deer flies; mosquitoes; tachinids; crane flies; and species of flies that exhibit mimicry. A final chapter is devoted to the non-Diptera—dragonflies, damselflies, fireflies and sawflies. An excellent flowing narrative.

Smart, Paul. *The International Butterfly Book.* (Illus.) Crowell, 1975. 275pp. $19.95. 75-15479. ISBN 0-690-00963-1. Index.

    SH-C-P More than one-third of this book is an introduction to butterflies. Topics include origin and classification, anatomy, life cycle, ecology, mobility, genetics, coloration, variation, mimicry and collecting. Half the book is devoted to butterfly families, mostly photographs identified as to species and the country of origin. Includes a systematic list of the world's butterflies and additional data.

Spoczynska, Joy O.I. *The World of the Wasp.* (Illus.) Crane, Russak, 1975. 188pp. $11.75. 74-13621. ISBN 0-8448-0560-2. Gloss.;index.

SH-C   Highly suitable for the advanced junior entomologist and the interested naturalist or as a reference volume for undergraduate university courses. The author provides a description of the natural history of various wasps and a useful appendix dealing with study methods and collecting.

Tyler, Hamilton A. *The Swallowtail Butterflies of North America.* (Illus.) Naturegraph, 1975. viii + 192pp. $8.95; $5.95(p). 75-30569. ISBN 0-87961-039-5; 0-87961-038-7. Gloss.;indexes;CIP;bib.

SH-C-P   This is primarily a handbook for identification of species in the family Papilonidae of North America. The author describes each species and its habitat.

Watson, Allan, and Paul E.S. Whalley. *The Dictionary of Butterflies and Moths in Color.* (Illus.) McGraw-Hill, 1975. xiv + 296pp. $39.95. 74-30433. ISBN 0-07-068490-1. Gloss.;CIP.

SH-C   The dictionary includes species from every major geographical region and from nearly all the recognized families. Following the excellent photographs is an alphabetical sequence of names, vernacular and Latin, which are keyed to the preceding photographs.

Worth, C. Brooke. *Mosquito Safari: A Naturalist in Southern Africa.* (Illus.) Simon&Schuster, 1971. 316pp. $8.95. 77-139670. ISBN 0-671-20827-6.

JH-SH   The enormous task of detecting viruses running wild in Africa is related by a member of a team in virus research which identified thousands of mosquitoes. Worth weaves a fine tale about the Zulu people as well as the scientific personnel with whom he worked and the projects he worked on.

596 VERTEBRATES

Alexander, R. McNeill. *The Chordates.* (Illus.) Cambridge Univ. Press, 1975. vi + 480pp. $28.50; $9.95(p). 74-76580. ISBN 0-521-20472-0; 0-521-09857-2. Index.

C-P   This important book treats the chordates with a phylogenetic approach, proceeding from the tunicates to the mammals. Alexander stresses those facts about these animals that have been determined by experimentation and has in each case noted the techniques applied. There is a strong emphasis on physics and engineering, with detailed presentations of many equations, graphs, records and mechanisms.

Kent, George C. *Comparative Anatomy of the Vertebrates, 3rd ed.* (Illus.) Mosby, 1973. ix + 414pp. $12.95. 72-91785. ISBN 0-8016-2649-8. Index;gloss.;bib.

C-P   This 3rd edition emphasizes phylogenetic aspects of vertebrate structure while integrating structural variation with ontogeny and adaptive significance. Most revised are chapters on skin, muscles and respiration. Chapter summaries, charts, and tables are continued.

Kent, George C. *Systemic Dissections of Vertebrates: A Laboratory Guide.* Mosby, 1975. viii + 187pp. $5.50. ISBN 0-8016-2653-6. Index.

C-P   Designed to be used with Kent's textbook (above), this guide will help the student dissect and study organ systems of lamprey, shark, bony fish, frog, salamander, turtle, alligator, bird and cat. Intended as a guide to structure only; function is described in the textbook.

Livaudais, Madeleine, and Robert Dunne. *The Skeleton Book: An Inside Look at Animals.* (Illus.) Walker, 1973. 31pp. $3.95. 72-81381. ISBN 0-8027-6125-9; 0-8027-6126-7.

    JH-SH-C-P   Introduces 13 chordate skeletons ranging from snake to human. Each plate is accompanied by a descriptive paragraph, with emphasis on overall structure and function. Although no evolutionary relationships are indicated, the adaptations are self-evident. The outstanding black-and-white photographs could illustrate university lectures.

McCauley, William J. *Vertebrate Physiology.* (Illus.) Saunders, 1971. xiv + 422pp. $9.75. 75-132177. ISBN 0-7216-5878-4.

    SH-C-P   A comparative approach of great flexibility, this text can be adapted to almost any vertebrate physiology course. The treatment includes enough material about nonmammalian vertebrates to make it worthwhile for every student majoring in biology or zoology.

Nixon, Marion. *The Oxford Book of Vertebrates: Cyclostomes, Fish, Amphibians, Reptiles, and Mammals.* (Illus.) Oxford Univ. Press, 1972. viii + 216pp. $14.75. ISBN 0-19-910009-8. Index.

    JH-SH   These superb descriptions of numerous members of each vertebrate class focus on geographical distribution, migratory and growth patterns, and breeding habits of all the British vertebrates except birds. There is a valuable taxonomy key. Best used as supplementary material for introductory biology or zoology.

Orr, Robert T. *Vertebrate Biology, 4th ed.* (Illus.) Saunders, 1976. viii + 472pp. $12.95. 75-19850. ISBN 0-7216-7018-0. Index;CIP.

    C   A general survey of the functioning of vertebrate organ systems with appropriate reference to morphology, classification, and vertebrate ancestry. It is concise, well written and easy to read. Discusses differences in the functioning of the classes of vertebrates, ecology, behavior (including population movements, dormancy, territorialism), and the methods that are used in studying migration and dormancy.

Phillips, Joy B. *Development of Vertebrate Anatomy.* (Illus.) Mosby, 1975. vii- + 473pp. $14.50. 74-14876. ISBN 0-8016-3927-1. Index;CIP.

    C   Phillips has done an exceptionally good job of combining anatomy and embryology at the vertebrate level. An evolutionary point of view accompanies the development of adult forms, with emphasis on morphogenesis, histogenesis and adaptation. Good beginning college biology text for students with good backgrounds. Also a good high school or public library reference.

Romer, Alfred S. *The Vertebrate Body, 4th ed.* (Illus.) Saunders, 1970. viii + 601pp. $10.25. 75-92143. ISBN 0-7216-7667-7.

    C   Discussions of vertebrates past and present, cells and tissues, embryology of the vertebrates, and development of all of the body systems combine to make an outstanding comparative anatomy text.

Sadleir, Richard M.F.S. *The Reproduction of Vertebrates.* (Illus.) Academic, 1973. xiii + 227pp. $4.95(p). 72-88371. ISBN 0-12-614250-5.

    C   This succinct, well-balanced account of the development of sexual reproduction in vertebrates from the lowest to the highest form interprets many aspects of evolution and behavior and the influence of various environments on the development of reproductive systems. For reference and supplemental reading.

597 FISHES

Blumberg, Rhoda. *Sharks.* (Illus.) Watts, 1976. 77pp. $3.90. 75-45120. ISBN 0-531-00846-0. Gloss.;index;CIP.

    JH   Many shark species are well illustrated and described in understandable terminology. There is a photograph of each of the more important species and a short description; the major distinguishing facts about the Elasmobranchii are noted.

Budker, Paul. *The Life of Sharks.* (Illus.) Columbia Univ., 1971. xvii + 222pp. $12.50. 71-148462. ISBN 0-231-03551-9.

    SH-C   A thoughtful, balanced discussion of sharks and their activities, and a clear presentation of their classification, morphology, and anatomy; feeding habits, myths and legend; and utilization of sharks for human benefit. A concise, authoritative book.

Caras, Roger. *Sockeye: The Life of a Pacific Salmon.* Dial, 1975. viii + 135pp. $7.95. 75-22260. ISBN 0-8037-7247-5. Gloss.;CIP.

    SH-C   Caras treats the typical life history of a male sockeye salmon, from birth to the death which follows spawning in the case of nearly all Pacific salmon. There are 22 chapters which correspond to stages or events in the life of the species and the individual.

Cousteau, Jacques-Yves, and Philippe Cousteau. *The Shark: Splendid Savage of the Sea.* (Illus.;trans.) Doubleday, 1970. 277pp. $7.95. 69-13004.

    SH-C   The narrative alternates between the words of Jacques and his son Philippe and presents the experiences of their research vessel's personnel with sharks in the Red Sea and the Indian Ocean. Illustrations are 124 color photographs by members of the crew. The text introduces the principal species of sharks in exciting encounters.

Fitch, John E., and Robert J. Lavenberg. *Tidepool and Nearshore Fishes of California.* (Illus.) Univ. of Calif. Press, 1976. 156pp. $8.95; $3.95(p). 74-84145. ISBN 0-520-02844-9; 0-520-02845-7. Gloss.;index.

    C   Data are arranged by family group, and the book includes identification keys, geographic range, life history, relationship to other fishes—living and fossil—and the meanings of the scientific names.

Hoar, W.S., and D.J. Randall (Eds.). *Fish Physiology, Vol. 6: Environmental Relations and Behavior.* (Illus.) Academic, 1971. xvi + 559pp. $30.00. 76-84233. ISBN 0-12-350406-6.

    SH-C   One of a six volume series which will be the authoritative reference on this subject for many years. This volume contains physiological reactions of fish to their environment and fish behavior. The chapters on freezing resistance, learning and memory make especially fascinating reading.

Lineaweaver, Thomas H., III, and Richard H. Backus. *The Natural History of Sharks.* (Illus.) Lippincott, 1970. 256pp. $6.95. 75-109174. Gloss.;bib.

    GA-C   A highly readable summary of the natural history of over 100 species of sharks, their behavior, feeding habits, and reproduction. Shark attacks on humans are a special concern of the authors; there is a final chapter on shark repellents, whose efficacy is questioned.

Netboy, Anthony. *The Salmon: Their Fight for Survival.* (Illus.) Houghton Mifflin, 1974. xix + 613pp. $15.00. 72-9022. ISBN 0-395-14013-7. Index;bib.

    C   Netboy discusses the importance of salmon to humankind; the evolution, life history and migrations of the salmon family; the fate of the two broad groups (the Atlantic and Pacific salmon) and aspects of the Asian stocks. A useful reference for students, teachers and fishery biologists.

Tinker, Spencer Wilkie, and Charles J. DeLuca.   *Sharks and Rays: A Handbook of the Sharks and Rays of Hawaii and the Central Pacific Ocean.* (Illus.) Tuttle, 1973. 80pp. $7.25. 72-83675. ISBN 0-8048-1082-6.

SH-C-P   A concise and interesting treatment of sharks and rays in typical handbook style. Common and scientific names are given, drawings and descriptive material covers the physical and reproduction characteristics, habitat and other unique traits of each creature.

Wheeler, Alwyne.   *Fishes of the World: An Illustrated Dictionary.* (Illus.) Macmillan, 1975. xiv + 366pp. $27.50. 75-6972. ISBN 0-02-626180-4. Gloss.;CIP.

GA   Wheeler covers many of the principal species of fish found in the world. Size, weight, range, a brief ecology and other useful facts are given for each. There are 501 beautifully reproduced color plates.

White, William, Jr.   *The Siamese Fighting Fish: Its Life Cycle: The Betta and Paradise Fish.* (Illus.) Sterling, 1975. 60pp. $5.95. 73-83442. ISBN 0-8069-3478-6; 0-8069-3479-4. Index.

SH   A useful book to introduce the inexperienced reader to a fish's life cycle and to instruct a beginning aquarist on keeping and breeding Siamese fish. Clearly written and interesting, but some aspects are oversimplified.

597.6 AMPHIBIANS

Blassingame, Wyatt.   *Wonders of Frogs and Toads.* (Illus.) Dodd, Mead, 1975. 80pp. $4.95. 74-25523. ISBN 0-396-07086-8. Index;CIP.

JH-SH   An unusually well-organized, concise introduction. Interest is created by discussions of amphibians in the geological time sequence, and superstitions, hibernation and habits of these animals. Can be used as a text or as supplementary reading.

Patent, Dorothy Hinshaw.   *Frogs, Toads, Salamanders, and How They Reproduce.* (Illus.) Holiday House, 1975. 142pp. $6.95. 74-26567. ISBN 0-8234-0255-X. Index;CIP.

JH   A wealth of interesting information is presented about amphibia. The origin and variety of amphibians, reproduction and development, and obscure and common groups are discussed. Written in a lecture format which helps the reader grasp important concepts.

Simon, Hilda.   *Frogs and Toads of the World.* (Illus.) Lippincott, 1975. 128pp. $6.95. 75-14095. ISBN 0-397-31634-8. Index;CIP.

JH-SH   Simon's clear and interesting book surveys the origins of amphibians, frog anatomy and development. Five groups are covered: primitive frogs and toads, toads, tree frogs, true frogs and maverick frog families. The unusual breeding habits of frogs are discussed.

598.1 REPTILES

Bellairs, Angus.   *The World of Reptiles.* (2 vols.;illus.) Universe Books, 1970. 590pp. $25.00 set. 70-99976. ISBN 0-87663-113-8. Bib.

C-P   Intended for the advanced student of zoology or the professional herpetologist; covers all aspects of the evolutionary history, geographical distribution, morphology, physiology, life history, embryology, and classification. Includes labeled drawings and diagrams and excellent photographs.

Conant, Roger. *A Field Guide to Reptiles and Amphibians of Eastern and Central North America.* (Illus.) Houghton Mifflin, 1975. xviii + 429pp. $10.00; $6.95(p). 74-13425. ISBN 0-395-19979-4; 0-395-19977-8. Gloss.;index;CIP.

    SH-C-P This field guide provides 331 excellent descriptions of species of amphibians and reptiles, and over 300 good quality distribution maps.

Harrison, Hal H. *The World of the Snake.* (Illus.) Lippincott, 1971. 160pp. $5.95. 72-146688. Index.

    JH-SH There are numerous excellent black-and-white photographs of various snakes, many identified with both common and scientific names, as well as clearly drawn anatomy diagrams. There is a good chapter on facts and fallacies and a checklist of over 100 American snakes in five families. Good for school and public libraries.

Klauber, Laurence M. *Rattlesnakes: Their Habits, Life Histories, and Influence on Mankind, 2nd ed.* (2 vols; illus.) Univ. of Calif. Press, 1972. 1533pp. $50.00 set. 78-1888573. ISBN 0-520-01775-7. Bib.

    C-P Deals exhaustively with rattlesnakes: description, classification, paleontology, zoogeography, phylogeny, geographic distribution, behavior, ecology, reproduction, control, enemies, poison apparatus, venoms, the treatment of envenomation, and the myth and folklore. Treats 31 species and 70 subspecies. Bibliography of 4085 titles.

Minton, Sherman A., Jr., and Madge R. Minton. *Giant Reptiles.* (Illus.) Scribner's, 1973. xiii + 345pp. $9.95. 72-9770. SBN 0-684-13267-2.

    JH-SH This pleasant and amusing book mixes biology with folklore and local customs. The authors' knowledge of their subject is such that even the specialist will find new facts and speculation. A mine of information.

Porter, Kenneth R. *Herpetology.* (Illus.) Saunders, 1972. xi + 524pp. $15.50. 75-188390. ISBN 0-7216-7295-7. Indexes;bibs.

    C This book for upper level courses discusses structure, evolution, distribution, basic physiology, moisture and temperature relations, reproductive behavior and associated adaptions, and more. Comprehensive but not thorough.

Simon, Hilda. *Snakes: The Facts and the Folklore.* (Illus.) Viking, 1973. 128pp. $6.95. 73-5154. SBN 670-65315-2.

    JH-SH Simon describes the evolution and distinctive anatomical features of snakes and traces the myths, folklore and position of the serpent in human history.

Stebbins, Robert C. *Amphibians and Reptiles of California.* (Illus.) Univ. of Calif. Press, 1972. 152pp. $2.75(p). 72-165229. ISBN 0-520-02090-1.

    SH-C-P This extremely fine guide contains concise, very readable information and superb illustrations. A valuable reference for both learning and observation.

598.13 –.14 Turtles, Tortoises, Crocodiles and Alligators

Blassingame, Wyatt. *Wonders of the Alligators and Crocodiles.* (Illus.) Dodd, Mead, 1973. 80pp. $3.95. 72-7747. ISBN 0-396-06734-4.

    JH-SH A good, short work for collateral reading. Attention is focused on the American alligator, but there are many other interesting items: prehistoric dinosaurs, observations of alligators in 16th century America, and the problem of endangered species. While readable and visually appealing, there is no sustained storyline. At the end, the author poses interesting problems for future naturalists.

Bustard, Robert. *Sea Turtles: Natural History and Conservation.* (Illus.) Taplinger, 1973. 220pp. $11.95. 72-2197. ISBN 0-8008-7018-2.

SH-C   Bustard presents sea turtle natural history for the green, loggerhead and flatback sea turtles. There are sections on the difference between turtles and reptiles; the nesting behavior of green, loggerhead, flatback and all other sea turtles; life history from egg to hatching, migration of adults and timing of egg laying; and conservation of sea turtles.

Ernst, Carl H., and Roger W. Barbour. *Turtles of the United States.* (Illus.) Univ. Press of Kentucky, 1973. x + 347pp. $22.50. 72-81315. ISBN 0-8131-1272-9.

SH-C-P   Over 1500 original publications are referenced in this major summary of the biology of U.S. turtles. Includes information on their structure, distribution, habitat, activity, reproduction, growth, food and feeding, economic importance, and conservation, plus illustrated taxonomic keys. Also covers recognition, geographic variation, confusing species, origin and evolution, care of turtles in captivity, parasites, commensals and symbionts.

Guggisberg, C.A.W. *Crocodiles: Their National History, Folklore and Conservation.* (Illus.) Stackpole, 1972. x + 203pp. $7.95. 73-179600. ISBN 0-8117-0460-2.

JH   Crocodiles' life cycle, ecological importance and future prospects are examined. This book could serve as a reference on characteristics and behavior of crocodiles. Stories are compiled here to strengthen support for preserving this species.

Neill, Wilfred T. *The Last of the Ruling Reptiles: Alligators, Crocodiles, and Their Kin.* (Illus.) Columbia Univ., 1971. xvii + 486pp. $15.95. 72-141239. ISBN 0-231-03224-2.

SH-C   This excellent review of crocodilian biology is a singular achievement that may be consulted by anyone interested in these reptiles. The author foresees rapid extinction of the crocodilian species. The text's five parts are "Little-Known Survivors From An Ancient Day," "Two Hundred Million Years of Crocodilian History," "A Little Light on Some Ancient Episodes," "Natural History of the American Alligator," and "The Modern Crocodilians."

Ricciuti, Edward R. *The American Alligator: Its Life in the Wild.* (Illus.) Harper &Row, 1972. 67pp. $4.95. 72-76502. SBN 06-024995-1; 06-024996-X. Index.

JH-SH   This excellent, well-written and highly accurate book introduces the alligator—its external anatomy, its relationship to man, its life style in four different localities, and its chances for survival. The book also describes the life history of an alligator by neatly integrating old observations with recent scientific investigation. A pleasurable reading experience.

Riedman, Sarah R., and Ross Witham. *Turtles: Extinction or Survival?* (Illus.) Abelard-Schuman, 1974. 156pp. $6.95. 73-6188. ISBN 0-200-00126-4. Index;CIP.

JH   This manual of turtle care and biology should help young turtle owners to care for their pets. Although the book does not specifically address problems of turtle conservation, it does provide young readers with an opportunity to learn more about turtles in general.

Scott, Jack Denton. *Loggerhead Turtle: Survivor from the Sea.* (Illus.) Putnam's, 1974. 64pp. $6.95. 73-83993. SBN GB-399-60869-9; TR-399-20379-6.

JH-GA   The author describes the life history of the loggerhead turtle, with particular emphasis on the hazards that adult and young turtles face during their lives, and shows how friends of the turtles are helping to protect the vulnerable eggs on several islands off the southeastern United States.

White, William. *A Turtle is Born.* (Illus.) Sterling, 1973. 96pp. $3.50. 72-95220. ISBN 0-8069-3528-6; 0-8069-3529-4.

JH-SH A concise report covering embryological development through the differences in life style of turtles and tortoises. A systematic approach to anatomy is combined with black-and-white drawings and photographs of dissected specimens.

598.2 BIRDS

Anderson, John M. *The Changing Worlds of Birds.* (Illus.) Holt, Rinehart&Winston, 1973. 122pp. $5.95. 75-182780. ISBN 0-03-091301-2; 0-03-091302-0. Bib.;gloss.

JH-SH A rather orthodox but well-done text covering all the traditional topics in introductory bird ecology. Also included are suggestions for birdwatching and attempts to reconcile the interests of hunters and birdwatchers. The author's literate style, sense of drama and wide familiarity with the subject combine to make this an excellent book.

Armstrong, Edward A. *The Life and Lore of the Bird: In Nature, Art, Myth, and Literature.* (Illus.) Chanticleer/Crown, 1975. 172pp. $15.95. 75-15261. ISBN 0-517-524341. Index.

JH-SH-GA A sampler of the history and current status of a variety of aspects of birds as seen by humans, especially the relationship between birds and humans. Fascinating for laypersons and as collateral or reference material in beginning biology classes.

Campbell, Bruce. *The Dictionary of Birds in Color.* (Illus.) Viking, 1974. 352pp. $22.50. 73-17954. SBN 670-27225-6. Gloss.

C A representative sample illustrated by over a thousand color photographs of the nearly 9000 species of birds. Includes valuable scientific information on the origin, anatomy and evolution of birds. A good general reference, appropriate for public libraries.

Dorst, Jean. *The Life of Birds, Vols. 1 & 2.* (Illus.;trans.) Columbia Univ. Press, 1974. 718pp. $35.00 set. 74-8212. ISBN 0-231-03909-3. Index;CIP.

GA-C These two volumes concentrate on all of the adaptations birds have made to various global environments. There are many diagrams, maps, charts and highly significant photographs.

Farner, Donald S., and James R. King (Eds.). *Avian Biology, Vol. 1.* (Illus.) Academic, 1971. xix + 586pp. $45.00. 79-178216. ISBN 0-12-249401-6. *Vol. 2.* 1972. 612pp. $40.00. ISBN 0-12-249402-4. *Vol. 3.* 1973. 573pp. $44.00. ISBN 0-12-249403-2.

C-P The interpretive skill of the authors makes for a lively and stimulating text, and even complex concepts are explained clearly. The coverage is superb, and the authors identify tentative and controversial content and direct attention to promising avenues of inquiry. An excellent reference for teachers.

Griffin, Donald R. *Bird Migration.* (Illus.) Dover, 1974. x + 180pp. $2.25(p). 74-76321. ISBN 0-48620529-0. Index;bib.

SH-C-P This republication of the 1964 edition includes a preface describing recent field research on the migratory mechanisms of birds. Griffin is a world-reknowned authority who presents information so that it is both attractive and understandable to the lay reader and useful to the professional biologist.

Gruson, Edward S. *Words for Birds: A Lexicon of North American Birds with Biographical Notes.* (Illus.) Quadrangle, 1972. xiv + 305pp. $8.95. 72-77537.

SH-C-P A lexicon on the origin of names of North American birds, those of Hawaii, and the exotic Asian Species sighted in the Aleutians and

Alaska. Given are common names, translations and origins of the scientific names, and brief biographies of the people for whom the birds were named.

Holden, Raymond P. *The Ways of Nesting Birds.* (Illus.) Dodd, Mead, 1970. 126pp. $3.69. 76-102728.

JH Written with whimsy, the book has one page on each of the 49 birds chosen to exemplify diverse reproductive habits. Teleological in tone, but it does impart an elementary understanding of the evolutionary and survival values of the elementary various nesting and food habits of these birds.

Kaufmann, John. *Birds in Flight.* (Illus.) Morrow, 1970. 96pp. $3.75. 79-101587.

JH Explains the special skeletal structures of birds and correlates variations in skeletal elements and musculature with the flight patterns of various bird species. Wing structure and all movements are carefully explained with diagrams, and the book provides a good primer on related aerodynamic principles.

Line, Les, (Ed.). *The Pleasure of Birds: An Audubon Treasury.* (Illus.) Lippincott, 1975. 191pp. $14.95. 75-17948. ISBN 0-397-01065-6. CIP.

C Some 25 articles originally published in *Audubon* magazine on such diverse subjects as tree nesting cavities, acronical flights of ibis returning to their roosts, and the wanton destruction of the environment and associated birdlife. Worthwhile reading for bird enthusiasts.

Luger, Harriet Mandelay. *Bird of the Farallons.* (Illus.) Young Scott (Addison-Wesley), 1971. 62pp. $4.50. 72-155914. ISBN 0-201-09124-0.

JH-SH Written to acquaint young people with the problems facing our wildlife which result from such pollutants as oil, but the vocabulary and sentence structure are difficult. However, the life of "Ding," a California murre, may inform future adults as to the destruction factor that often accompanies "progress."

Murton, R.K. *Man and Birds.* (Illus.) Taplinger, 1974 (c.1971). xx + 364pp. $8.95. 75-183134. ISBN 0-8008-5083-1. Index.

C-P Murton skillfully sets the ecological stage for human interaction with birds and then traces avian behavioral changes as they have been adapted to humankind's evolving cultures. Recommended for college libraries.

Scott, Peter, (Ed.). *The World Atlas of Birds.* (Illus.) Random House, 1974. 272pp. $29.95. 74-8575. ISBN 0-394-49483-0. Gloss.;index;CIP.

SH-C-P This guide to birds of the world concentrates on Old World species and can be readily understood by the layperson. Includes numerous photographs and an introduction by Roger Tory Peterson.

Short, Lester L. *Birds of the World.* (Illus.) Grosset&Dunlap, 1975. 159pp. $2.95(p). 75-14680. ISBN 0-448-12144-1; 0-448-13327-0(p). Index.

JH-SH-GA Many interesting facts about various birds, including 139 colorful bird pictures plus a short history of each bird. Easy enough for junior high school but not boring for adults.

Thomas, Arline. *Bird Ambulance.* (Illus.) Scribner's, 1971. xi + 131pp. $6.95. 77-140775.

JH-SH Anyone interested in birds, animals or ecology will be moved by this truly enjoyable book. Each chapter deals with a migratory or predatory bird, both city dwellers and birds of the fields, woods and marshes. Nicely rendered black-and-white photographs; pertinent and lucid.

Van Tyne, Josselyn, and Andrew J. Berger. *Fundamentals of Ornithology, 2nd ed.* (Illus.) Wiley, 1976. xv + 808pp. $22.50. 75-20430. ISBN 0-471-89965-8. Index;CIP.

C-P   Not broad or even well balanced, but the book does provide an excellent discussion of voice and sound production, and a synopsis of the birds of the world by family. An excellent resource for high school and college biology teachers and advanced students, it could serve as a text for an advanced ornithology course. Excellent illustrations.

598.29 BIRDS—GEOGRAPHICAL TREATMENT

Bond, James. *Birds of the West Indies, 2nd ed.* (Illus.) Houghton Mifflin, 1971. 256pp. $8.95.

SH-C-P   A competent and highly useful field guide. Nine beautifully rendered color plates of some 90 species and abundant line sketches will enable even the amateur ornithologist to identify readily most birds encountered in this area.

Bruun, Bertel. *Birds of Europe.* (Illus.) Golden, 1971 (c.1969). 320pp. $16.95. 74-147881.

SH-C   This coffee-table book covers all the birds known to occur regularly in Europe. The hundreds of small insets are merely adequate, but the full-page color plates are magnificent. The text is chatty, informative, sometimes anthropomorphic, and it should delight the lay reader and perhaps even the professional.

Clement, Roland C. *American Birds.* (Illus.) Bantam, 1973. 159pp. $1.45(p). 77-77600. Index.

JH-SH-C   This helpful, pocket-sized volume of common birds begins with a concise introduction to the how and why of classification, where to look for birds, and a discussion of the excitement of becoming adept enough at birding to start one's own life-list. The excellent photographs are from the National Audubon Society. The descriptive material is adequate, and the index includes both common and group names.

de Schauensee, Rodolphe Meyer. *A Guide to the Birds of South America.* (Illus.) Livingston, 1970. xiv + 470pp. $20.00. 76-113640. ISBN 0-87098-027-0.

SH-C-P   Each bird family and member of the family is briefly described: plumage coloration, bill size and shape, body length, and geographic range. Some 676 species are illustrated, most grouped into 50 plates (31 in color). This is the first time all the birds native to South America have been brought together in one volume.

Harrison, Hal H. *A Field Guide to Birds' Nests of 285 Species Found Breeding in the United States East of the Mississippi River.* (Illus.) Houghton Mifflin, 1975. xxviii + 257pp. $8.95. 74-23804. ISBN 0-395-20434-8. Index;CIP.

SH-C   This volume is composed almost entirely of individual color photographs of nests and clutches of eggs of 222 of the 285 species of birds included. Breeding range, habitat, nest and eggs are described for each species. An excellent guide for amateurs and professionals.

Johnsgard, Paul A. *North American Game Birds of Upland and Shoreline.* (Illus.) Univ. of Nebraska Press, 1975. xxix + 183pp. $11.95; $6.95(p). 74-15274. ISBN 0-8032-0701-8; 0-8032-5811-9. Gloss.;index;CIP.

C   This intermediate guide includes all the native and established exotic gallinaceous game birds found above Mexico, plus all major nonwaterfowl, migratory game birds common to this region. Each species account includes information on identification, sex and age determination, geographic range and basic biology.

Leck, Charles. *Birds of New Jersey: Their Habits and Habitats.* (Illus.) Rutgers Univ. Press, 1975. xvii + 190pp. $12.50. 75-2112. ISBN 0-8135-0803-7. CIP.

SH-C   The author divides the state into three regions, then discusses 16 typical habitats within these regions from an ecological point of view. Three very useful appendixes are included.

Matthiessen, Peter. *The Wind Birds.* (Illus.) Viking, 1973. 159pp. $9.95. 72-11906. SBN 670-77096-5.

SH-C   "Wind birds" are shorebirds, primarily sandpipers and plovers. A nontechnical (almost poetic and occasionally wryly humorous) discussion of comparative behavior, avian physiology, plummages and protective coloration, structural adaptations, taxonomy and evolution.

Russell, Franklin. *The Sea Has Wings.* (Illus.) Dutton, 1973. 189pp. $10.00. 72-94696. ISBN 0-87690-097-X.

GA   This is a series of impressions of the total environment of the islands off Maine and the Maritime provinces. The birds, the sea, and the rocky islands are the principal characters.

Saunders, David. *Sea Birds.* (Illus.) Grosset&Dunlap, 1973. 159pp. $3.95. 70-175347. ISBN 0-448-00813-0; 0-448-04180-4. Index.

JH-SH   Begins with an excellent taxonomic introduction to the four orders of sea birds. Each group is discussed in terms of distribution, size, reproduction, diet, migration, etc. Especially interesting are the anecdotes and descriptions of recent research. An excellently illustrated reference.

Sutton, George Miksch. *At a Bend in a Mexican River.* (Illus.) Eriksson, 1972. xvii + 184pp. $14.95. 72-83709. ISBN 0-8397-0780-0. Index.

GA   Sutton, a well-known ornithologist, author, and artist, writes about three expeditions to collect and paint bird species in Mexico. Interestingly told in diary form, with many superb paintings and drawings by the author and excellent photography. There is also a separate list of the scientific names of birds mentioned in the text.

598.33 GULLS AND TERNS

Costello, David F. *The World of the Gull.* (Illus.) Lippincott, 1971. 160pp. $5.94. 74-159726. Index;bib.

SH-C   The author emphasizes gulls' versatility in gathering food, defending their nests, adapting to temperature changes, etc. The book contains 87 photographs and a distributional checklist of species (without diagnostic features) and an extended bibliography.

Graham, Frank, Jr. *Gulls: A Social History.* (Illus.) Random House, 1975. vi + 169pp. $8.95. 75-10265. ISBN 0-394-49333-8. Index;CIP.

SH-C   Since this is a social history, it deals primarily with human interactions with gulls. A list of the 44 species of gulls found throughout the world is included.

Hay, John. *Spirit of Survival: A Natural and Personal History of Terns.* (Illus.) Dutton, 1974. 175pp. $7.95. 73-20402. ISBN 0-87690-116-X. CIP.

SH-C   This well-written book is enjoyable reading and a fine reference book as well. Hay presents a well-researched and factual account of the tern population and the effects of human encroachment on the terns' breeding grounds.

598.4 ANSERIFORMS AND RELATED ORDERS

Fegely, Thomas D. *Wonders of Geese and Swans.* (Illus.) Dodd, Mead, 1976. 96pp. $4.95. 75-38360. ISBN 0-396-07307-7. Index;CIP.

JH-SH Of interest to the beginning birdwatcher as well as to the advanced amateur, this book covers physical and behavioral characteristics of the various species and subspecies of geese and swans. High quality photographs, excellent range maps, and easy and enjoyable reading.

Halle, Louis J. *The Storm Petrel.* (Illus.) Princeton Univ. Press, 1970. xii + 268pp. $6.00. 76-100356. ISBN 0-691-09349-0.

JH-SH An attractive, well-written book which includes a discussion of human intervention in the world of nature. Describes birds on the coasts of the Shetland Islands north of Scotland and many other shore birds.

Hester, F. Eugene, and Jack Dermid. *The World of the Wood Duck.* (Illus.) Lippincott, 1973. 160pp. $5.95. 72-14117. ISBN 0-397-00763-9.

JH-SH A popular history of a beautiful duck, which also provides an interesting view of some of the work of a wildlife biologist and of careers in conservation and management. Plans for nest boxes are provided and their uses to help preserve the species are explained.

MacSwiney, Marquis of Mashanaglass. *Six Came Flying.* (Illus.) Knopf, 1972. 270pp. $6.95. 71-171150. ISBN 0-394-47282-9.

SH-GA A personal story about the author and a small group of wild swans which settles on his estate, with detailed examination of the relationship between the human and the birds. Gives very clear explanations of the birds' behavior, including territoriality, courtship, etc.

Pettingill, Olin Sewall, Jr. *Another Penguin Summer.* (Illus.) Scribner's, 1975. 80pp. $10.00. 75-13493. ISBN 0-684-14331-3. CIP.

JH-SH An account of the author's study of penguins on the Falkland Islands. Pettingill discusses locomotion; vocalization; social, feeding, breeding and parental behavior. Five species are described: Gentoo, Rockhopper, Magellanic, King and Macaroni.

Scott, Jack Denton, and Ozzie Sweet. *That Wonderful Pelican.* (Illus.) Putnam's, 1975. 63pp. $6.95. 74-21064. SBN GB-399-60939-3; TR-399-20449-0.

JH-SH The book deals essentially with the brown pelican on the coast of Florida. Exceptional photographs; useful as a reference for junior high school students and as collateral reading at the high school level.

Scott, Peter, and The Wild Fowl Trust. *The Swans.* (Illus.) Houghton Mifflin, 1972. x + 242pp. $15.00.

SH-C-P A definitive work on the swans of the world. All eight species are treated in detail: their classification, distribution, natural history and related topics. The swan in art and mythology, its exploitation and its conservation are also discussed. Excellent photographs and innumerable sketches.

Simpson, George Gaylord. *Penguins: Past and Present, Here and There.* (Illus.) Yale Univ. Press, 1976. xi + 150pp. $10.00. 75-27211. ISBN 0-300-01969-6. Index;bib.

SH-C-P An account of the structural and other characteristics that all penguins have in common. In addition, six genera are discussed. Focuses on distribution, evolution, nomenclature, biological characteristics, ecology, behavior, and relationship of penguins and humans. Includes nine area maps.

Van Wormer, Joe.   *The World of the Swan.* (Illus.) Lippincott, 1972. 160pp. $5.95. 77-39150. ISBN 0-397-00845-7. Bib.

JH-SH-C   Seven species of swans are presented, with an emphasis on the trumpeter, mute, whistler and whooper. Natural history and life activities are discussed. There is a chapter for each of the seasons, and one on human interaction with the swan. Black-and-white photographs.

598.899 SWIFTS AND HUMMINGBIRDS

Kaufmann, John.   *Chimney Swift.* (Illus.) Morrow, 1971. 63pp. $3.75. 77-127639.

JH   A great deal of useful information on chimney swifts is provided along with occasional allusions to other species. The text covers migration, nesting, wintering, and other pertinent data.

Skutch, Alexander F.   *The Life of the Hummingbird.* (Illus.) Crown, 1973. 95pp. $9.95. 73-81514. ISBN 0-517-50572-X. Bib.

SH-GA-C   Shows the wide diversity in the morphology, habits, and habitats among the more than 300 species of hummingbirds in the Western Hemisphere. A delightful, light treatment in which the author describes many situations that suggest interesting research problems in behavior, distribution, color physiology, metabolism and more.

598.91 BIRDS OF PREY

Brandon-Cox, Hugh.   *Summer of a Million Wings: Arctic Quest for the Sea Eagles.* (Illus.) Taplinger, 1975. 184pp. $8.95. 74-10365. ISBN 0-8008-7492-7.

JH-SH-C   This is the story of a typical breeding season of birds on the island of Vaeroy, north of the Arctic Circle. Beautifully written, excellent illustrations.

Brown, Leslie.   *African Birds of Prey.* (Illus.) Houghton Mifflin, 1970. 320pp. $8.95.

C-P   Both the serious student of ornithology and the hobbyist will find much of interest in the wealth of information about the feeding, nesting, and breeding and other habits of 120 species of predatory African birds.

Callahan, Philip S.   *The Magnificent Birds of Prey.* (Illus.) Holiday House, 1974. 190pp. $6.95. 74-7571. ISBN 0-8234-0248-7. Gloss.;index;CIP.

SH-C   Well written and enthusiastic; contrasts ancient people's devotion to raptorial birds with the modern desire to slaughter them. Chapters on form and flight and the history of falconry. Recommended as collateral reading in biology.

Everett, Michael.   *Birds of Prey.* (Illus.) Putnam's, 1976. 128pp. $12.95. 75-19596. SBN 399-11675-3. Index.

JH-SH-GA-C   Discusses the evolution of the Falconiformes (falcons, hawks, harriers, hites, eagles, vultures and buzzards), their taxonomy and anatomy, and hunting, breeding and migration. Includes a detailed list of all extant species. Magnificently illustrated.

Harwood, Michael.   *The View from Hawk Mountain.* (Illus.) Scribner's, 1973. 191pp. $6.95. 73-38905. SBN 684-13378-4.

SH-C   An account of the fortunes and misfortunes of the stream of migrating hawks (and other birds) which passes a steep ridge in Pennsylvania known as Hawk Mountain. Harwood sketches the struggle between the shooters and the watchers at Hawk Mountain. Not really a reference, but a source of much information not given in technical accounts of birds.

Kaufmann, John, and Heinz Meng. *Falcons Return.* (Illus.) Morrow, 1975. 128pp. $5.95. 74-32457. ISBN 0-688-22027-4; 0-688-32027-9. Index.

JH-SH-GA The tragic story of the decline of the endangered duck hawk and last minute rescue efforts are interestingly recounted. Intertwined with the story are descriptions of the sport of falconry and breeding programs. Useful resource in environmental subjects and conservation.

Lavine, Sigmund A. *Wonders of the Hawk World.* (Illus.) Dodd, Mead, 1972. 64pp. $3.95. 70-38564. ISBN 0-396-06509-0.

JH-SH Covers the physical appearance, habits and behavior, and mythology of hawks, and is global in scope. Excellent for the student who already has some knowledge of hawks and as a reference.

Turner, John F. *The Magnificent Bald Eagle: America's National Bird.* (Illus.) Random House, 1971. 83pp. $3.50. 73-158383. ISBN 0-394-82061-4.

JH Combines biological facts, natural history, and pertinent issues of environmental stewardship in a lucid, simple yet accurate essay. The numerous photos are a significant asset.

598.97 OWLS

Burton, John A., (Ed.). *Owls of the World: Their Evolution, Structure and Ecology.* (Illus.) Dutton, 1973. 216pp. $17.95. 73-8271. SBN 0-525-17432-X.

C-P Summarizes the folklore, paleo-ornithology, physiology, and various individual genera of owls. Much remains to be learned, and Burton offers starting points for much additional research.

Cameron, Angus, and Peter Parnall. *The Nightwatchers.* (Illus.) Four Winds, 1971. 112pp. $8.95. 70-161023.

JH-SH-C An extraordinary collection of anecdotes and descriptions of owls which reveals some little-known facts. Excellent drawings; for students of folklore or art as well as of wildlife.

Walker, Lewis Wayne. *The Book of Owls.* (Illus.) Knopf, 1974. xiii + 255pp. $12.50. 73-20746. ISBN 0-394-49218-8. Index;CIP.

SH-C Descriptions and life histories of the principal types of owls of North America.

599 MAMMALS

Boorer, Michael. *Mammals of the World.* (Illus.) Grosset&Dunlap, 1971. 159pp. $3.95. 70-134998. ISBN 0-448-00860-2.

JH-SH Written in exact scientific terminology, introductory sections cover mammalian evolution, structure, diet, reproduction, growth, senses, distribution and taxonomy. Each of the eighteen living mammalian orders is carefully reviewed.

Dorst, Jean, and Pierre Dandelot. *A Field Guide to the Larger Mammals of Africa.* (Illus.) Houghton Mifflin, 1970. 287pp. $8.50.

JH-SH-C-P Each species is depicted by a color illustration and distribution map, and there are diagrams for distinguishing between closely related and confusing species. The common and scientific name of each is followed by a brief description, the preferred habitat, distribution and habits. A unique, outstanding reference.

Jenkins, Marie M. *Kangaroos, Opossums, and Other Marsupials.* (Illus.) Holiday House, 1975. 160pp. $6.95. 75-10798. ISBN 0-8234-0264-9. Index;CIP.

JH   Jenkins begins with the evolution of mammals and continues with a description of the unique characteristics of marsupials, their life cycles and habits. The theory of "drifting continents" is used to show the possible routes by which marsupials dispersed to different areas.

Laycock, George. *People and Other Mammals.* (Illus.) Doubleday, 1975. 143pp. $4.95. 74-4874. ISBN 0-385-00179-7; 0-385-00227-0. Index;CIP.

JH   This book focuses on the natural history of many mammals, including shrews, moles, bats, rabbits, porpoises, whales, bears, wolverines, badgers, sea otter, bobcats, cougars, seals, white-tailed deer, caibou and musk-oxen.

Walker, Ernest P., et al. *Mammals of the World, 3rd ed.* (2 Vols.;illus.;rev. by John L. Paradiso.) Johns Hopkins Univ. Press, 1975. liii + 2142pp. $37.50 set. 74-23327. ISBN 0-8018-1657-2. Indexes;CIP.

C-P   This set is complete in coverage and serves as the standard English-language reference on living mammals. There is a brief overall description of a mammalian order followed by a more extensive description of each family in the order, followed by a brief natural history on each genus in the family.

Williamson, H.D. *The Year of the Koala.* (Illus.) Scribner's, 1975. xiii + 209pp. $8.95. 75-12706. ISBN 0-684-14351-8. Gloss.;index;CIP.

JH-SH-C   Williamson traces the lives of two koalas for one year. He stresses the conflict between population growth, urban development and loss of habitat for an animal with highly specialized eating habits. Information about anatomy, physiology and behavior is supplied.

Young, J.Z., with M.J. Hobbs. *The Life of Mammals: Their Anatomy and Physiology, 2nd ed.* (Illus.) Clarendon/Oxford Univ. Press, 1976. xv + 528pp. $22.50. ISBN 0-19-857156-9; 0-19-857158-5(p). Indexes;bib.

C-P   This is a detailed textbook (for the benefit of medically-oriented students) with emphasis on rabbits, rats and humans. Structure, function and development are interwoven so that all systems are examined with regard to internal consistency and external interactions. Should stand as a standard text and basic reference for years to come.

599.09 MAMMALS—GEOGRAPHICAL TREATMENT

Fletcher, Colin. *The Winds of Mara.* (Illus.) Knopf, 1973. 343pp. $7.95. 72-2237. ISBN 0-394-47091-5. Bib.

SH-C   A scientifically accurate but nontechnical story of the great African mammals of the savanna in Kenya. One sees at first hand the Pleistocene climax of biological evolution with its elephants, giraffes, zebras, rhinos, various gazelles, lions, hyenas, etc. More informative than many formal books on East African mammals.

Hoffmeister, Donald F. *Mammals of Grand Canyon.* (Illus.) Univ. of Ill. Press, 1971. 183pp. $7.50; $1.95(p). 75-141517. ISBN 0-252-00154-0; 0-252-00155-9.

JH-SH   The species descriptions are accompanied by excellent drawings of 63 of the 74 species that have been found. Mammals as widely different as skunks, rabbits, mice, voles, squirrels, coyotes, mountain lions, deer, and bighorn sheep occupy the different topographic areas (desert, plateau, tree zones, mountain grasslands, etc.). A generally useful reference.

Morcombe, Michael. *Australian Marsupials and Other Native Mammals.* (Illus.) Scribner's, 1974 (c.1972). 100pp. $7.95. 73-9324. ISBN 684-13597-3. Index.

   JH-SH-C Excellent color photographs illustrate an informative text describing each major grouping of Australian mammal. Distributional maps indicate the unique form of adaptation and specialization of Australian mammals.

Orr, Robert T. *Marine Mammals of California.* (Illus.) Univ. of Calif. Press, 1972. 64pp. $1.95(p). 78-165233. ISBN 0-520-0277-4. Index.

   SH-C One of a series of California Natural History Guides, this brief yet complete text is supplemented with excellent photographs and fair drawings and provides a resumé of protective measures being taken for each species of California marine mammals.

Russell, Franklin. *Season on the Plain.* Reader's Digest Press (distr.: Dutton), 1974. 313pp. $8.95. 73-20465. ISBN 0-88349-024-2. CIP.

   JH-SH Franklin intertwines stories about the lives of a baboon, a leopard, a lion and several other animals living on the plains of Africa. He discusses the ecological aspects of drought, rain, vegetable growth and animal population explosions.

Schaefer, Jack. *An American Bestiary.* (Illus.) Houghton Mifflin, 1975. xxvii + 287pp. $10.00. 75-5630. ISBN 0-395-20710-X. CIP.

   SH-C Schaefer discusses 32 species of mammals, providing information about taxonomy, evolutionary history, natural history and ethnology of several of the species. The importance of wildlife survival is stressed. Nonscientists especially will appreciate this book.

Summers, Gerald. *An African Bestiary.* (Illus.) Simon&Schuster, 1974. 222pp. $7.95. 74-6127. ISBN 0-671-21813-1. CIP.

   JH-SH Summers provides a wealth of information on the birds, mammals and insects of Africa. It is a storybook that even a professional zoologist can read for enjoyment and information on the natural history of Africa.

599.32 RABBITS AND RODENTS

Barkalow, Frederick S., Jr., and Monica Shorten. *The World of the Gray Squirrel.* (Illus.) Lippincott, 1973. 159pp. $5.95. 72-2920. ISBN 0-397-00749-3. Bib.

   JH-GA Describes the many kinds of squirrels, including the American red squirrel, the fox squirrel and chipmunk. Likenesses and differences of various species and subspecies are compared before attention is centered on the gray squirrel. The text and photographs are detailed but not technical. Among the best publications on the topic.

Barnett, S.A. *The Rat: Study in Behavior, rev. ed.* (Illus.) Univ. of Chicago Press, 1976. xiv + 313pp. $20.00. 74-33509. ISBN 0-226-03740-1. Gloss.;indexes;CIP.

   C-P Barnett surveys the rat's feeding and reproductive behavior, activity patterns, conditioning and learning, emotions, instinct and social behavior. He includes recent findings from three disciplines: experimental psychology, ethology and behavioral physiology. A fine reference work.

Costello, David F. *The World of the Prairie Dog.* (Illus.) Lippincott, 1970. 106pp. $5.82. 70-110650. Index.

   SH-C An amiable, anecdotal, nontechnical account of an interesting rodent during a typical year. Provides a framework for considering the ecology of a once abundant rodent which now occupies only a restricted part of its former habitat.

Lockley, R.M.   *The Private Life of the Rabbit: An Account of the Life History and Social Behavior of the Wild Rabbit.* (Illus.) Macmillan, 1974 (c.1964). 152pp. $6.95. 74-8855. ISBN 0-02-573900-X. Index;CIP.

    JH-SH-C   A British writer's description of rabbit life based on his own observations and experiments dealing with productivity and behavior of the wild rabbit. Also includes an excellent account of myxomatosis and its use as a biological control agent.

MacClintock, Dorcas.   *Squirrels of North America.* (Illus.) Van Nostrand Reinhold, 1970. vi + 184pp. $7.95. 74-110059.

    JH-SH   This unique, handy reference covers all squirrel species north of Panama. Although weak on ecological aspects, it gives accurate descriptions and general life history information. Includes distribution maps and excellent drawings.

Shuttlesworth, Dorothy E.   *The Story of Rodents.* (Illus.) Doubleday, 1971. 95pp. $4.50. 71-103922.

    JH-SH   A delightful, well-organized account of the natural history of a number of representatives of a disparate mammalian order. Gives an understanding of rodent adaptations in their environmental niches. Accurate facts, simply presented.

599.4 BATS

Turner, Dennis C.   *The Vampire Bat: A Field Study in Behavior and Ecology.* (Illus.) Johns Hopkins Univ. Press, 1975. x + 145pp. $12.00. 74-24396. ISBN 0-8018-1680-7. Index;CIP,bib.

    C-P   This book details studies done on vampire bats to discover how, when and why they prey on cattle and how they transmit rabies to their victims, The importance of these studies to bat control programs is discussed.

Wimsatt, William A., (Ed.).   *The Biology of Bats: Vols. 1 & 2.* (Illus.) Academic Press, 1970. xi + 406pp; xv + 477pp. Vol. 1, $25.00; Vol. 2, $26.00. 77-11710. ISBN 0-12-758001-8; 0-12-758002-6.

    C-P   This set by many contributors is a broad assemblage providing in-depth information on basic morphological, physiological, behavioral, ecological and bioeconomic characteristics of bats. The books also treat bats in relation to humans, including bats' role in the spread of infectious disease. Useful to students, professional mammalogists, ecologists, and public health workers.

Yalden, D.W., and P.A. Morris.   *The Lives of Bats.* (Illus.) Quadrangle/N.Y. Times, 1976. 247pp. $9.95. 75-20692. ISBN 0-8129-0600-4. Index;CIP.

    SH-C-P   Condensed accounts of structure and origin of bats, their flight and wing anatomy, food and feeding, hibernation, reproduction, marking and populations, seeing with sound, bats and humans, and a review of the 17 bat families. Readable and useful.

599.5 WHALES AND DOLPHINS

Caldwell, David K., and Melba C. Caldwell.   *The World of the Bottlenosed Dolphin.* (Illus.) Lippincott, 1972. 160pp. $5.95. 71-159728. ISBN 0-397-00734-5.

    JH-SH   Most of the text deals with behavioral studies of the intelligence and sensory-communication processes of dolphins. Other topics include classification, natural history and life cycle, color patterns, capture and adaptation to captivity, and economic importance. Highly recommended as a school and public library reference.

Ciampi, Elgin. *Those Other People the Porpoises.* (Illus.) Grosset&Dunlap, 1972. 163pp. $5.95. 78-130852. ISBN 0-488-21406-7. Index.

   JH-SH-GA   A brief, simple, moving account of the ways of porpoises, which contains a surprising depth of factual material. Some 80 full-page, black-and-white photographs.

Cook, Joseph J., and William L. Wisner. *Blue Whale: Vanishing Leviathan.* (Illus.) Dodd, Mead, 1973. 80pp. $3.95. 72-7750. ISBN 0-396-06739-5.

   JH-SH   A brief compendium of the evolutionary history and the adaptations of whales in general, and the behavior and life style of the blue whale and its relationships in the ecosystem. Requires knowledge of specialized terms or willingness to use a technical dictionary.

Cousteau, Jacques-Yves, and Philippe Diole. *Dolphins.* (Illus.;trans.) Doubleday, 1975. 304pp. $12.95. 74-9481. ISBN 0-385-00015-4. Gloss.;index;CIP.

   JH-SH-C   Cousteau tells the fascinating story of his encounters with dolphins. Popular illusions and myths are explored along with scientific investigations of dolphin behavior. Numerous photographs.

Eckert, Allan W. *In Search of a Whale.* (Illus.) Doubleday, 1970. $5.95. 69-15172.

   JH   The story of Marlin Perkins's expedition to capture pilot whales for exhibition and scientific studies at Marineland of the Pacific is a well-written action narrative with much solid natural history material concerning ocean fishes and marine mammals in general and pilot whales in particular.

Hoke, Helen, and Valerie Pitt. *Whales.* (Illus.) Watts, 1973. 90pp. $3.95. 72-11769. ISBN 0-531-00779-0. Index;gloss.;bib.

   JH   The history of whaling and the biology, classification, and conservation of whales, dolphins, and porpoises. A section on where to see cetaceans and whaling exhibits adds to the text. A very easy-to-read book with 30 pages of pictures.

Ommanney, Francis D. *Lost Leviathan.* (Illus.) Dodd, Mead, 1971. 280pp. $7.95. 72-155066. ISBN 0-396-06253-9.

   SH-C   Biological information on whales, their characteristics, life histories, migratory habits, statistics on size and age, and the biology and distribution of krill on which baleen whales feed are some of the background sections. A history of whaling is provided, and descriptions of the author's experiences at whaling stations and on factory ships are excellent. Good recreational reading and a sound reference volume.

Riedman, Sarah R., and Elton T. Gustafson. *Home Is the Sea: For Whales.* (Illus.) Abelard-Schuman, 1971. 264pp. $4.95. 73-160111. ISBN 0-200-71859-2. Index;bib.

   JH-SH-C   A well-balanced introduction to cetacean biology—evolution, anatomy, physiology, behavior, and economic implications of the whales, dolphins, and porpoises of the world. The text and 102 figures, many diagrammatic, answer common questions. A table gives characteristics and distribution of 34 of the approximately 80 known kinds of cetaceans.

Small, George L. *The Blue Whale.* (Illus.) Columbia Univ. Press, 1971. xiii + 248pp. $9.95. 76-134986. ISBN 0-231-03288-9. Bib.

   C-P   Provides a resumé of the development of the whaling industry, a review of the biology and natural history of the blue whale, and a critical discussion of national whaling policies and the development of international cooperation to manage and conserve the whaling stocks. A unique study and analysis for all reference collections.

599.61 ELEPHANTS

Douglas-Hamilton, Iain, and Oria Douglas-Hamilton. *Among the Elephants.* (Illus.) Viking, 1975. 285pp. $14.95. 74-7502. SBN 670-12208-4. Index.

JH-SH   A mixture of scholarly and personal observations about the lives of African elephants makes this book both informative and enjoyable. Aspects of elephant reactions to humans and other animals are clearly presented. A unique study; forward by Niko Tinbergen.

599.7 HOOFED MAMMALS

Geist, Valerius. *Mountain Sheep: A Study in Behavior and Evolution.* (Illus.) Univ. of Chicago Press, 1971. xv + 383pp. $14.50. 77-149596. ISBN 0-226-28572-3.

C-P   The author has devoted 4 years to a rigorous behavioral study of these ungulates in their native habitats, and his findings deal with them from taxonomic, biological, physiological, social, and ecological standpoints. The book is clearly written, well-documented, extremely enjoyable, and an excellent reference.

Haines, Francis. *The Buffalo.* (Illus.) Crowell, 1970. 242pp. $7.95. 76-94794.

JH-SH-C   A detailed and stimulating account of the buffalo's role in the lives of both the American Indian and the early settlers in America. The author traces the buffalo from their arrival in North America to their near demise. Indian origins and cultural patterns are well described. Photographs include ones by George Catlin.

Haines, Francis. *Horses in America.* (Illus.) Crowell, 1971. 213pp. $7.95. 74-139096. ISBN 0-690-40253-8.

JH-SH-C   The horse was an inseparable part of the growth and development of this country. For those who love horses and for those who have taken them for granted, this book is a must.

Hiser, Iona Seibert. *Collared Peccary—The Javelina.* (Illus.) Steck-Vaughn, 1971. 30pp. $2.95. 70-150341. ISBN 0-8114-7726-6.

JH   A brief, nontechnical presentation of the physical characteristics, food habits and living conditions of the javelina, as well as its relations with other animals, its environment, and with humans. There is limited information on reproductive habits, length of life, diet, and diseases. Numerous line drawings.

Ipsen, D.C. *The Elusive Zebra.* (Illus.) Addison-Wesley, 1971. 112pp. $3.75. 70-127786. ISBN 0-201-01367-1.

JH-SH   This fascinating history attempts to rationalize color and pattern in animals. The author cites examples in other species which are consistent with each of several theories and then shows how none appears to explain the color and pattern of the zebra. Experimental methods are described in which other animals are also used to test these hypotheses. Delightful!

Lavine, Sigmund A., and Vincent Scuro. *Wonders of the Bison World.* (Illus.) Dodd, Mead, 1975. 64pp. $4.95. 75-8732. ISBN 0-396-07146-5. Index;CIP.

JH   Evolutionary changes in the American bison are traced and the characteristics of several species are described in detail. The author tells about the animal's habits, erratic behavior and great strength. The book is illustrated with copies of old prints.

MacClintock, Dorcas. *A Natural History of Giraffes.* (Illus.) Scribner's, 1973. x + 134pp. $5.95. 72-9580. SBN 684-13239-7.

SH-C   Describes in detail the anatomy and functional physiology of giraffes; how individual giraffes live in the wild and how they loosely herd with other giraffes; behavior; reproduction, birth, the characteristics and growth of the calf; and how giraffes deal with predators, other animals, and humans. Every typical pose of the giraffe is illustrated with accuracy and clarity in outstanding silhouettes by Ugo Mochi.

Mochi, Ugo, and T. Donald Carter. *Hoofed Mammals of the World.* (Illus.) Scribner's, 1971. xx + 268pp. $9.95. 75-169790. ISBN 0-684-12382-7.

JH-SH-C   An extraordinary combination of art and natural history, illustrated by incredible "skiograms" which are absolutely correct in pattern of outline and are exactly to scale. Hoofed mammals are described with sensitivity, beautiful simplicity, absolute precision, and enormous interest.

Ryden, Hope. *America's Last Wild Horses.* (Illus.) Dutton, 1970. 311pp. $8.95. 72-87198. ISBN 0-525-05477-4.

SH-C   A slick, well-illustrated potpourri of materials on the horse, from Eohippus to the Appaloosa. Both a plea to save the remaining wild horses and a pleasant collage. Fact and anecdote are indiscriminately mixed, with sources varying from technical documents to personal recollections.

Stadtfeld, Curtis K. *Whitetail Deer: A Year's Cycle.* (Illus.) Dial, 1975. vii + 163pp. $7.95. 74-14724. ISBN 0-8037-6101-5. CIP;bib.

JH-SH-C   The niche of the whitetail deer in the Michigan forest habitat is described; one year's events are the basis for analyzing the complex problems of overpopulation in the absence of predation and the thermodynamics of the forest ecosystem. For both amateur and expert.

Van Wormer, Joe. *The World of the Moose.* (Illus.) Lippincott, 1972. 160pp. $5.95. 79-38995. ISBN 0-397-00846-5. Index;bib.

SH-C   Combines very readable text with excellent photography to introduce the reader to the year-round life of the moose. Extensive information about the American elk is followed by four chapters, each devoted to a season. Includes detailed descriptions of behavior patterns and a section on classification.

599.74 CARNIVORES

Bueler, Lois E. *Wild Dogs of the World.* (Illus.) Stein&Day, 1973. 274pp. $8.95. 72-96435. ISBN 0-8128-1568-6.

C-P   Discusses wolves, coyotes, jackals, dingoes, domestic dogs and foxes as well as the more unusual species such as dholes, bush dogs, raccoon dogs, bat-eared foxes, and maned wolves. For classroom use, collateral reading or reference in zoology or ecology.

Cook, David, and Valerie Pitt. *A Closer Look at Dogs.* (Illus.) Watts, 1975. 30pp. $4.90. 75-4390. ISBN 0-531-02425-3; 0-531-01100-3.

JH-SH   The authors, explaining that the behavior of dogs can best be understood in its social context, present an accurate, simplified and extremely brief view of wolf society. Excellent illustrations.

Eaton, Randall L. *The Cheetah: The Biology, Ecology, and Behavior of an Endangered Species.* (Illus.) Van Nostrand Reinhold, 1974. xii + 178pp. $12.95. 73-12148. ISBN 0-442-2229-7. Indexes;CIP.

JH-SH   A comprehensive summary of current literature on the basic behavior and psychology of the cheetah which also provides the reader with casual information on cheetah biology.

294   CARNIVORES

Ewer, R.F. *The Carnivores.* (Illus.) Cornell Univ. Press, 1973. xv + 494pp. $21.50. 72-6263. ISBN 0-8014-0745-1. Bib.

C-P   Half of the text deals with functional anatomy; the other half with feeding, social and reproductive behavior. Lists 239 wild and domesticated carnivores, together with their distribution, systematic and common names, and taxonomic notes. Includes references to the publications of more than 900 authors. A clear, comprehensive book.

Gilbert, Bil. *The Weasel: A Sensible Look at a Family of Predators.* (Illus.) Pantheon, 1970. 201pp. $4.95. 68-12655.

SH-C   The author begins with a brief discussion of the evolution of the family and proceeds to describe each group in detail, along with geographical ranges, habits, and some of their biology. Drawings of each animal are included. Emphasis on predator-prey relationships introduces a very fundamental biological concept. Animals included are weasels, minks, badgers, skunks, martens, fishers, wolverines, otters and ferrets.

Gray, Robert. *Cougar: The Natural Life of a North American Mountain Lion.* (Illus.) Grosset&Dunlap, 1972. 150pp. 74-182012. ISBN 0-448-21439.

JH-GA   Based on a 5-year study by the Wildlife Society, this book describes a beautiful animal which in the course of evolution has developed the predator capability of quiet mobility in a complex environment. Through the perceptions and understanding of a small boy, this book tells the life story of both the cougar and the whole cat family. The wild stories and misinformation are dispelled, and we realize that we too are part of the natural processes of the earth.

Hopf, Alice L. *Wild Cousins of the Cat.* (Illus.) Putnam's, 1975. 159pp. $6.95. 74-21077. SBN GB-399-60932-6; TR-399-20442-3. Index.

SH-C   Briefly reviews 35 varieties of the big cat family, nearly all of which are on the endangered species list. The photographs show the animals in typically majestic poses.

Hopf, Alice L. *Wild Cousins of the Dog.* (Illus.) Putnam's, 1973. 128pp. $3.86. 72-189235. SBN 399-20323-0; 399-60812-5. Bib.

JH-SH   Presents descriptions and natural histories of the wolf, Arctic fox, coyote, maned wolf, bush dog, black-backed jackal, African wild dog, bat-eared fox, dingo, red fox, raccoon dog, and Antarctic wolf. Capsulizes recent information on canine behavior and ecology. Useful as collateral reading, reference or in the classroom.

Koch, Thomas J. *The Year of the Polar Bear.* (Illus.) Bobbs-Merrill, 1975. 150pp. $8.95. 74-17660. ISBN 0-672-52062-1. Index.

JH-SH-C   This is a natural history account of the polar bear of Southampton Island, which guards the northern approaches to Hudson Bay. Color plates delineate the environment and this endangered species. Pleasant reading.

Kruuk, Hans. *The Spotted Hyena: A Study of Predation and Social Behavior.* (Illus.) Univ. of Chicago Press, 1972. xiii + 335pp. $15.00. 70-175304. ISBN 0-226-45507-6.

SH-C   Documents the author's careful study of the total life pattern of the hyena. Shows that this animal is not a scavenger living on leftover kills, but a successful predator in its own right. Also discusses the social interactions of the 385 adult hyenas living in the Ngorongoro Crater. A fascinating, scholarly and highly readable report.

Roberts, Charles G.D. *Red Fox.* (Illus.) Houghton Mifflin, 1972. x + 187pp. $4.95. 77-187419. ISBN 0-395-13735-7.

JH-SH   This realistic picture of a red fox's life in the eastern Canadian backwoods at the turn of the century was first published in 1905. To the lucid and sensitive words of the late Sir Charles G.D. Roberts, the present publisher has added lovely scratchboard drawings.

Schaller, George B.   *The Serengeti Lion: A Study of Predator-Prey Relations.* (Illus.) Univ. of Chicago Press, 1972. xii + 480pp. $12.50. 78-180043. ISBN 0-226-73639-3.

C-P   A well-rounded study of the lion, both in Africa and in the Gir Forest of India. Includes a careful analysis of the Serengeti National Park as a component of a larger ecological unit in northern Tanzania and discussions of the interactions of lion prides, the role of nomadic lions, and behavior within the group. Considers the other major predators in the Serengeti and examines both the dynamics of social systems among predators and the dynamics of predation.

Scheffer, Victor B.   *The Year of the Seal.* (Illus.) Scribner's, 1970. xi + 207pp. $7.95. 74-123840. Index;bib.

SH-C   Fiction based on fact. The life of seals is described from a seal's point of view; the accounts of the biologists' activities, the biology of seals, commercial sealing and conservation reflect a human point of view.

Thomas, Harold E.   *Coyotes: Last Animals on Earth?* (Illus.) Lothrop, Lee&Shepard, 1975. 125pp. $5.25. 74-26969. ISBN 0-688-41699-3; 0-688-51699-8. Index;CIP.

JH-SH-GA   Coyote behavior, characteristics, and environmental relationships are described, and anecdotes from Indian lore and legend are included. Interesting, instructive and entertaining.

van Lawick, Hugo.   *Solo: The Story of an African Wild Dog.* (Illus.) Houghton Mifflin, 1974. 159pp. $6.95. 73-19747. ISBN 0-395-18321-9.

JH-SH-C   *Solo* retells the fascinating story portrayed in the documentary TV film, and the reader is offered the opportunity through the succinctly written narrative and ample illustrations to follow along with van Lawick in his observations of the Genghis pack and a sequence of events in the life of one wild dog in particular.

van Lawick-Goodall, Hugo, and Jane van Lawick-Goodall.   *Innocent Killers.* (Illus.) Houghton Mifflin, 1970. 222pp. $10.00. 78-132786. ISBN 0-395-12109-4.

JH-SH-C   The hyenas, jackals, and wild dogs in Tanzania are described, including their hunting and acquisition of food, relationships between these three species and to other species such as lions, rearing of offspring, dominance, hierarchy, territorial defense, activity patterns, and feeding behavior. Includes some of the best photographs of these species in existence.

Young, Stanley Paul.   *The Last of the Loners.* (Illus.) Macmillan, 1970. 316pp. $9.95. 75-101728.

SH-C   Beginning with a history of the wolf in the American West, their evolutionary origin and geographic distribution, Young continues with a history of bounty hunting and other methods of control used in Europe and America and nine excellent stories of these "last loners."

599.8 PRIMATES

See also 156 Comparative Psychology.

Bourne, Geoffrey H.   *The Ape People.* (Illus.) Putnam's, 1971. 364pp. $7.95. 77-135253.

SH-C   Not a scientific treatise but a factual and anecdotal report based on Bourne's own experiences in working with nonhuman primates, with detailed mention of the work of others. Well written, with a collection of well-chosen photographs, it is a nontechnical yet authoritative review of clinical-medical, physiological, and psychological studies.

Bourne, Geoffrey H. *Primate Odyssey.* (Illus.) Putnam's, 1974. 479pp. $9.95. 73-78596. SBN 399-11200-6. Index.

SH-C-P   Bourne presents a survey of the world's primates, their evolution, biology, behavior, and status, as well as a history of their discovery by humans and subsequent scientific studies. Bourne enriches his text with anecdotes and examples from his own work and travels. Recommended both for pleasant reading and for reference.

Bourne, Geoffrey H., and Maury Co en. *The Gentle Giants: The Gorilla Story.* (Illus.) Putnam's, 1975. 319pp. $12.50. 75-25753. SBN 399-11528-5. Index;CIP.

SH-C   In a witty, anecdotal style, the authors trace the first discoveries of gorillas and discuss the public image of the creatures. Detailed studies and stories about gorillas, both in the wild and in captivity, are presented. Growth patterns, love life, "speech" habits, learning feats and more are discussed.

Clark, W.E. Legros. *The Antecedents of Man: An Introduction to the Evolution of the Primates, 3rd ed.* (Illus.) Quadrangle, 1971. ix + 395pp. $8.95. 73-162309. ISBN 0-8129-0224-6.

C-P   Long recognized as one of the most relevant introductory books depicting the evolutionary history of the primates, this edition offers recent paleontological evidence and includes taxonomic and interpretive revisions. Recent discoveries and interpretations combine to make a stimulating and provocative finale to a most important reference work.

Gardner, Richard. *The Baboon.* (Illus.) Macmillan, 1972. 151pp. $4.95. 73-102963. Bib.

JH-SH-GA   Covers just about everything one might wish to know about the baboon: legends, evolutionary history, varieties, behavior, life cycle, complex social organization, relationship to humans, and baboon communication.

Hahn, Emily. *On the Side of the Apes.* (Illus.) Crowell, 1971. 239pp. $7.95. 78-146282. ISBN 0-690-59992-7.

SH-GA-C-P   Transmits the enthusiasm felt by those scientists who work with primates to the lay reader while giving an account stimulating to the primatologist. The vignettes of the seven regional primate centers established under the auspices of the National Institutes of Health make up the bulk of the book. One learns as much about the primatologists as about the primates, and the studies range from behavior to virology, atherosclerosis, neurophysiology and more.

MacKinnon, John. *In Search of the Red Ape.* (Illus.) Holt, Rinehart&Winston, 1974. 222pp. $8.95. 73-15457. ISBN 0-03-012496-4. Index;CIP.

JH-SH   The reader accompanies the author into the rain forests of Borneo and Sumatra to observe free-ranging orangutans. MacKinnon describes the progressive changes in behavior as a youngster frees itself from dependence upon its mother, reaches sexual maturity and is involved in mating activity. Distribution maps are included.

Napier, Prue. *Monkeys and Apes.* (Illus.) Bantam, 1973. 159pp. $1.45(p).

JH-SH-C   Half the book is a descriptive section which covers generalities and numerous species of both fossil and extant primates. Other chapters deal with evolution, structure, locomotion, diet, reproduction and social behavior. For the lay reader.

Napier, Prue, and John Napier. *Monkeys and Apes.* (Illus.) Time-Life Films (distr.: Little, Brown), 1976. 128pp. $7.95. 75-40056. ISBN 0-913948-03-9. Index.

JH-SH   This is a book to read for enjoyment, for the several intriguing excerpts from well-known books, and for its color photographs. Not a text or reference, but a discussion of numerous species, with emphasis on appearance, behavior and ecology.

Struhsaker, Thomas T. *The Red Colobus Monkey.* (Illus.) Univ. of Chicago Press, 1976. xiv + 311pp. $25.00. 74-21339. ISBN 0-226-77769-3. Index;CIP.

C-P   A detailed, readable account of one troop of forest-living red colobus monkeys in equatorial Africa, a species in danger of extinction. Emphasis is on social structure, intragroup relationships, vocalizations and ecology. Information is documented by tables, field notes, references and photographs. Much of the data is statistically analyzed. This is an important study for primatologists, behaviorists, ecologists, conservationists and environmentalists.

van Lawick-Goodall, Jane. *In the Shadow of Man.* (Illus.) Houghton Mifflin, 1971. xx + 297pp. $10.00. 71-162007. ISBN 0-395-12726-2.

SH-GA-C-P   The modern scientific study of animal behavior (ethology) has never been better illustrated than in this intimate and detailed account of a large group of wild chimpanzees living on the shores of Lake Tanganyika in the heart of Africa. The book ends with a thought-provoking comparison of chimpanzee and human behavior.

Technology

603 ENCYCLOPEDIAS

Kerrod, Robin. *Concise Color Encyclopedia of Science.* (Illus.) Crowell, 1975. 256pp. $9.95. 73-13688. ISBN 0-690-00683-7. Index;CIP.

JH-SH-GA   This is essentially an encyclopedia of technology, covering such areas as technological products, manufacturing and industrial processes, engineering construction and processes, farming and food preservation.

Lodewijk, T., et al. *The Way Things Work: An Illustrated Encyclopedia of Technology.* (Illus.) Simon&Schuster, 1973. 288pp. $9.95. 73-13009. SBN 671-65212-5.

JH-SH   This is a handy reference, and this particular edition is written especially for junior high school use.

609 HISTORY OF TECHNOLOGY

Baldwin, Gordon C. *Inventors and Inventions of the Ancient World.* (Illus.) Four Winds, 1973. 251pp. $6.50. 73-76461.

JH-SH   In lean prose and with fascinating asides, Baldwin describes humanity's change from hunter to farmer, seafarer and city dweller. Fire, plant cultivation, domesticated animals, housing, weaving, metal working, the wheel, the emergence of cities, writing, and money are all discussed and related to geography, climate and human temperament.

Bernal. J.D. *The Extension of Man: A History of Physics Before the Quantum.* (Illus.) MIT Press, 1972. 317pp. $12.50. 72-17898-2. ISBN 262-02-086-6.

SH-C   The story of humans' extension of their natural powers through technology
and later through physics, beginning with simple prehistoric tools and progressing to the end of the 19th century. The book is copiously illustrated with fine
prints. An off-hand, anecdotal style may be somewhat disconcerting. A much better
history of technology than of physics.

Biwas, Asit K.   *History of Hydrology.* (Illus.) American Elsevier, 1970. xii + 336pp.
$16.00. 69-18384. ISBN 0-0444-10025-3.

C   An excellent introduction to the effort to understand and control the planet's
water, the book begins with a description of ancient river-based civilizations
and traces the socioeconomic importance of river management through successive
cultures. Excellent collateral reading for hydraulic engineers and in history of civilization courses.

Bulliet, Richard W.   *The Camel and the Wheel.* (Illus.) Harvard Univ. Press, 1975.
xiv + 327pp. $16.00. 75-571. ISBN 0-674-09130-2. Index.

SH-C   A clear, vivid, powerfully argued and carefully researched example of how
history should be taught. Deals with how and why the camel replaced the
wheel in North Africa and the Middle East from about the third through sixth century
A.D. Discusses the camel's origins, domestication and breeding; camel saddle design;
trade routes; and a great deal more, including the effects on the state of mind of living
in a wheelless culture—effects that still linger today.

Chubb, Thomas Caldecot.   *Prince Henry the Navigator and the Highways of the Sea.*
(Illus.) Viking, 1970. 160pp. $4.53. 78-106923. ISBN 0-670-57623-9.

JH-SH   The methods of navigation, ship building, and the cartographic innovations which made the 15th-century Portuguese "geographic explosion"
possible are described easily as part of the narrative, so as to make the technical
details understandable.

Daumas, Maurice.   *Scientific Instruments of the Seventeenth and Eighteenth Centuries.*
(Illus.;trans. and edited by Mary Holbrook.) Praeger, 1972. vi + 361pp. $38.50. 77-
112019.

SH-C   The treatment of the instrument problems of the early 1700–1800s is skillful
and scholarly. Daumas divides the subject chronologically and by economic, social, industrial, and technical factors. A wealth of historical background
and reference material on the work of the early scientific craftsmen.

Daumas, Maurice, (Ed.).   *A History of Technology and Invention: Progress Through the
Ages.* (2 vols.;illus.) Crown, 1970. 596pp. and 694pp. $10.00ea. 71-93403.

SH-C   Treats the origins of technologies, broadly defined, from prehistoric time
through the early historic period, discusses the beginnings of the industrial
era, and ends with the middle of the 18th century. Emphasizes the development of
technology almost independently of science, in the context of social and other conditions permitting or forcing innovation.

De Camp, L. Sprague.   *The Ancient Engineers.* (Illus.) MIT Press, 1970. 408pp.
$2.95. 76-95278.

SH-C   The engineering technology of Egypt, Mesopotamia, Greece, Hellena,
Rome, the Orient, and Europe is detailed in this well-developed and closely
written overview of eight significant historical periods.

Guye, Samuel, and Henri Michel.   *Time & Space Measuring Instruments from the 15th
to the 19th Century.* (Illus.;trans.) Praeger, 1971. 289pp. $28.50. 77-111070.

C-P   Covers the evolution of time-keeping devices, the evolution and refinements
of clock mechanisms and the development of more precise escapements, and

shows the evolution and design of measuring instruments. Splendid examples are described and illustrated, and the beauty and ornamentation of these handcrafted objects is spectacular.

Hodges, Henry. *Technology in the Ancient World.* (Illus.) Knopf, 1970. xvi + 287pp. $10.00. 71-79353.

SH-C   A lucidly written history of technology from the Paleolithic era to the decline of Rome. The book centers on developments in the Near East, with emphasis on tools, inventions, and processes in relation to environment and social organization. Fine illustrations and maps.

Hughes, Thomas Parke. *Elmer Sperry: Inventor and Engineer.* (Illus.) Johns Hopkins, 1971. xvii + 348pp. $15.00. 71-110373. ISBN 0-8018-1133-3.

SH   The career and accomplishments of a genial inventor and successful entrepreneur are detailed and commented upon. Beginning in 1880, Sperry had built feedback, servomechanisms, and computation devices into his inventions which were recognized decades later as the essence of cybernetics.

Hutchings, David W. *Edison at Work: The Thomas A. Edison Laboratory at West Orange, New Jersey.* (Illus.) Hastings House, 1969. 94pp. $4.95. 78-98058. ISBN 0-8038-1893-9.

JH   An excellent, factual description of the exhibits in the laboratories built in 1887. Presents the problems Edison and his staff met and solved in developing the first incandescent electric lamp, the phonograph, motion picture machine, and improvements in the telephone, telegraph, and storage batteries.

Kenner, Hugh. *Bucky: A Guided Tour of Buckminster Fuller.* (Illus.) Morrow, 1973. 338pp. $7.95. 79-182966. ISBN 0-688-00141-6; 0-688-05141-3(p).

SH-C   There are clear explanations, detailed drawings and directions for building such things as Tensegrity Spheres as well as rich descriptions and an in-depth view of Bucky Fuller.

Knauth, Percy, et al. *The Metalsmiths.* (Illus.) Time-Life (distr.: Little, Brown), 1974. 160pp. $7.95. 73-89680. Index;bib.

JH-SH   Centers on the working of copper, bronze, iron and gold, and includes references to silver, platinum and steel. Archeological discoveries are used to reconstruct the development of metalworking techniques. An excellent, unique presentation of well-chosen illustrations and well-written text.

Pacey, Arnold. *The Maze of Ingenuity: Ideas and Idealism in the Development of Technology.* (Illus.) Holmes&Meier, 1975. 350pp. $12.95. 74-18380. ISBN 0-8419-0181-3. Index;CIP.

C   Pacey traces the influence of noneconomic ideals and impulses in the development of technology in Europe from 1100 to 1870 and then applies his findings to a discussion of ideals and objectives of technology in the 1970s.

Robbins, Michael S. *Electronic Clocks and Watches.* (Illus.) Sams/Bobbs-Merrill, 1975. 208pp. $6.50(p). 74-33836. ISBN 0-672-21162-9. Index.

C   Time scales are described with excellent illustrations in this tutorial review. Also gives a brief history of timekeeping and methods of measuring time, and units of time. Manufacturers' data, suppliers and manufacturers are identified.

Roller, Duane H.D., (Ed.). *Perspectives in the History of Science and Technology.* (Illus.) Univ. of Oklahoma Press, 1971. x + 310pp. $9.95. 77-144163. ISBN 0-8061-0952-1.

C-P   Papers and commentaries on the kinds of problems and the concerns of contemporary historians of science and technology.

Rowland, K.T. *Eighteenth Century Inventions.* (Illus.) Barnes&Noble, 1974. 160pp. $10.00. ISBN 06-496015-3. Index;bib.

JH-SH-C Rowland describes each invention briefly and provides some insight into the inventor's background. The inventions range from aeronautics to water and wind power devices. May be used as a reference in design, graphics, home economics and engineering.

Sandström, Gösta E. *Man the Builder.* (Illus.) McGraw-Hill, 1970. 280pp. $16.00. 78-97121. Gloss.;index.

SH-C An interesting history of engineering technology and human enterprise. Roads, pyramids, houses, religious and governmental structures, aqueducts and waterways, tunnels, and bridges show the origin and evolution of design, of managerial and organizational skills, and of the invention of equipment and tools. A supplement describes certain prehistorical developments.

Sinclair, Angus. *Development of the Locomotive Engine.* (Illus.) MIT Press, 1970. ix + 708pp. $16.00. 72-95285. ISBN 0-262-19068-0.

SH First published in 1907, this classic book contains immense detail concerning specifications of the many locomotive designs and hundreds of pictures or line drawings. The afterword describes the final days of the steam locomotive and brings the history up to 1969.

Sinclair, Bruce. *Philadelphia's Philosopher Mechanics: A History of the Franklin Institute, 1824-1965.* (Illus.) Johns Hopkins Univ. Press, 1974. xi + 353pp. $15.00. 74-6843. ISBN 0-8018-1636-X. Index;CIP.

C The history of the Franklin Institute is described in an interesting way and the book is a useful reference for students of the history of technology in the United States.

610 Medicine

Stedman, Thomas L. *Stedman's Medical Dictionary, 23rd ed.* (Illus.) Williams&Wilkins, 1976. lii + 1730pp. $26.50. 78-176294. ISBN 0-683-07924-7. Index.

C-P A widely used authoritative medical dictionary that contains anatomical, bacteriological, chemical, dental, pharmacological, veterinary and other special terms. Includes discussions of medical etymology and of recent pharmacological preparations, and biographical notes. Excellent color plates show the muscular, skeletal and circulatory systems. Should be included in general reference collections.

610.09 MEDICINE—HISTORY AND BIOGRAPHY

Baldry, P.E. *The Battle Against Heart Disease.* (Illus.) Cambridge Univ. Press, 1971. 189pp. $10.00. 75-108098. ISBN 0-521-97490-8.

SH-GA The historical developments of our understanding of the structure and function of the cardiovascular system and of the diagnosis and treatment of cardiovascular diseases are traced from the Greeks to contemporary cardiovascular surgery. Presents portraits of the pioneers in this field.

Bedeschi, Giulio. *Science of Medicine.* (Illus.) Collins/Watts, 1975. 127pp. $6.90. 74-12752. SBN 0-531-02122-X. Index.

SH-GA Bedeschi provides a synopsis of the history of medicine and of the scientific contributions and discoveries which are the milestones in the field. A useful book for high school libraries.

Bickel, Lennard. *Rise Up to Life: A Biography of Howard Walter Florey Who Gave Penicillin to the World.* (Illus.) Scribner's, 1973. xix + 314pp. $9.95. 73-1051. SBN 684-13429-2.

SH-C-P   Bickel handles nicely the events of Florey's work, including the specula-
tion that purports to show that the discovery and isolation of penicillin was the result of serendipity. The book is well written; it shows how science is practiced.

Brody, Saul Nathaniel. *The Disease of the Soul: Leprosy in Medieval Literature.* (Illus.) Cornell Univ. Press, 1974. 223pp. $9.50. 73-8407. ISBN 0-8014-0804-0. Index;bib.

SH-C   Brody spells out the reasons why leprosy has been, and still often is,
associated with moral defilement and uncleanness. A fascinating account as seen in both lay and scientific literature.

Clapp, Patricia. *Dr. Elizabeth: The Story of the First Woman Doctor.* Lothrop, Lee&Shepard, 1974. 156pp. $4.95. 73-17702. ISBN 0-688-40052-3; 0-688-50052-8(p). CIP.

JH   Writing as though it were a first person account of about one-fourth of the life
of the subject, Clapp recounts the difficulties Blackwell encountered at each stage of her career and the determination with which she overcame those difficulties to achieve her goals.

Cohn, Victor. *Sister Kenny: The Woman Who Challenged the Doctors.* (Illus.) Univ. of Minn. Press, 1976. 302pp. $16.50. 75-15401. ISBN 0-8166-0755-9. Index.

SH-C-P   The story of a nurse who, through common sense, developed an empiri-
cally effective method for treating polio. Her conflict with the medical profession in Australia and America is vividly described. The book is nontechnical, but footnoted to the medical literature. There is a bibliography of Kenny's publications and of pertinent articles on polio.

Edelson, Edward. *Healers in Uniform.* Doubleday, 1971. 177pp. $3.95. 78-131072.

JH-SH-C   This popular account of the achievements of a selected group of mili-
tary physicians describes significant contributions of military medicine to public health, and preventive and aerospace medicine. Well written; offers good insight into the controversies and difficulties surrounding the advancements of military medicine.

Forster, Robert, and Orest Ranum (Eds.). *Biology of Man in History.* (Illus.;trans.; selections from the *Annales, Economies, Sociétés, Civilisations,* 1975.) Johns Hopkins Univ. Press, 1975. x + 205pp. $12.00; $2.95(p). 74-24382. ISBN 0-8018-1690-4; 0-8018-1691-2. CIP.

C-P   This collection on sex, disease, hunger, and research methods reflects the
increasing emphasis on the integration of the social and biologically oriented sciences and ranges from the early Byzantine empire to the present and from the Levant to Western Europe.

Glemser, Bernard. *Mr. Burkitt and Africa.* (Illus.) World, 1970. xii + 236pp. $7.95. 70-112435.

SH-C   Denis Burkitt, an Irish surgeon, diagnosed a lymphoma as a specific dis-
ease entity affecting mainly children in certain parts of Africa. The book contains a brief description of Burkitt's early life and considerable detail about African lymphoma; provides an insight into the conflicts surrounding the proposed treatment of the malignancy and the personalities of the men behind these discoveries.

Hechtlinger, Adelaide. *The Great Patent Medicine Era or Without Benefit of Doctor.* (Illus.) Grosset&Dunlap, 1970. 248pp. $14.95. 70-122554.

JH-SH-C   An attractive and carefully prepared book with many excellent colorful illustrations from patent medicines. Of value in social studies, sociology and history classes.

Huxley, Elspeth. *Florence Nightingale.* (Illus.) Putnam's, 1975. 254pp. $15.00. 74-19237. SBN 399-11480-7. Index.

JH-SH-C   Huxley shows how a well-born, intelligent woman broke the bondage common to 19th century women of that class. Recommended for health professionals and for students considering health careers.

Kaufman, Martin. *Homeopathy in America: The Rise and Fall of a Medical Heresy.* (Illus.) Johns Hopkins Univ. Press, 1971. x + 205pp. $10.00. 79-149741. ISBN 0-8018-1238-0.

C-P   A thorough history of homeopathy — its inception; early misuses; gradual, partial acceptance; its long struggle with the American Medical Association; and its eventual demise.

Kennedy, David M. *Birth Control in America: The Career of Margaret Sanger.* Yale, 1970. xi + 320pp. $8.75. 79-99827. ISBN 0-300-01202-0.

SH-C   Kennedy capably reviews the earlier phases of birth control in America and relates in detail the determination and perseverance of Margaret Sanger, pioneer and leader in the field. Despite ridicule, defamation and imprisonment, she lived to be accepted and receive long-overdue accolades. Many original sources are cited in a readable, informative book.

King, Lester S. *The Road to Medical Enlightenment 1650–1695.* American Elsevier, 1970. x + 209pp. $11.50. 79-111293. ISBN 0-444-19685-4.

C-P   The author reveals an impressive scholarship and a delightful, stylistic facility, focusing on key figures who represent an intellectual history of medicine. Among those included are Lazar Riverius, Jean-Baptiste van Helmont, Robert Boyle, Franciscus de la Boe and Friedrich Hoffman.

King, Lester S., (Ed.). *A History of Medicine: Selected Readings.* (Illus.) Penguin, 1971. 316pp. $5.75. ISBN 0-14-08-0220-7.

C   The book is an anthology from the writings of two dozen physicians and spans the period from 400 B.C. to A.D. 1967. Included are Hippocrates, Galen, Harvey, John Hunter, Oliver Wendell Holmes, and Walter Reed. The book provides sufficient material for seminars in the history of science.

Majno, Guido. *The Healing Hand: Man and Wound in the Ancient World.* (Illus.) Harvard Univ. Press, 1975. xxiii + 571pp. $25.00. 74-80730. ISBN 0-674-38330-3. Index.

SH-C-P   A short description of prehistoric treatment of wounds is followed by a discussion of Mesopotamian and Egyptian medicine and of the techniques and treatments used by the physicians of ancient Greece, Arabia, India, Alexandria, and Imperial Rome. An immensely enjoyable book.

Marks, Geoffrey, and William K. Beatty. *The Story of Medicine in America.* (Illus.) Scribner's, 1974. xi + 416pp. $10.00. 73-1369. SBN 684-13537-X.

SH-C   Covers medicine in America from 1600 to the present, with quotes from medical practitioners and the medical literature. For those interested in either medicine or history.

Marks, Geoffrey, and William K. Beatty. *Women in White.* (Illus.) Scribner's, 1972. 239pp. $6.95. 70-38281. ISBN 0-684-12843-8.

SH-C Chronicles the role of women throughout the evolution of the healing arts. Describes ancient healers, practitioners of the Middle Ages, midwifery, the struggle by women for the right to receive medical training, to practice healing arts, to enter graduate and medical schools and to be licensed to practice medicine; and discusses women in related fields such as nursing, social work and medical research.

Rather, L.J. *Addison and the White Corpuscles: An Aspect of Nineteenth-Century Biology.* Univ. of Calif. Press, 1972. viii + 236pp. $8.50. 71-149940. ISBN 0-520-01972-5.

SH-C-P An interesting discussion of William Addison's long-overlooked discovery that white blood cells pass through the walls of blood vessels into tissues during the inflammatory process. The story is told in the context of 19th century medical and biological thought. A worthwhile reference.

Rothstein, William G. *American Physicians in the Nineteenth Century: From Sects to Science.* Johns Hopkins Univ. Press, 1972. xv + 362pp. $15.00. 77-186517. ISBN 0-8018-1242-9.

C-P A sociological analysis of a period when sects of physicians were committed to "heroic" and dangerous procedures and struggled among themselves over admission to medical societies, licensing, and standards of medical education. Only when the closing decades brought scientific understanding and technological improvements to medical practice did the profession find a basis for unity. References and quotations from primary sources make this a useful monograph.

610.28 BIOMEDICAL ENGINEERING

Cromwell, Leslie, et al. *Biomedical Instrumentation and Measurements.* (Illus.) Prentice-Hall, 1973. xiv + 446pp. $16.95. 73-6. ISBN 0-13-077131-7. Index;gloss.

C An introductory level text for engineering and electronics students. Describes the setting in which medical instruments are used and the scope and limitations of the major instrument types. Includes notes on medical terminology, a summary of the major physiological measurements, problems, and exercises.

Cromwell, Leslie, et al. *Medical Instrumentation for Health Care.* (Illus.) Prentice-Hall, 1976. xiii + 418pp. $18.95. 75-35505. ISBN 0-13-572602-6. Index;gloss.;CIP.

C-P This well-written review of biomedical instrumentation, primarily for the critical-care nurse, deals with the principles of electronics, the measurement of physiological variables, major areas of clinical application, and the principles of safety, care and utilization of medical instrumentation. Familiarity with medicine and electronics is required.

Rushmer, Robert F. *Medical Engineering: Projections for Health Care Delivery.* (Illus.) Academic, 1972. xiii + 391pp. $17.50. 77-182648.

C-P The first portion summarizes the current state of health care delivery in the United States and identifies problems for which solutions may be forthcoming from biomedical engineering. The second portion relates recent discoveries in engineering and physical-chemical technology which have applicability for improving diagnostic, therapeutic and administrative techniques in medicine.

610.6 MEDICAL ORGANIZATIONS AND PERSONNEL

Dimond, E. Grey. *More than Herbs and Acupuncture.* Norton, 1975. 223pp. $7.95. 74-11166. ISBN 0-393-06400-X. CIP.

C The book deals with the results of Dimond's study of Chinese medicine today. Such topics as methods of selecting and training students, the medical curriculum and the return of the medical student to the service of the people are described and analyzed. Accounts of operations performed under acupuncture are included. An excellent source of information on current Chinese attitudes both inside and outside the field of medicine.

MacNab, John. *The Education of a Doctor: My First Year on the Wards.* Simon&Schuster, 1971. 222pp. $6.95. 71-156158. ISBN 0-671-21017-3.

   SH-C   In an unusual book, the author interjects personal philosophical observations about his professors, clinicians, and fellow students that are perceptive jewels. Replete with valuable information to those planning to study medicine, it exposes the bare bones of medical education in an entertaining manner.

Mumford, Emily. *Interns from Students to Physicians.* Harvard Univ. Press, 1970. ix + 298pp. $8.50. 74-99519. ISBN 0-674-45925-3. Index.

   C-P   A careful and accurate description of the socialization processes to which young physicians are exposed. The internship experience is related to both prehospital training and to subsequent professional growth and development. Ample documentation; an excellent set of records.

610.7 NURSING

Cafferty, Kathryn W., and Leone K. Sugarman. *Steppingstones to Professional Nursing, 5th ed.* (Illus.) Mosby, 1971. xii + 361pp. $9.25. 76-158487. ISBN 0-8016-0929-1.

   C   The development of nursing is explored and presented in a concise, interesting, colorful manner, which makes the reading of history most enjoyable.

Hamilton, Persis Mary. *Basic Maternity Nursing, 2nd ed.* (Illus.) Mosby, 1971. viii + 244pp. $6.90. 70-155324. ISBN 0-8016-2029-5.

   C   The audience for this basic text comprises students of licensed practical or vocational nursing. The basic physiological processes involved in reproduction are discussed, and the nursing responsibilities are outlined. Less attention is paid to the psychological problems that may be related to pregnancy and delivery.

Redman, Barbara Klug. *The Process of Patient Teaching in Nursing, 2nd ed.* (Illus.) Mosby, 1972. viii + 178pp. $7.95. 71-188732. ISBN 0-8016-4097-0.

   SH-C-P   After an introduction to the psychology of learning, chapters are organized around the teaching-learning process: the place of teaching in nursing; readiness for health education; objectives of health education; and evaluation of health teaching. There are case studies, questions and answers, chapter summaries, and numerous references.

611–612 ANATOMY AND PHYSIOLOGY

Anthony, Catherine Parker, with Norma Jane Kolthoff. *Textbook of Anatomy and Physiology, 9th ed.* (Illus.) Mosby, 1975. viii + 598pp. $12.50. 74-20991. ISBN 0-8016-0254-8. Gloss.;index;CIP.

   C-P   Recommended for a 1-semester course in anatomy and physiology for nursing and pharmacy schools. The section on anatomy is meager, but the author draws excellent correlations between the structure and function of the major parts of the human body. The charts and tables are clear and precise.

Basmajian, J.V. *Grant's Method of Anatomy: By Regions Descriptive and Deductive, 8th ed.* (Illus.) Williams&Wilkins, 1971. xviii + 707pp. $15.50. 70-147079. ISBN 0-683-00431-X. Index;bib.

   C-P   A standard tool of medical pedagogy. The current edition attempts to relate the subject matter and its presentation to the needs of medical students in the 1970s. Profusely illustrated with clear diagrams and accurately labeled line drawings.

Extensive index, reading references, and many of the names of structures are followed by the derivation of their principal roots.

Copenhaver, Wilfred M., Richard P. Bunge, and Mary Bartlett Bunge. *Bailey's Textbook of Histology, 16th ed.* (Illus.) Williams&Wilkins, 1971. xv + 745pp. $19.50. 79-127303. ISBN 0-683-02073-0. Index;bib.

   C-P The treatment is exhaustive and accurate as befits a standard teaching and reference text. Highly recommended for medical collections.

Crouch, James E., and J. Robert McClintic. *Human Anatomy and Physiology, 2nd ed.* (Illus.) Wiley, 1976. xxiii + 892pp. $17.50. 76-868. ISBN 0-471-18918-9. Gloss.;index;CIP.

   C The presentation is clear and well organized, with good artwork and photographs; excellent for a broad spectrum of undergraduate level courses. Extensive use of tables and concise chapter summaries add to the value of the book.

Francis, Carl C., and Alexander H. Martin. *Introduction to Human Anatomy, 7th ed.* (Illus.) Mosby, 1975. xii + 512pp. $12.95. 75-2455. ISBN 0-8016-1646-8. Gloss.;index;CIP.

   C-P A readable, lucid description of human anatomy is presented, emphasizing a "systems" rather than a "regional" approach for students in paramedical fields.

Griffiths, Mary. *Introduction to Human Physiology.* (Illus.) Macmillan, 1974. xx + 555pp. $12.95. 73-1960. ISBN 0-02-347220-0. Index;CIP.

   C A simple approach to physiology for beginning college students with little or no science background. Provides an up-to-date treatment of the subject of endocrines, and good illustrations, some using electron micrographs.

Guyton, Arthur C. *Function of the Human Body, 4th ed.* (Illus.) Saunders, 1974. viii + 473pp. $10.50. 73-81832. ISBN 0-7216-4377-9. Bib.

   C This text will be extremely valuable in teaching a 1-semester course in physiology or anatomy to students in any discipline. The material is current and the physiology of the basic organ systems is covered in excellent detail.

Jacob, Stanley W., and Clarice Ashworth Francone. *Elements of Anatomy and Physiology.* (Illus.) Saunders, 1976. x + 251pp. $6.95(p). 75-28795. ISBN 0-7216-5088-0. Index;CIP.

   C A well-balanced selection of topics in clear text on human gross anatomy. Short discussions of related physiology and histology (but lacks adequate discussion of some crucial topics). Superb drawings and learning exercises.

Landau, Barbara R. *Essential Human Anatomy and Physiology.* (Illus.) Scott, Foresman, 1976. 598pp. $14.95. 75-22138. ISBN 0-673-05989-8. Gloss.;index;CIP.

   C-P The author begins with selected basic chemical and physical principles, introduces aspects of cell structure and function, and covers anatomy and physiology of various organ systems. An excellent text for students who plan a future in the field or for advanced students who need a review.

Nilsson, Lennart, with Jan Lindberg. *Behold Man: A Photographic Journey of Discovery Inside the Body.* (Illus.;trans.) Little, Brown, 1974. 254pp. $25.00. 73-14087. ISBN 91-0-038093-8.

   JH-SH-C-P Beautifully illustrated with macro- and microscopic views of inside the human body, this book depicts various anatomical structures and processes. Should be most appreciated by health professionals, especially those who are amateur photographers.

Schroeder, Charles R.   *The Human Body: Its Structure and Function.* (Illus.) Brown, 1971. ix + 60pp. $0.95. 79-156097. ISBN 0-697-07336-X. Index;bib.

JH-SH   A lucid presentation with extremely clear and pertinent line drawings. Ideal both for high schools and for general readers. Each section has review questions on central concepts.

Stevens, Leonard A.   *Neurons: Building Blocks of the Brain.* (Illus.) Crowell, 1974. 87pp. $5.50. 74-4399. ISBN 0-690-00403-6. Index;CIP.

JH-SH   A brief review of the development of the scientific observation and description of the neuron. Useful as collateral reading for a human biology course.

Stonehouse, Bernard, et al.   *The Way Your Body Works.* (Illus.) Crown, 1974. 96pp. $9.95. 74-80724. ISBN 0-517-51678-9. Gloss.;index.

SH-GA   This atlas-type book provides a description of the structure and function of humans, primarily through lifelike illustrations of the systems of the human body. A worthwhile supplement for both the scientist and layperson.

612.6 REPRODUCTION, DEVELOPMENT, MATURATION

Berenstain, Stan, and Jan Berenstain.   *How to Teach Your Children About Sex . . . Without Making a Complete Fool of Yourself.* (Illus.) McCall, 1970. 64pp. $2.95. 79-122114. ISBN 0-8415-0039-8.

GA   This is a humorous cartoon book that parents, elementary school teachers, pediatricians, nurses and others will enjoy. It explores the various approaches and techniques of sex education in the home or at school and demonstrates that what is taught is not always what is learned.

Hettlinger, Richard D.   *Growing Up with Sex.* (Illus.) Seabury, 1971. ix + 162pp. $4.95; $2.25(p). 78-157067.

JH-SH   Written for youngsters who exhibit an uncomfortable mixture of sophistication and naïveté. Discusses the physical aspects of adolescent development: sex urges; masturbation and other forms of physical gratification, including intercourse; homosexuality; influences on sex attitudes and practices; sex and religion; premarital sex; sex and love; teenage marriage; and the sex organs and reproduction, birth control, venereal disease and abortion.

Johnson, Eric W.   *Love and Sex in Plain Language.* (Illus.) Lippincott, 1974. 143pp. $4.95. 73-14797. ISBN 0-397-00988-7.

JH-GA   Johnson provides information about sexual development, the reproductive system, intercourse, heredity, conception, and sexual behavior. He discusses myths about masturbation, homosexuality and birth control in such a way as to offer young people a choice.

Katchadourian, Herant A., and Donald T. Lunde.   *Fundamentals of Human Sexuality, 2nd ed.* (Illus.) Holt, Rinehart&Winston, 1975. xii + 595pp. $15.00. 74-23704. ISBN 0-03-089523-5; 0-03-014136-2. Index;CIP.

C   This is a highly descriptive textbook with a great deal of anatomical and physiological information. Topics include reproductive behavior, childbirth, contraception, psychosexual development, autoeroticism, sexual deviation and sexual disorders.

Krogman, Wilton Marion.   *Child Growth.* (Illus.) Univ. of Mich. Press, 1972. 231pp. $7.95; $2.95(p). 71-163623. ISBN 0-472-00119-1; 0-472-05019-2.

C-P   An immense variety of growth changes illustrate a series of concepts in the field. Covers biological and behavioral growth patterns and discusses both intrinsic factors (genetic, endocrine) and extrinsic factors (nutrition, socioeconomic, etc.) in great depth and breadth.

Lerner, Marguerite Rush. *Color and People: The Story of Pigmentation.* (Illus.) Lerner, 1971. 55pp. $4.50. 70-128800. ISBN 0-8225-0625-4.

   JH-SH   Excellent photographs, concise writing and well-formed concepts make this small monograph a choice introduction to the basic biology of skin color. Treatment is from the technical point of view of a dermatologist and does not include the social, psychological or literary aspects.

Navarra, J.G., Joseph S. Weisberg, and Frank M. Mele. *From Generation to Generation: The Story of Reproduction.* (Illus.) Natural History Press, 1970. 115pp. $3.95. 72-97673.

   JH-SH   The authors, all science teachers, compare the human story with that of the animals. The progress from asexual to sexual reproduction is explained as nature's means to help its creatures to adapt to an ever-changing environment. Useful in sex education as well as science classes.

Stein, Sara Bonnett. *Making Babies: An Open Family Book for Parents and Children Together.* (Illus.) Walker, 1974. 46pp. $4.50. 73-15267. ISBN 0-8027-6171-2.

   SH   Designed as a sex education book for children, this book is factually accurate and its illustrations are excellent. Will also be helpful in teaching parents how to educate their children.

Swanson, Harold D. *Human Reproduction: Biology and Social Change.* (Illus.) Oxford Univ. Press, 1974. 392pp. $10.00; $4.95(p). 73-92868. ISBN 0-19-501771-4. Index;gloss.

   C   Swanson is concerned that the next generation will face dilemmas caused by unprecedented reproductive and genetic choices. A timetable and a discussion of coming technologies are included. The book is designed for a course in human reproduction for nonbiologists.

Weiss, Elizabeth. *The Female Breast.* (Illus.) Bantam, 1975. ix + 164pp. $1.95(p). Bib.

   SH-C   Most of the text is devoted to the normal structure and physiology of the breast. There is discussion of variations in anatomy and development, with emphasis on cultural and psychological meanings of the breast, and chapters on breast feeding and breast cancer.

612.8 NEUROPHYSIOLOGY AND SENSORY PHYSIOLOGY

Luce, Gay Gaer. *Body Time: Physiological Rhythms and Social Stress.* (Illus.) Pantheon, 1971. xii + 394pp. $6.95. 70-162394. ISBN 0-394-36891-0.

   SH-C   Describes the results of many studies on the time patterns of tree growth, disease resistance in humans, patterns of sleep, hormone levels, other bodily functions, etc. The message is that technology has greatly changed our environment but that we have paid too little attention to the effects of these changes on our innate biological rhythms.

Melzack, Ronald. *The Puzzle of Pain.* (Illus.) Basic, 1974. 232pp. $9.50. 73-81726. SBN 465-06779-4. Gloss.;bib.

   C-P   A careful exposition of the physiology and pathology of pain and painful stimuli, and a less extended treatment of methods of pain control, drug action and the psychiatric-psychological literature on placebos. Recommended for the specialist or serious student desiring an overview and introduction to the field.

Pool, J. Lawrence. *Your Brain and Nerves.* (Illus.) Scribner's, 1973. xi + 211pp. $7.95. 79-143957. SBN 684-13268-0.

   SH-C   Nontechnical coverage of the anatomy and function of the nervous system, including the brain, skull, spinal cord, the spine and peripheral nerves. Describes injuries to and diseases of the nervous system and their diagnosis and treatment.

Silverstein, Alvin, and Virginia B. Silverstein. *The Nervous System: The Inner Networks.* (Illus.) Prentice-Hall, 1971. 64pp. $4.50. 76-119514. ISBN 0-13-610964-0.

   JH   The basic features of the nervous system are presented, and concepts of nerve cells, pathways and reflexes are clearly developed by comparison to everyday occurrences and familiar objects. There is judicious use of illustrations; key words are italicized, and new words appear in bold type followed by the phonetic spelling. Touches on comparative anatomy, neurological research and animal behavior, and diseases of the mind.

Stiller, Richard. *Pain: Why It Hurts, Where It Hurts, When It Hurts.* (Illus.) Nelson, 1975. 162pp. $6.95. 75-6522. ISBN 0-8407-6430-8. Index;CIP.

   SH-C-P   Stiller starts with pain as a part of life and concludes with an objective summary of how "to win the battle against pain without losing one's sense of life, one's ability to feel." Four chapters are devoted to vanquishing pain through drugs, electroanalgesia, acupuncture and hypnosis and biofeedback.

Strughold, Hubertus. *Your Body Clock.* (Illus.) Scribner's, 1971. 95pp. $5.95. 70-162753. ISBN 0-684-12557-9.

   C-P   A valuable aid for plane travelers who cross many time zones in a short time: it will help them understand their body responses and will suggest what they can do to minimize discomfort. The chapter dealing with the structure and function of the physiologic clock explores in detail the deep sleep patterns of humans and animals. Contains numerous facts on human body rhythms.

612.82 THE BRAIN AND SLEEP PHENOMENA

Calder, Nigel. *The Mind of Man.* (Illus.) Viking, 1971. 288pp. $8.95. 71-145661. ISBN 0-670-47640-4.

   SH   A well-written view of brain research on normal and pathological humans, with descriptions of brain function in speech and sleep, control of visceral functions and emotions, group behavior, perception, and motor movements.

Clarke, Edwin, and Kenneth Dewhurst. *An Illustrated History of Brain Function.* (Illus.) Univ. of Calif. Press, 1973. 154pp. $14.00. 72-87206. ISBN 0-520-02316-1.

   GA-C-P   Much of this historical presentation of brain neurology deals with the elusive problem of cortical localization of brain function. Phrenology is discussed. Suitable for the general reader except for some difficult sections.

Dement, William C. *Some Must Watch While Some Must Sleep.* (Illus.) Freeman (distr.: Scribner's), 1974. 148pp. $5.95. 74-7334. ISBN 0-7167-0769-1; 0-7167-0768-3(p). Gloss.;index;CIP.

   SH-GA-C-P   An introduction to sleep research, covering dreams, sleep disorders and disturbances, REM and NREM sleep, sleep disturbances in mental illness and creativity or problem solving while sleeping.

Freedman, Russell, and James E. Morriss. *The Brains of Animals and Man.* (Illus.) Holiday House, 1972. 160pp. $4.95. 71-151754. ISBN 0-8234-0205-3.

   SH-C-P   Concepts are developed in a way that takes the reader from familiar objects and everyday occurrences to the realm of experimentation and scientific observation. The structure of the nervous system, stimulus and response, localization of function and electrical activity of the brain are explained. Discussions of brain waves, sleep and dreams, memory and intelligence, and remote control of behavior are included.

Hartmann, Ernest L. *The Functions of Sleep.* (Illus.) Yale Univ. Press, 1973. ix + 198pp. $8.50; $2.95(p). 73-79983. ISBN 0-300-01700-6; 0-300-01701-4.

C-P   A remarkable, short compendium on sleep research and the varying theories about the nature of sleep. Both biochemical and behavioral aspects are competently reviewed. An interdisciplinary approach is successfully achieved in the discourse on dreams, sleep pathology, REM sleep, the effects of the body's internal chemistry, drugs, psychic states and other numerous factors.

McGuigan, F.J., and R.A. Schoonover (Eds.). *The Psychophysiology of Thinking: Studies of Covert Processes.* (Illus.) Academic, 1973. xvi + 511pp. $26.50. 72-82658. ISBN 0-12-484050-7.

C-P   Discusses physiological measures as useful correlates of mental activity during sleep. Jacobson's significant chapter dispels simplistic beliefs about the relationship of muscle activity to thought and directly relates his work to that on sleep. Other papers cover current sleep research, effects of separated brain hemispheres, experimentally produced perceptions in the absence of stimulation, relation of brain activity changes to visual attention, the central processes controlling speech production during sleep and while awake, etc.

Oatley, Keith. *Brain Mechanisms and Mind.* (Illus.) Dutton, 1972. 216pp. $7.95. 74-172077. ISBN 0-525-07050-8; 0-525-03050-6(p).

JH-SH-C   The text focuses on the structural aspects of brain function, with chapters on the neurone, vision, learning, memory, language, and such theoretical aspects as artificial brains and future challenges. Multiple illustrations.

Penfield, Wilder. *The Mystery of the Mind: A Critical Study of Consciousness and the Human Brain.* (Illus.) Princeton Univ. Press, 1975. xxix + 123pp. $8.95. 74-25626. ISBN 0-691-08159-X. Index;CIP.

GA-C-P   The author discusses the relation between brain structure, consciousness and concepts of minds. He relates case material from his published reports and presents some simplified anatomical drawings.

Pines, Maya. *The Brain Changers: Scientists and the New Mind Control.* Harcourt Brace Jovanovich, 1973. 248pp. $7.95. 73-8721. ISBN 0-15-113700-5.

C   Pines gives a clear, if patchy, report on current research on brain manipulations. Several means of brain control are described: manipulating the environment of the young, drugs, biofeedback and direct electrical stimulation of pain and pleasure centers. The applications of these controls to intelligence, memory and violence are discussed.

Silverstein, Alvin, and Virginia B. Silverstein. *Sleep and Dreams.* (Illus.) Lippincott, 1974. 159pp. $5.50. 73-13825. ISBN 0-397-31325-X.

SH   The main focus is sleep and the array of new knowledge made available through the widespread use of the REM electroencephalographic, all-night monitoring technique. Covers animal and human experimentation, hibernation and estivation, and how this new knowledge applies to the many unsolved sleep problems.

Valenstein, Elliott S. *Brain Control: A Critical Examination of Brain Stimulation and Psychosurgery.* (Illus.) Wiley, 1973. xx + 407pp. $10.95. 73-13687. ISBN 0-471-89784-1.

C   An accurate picture of the state of brain manipulation experiments. Valenstein sees brain research as being useful only as a research tool to study how the brain is organized and how it carries out its functions. The book is clear and thorough.

613 GENERAL AND PERSONAL HYGIENE

Bucher, Charles A., et al. *The Foundations of Health, 2nd ed.* (Illus.) Prentice-Hall, 1976. viii + 408pp. $11.95. 75-28002. ISBN 0-13-329896-5. Glosses;index;CIP;bib.

SH-C   An established textbook, modern, serious, philosophical, and comprehensive, making it a useful reference for teachers. Sections on life and the mind, human sexuality, chemical alteration of behavior, anatomy and physiology, nutrition, health services, health frauds, environmental health and disease, and drugs, alcohol and tobacco.

Cain, Arthur H. *Young People and Health.* John Day, 1973. xix + 171pp. $5.95. 72-12083. ISBN 0-381-98241-6.

JH-SH   For anyone concerned with maintaining good mental and physical health. Diet is emphasized along with some techniques useful in dieting. Normal weight charts and simple diet and calorie charts are included. Problems associated with alcohol and tobacco abuse are related and advice on ways to quit is provided.

Frank, Arthur, and Stuart Frank. *The People's Handbook of Medical Care.* Random House, 1972. 494pp. $8.95. 72-4593. ISBN 0-394-47925-4.

SH-GA-C   Attempts to present medical problems in simple, clear language for lay audiences. Chapter topics include timely medical care problems (drug addiction, rape, abortion, free clinics); the business of medicine (costs); recognizing diseases; medical emergencies; medical rights; social and political medicine; and food, nutrition and obesity.

Gregg, Walter H. *Physical Fitness Through Sports and Nutrition.* (Illus.) Scribner's, 1975. x + 112pp. $5.95. 74-4847. ISBN 0-684-13903-0.

JH-SH   This guide to attaining self-confidence and self-discipline through good nutrition and exercise explains the physiology of bodily motion. Segregated illustrations.

Miller, Benjamin F., and Lawrence Galton. *The Family Book of Preventive Medicine: How to Stay Well All the Time.* Simon&Schuster, 1971. 704pp. $12.95. 70-139644. ISBN 0-671-20812-8.

SH-GA-C   An interesting volume intended for family use as an aid to healthful living. There are six parts: the nature of preventive medicine; preventive therapy; care of one's body; mental health; family care (including healthy adjustments in marriage and preventive medicine for children); and disease scenarios, listing 63 diseases. A useful reference.

Miller, Benjamin F., Edward B. Rosenberg, and Benjamin L. Stackowski. *Investigating Your Health.* (Illus.) Houghton Mifflin, 1971. xii + 564pp. $5.55. 74-111257.

JH-SH   Opportunities are provided for the students to conduct experiments, to make surveys, and to evaluate evidence and opinions on health problems. Individual, family, and community interests are considered in relation to physical systems and functions; fitness, rest and recreation; nutrition and grooming; emotional and psychological development; stimulants and drugs; disease processes, care and prevention; safety and first aid; environmental issues; and health services and careers. Teachers' edition available.

Vickery, Donald M., and James F. Fries. *Take Care of Yourself: A Consumer's Guide to Medical Care.* (Illus.) Addison-Wesley, 1976. xvi + 269pp. $5.95(p). 76-1378. ISBN 0-201-02401-2; 0-201-02402-0. Index;CIP.

GA   This very readable, practical, well-organized consumer's guide is devoted to self-treatment advice for common complaints, including indicators for emergency care. Sections on home pharmacies, protecting one's own health, locating physicians, and choosing hospitals.

613.8 DRUGS, ALCOHOL AND TOBACCO

Brecher, E.M., and Editors of Consumer Reports. *Licit and Illicit Drugs: The Consumers Union Report on Narcotics, Stimulants, Depressants, Inhalants, Hallucinogens, and Marijuana—Including Caffeine, Nicotine, and Alcohol.* (Illus.) Little, Brown, 1972. xv + 623pp. $12.50. 75-186972. ISBN 0-326-15340-0.

GA-C-P   A controversial, authoritative and extremely well-written book in which the use of drugs by our society is carefully examined. Analyzes the principal areas of drug abuse in America today. Their conclusions and recommendations may appear as heresy to many people over 40, but when viewed in a thoroughly objective manner, they certainly make sense.

Girdano, Daniel A., and Dorothy Dusek Girdano. *Drug Education: Content and Methods.* (Illus.) Addison-Wesley, 1972. viii + 288pp. $5.95(p). 75-171431.

C   Directed to public school teachers, this book provides both content (social, psychological, pharmacological) and methods (classroom projects, discussion topics, etc.) for drug education. The methods emphasize student participation. Teachers using this text need some background in physiology and pharmacology.

Goodwin, Donald. *Is Alcoholism Hereditary?* Oxford Univ. Press, 1976. x + 171pp. $7.95. 75-32346. ISBN 0-19-502009-X. Index.

GA   The author believes that the severest forms of alcoholism may be hereditary. Based on a unique study of adopted men with a biologic parent who had been hospitalized for alcoholism, the book includes a somewhat negative appraisal by two independent workers, and a discussion of the entire problem of alcoholism. Concise, clear, engaging style. Footnotes provided for further study.

Health, Education and Welfare. *Alcohol and Health: Report from the Secretary of Health, Education and Welfare.* (Illus.) Scribner's, 1973. xxv + 372pp. $3.95(p). 72-11115. SBN 684-13271-0.

SH-C-P   This report to Congress is highly readable and should help clarify many misconceptions; however, it does not supply much new data. Suggests how our alcohol problems could be handled and what the scientific problems are. Gives an interesting and clear view of the historical use of alcohol and its effect in the body. A valuable book.

Hofmann, Frederick G., with Adele D. Hofmann. *A Handbook on Drug and Alcohol Abuse: The Biomedical Aspects.* Oxford Univ. Press, 1975. xv + 329pp. $10.95; $6.95(p). 74-83987. ISBN 0-19-501922-9. Index.

SH-GA-C   After careful introduction to drug terminology, drug abuse and often encountered misconceptions, the common areas of drug concern are discussed, including clinical signs used to identify drug abuse. Although some knowledge of biology is helpful, it will inform and stimulate even the lay reader. Excellent background reading for lecture courses.

Milgram, Gail Gleason. *The Teenager and Smoking.* (Illus.) Richards Rosen, 1972. 114pp. $3.99. 70-181421. ISBN 8239-0255-2.

JH   Readable, well-organized information on smoking both tobacco and marijuana. The section on tobacco includes its history and uses, characteristics of smokers, difficulties in stopping smoking, and medical information on smoking. The section on marijuana discusses the traits of drug users, the history and laws governing the use of marijuana, and the effects of use. Each section ends with a chapter on deciding whether or not to use the drug discussed. Both sections are followed by summarizations in convenient question-answer form.

613.94 BIRTH CONTROL

Arnstein, Helene S. *What Every Woman Needs to Know About Abortion.* Scribner's, 1973. 144pp. $5.95; $2.45(p). 73-5168. ISBN 0-684-13547-7; 0-684-13559-0.

Planned Parenthood of New York City. *Abortion: A Woman's Guide.* (Illus.) Abelard-Schuman, 1973. xii + 147pp. $5.95. 72-9549. ISBN 0-200-04003-0; 0-200-04008-1(p).

JH-SH-C   Both books describe pregnancy testing, methods of abortion in clinics and in hospitals, and contraception; both make some attempt to help the prospective patient sort out her motives. Either would be useful for classes in sex education.

Luker, Kristin. *Taking Chances: Abortion and the Decision Not to Contracept.* (Illus.) Univ. of Calif. Press, 1975. xii + 207pp. $10.95. 74-22965. ISBN 0-520-02872-4. Index.

C   Luker examines the use of abortion as a preferred alternative to contraception. Teachers, counselors, nurses, and administrators should find this book stimulating.

Pierson, Elaine C. *Sex Is Never an Emergency: A Candid Guide for College Students, 2nd ed.* Lippincott, 1971. vii + 56pp. $0.95(p) (12 copies, minimum). 78-148244.

C   The purpose of this excellent book is "to prevent unwanted pregnancies and, secondarily, to help students be more comfortable with their level of sexuality, whatever that level is." In a question-and-answer plus discussion format, it covers a vast amount of information in complete candor without obvious value judgment.

Wylie, Evan McLeod. *The New Birth Control: A Guide to Voluntary Sterilization.* Grosset&Dunlap, 1972. 215pp. $6.95. 70-183035. ISBN 0-448-01221-9.

SH-GA-C   Gives the pertinent facts about sterilization (vasectomy and laparoscopy) in language which can easily be understood by the intelligent layperson and offers some viewpoints of interest to health professionals. Lists addresses of clinics which sponsor or perform the operations and gives details about possible insurance coverage in each state.

613.95 SEX HYGIENE

Blanzaco, Andre. *VD: Facts You Should Know.* (Illus.) Lothrop, Lee&Shepard, 1970. 63pp. $3.95. 78-120168.

JH   All the information on syphilis and gonorrhea that a layperson needs is presented here in language understandable to a 12-year-old. The nature of the problem and male and female reproductive systems are discussed. On almost every page is a multiple choice test, with answers below. The book ends with an excellent history of venereal disease.

Brooks, Stewart M. *The V.D. Story.* (Illus.) Littlefield, 1972. 157pp. $1.95(p).

SH-C   A very timely subject presented in a readable and interesting manner. Should serve as an excellent compendium for both students and their teachers and parents.

Helmer, Robert. *The Venus Dilemma.* (Illus.) Nash, 1974. 172pp. $5.95. 73-83533. ISBN 0-8402-1328-X.

Sgroi, Suzanne M. *VD: A Doctor's Answers.* (Illus.) Harcourt Brace Jovanovich, 1974. 182pp. $6.50. 73-5245. ISBN 0-15-293350-6. Gloss.;index;CIP.

JH-SH-C   These two primers are accurate and useful; both discuss the epidemiology, pathophysiology, complications, prevention and treatment of VD. Sgroi presents an easily readable, brief but complete summary; most of her book is a directory of VD treatment centers. Helmer's is in greater depth, but loosely organized, verbose, and includes his own opinions.

Hyde, Margaret O. *V.D.: The Silent Epidemic.* (Illus.) McGraw-Hill, 1973. 64pp. 73-1440. ISBN 0-07-031637-6; 0-07-031638-4.

JH-SH-C   Hyde's factual approach to venereal disease avoids scare tactics and preaching and provides a very readable account of the major venereal diseases, their symptoms and modes of spread. Readers are encouraged to seek medical advice quickly. Brief summaries of control methods and research programs are presented.

Johnson, Eric W. *V.D.: Venereal Disease and What You Should Do About It.* (Illus.) Lippincott, 1973. 127pp. $4.75. 73-5515. ISBN 0-397-31447-7.

SH-C   Johnson presents well-documented information within a problem-solving format. Topics range from how venereal disease spreads to specific facts and figures on syphilis and gonorrhea.

614 PUBLIC HEALTH

Bergwall, David F., et al. *Introduction to Health Planning.* (Illus.) Information Resources Press, 1974. vii + 221pp. $15.00. 73-87159. ISBN 0-87815-012-9.

C-P   This introductory text is for laypersons interested in planning for the health needs of society. History, goals and requirements for such planning are set forth in uncomplicated language. Chapters on cost-benefit analysis and evaluation of planning are commendable in their concise definition of aims, methodologies, virtues and pitfalls.

Busvine, James R. *Arthropod Vectors of Disease.* (Illus.) Arnold (distr.: Crane, Russak), 1975. 68pp. $3.00(p). ISBN 0-7131-2500-4; 0-7131-2501-2.

SH-C   The opening chapters deal with vectors and the pathogens involved. In subsequent chapters each arthropod-borne disease is covered, including a brief history, its importance, nature of the pathogen, characteristics of the vector and treatment and control of the disease. An interesting monograph.

Chaney, Margaret S., and Margaret L. Ross. *Nutrition, 8th ed.* (Illus.) Houghton, Mifflin, 1971. viii + 486pp. $9.50. 76-151636. ISBN 0-395-12425-5.

C   The focus of this very readable introductory text is normal nutrition in humans and the interrelationships among nutrients. Some knowledge of organic chemistry and physiology is assumed. Discusses the importance of nutrition; the agencies concerned with the development of nutritional standards; the functions of food in the body; energy requirements; specific components of foodstuffs; varying nutritional needs during the life cycle; and a number of world-wide nutritional problems.

Fox, John P., Carrie E. Hall and Lila R. Elveback. *Epidemiology: Man and Disease.* (Illus.) Macmillan, 1970. ix + 339pp. $12.95. 76-80305.

C-P   Develops the logic behind concepts such as vital statistics, mortality rates, morbidity rates, cause-specific death rates, etc. Difficulties and errors in sampling techniques are outlined, as are the nature and practice of epidemiology.

Hunter, Beatrice Trum. *Food Additives and Federal Policy: The Mirage of Safety.* Scribner's, 1976. x + 322pp. $9.95. 75-20299. ISBN 0-684-14426-3. Index;CIP.

SH-C-P This is a panoply of research facts, expert statements and events about the known, unknown and suspected toxicity of food additives. Shows consumers why and how animal safety studies on these chemicals are performed, why the tests are largely inadequate and why the public is not effectively protected from these substances. Useful for reference.

Lieberman, E. James, (Ed.). *Mental Health: The Public Health Challenge.* Amer. Public Health Assoc., 1975. xx + 293pp. $6.00. 74-34564. ISBN 0-87553-075-3. Index;CIP.

C These 48 articles by knowledgeable experts reflect the many complex considerations in the contemporary public mental health field and provide a broad state-of-the-field report. Should be useful to mental health workers, legislators, students, social planners and concerned citizens.

Retherford, Robert D. *The Changing Sex Differential in Mortality.* Greenwood, 1975. xi + 139pp. $11.00. 74-19808. ISBN 0-8371-7848-7. Index;CIP.

C-P Retherford explores causes of sex differences in mortality and examines changes in cause of death and the effects of altered age structures, increased cigarette smoking, lowered age at marriage, reduction of occupation hazards and other factors. Documents effects of smoking patterns on mortality rates.

Roe, Francis J.C., (Ed.). *Metabolic Aspects of Food Safety.* (Illus.) Academic, 1970. xxiii + 612pp. $22.00. 72-142181. ISBN 0-12-592550-6. Indexes.

C-P With chapter summaries and extensive documentation, this symposium volume achieves textbook status. The major thrust is a detailed discussion of the kidneys, liver and the function of the small intestines—the major sites of formation of metabolites from food additives. Useful reference for college biology classes.

Schroeder, Henry A. *The Trace Elements and Man: Some Positive and Negative Aspects.* (Illus.) Devin-Adair, 1973. ix + 171pp. $7.95. 72-85732.

C-P A valuable book on the role of minerals in human health which covers both negative and positive aspects of mineral intake, health and disease. Industrial pollution, the treatment of foods, soil deficiencies and nonessential minerals, and fluorides in the diet are discussed. A reference work especially useful for its relatively extended treatment of lesser known elements.

Silverstein, Alvin, and Virginia Silverstein. *The Chemicals We Eat and Drink.* (Illus.) Follett, 1973. 112pp. $4.95. 72-85576. ISBN 0-695-80372-7; 0-695-40372-9.

JH Provides a brief account of the food we eat and then discusses the various chemicals that are added, advertently or inadvertently. The authors also discuss the natural poisons found in common foods, aflatoxins produced by fungi, marine toxins, and excessive amounts of vitamin D.

Snow, Keith R. *Insects and Disease.* (Illus.) Halsted/Wiley, 1974. x + 208pp. $9.50. 73-15433. ISBN 0-470-81017-3.

SH-C A brief account of medical entomology, useful in parasitology, entomology and general zoology; biology of the insects and biology of the pathogens. Contains truly noteworthy accounts of the biology of mosquitoes, flies, fleas, lice and bugs, and many good, clear diagrams.

West, Geoffrey P. *Rabies in Animals and Man.* (Illus.) Arco, 1973. 168pp. $6.95. ISBN 0-668-02875-0.

SH-C The history, epidemiology, and epizootiology of rabies are remarkably well covered, and the rabies virion, its microbiological properties, morphology, and the development of safe and effective vaccines are described.

614.7 ENVIRONMENTAL SANITATION AND COMFORT

Berland, Theodore. *The Fight for Quiet.* Prentice-Hall, 1970. 370pp. $8.50. 74-121724. ISBN 0-13-314591-3. Gloss.

SH-C This readable collection of fact and opinion for people who dislike noise includes descriptions of sound, noise and some effects of noise upon humans, sources of noise at home, at work, and in travel, together with suggestions for eliminating or minimizing noise, and a list of anti-noise organizations in the United States.

Epstein, Samuel S., and Marvin S. Legator. *The Mutagenicity of Pesticides: Concepts and Evaluation.* (Illus.) MIT Press, 1971. xvii + 220pp. $12.50. 70-130274. ISBN 0-262-05008-0.

C-P Based on the report of an HEW advisory panel, this monograph is both a guide to the literature and a compendium of information on some 390 widely used pesticides with respect to their mutagenic hazards. Concise and clearly written, the chapters cover the potential health hazards of mutagenic pesticides, methods for testing for mutagenic activity, general considerations of structure-activity relations, and recommendations.

Grey, Jerry. *Noise, Noise, Noise!* (Illus.) Westminster, 1976. 104pp. $6.50; $4.50(p). 75-23302. ISBN 0-664-32575-0; 0-664-34010-5. Gloss.;index;CIP;bib.

SH-C Grey begins with a definition of noise pollution and an explanation of how it is generated. He describes the common noise sources and the effect of noise on humans.

Lawrence, Thomas Gordon, et al. *Your Health and Safety in a Changing Environment.* (Illus.) Harcourt Brace Jovanovich, 1973. xii + 692pp. $6.54. ISBN 0-15-369525-0. Index;gloss.

JH-SH A carefully constructed text with examples, exercises and appropriate references for further study. The material is personalized and presented without condescension. Topics range from general concepts of health and environment through anatomy, physiology, genetics and social development to community concerns and safety. Factual statements deal with problems of current interest to both high school students and health professionals. The book includes excellent illustrations.

Lipscomb, David M. *Noise: The Unwanted Sounds.* (Illus.) Nelson-Hall, 1974. 331pp. $15.00. 73-89466. ISBN 0-911012-97-4. Index;CIP.

SH-C A complete survey of noise, its incidence and effects on humans. Based on many years of research, this book discusses temporary and permanent hearing loss and other physiological and psychological effects of noise. Also addresses issues in noise reduction.

Purdom, P. Walton, (Ed.). *Environmental Health.* (Illus.) Academic, 1971. xiv + 584pp. $19.50. 70-154403. ISBN 0-12-567850-9.

C Both a textbook and a handbook on pollution in the environment. Readable yet precise, topics covered are disease, food, water, air, solids, radiation, occupational environment, housing, accidents, and planning. History is reviewed and techniques are explained.

Taylor, Rupert. *Noise, 2nd ed.* (Illus.) Penguin, 1975. 274pp. $3.50(p). ISBN 0-1402-1233-7. Gloss.;index.

C Taylor discusses the generation of sound, the physics and physiology of its perception, the effects of noise on the hearer, and some methods of minimizing noise. An ingenious and effective presentation.

Waldbott, George L. *Health Effects of Environmental Pollutants.* (Illus.) Mosby, 1973. x + 316pp. $7.50(p). 72-11656. ISBN 0-8016-5330-4.

> **SH-C**   This well-organized and simplified review by a physician deals with the sources of pollutants, their measurement and toxic effects, and the physiological defense mechanisms of the body, and describes the effects of specific pollutants.

615 PHARMACOLOGY AND THERAPEUTICS

Arnow, L. Earle. *Health in a Bottle: Searching for the Drugs that Help.* (Illus.) Lippincott, 1970. 272pp. $5.95. 76-85423.

> **C**   A concise, vibrant and kaleidoscopic view of the mechanisms involved in the functioning of the ethical pharmaceutical industry. Neatly covers the considerations of management thinking in relation to the search for new therapeutic agents.

Girdano, Dorothy Dusek, and Daniel A. Girdano. *Drugs: A Factual Account.* (Illus.) Addison-Wesley, 1973. viii + 216pp. $3.95(p). 75-171431.

> **SH-C-P**   The authors—health educators—aim to counteract fear-provoking propaganda. The information is superficial, but ample references are included. Discusses the advertising and widespread abuse of over-the-counter drugs. Could satisfy a real need where Boards of Education require instruction on drugs but have not provided adequate instructional materials.

Marks, Geoffrey, and William K. Beatty. *The Medical Garden.* (Illus.) Scribner's, 1971. xii + 179pp. $6.95. 74-167777. ISBN 0-684-12383-5.

> **SH-C**   In a free-flowing style, the authors discuss opium, cocaine, quinine, aspirin, colchicine, digitalis, and penicillin. They cover origins, preparation, syntheses in the laboratory, medical usage, contraindications, and folklore. Pleasant and informative collateral reading.

Massett, Larry, and Earl W. Sutherland, III. *Everyman's Guide to Drugs and Medicines.* Luce (distr.: McKay), 1975. 262pp. $9.95. 74-7833. ISBN 0-88331-069-4. Indexes;CIP.

> **SH-C**   The authors discuss how the body works and how drugs act against diseases. There are discussions of generic versus brand-name drugs, side effects and drug interactions. A number of useful references on both price and drug evaluation are given.

Silverman, Milton, and Philip R. Lee. *Pills, Profits, and Politics.* Univ. of Calif. Press, 1974. xviii + 403pp. $10.95. 73-89166. ISBN 0-520-02616-0. Index.

> **SH-C-P**   The authors discuss modern drugs, their usefulness, their side effects, and their economic ramifications. They provide an outline of the problems of drug research and development, promotion, and factors involved in pricing. Written in nontechnical language.

615.7 PHARMACODYNAMICS

Aaronson, Bernard, and Humphry Osmond (Eds.). *Psychedelics: The Uses and Implications of Hallucinogenic Drugs.* (Illus.) Doubleday, 1970. 512pp. $2.45(p). 70-103788.

> **C-P**   An anthology of 32 articles which discuss drugs that alter sensory perception and the interpretation of stimuli from clinical, anthropological, religious, psychiatric and therapeutic points of view.

Edwards, Gabrielle I. *The Student Biologist Explores Drug Abuse.* (Illus.) Richards Rosen, 1975. xi + 112pp. $4.80. 73-93545. SBN 8239-0298-6. Gloss.

JH-SH   Topics include hallucinogens, stimulants, sedatives, narcotics, volatile substances, tobacco and alcohol. There is a simplified discussion of how drugs act and the harmful effects that may occur from their use.

Grinspoon, Lester. *Marihuana Reconsidered.* Harvard Univ. Press, 1971. ix + 443pp. $9.95. 75-150009. ISBN 674-54835-3.

SH-C-P   A comprehensive account of marijuana, its properties, dangers and utilities. The author, a clinical psychiatrist, has compiled information on the history, legalities, chemistry and pharmacology of the drug. Although somewhat simplistic, it is an excellent book.

Grinspoon, Lester, and Peter Hedblom. *The Speed Culture: Amphetamine Use and Abuse in America.* Harvard Univ. Press, 1975. viii + 340pp. $15.00. 74-27257. ISBN 0-674-83192-6. Index.

C   Actual cases, including commentary by drug users, give the reader a full understanding of amphetamine use. The authors describe amphetamine use in general and the history, pharmacology and effects of the drugs.

Hyde, Margaret O., (Ed.). *Mind Drugs, 3rd ed.* McGraw-Hill, 1974. 190pp. $4.95. 74-9668. ISBN 0-07-031633-3; 0-07-031634-1(p). Index;CIP.

SH-C-P   Third edition of an excellent book describing difficulties in treating new varieties of "street drugs," including bootleg hallucinogens and heroin. Recommended for both students and drug counselors.

Julien, Robert M. *A Primer of Drug Action.* (Illus.) Freeman, 1975. x + 290pp. $10.00; $4.95(p). 74-23271. ISBN 0-7167-0756-X; 0-7167-0755-1. Gloss.;index;CIP;bib.

SH-C-P   An introduction to drug action and interaction for those with little scientific background. For complete understanding, however, guidance in class or an elementary knowledge of organic chemistry and biological terminology would be beneficial. The emphasis is on the psychoactive drugs.

Kendall, Edward C. *Cortisone.* (Illus.) Scribner's, 1971. 175pp. $7.95. 72-123853.

JH-SH-C   A narrative by one of the co-winners of the 1950 Nobel Prize in medicine and physiology concerning his voyage of scientific discovery. Admirable collateral reading in chemistry, biochemistry, or history of science or for pleasure.

Klein, Aaron E. *Trial by Fury.* (Illus.) Scribner's, 1972. 175pp. $6.95. 72-1177. SBN 684-12997-3.

JH-SH-C   Traces the development of the Salk and Sabin polio vaccines from the first unsuccessful experiments to the final mass production of the vaccines and the ultimate apparent eradication of infantile paralysis. Provides an excellent study of the "ivory tower" of medical research, the chronic need for money; the petty jealousies between scientists; the unwritten code of ethics in the scientific community; and the influence on medical research of politics, the news media, and the emotional climate of the nation.

Lieberman, Mark. *The Dope Book: All About Drugs.* Praeger, 1971. 141pp. $5.95. 74-122090.

JH-SH-C   Provides information about physiology, pharmacology, and pathology in relation to abused drugs; a history of some of the mind-altering drugs; and the legal aspects of drug abuse. Without lecturing, the author presents excellent arguments against the use of everything from coffee to LSD.

Wells, Brian. *Psychedelic Drugs.* Penguin, 1974. 250pp. $1.95(p). ISBN 0-14-00-3800-0.

   C   Describes the major types of psychedelic drugs, their use and abuse, primary
   psychology and physiological effects. Cannabis, mescaline and LSD are covered, along with organic or genetic social and societal complications which may
result from sustained use.

615.8 PHYSICAL AND OTHER THERAPIES

Duke, Marc. *Acupuncture.* (Illus.) Pyramid, 1972. 223pp. $6.95. 70-189541. ISBN 0-515-09303-3.

   SH-GA-C-P   A readable and lucid introduction to Oriental medicine and the philosophy on which it is based. Outlines briefly the history of Chinese
medicine, the traditional system of anatomy and its roots in Taoist philosophy. While
intended for the general reader, it might also have value for the physician.

Hilgard, Ernest R., and Josephine R. Hilgard. *Hypnosis in the Relief of Pain.* (Illus.)
Kaufmann, 1975. ix + 262pp. $12.50. 75-19490. ISBN 0-913232-16-5. Index;CIP.

   C-P   The authors document the experimental and clinical research on hypnosis as
   it is used in the relief of pain and suffering. A summary of the literature is
included.

Kavaler, Lucy. *The Wonders of Cold: Cold Against Disease.* (Illus.) Day, 1971. 158pp.
$6.50. 69-10810.

   JH-SH-C   Describes the current use of cold as a surgical tool for removing dis-
   eased tissues or to cool the surgical patient as protection from the rigors
of surgery; to preserve blood, body tissues, and various organs for transplant; and as
treatment for fever, pain and various symptoms of disease. The dubious use of cold
to insure immortality (cryonics) is also discussed. Historical aspects and research are
described.

Nolen, William A. *Healing: A Doctor in Search of a Miracle.* Random House, 1974.
308pp. $8.95. 74-9084. ISBN 0-394-49095-9. CIP.

   JH-SH-C   Nolen examines two successful faith healers now working in the United
   States and some "psychic surgeons" in the Philippine Islands.

Pauling, Linus. *Vitamin C and the Common Cold.* Freeman, 1970. 122pp. $3.95.
76-140232. ISBN 0-7167-0159-6.

   SH-C   Pauling's thesis is that the recommended human daily allowance of vitamin
   C is perhaps 100 times less than optimum. Pauling proposes that 1 or 2
grams of vitamin C taken daily will reduce the incidence of the common cold and
have other beneficial effects such as improving mental ability. This controversial
thesis is supported through a critical review of published laboratory reports presented in a delightful, lucid style.

615.9 TOXICOLOGY

Arnold, Robert E. *What to Do About Bites and Stings of Venomous Animals.* (Illus.)
Macmillan, 1973. 122pp. $5.95. 72-77647.

   SH-C-P   Insects, arachnids, marine animals (fish, mollusk and jellyfish) and reptiles are covered. The discussions on first aid and definitive treatments
are excellent. There are three appendixes: snake descriptions (the major New World
vipers); antivenin sources (addresses); and the annual antivenin inventory report (the
location and amount of antivenin stored in the U.S.).

Brooks, Stewart M. *Ptomaine: The Story of Food Poisoning.* (Illus.) A.S. Barnes, 1974. 136pp. $6.95. 73-10512. ISBN 0-498-01355-3. Index;CIP.

JH-SH Presents well-documented case studies in food poisoning and how it occurs. A well-written and entertaining presentation of an important topic.

Bucherl, W., and Eleanor E. Buckley, (Eds.). *Venomous Animals and Their Venoms.* (Illus.) Academic, 1971. xxiv + 687pp. $35.00. 66-14892. Indexes.

C-P Volume 2 of a three-volume treatise on venomous animals is the work of 53 highly qualified scientists from 32 countries. It is brief on natural history, but thorough on collection, biochemistry, pharmacology, antigens, clinical symptoms, and therapy. Coverage is comprehensive and technical, supported by tables, charts, maps, line drawings, and halftones.

Schroeder, Henry A. *The Poisons Around Us: Toxic Metals in Food, Air, and Water.* Indiana Univ. Press, 1974. xii + 144pp. $6.95. 73-15283. ISBN 0-253-16675-6. Index;CIP;bib.

SH-C-P An excellent, nontechnical, easy-to-read, informative book which puts the "mercury-in-fish" scare and other similar problems in proper perspective. A useful reference.

616 Diseases

Berland, Theodore, and Mitchell A. Spellberg. *Living with Your Ulcer.* (Illus.) St. Martin's, 1971. ix + 178pp. $5.95. 77-145432.

C-P This guide for the ulcer patient covers etiology (with anatomical sketches), symptoms, and methods of diagnosis and treatment. The follow-up care of an ulcer after hospitalization, with an explanation of diet, is adequately outlined, as is the ulcer surgery required in complicated cases. A good book, medically sound, which discusses the patient as a whole, not just as a medical problem.

Ferguson, L. Kraeer, and John H. Kerr. *Explain It to Me, Doctor.* (Illus.) Lippincott, 1970. xv + 464pp. $10.00. 69-14856.

JH-SH-C Explains in nontechnical language the causes of disease as well as the reasons behind approved methods of treatment and what to expect from proper treatment. Covers most common adult diseases (except mental problems and common contagious diseases). Various doctors provide chapters on their special interests. Useful as a family reference.

Friedman, Arnold P., et al. *The Headache Book.* Dodd, Mead, 1973. 180pp. $5.95. 73-2131. ISBN 0-396-06806-5.

SH-C A clear, concise book for the lay reader, detailing the causes, symptoms, and treatment of each type of headache. The folkloric background concerning headaches is discussed briefly. Not a reference but an interesting reader.

Fuller, John G. *Fever! The Hunt for a New Killer Virus.* Reader's Digest Press (distr.: Dutton), 1974. 297pp. $8.95. 73-21818. ISBN 0-88349-012-9.

SH Fuller traces the appearance of the Lassa fever virus in a remote area of Nigeria, and takes the reader into hospitals and laboratories to give a vivid and readable account of the methods required to establish the causative agent of a disease as a virus. Sample handling, tissue culture, animal inoculation, virus titration and the use of passive immunization are described so that the nontechnical audience can appreciate the need for highly skilled and competent health personnel.

Garcia, Lynne Shore, and Lawrence R. Ash. *Diagnostic Parasitology: Clinical Labora-*

tory Manual. (Illus.) Mosby, 1975. viii + 112pp. $7.50(p). ISBN 0-8016-1740-5. Gloss.; index.

C-P   This well-organized, well-written manual covers techniques of collecting, handling and finding parasites. About one-third of the book is devoted to parasite identification, and chapters cover culture methods, animal inoculation, serology and other topics.

Hamburger, Jean. *The Power and the Frailty: The Future of Medicine and the Future of Man.* (Trans.) Macmillan, 1973. xv + 140pp. $4.95. 72-12750.

C-P   Discusses new methods of protecting people from disease and suggests that academicians, physicians and chemists work in freedom, but explore each other's disciplines and thus contribute more to general medical knowledge. A literary find.

Jayson, Malcolm I.V., and Allan St. J. Dixon. *Understanding Arthritis and Rheumatism: A Complete Guide to the Problems and Treatment.* (Illus.) Pantheon, 1974. xxi + 228pp. $7.95. 73-18724. ISBN 0-394-49177-7. Index;CIP.

GA   This is a fairly extensive, relatively simple and reasonably complete book on arthritis and rheumatism, particularly appropriate for public library collections.

Lance, James W. *Headache: Understanding, Alleviation.* (Illus.) Scribner's, 1975. 232pp. $8.95. 75-11809. ISBN 0-684-14372-0. Gloss.;index;CIP.

SH-C   There is a brief history of headaches through the ages and a discussion of various types of headaches. The mechanisms of headaches and their relationship to brain tumors, sexual intercourse and head injury are presented.

Lowell, Anthony M., Lydia B. Edwards, and Carroll E. Palmer. *Tuberculosis.* (Illus.) Harvard Univ. Press, 1969. xxvi + 230pp. $6.00. 79-82296. ISBN 0-674-91135-0.

C-P   Deals with morbidity, mortality and the control of tuberculosis in the whole population, gives a detailed study of tuberculosis among Navy recruits and a brief but well-documented history of the disease.

McQuade, Walter, and Ann Aikman. *Stress: What It Is, What It Can Do to Your Health, How to Fight Back.* (Illus.) Dutton, 1974. 243pp. $8.95. 73-79558. ISBN 0-525-21114-4. Index;CIP;bib.

SH-C   The authors discuss techniques for helping people to better tolerate stress—hygiene, diet, drugs, encounter groups, etc. The stress-connected depression of the immune system is viewed as an important factor in the origin of cancer.

Nourse, Alan E. *Viruses.* (Illus.) Watts, 1976. 65pp. $3.90. 75-19142. ISBN 0-531-00839-8. Index;CIP;bib.

JH   An informative delight for both youngsters and adults with some science background. Covers the 18th century to present-day knowledge and research. Discusses the discovery and study of viruses, viral diseases and their effect on humans, and the search for vaccines.

Selye, Hans. *The Stress of Life, rev. ed.* (Illus.) McGraw-Hill, 1976. xxvii + 515pp. $8.95. 75-12746. ISBN 0-07-056208-3. Gloss.;index;CIP.

SH-C   The puzzling, ambivalent nature of stress as an active force and, at the same time, as a biological state created by force is explained, as is the nonspecific character of stress. The specific quality of the triad of the general adaptation syndrome is not discussed.

616.079 IMMUNOLOGY

Barrett, James T. *Textbook of Immunology: An Introduction to Immunochemistry and Immunobiology, 2nd ed.* (Illus.) Mosby, 1974. xi + 417pp. $12.50. 73-16095. ISBN 0-8016-0502-4. Index;gloss.;CIP.

C-P   An encyclopedic approach. There are many references to historical development and applications to the detection and development of various diseases. May be used for a comprehensive course in immunology; will also serve as a fine reference book.

Burrell, Robert. *Experimental Immunology, 4th ed.* (Illus.) Burgess, 1974. vi + 98pp. $5.75(p). SBN 8087-0274-2.

SH-C   Designed as a manual for laboratory courses in immunology. Serological and chemical aspects are examined and experiments and demonstrations are clearly presented. An appendix lists the materials necessary, together with sources of supply.

Good, Robert A., and David W. Fisher. *Immunobiology: Current Knowledge of Basic Concepts in Immunology and Their Clinical Applications.* (Illus.) Sinauer, 1971. xii + 305pp. $10.95. 77-155366. ISBN 0-87893-201-1.

C-P   A very clear and comprehensive review. Both the normal and abnormal mechanisms of cellular and humoral immune systems and resulting disease states, with clinical applications of prophylaxis and therapy, are presented in moderate detail and in a simple manner that is aided by many clear illustrations and diagrams. There are very few books that present an entire discipline so beautifully and accurately.

Lewin, Roger. *In Defense of the Body: An Introduction to the New Immunology.* (Illus.) Anchor/Doubleday, 1974. 146pp. $2.50(p). 74-4900. ISBN 0-385-03790-X. Index.

SH-C   This thorough, enjoyable text presents recent developments in immunology. Topics range from the anatomy of the immune system and antibodies to genetic aspects and organ transplants. A valuable contribution to any course on human biology, particularly since it provides insights into the processes of scientific discovery.

Schmeck, Harold M., Jr. *Immunology: The Many-Edged Sword.* Braziller, 1974. 143pp. $6.95; $2.95(p). 73-80335. ISBN 0-8076-0711-8; 0-8076-0712-6.

GA-C   Schmeck has done a moderately good job of capturing some of the excitement of immunology in this collection of short essays. Recommended to the general reader or the nonspecialist undergraduate.

616.1 DISEASES OF THE CARDIOVASCULAR SYSTEM

Diethrich, Edward B., and John J. Fried. *Code Arrest: A Heart Stops: A Doctor Reports on the Battle Against the Heart Disease Epidemic.* Saturday Review Press/Dutton, 1974. 224pp. $6.95. 73-18156. ISBN 0-8415-0318-4. Gloss.;index;CIP.

Franklin, Marshal L., et al. *The Heart Doctors' Heart Book.* (Illus.) Grosset&Dunlap, 1974. xvi + 368pp. $8.95. 73-15131. ISBN 0-448-11524-7. Index.

SH-C   Both books describe, in readable fashion, what heart disease is, its contributing factors, how it affects the stricken patient, and the treatment. *Code Arrest* has the stronger emphasis on surgery, but without illustrations. Neither book is suited for school libraries, but Franklin's is recommended for patients and their families.

Finnerty, Frank A., Jr., and Shirley Motter Linde. *High Blood Pressure: What Causes It, How to Tell If You Have It, How to Control It for a Longer Life.* McKay/Pavillion, 1975. ix + 299pp. $9.95. 74-83690. ISBN 0-679-50512-1. Gloss.;index.

SH-GA-C   The authors discuss hypertension as it relates to the American public. The physiology is clearly presented, and diagnosis and treatment are emphasized. Recommended for anyone with hypertension.

Galton, Lawrence. *The Silent Disease: Hypertension.* (Illus.) Crown, 1973. xiii + 210pp. $5.95. 72-96654. ISBN 0-517-503573. Index;gloss.;bib.

GA   Galton provides information on the structure and function of the cardiovascular system and explains some of the causes of arterial hypertension. Methods of treatment are explained and future methods are predicted. Communicates important information in an understandable and accurate way.

Linde, Shirley Motter. *Sickle Cell: A Complete Guide to Prevention and Treatment.* (Illus.) Pavilion, 1972. 188pp. $2.00(p). Bib.

C-P   A very complete and readable presentation, both technically and scientifically accurate, particularly useful for medical and social public health workers. Describes symptoms, diagnostic tests, and steps the patient may take to avoid a crisis. The book is an excellent reference and could be used by the general practitioner. Contains an appendix of resource agencies.

Riedman, Sarah R. *Heart.* (Illus.) Golden/Western, 1974. 160pp. $1.95(p). 72-95511. Index.

JH-SH-GA   Riedman devotes the first half of her book to normal anatomy and functioning, and the latter half to disease detection, treatment, research and education. She explores the relationships between heart disease and diet, high blood pressure, smoking, etc.

Vinsant, Marielle O., et al. *A Commonsense Approach to Coronary Care: A Program.* (Illus.) Mosby, 1972. $5.95. 72-86519. ISBN 0-8016-5238-3.

C-P   Intended for nurses, this question-and-answer format concentrates on problems incident to acute myocardial infarction, rather than its psychological and rehabilitative aspects. Presumes a thorough knowledge of normal anatomy and physiology. Best used as a teaching and reference manual for those already familiar with the principles and problems involved in coronary care, and also useful in training ambulance technicians.

616.8 DISEASES OF THE NERVOUS SYSTEM

Ancowitz, Arthur. *Strokes and Their Prevention: How to Avoid High Blood Pressure and Hardening of the Arteries.* (Illus.) Van Nostrand Reinhold, 1975. 255pp. $7.95. 74-13360. ISBN 0-442-20330-6. Index;CIP.

SH-GA-C   Ancowitz explains stroke and cites the principles of prevention. He describes accurately and with considerable detail the predisposing circumstances, the warning events, the immediate sequellae, and the long-term prognosis for stroke victims.

Bruch, Hilde. *Eating Disorders: Obesity, Anorexia Nervosa, and the Person Within.* Basic Books, 1973. x + 396pp. $12.50. 72-89189. SBN 465-01784-3.

C-P   Treatment of an overpopularized subject in a calm, scientific, yet highly readable manner. Bruch covers the physio-psychological setting; the general pathology of weight, both developmentally (in childhood, adolescence, etc.) and as it occurs with other psychological illnesses, especially schizophrenia; *anorexia nervosa;* and finally, treatment for these conditions.

Cooper, I.S.   *The Victim Is Always the Same.* (Illus.) Harper&Row, 1973. xix + 160pp. 72-9111. ISBN 0-06-010856-8.

SH-C   By a world renowned neurosurgeon who pioneered a group of operations upon the human brain that relieved uncontrollable muscular movements of dystonia musculorum, this book is the skillfully developed story of two youngsters, their suffering and struggle to regain health. In addition, the book addresses profound questions of medical ethics and responsibility in experimental procedures involving human beings.

Silverstein, Alvin, and Virginia B. Silverstein.   *Epilepsy.* (Illus.) Lippincott, 1975. 64pp. $5.50; $1.95(p). 74-31382. ISBN 0-397-31615-1; 0-397-31624-0. Index;CIP.

JH-SH-GA   After giving a brief history of epilepsy, the authors detail its diagnosis, treatment and what happens in the body and brain during a seizure. Helpful to epileptic patients, their families and the general public.

Temkin, Owsei.   *The Falling Sickness: A History of Epilepsy from the Greeks to the Beginnings of Modern Neurology, 2nd ed.* (Illus.) Johns Hopkins Univ. Press, 1971. xv + 467pp. $15.00. 70-139522. ISBN 0-8018-1211-9.

C-P   Temkin applies the skills of physician, historian, and scientist in unravelling the long history of epilepsy. Of particular interest is the changing definition of "rational" treatment. The numerous literature citations from the Hippocratic era serve as a model for writing the history of a disease from superstition to modern outlooks.

616.89 PSYCHIATRY

Evans, Richard I.   *R.D. Laing: The Man and His Ideas.* Dutton, 1976. lxxv + 170pp. $12.95; $3.95(p). 75-14294. ISBN 0-525-18765-0; 0-525-47411-0. Index;CIP.

C-P   Written in dialogue form, with an introduction, this volume presents Laing's major ideas and personal style. Laing focused attention on the weaknesses of institutionalized psychiatry and was involved in the problem of classification, the nature of schizophrenia, the impact of Eastern thought, and innovative attempts at communal therapeutic arrangements.

Foudraine, Jan.   *Not Made of Wood: A Psychiatrist Discovers His Own Profession.* (Trans.) Macmillan, 1974. 414pp. $9.95. 73-6057. ISBN 0-02-540200-5. Index;CIP.

C   Foudraine attempts to strip psychiatry of its mysterious properties and expose psychiatrists' own problems in dealing with their humanity. He advocates interpersonal relationships with patients in order to allow for patients' emotional exposure to an understanding individual. Technical terms are translated into lay language.

Frank, Jerome D.   *Persuasion and Healing: A Comparative Study of Psychotherapy, rev. ed.* Johns Hopkins Univ. Press, 1973. xx + 378pp. $12.50. 72-4015. ISBN 0-8018-1443-X.

C-P   Maintains that the success of psychotherapy depends on the position of the psychotherapist as a socially sanctioned expert and healer. Examines several other areas of interpersonal contact which share the common features of being meaningful, hopeful, socially sanctioned and clearly defined; and describes experimental studies of persuasion and the placebo effect, the interactions between psychotherapist and patient, and the major forms of psychotherapy.

Freeman, Lucy, (Ed.).   *Karl Menninger, M.D.: Sparks.* Crowell, 1973. xiv + 290pp. $7.95. 73-4698. ISBN 0-690-75825-1.

SH-C   This compilation of unpublished writings of one of America's most noted psychiatrists introduces readers to his personality, thinking and professional work, as well as to the history of psychiatry, colleagues Freud and Schweitzer, other well-known psychiatrists and friends, the background history of the Menninger family and the community of Topeka, Kansas, the site of the famous family clinic.

Group for the Advancement of Psychiatry. *The VIP with Psychiatric Impairment.* Scribner's, 1973. 78pp. $4.95; $1.95(p). 76-37208. SBN 684-12778-4; 684-13259-1.

C-P   The role of the psychiatrist *vis-à-vis* the VIP is dealt with in the most general
          terms. There is brief treatment of problems of particular individuals (Gov. Long, Secretary Forrestal, certain members of the Judiciary), the grave problem of psychiatric determination of reliability of judgment and mental competence, and conclusions and recommendations. Unfortunately the treatment is very disappointing, but it is virtually the only responsible statement available on a problem of the greatest importance to all citizens in a democracy.

Havens, Leston L. *Approaches to the Mind: Movement of the Psychiatric Schools from Sects Toward Science.* Little, Brown, 1973. xiii + 385pp. $10.00. 73-1419. ISBN 0-316-35045.

C-P   Intimate sketches of some of the great contributors to human psychology in
          the last 150 years. Shows the development and the interrelationships between these individuals and the schools and approaches they founded, as well as the main psychiatric theories in use today.

Howells, John G., (Ed.). *Modern Perspectives in the Psychiatry of Old Age.* (Illus.) Brunner/Mazel, 1975. viii + 630pp. $20.00. 74-78715. SBN 87630-097-2. Indexes;bibs.

C-P   Sections on the theoretical foundations of the psychiatry of old age; histori-
          cal overview; treatment modalities, chiefly in the U.S., the United Kingdom, the U.S.S.R. and Japan; and training of psychogeriatricians.

Howells, John G., (Ed.). *World History of Psychiatry.* (Illus.) Brunner/Mazel, 1975. xxv + 770pp. $25.00. 73-81475. SBN 87630-082-4. Indexes.

SH-C-P   This reference book contains a chapter on every major western and
               eastern nation, as well as third world and underdeveloped nations. It reviews medical, scientific and humanistic developments within national ideologies and traditions. Recommended to students in all branches of the behavioral sciences.

Kaplan, Berton H., et al. (Eds.). *Further Explorations in Social Psychiatry.* (Illus.) Basic, 1976. xxi + 420pp. $20.00. 74-25914. ISBN 0-465-02589-7. Index;CIP;bibs.

C-P   These essays by psychiatrists, psychologists, sociologists and an-
          thropologists provide an excellent overview of the current status of social psychiatry. They range from elementary discussions to technical articles.

Kaplan, Helen Singer. *The New Sex Therapy: Active Treatment of Sexual Dysfunctions.* (Illus.) Brunner/Mazel, 1974. xvi + 544pp. $17.50. 73-87724. SBN 87630-083-2. Index;bibs.

C-P   Singer explains the psychodynamics of rapid treatment of sexual dysfunc-
          tions, offering an unbiased account of the concepts inherent in such treatment. Suitable as a text for the sex counselor or any professional dealing with human sexuality.

Lieberman, Morton A., et al. *Encounter Groups: First Facts.* Basic, 1973. x + 495pp. $15.00. 72-89174. SBN 465-01968-4.

C-P   A careful, comprehensive study of 10 approaches to group therapy. Profes-
          sional therapists led the groups of students, and a broad battery of assessment devices was employed prior to, immediately after and 6 months after the group meetings ended. This is an important book which presents hard data on the effectiveness of various procedures.

Seligman, Martin E.P. *Helplessness: On Depression, Development, and Death.* (Illus.) Freeman (distr.: Scribner's), 1975. xv + 250pp. $8.95. 74-23125. ISBN 0-7167-0752-7; 0-7167-0751-9(p). Indexes;CIP.

C   An investigation into the concept of helplessness and its resulting anxiety and depression. The author suggests that these responses are learned and can cause long-term character changes. The ideas presented have provocative implications about society and the individual.

Shepard, Martin. *Fritz.* Saturday Review Press/Dutton, 1975. xvi + 235pp. $9.50; $3.95(p). 74-16380. ISBN 0-8415-0354-0; 0-8415-0358-3. Index;CIP.

C-P   This is a splendid biography of Frederick Perls, the founder of Gestalt psychotherapy. Details interviews with colleagues, family and lovers. Workers in psychotherapy and counseling will be most interested, but the book can be appreciated by anyone concerned with the human condition.

Sugar, Max, (Ed.). *The Adolescent in Group and Family Therapy.* Brunner/Mazel, 1975. xvii + 286pp. $13.50. 73-92809. ISBN 87630-092-1. Index.

C-P   The collection offers a wide variety of situations in which group and family therapy can be effective in helping the adolescent with certain phase-specific and situational problems. Useful to both professionals and paraprofessionals.

Szasz, Thomas S. *The Manufacture of Madness: A Comparative Study of the Inquisition and the Mental Health Movement.* Harper&Row, 1970. xxvii + 383pp. $8.95. 71-83626.

C-P   This interesting and difficult book attempts "to display the forms in which the perennial scapegoat principle manifests itself in the modern world," and compares witches and their persecution by priests to contemporary ideas about madmen and their "persecution by physicians." The arguments presented are not simplistic and challenge the standard interpretations of the history of psychiatry.

Tennov, Dorothy. *Psychotherapy: The Hazardous Cure.* Abelard-Schuman, 1975. xviii + 314pp. $8.95. 75-9856. ISBN 0-200-04028-6. Index;CIP.

C-P   Tennov, arguing that psychotherapy is hazardous to one's health, criticizes psychotherapists for their narrow-minded dependence on and adherence to Freudian concepts of mental illness. Unfortunately, the author touches only briefly on alternatives for treatment of mental illness.

616.994 CANCERS

Bouchard, Rosemary, and Norma F. Owens. *Nursing Care of the Cancer Patient, 2nd ed.* (Illus.) Mosby, 1972. xiii + 290pp. $12.50. 72-79787. ISBN 0-8016-0715-9.

C-P   The relevant pathophysiology is simply stated, although a beginning background in anatomy, physiology and basic nursing techniques is assumed. Reasonable suggestions for the psychological support of the patient are made. A straightforward presentation of treatment and prognosis.

Goodfield, June. *The Siege of Cancer.* Random House, 1975. xi + 240pp. $8.95. 75-10298. ISBN 0-394-49119-X. Gloss.;CIP.

SH-C   Provides a basic foundation in the problems associated with modern cancer research, considering cancer from a biological perspective and as a complex genetic problem which results from the dysfunction of any of a number of different processes which control cell growth.

Harris, R.J.C., (Ed.). *What We Know About Cancer.* (Illus.) St. Martin's, 1972. 240pp. $6.95. 70-188547. Gloss.

C-P   A collection of reviews to promote education about cancer and its prevention, treatment and aftercare. Examples are largely from Great Britain. A lucid, coordinated summary of both the biological and the psychosocial aspects relating to the patient and his or her family.

Johnson, F. Leonard, and Marc Miller.   *Shannon: A Book for Parents of Children with Leukemia.* (Illus.) Hawthorn, 1975. xii + 132pp. $6.95. 74-18689. ISBN 0-8015-6776-9. Index.

SH-C   The touching story of a child with leukemia, told in alternating chapters by her physician and grandmother. Both the personal and medical aspects are presented in the diagnosis, therapy and adjustment to the disease. Appropriate for families or even an older child with leukemia and recommended for public libraries.

Maugh, Thomas H., II, and Jean L. Marx.   *Seeds of Destruction: The* Science *Report on Cancer Research.* (Illus.) Plenum, 1975. xiv + 251pp. $17.95. 75-15860. ISBN 0-306-30836-3. Gloss.;index;CIP.

GA-C-P   One of the most informative, authoritative, well-balanced and readable books on cancer research, covering cancer etiology, biochemistry, and therapy, and specific cancers. Blends scientific fact, historical background, current findings and lucid, unassuming interpretation. For the general reader as well as a useful reference source for cancer researchers.

Pilgrim, Ira.   *The Topic of Cancer.* Crowell, 1974. xii + 255pp. $6.95. 74-7500. ISBN 0-690-00516-4. Gloss.;index;CIP.

SH-GA-C   The author, an active participant in experimental cancer research, has provided a frank discussion of the politics, accomplishments and unsolved problems of cancer research. His sensitive, humanistic survey of current diagnosis and treatment should be of interest to patients and families.

Rosenbaum, Ernest H.   *Living with Cancer.* Praeger, 1975. ix + 214pp. $10.00. 75-4342. ISBN 0-275-22600-X. Gloss.

SH-C-P   Rosenbaum presents the reactions of 11 terminally ill patients. Sources of outside aid for patients are indicated.

Silverstein, Alvin, and Virginia Silverstein.   *Cancer.* (Illus.) John Day, 1972. 96pp. $4.89. 74-155010. Index;gloss.

JH   An explanation of cancer: what it is, how the various kinds are treated, what natural resources the body brings to bear, and what research is doing to help combat it. An excellent section on "what you can do about cancer" provides suggestions for children whose parents smoke. Valuable for biology and health classes and as a reference.

617 SURGERY, DENTISTRY, AUDIOLOGY, TRANSPLANTS

Deaton, John G.   *New Parts for Old: The Age of Organ Transplants.* (Illus.) Franklin, 1974. 160pp. $7.40. Index.

JH-SH   Deaton provides an informative survey of various aspects of organ transplantation, including sources of donor organs, the body's immunity to and rejection of transplanted organs, alternatives to transplantation, and speculation on the future of transplants.

Galton, Lawrence.   *The Laboratory of the Body: Dental Research, Science's Newest Frontier.* (Illus.) Pyramid, 1972. 226pp. $6.94; $0.95(p). 73-189762.

GA-C   Describes many intriguing areas of oral research and links observable problems in the oral cavity to those which may be hidden in the rest of the body. Topics range from the formation of collagen to dental plaque; from dental decay to cellular transport mechanisms in saliva formation; and facial growth in relation to clefts and orthodontics. Well documented, but may be heavy going.

Gaylin, Willard M., et al. *Operating on the Mind: The Psychosurgery Conflict.* Basic, 1976. viii + 216pp. $11.95. 74-79276. ISBN 0-465-05288-6. Index;CIP.

C-P   The contributors represent medicine, law, philosophy and sociology. Each author illuminates and clarifies the many complex issues involved in the use of this controversial behavioral modification technique. They conclude that regulations should be drafted which specify and restrict the circumstances and conditions under which psychosurgery may be performed.

Lantner, Minna, and Gerald Bender. *Understanding Dentistry.* (Illus.) Beacon, 1969. viii + 214pp. $6.00. 70-84802.

JH-SH   This book, presented with the approval of the Ethics Committee of the American Dental Association, contains clear, useful, and accurate information on dental care, dentists, and the practice of dentistry in the United States.

Mowbray, A.Q. *The Transplant.* McKay, 1974. 266pp. $8.95. 74-82987. ISBN 0-679-50522-9.

SH-C   The easily readable, journalistic style gives the reader an eyewitness account of a kidney transplant performed on a 17-year-old girl. Explores psychic and social problems facing the kidney victim's family, and also traces historical development and current state-of-the-art of renal transplants.

Richardson, Robert G. *The Scalpel and the Heart.* (Illus.) Scribner's, 1970. vii-+ 323pp. $8.95. 77-106553.

SH-C   Accurate, enjoyable and useful. Deals with the aspects of heart surgery and organized on the basis of a broad category of those heart diseases amenable to surgery. Progression is chronological.

Roberts, Elizabeth H. *On Your Feet.* (Illus.) Rodale, 1975. 192pp. $8.95. 75-6712. ISBN 0-87857-099-3. Gloss.;index;CIP.

SH-C-P   A guide to footcare from before birth into old age. The development and anatomy of feet are well described, function and movement are explained, and a description of the pathology of feet is provided.

Rosenthal, Richard. *The Hearing Loss Handbook.* (Illus.) St. Martin's, 1975. xii + 222pp. $8.95. 75-9496. Index;CIP.

SH-GA-C   Rosenthal justifiably criticizes some methods of medical examination and treatment of patients as well as some clinical audiological procedures and recommendations. He is quite harsh with hearing-aid dealers and companies, and describes methods for achieving better success with amplification devices and habilitative or rehabilitative processes. Useful for anyone with a hearing loss.

Thompson, Thomas. *Hearts: Of Surgeons and Transplants, Miracles and Disasters Along the Cardiac Frontier.* (Illus.) McCall, 1971. xii + 304pp. $7.95. 70-154248. ISBN 0-8415-0123-8.

C   A first-person account of desperately ill patients racing cross-country to acquire new hearts, of desperate efforts of doctors to secure the survival of their patients, and of the sudden realization that death was no longer easily defined. All this serves as a background to the bitter competition between two superb surgeons—Michael Ellis DeBakey and Denton Arthur Cooley.

Woodforde, John. *The Strange Story of False Teeth.* (Illus.) Universe Books, 1970. 137pp. $4.95. 73-97597. ISBN 0-87663-118-9. Index.

JH-SH   A lively historical and descriptive account of dentures, denture wearers, and the materials and methods of their fabrication. Carefully researched, with nearly a hundred well-chosen illustrations.

618 PEDIATRICS

Boston Children's Medical Center and Richard I. Feinbloom. *Child Health Encyclopedia: The Complete Guide for Parents.* Delacorte/Lawrence, 1975. xxvii + 561pp. $15.00. 75-5773. ISBN 0-440-01385-2. CIP.

GA   A wealth of useful information on the rearing of children, health maintenance and symptoms, illnesses and hazards commonly encountered. A sound investment for any family.

Kappelman, Murray M. *What Your Child Is All About.* Reader's Digest Press, 1974. ix + 223pp. $7.95. 74-8945. ISBN 0-88349-043-9. Index;CIP.

SH-GA-C   A thoughtful pediatrician has written a book full of useful information, observations and advice. It covers such areas as normal and problem children, divorce, ear problems, bed wetting and tonsillectomy. Useful reference not only for parents but also for high school and college students studying child development or family related subjects.

Shneour, Elie A. *The Malnourished Mind.* Anchor/Doubleday, 1974. vxiii + 196pp. $6.95. 73-9175. ISBN 0-385-03909-3. Index.

SH-C   Shneour writes about the nutrition of the unborn, discussing hunger, malnutrition and starvation. He also outlines some general food and nutritional principles and infant feeding practices. His thesis is that early malnutrition is probably the most important element in shaping human cognitive potential.

Zimmerman, David R. *Rh: The Intimate History of a Disease and Its Conquest.* (Illus.) Macmillan, 1973. xviii + 371pp. $8.95. 72-90280. Index;gloss.

C   This book documents, through interviews with principals and references to primary publications, the events leading to the conquering of erythroblastosis fetalis, a disease which jeopardizes the life of Rh-positive babies before delivery from Rh-negative mothers. Demonstrates that major scientific advances often represent the synthesis of many small, trivial or abstruse discoveries and shows the effects of the pressures of competitive research on scientists.

619 EXPERIMENTAL MEDICINE

Beveridge, W.I.B. *Frontiers in Comparative Medicine.* Univ. of Minn. Press, 1972. 104pp. $4.75. 72-79500. ISBN 0-8166-0643-9.

C-P   Beveridge views comparative medicine and discusses the contributions to human medicine resulting from studies of disease in animals. Describes current comparative studies in cancer, cardiovascular diseases, neuropathology, immunopathology, congenital defects, mental health, reproduction, population regulation, and the possible origin of human influenza pandemics. Altogether fascinating!

620 Engineering

Angrist, Stanley W. *Closing the Loop: The Story of Feedback.* (Illus.) Crowell, 1973. 65pp. $4.95. 73-3. ISBN 0-690-19644-X.

JH-SH-GA   This is an introduction to the concept and essential elements of feedback. The emphasis is on negative feedback, but stability and equilibrium are carefully explained. The evolution of applications (including biofeedback) is well traced.

Daitch, Paul B. *Introduction to College Engineering.* (Illus.) Addison-Wesley, 1973. ix + 288pp. $7.50(p). 72-2651.

C   This book links engineering design concepts and related calculations to the broader nontechnical aspects of contemporary problems. Applications include use of simultaneous linear equations, analysis and design of static structures, electrical circuits, feedback control, thermal energy, and heat conduction. Includes FORTRAN matrix programs.

Hill, Robert W. *The Chesapeake Bay Bridge-Tunnel: The Eighth Wonder of the World.* (Illus.) John Day, 1972. 47pp. $4.89. 75-155013.

JH-SH-GA   An excellent and thorough description of the process by which suspension bridges are built. Includes maps and a brief description of the historical necessity for the structure. Illustrated thoroughly.

Hilson, Barry. *Basic Structural Behaviour via Models.* (Illus.) Halsted/Wiley, 1973. xiii + 113pp. $8.50. 72-11072. ISBN 0-470-39966-X.

SH-C   A systematic presentation of the behavior of structures, using simple models of balsam wood and cartridge paper. The author states the problem, describes the construction of the study model and test procedure, evaluates the results, and offers thought-provoking questions to develop an intuitive feeling for how a given form responds to an applied load.

Munch, Theodore W. *Man the Engineer—Nature's Copycat.* (Illus.) Westminster, 1974. 125pp. $6.50; $4.50(p). 74-8823. ISBN 0-664-32555-6; 0-664-34007-5. Index;CIP.

JH-SH   This is a clear, lucid but brief description of many similarities between animal and human engineering.

Popov, E.P., et al. *Mechanics of Materials, 2nd ed.* (Illus.) Prentice-Hall, 1976. xiii + 590pp. $18.95. 75-34466. ISBN 0-13-5713560-0. Index;CIP.

C-P   For undergraduate engineers in the first mechanics/strength of materials course. Examples are given in both English and SI units. A very thorough text.

Sullivan, George. *How Do They Find It?* (Illus.) Westminster, 1975. 160pp. $6.50. 74-26713.

JH-SH   The volume includes well-organized information on all kinds of searches: for water, for metal in the earth and underwater, for ships and fishes, for predicting weather changes and even for finding airport runways in storms and at night. A good bibliography for in-depth reading is provided.

Young, Donald F. *Introduction to Applied Mechanics: An Integrated Treatment for Students in Engineering, Life Science, and Interdisciplinary Programs.* (Illus.) Iowa State Univ. Press, 1972. viii + 269pp. $8.50. 74-153162. ISBN 0-8138-1075-2.

C   A simple, concise treatment in which the coverage of dimensional analysis is especially good. Continuum mechanics, vibrations, materials, and fluid mechanics are covered in standard fashion, but with appropriate examples.

620.8 HUMAN FACTORS ENGINEERING

Chapanis, Alphonse, (Ed.). *Ethnic Variables in Human Factors Engineering.* (Illus.) Johns Hopkins Univ. Press, 1975. xvii + 290pp. $17.50. 74-24393. ISBN 0-8018-1668-8. Index.

C-P   These papers highlight the problems created by the increased mobility of the world's inhabitants and their differences in physical size, language, and other cultural variables. For engineering and for collateral reading in anatomy, biomechanics, cultural anthropology, etc.

Diffrient, Niels, et al. *Humanscale 1/2/3: A Portfolio of Information: Manual and 3 Pictorial Selectors.* (Illus.) MIT Press, 1974. Portfolio and Manual (32pp.). $25.00. 74-5041. ISBN 0-262-54027-4.

SH-C-P   Brings together information on the sizes of people, seating considerations and requirements for the handicapped and elderly. Although chiefly for architects, engineers, designers, and planners, this is an excellent basic reference on human variations.

621.3 ELECTRICAL, ELECTRONIC AND ELECTROMAGNETIC ENGINEERING

Beesley, M.J. *Lasers and Their Applications.* (Illus.) Taylor&Francis (distr.: Barnes &Noble), 1971. xii+234pp. $18.00. ISBN 0-389-04171-8.

C-P   Aimed at prospective users, this text explains the theory of emission and absorption of photons, laser theory and the operation of gas and solid-state lasers, scientific and technological applications, and holography.

Clifford, Martin. *Basic Electricity and Beginning Electronics.* (Illus.) Tab, 1973. 256pp. $7.95; $4.95(p.). 72-94809. ISBN 0-8306-3628-5; 0-8306-2628-X.

JH-SH-GA   The main emphasis is on DC circuits from a component-function standpoint, although AC circuits are also covered. Clifford discusses vacuum tubes and transistors; the only math used is simple algebra. Suitable as a reference and for hobbyists.

MacKay, R. Stuart. *Bio-Medical Telemetry: Sensing and Transmitting Biological Information from Animals and Man, 2nd ed.* (Illus.) Wiley, 1970. xiv+533pp. $14.95. 74-121909. ISBN 0-471-56030-8.

C-P   This well-established textbook covers a unique topic of importance in biologic investigation and experimental medicine. The major emphasis is on instrumentation and the technical aspects of telemetry. Many applications are covered.

Marshall, J.L. *Lightning Protection.* (Illus.) Wiley, 1973. xiii+190pp. $14.95. 73-4415. ISBN 0-471-57305-1.

SH-C-P   Discusses the physics of lightning, its distribution throughout the world, the ability of various soils to ground lightning and methods of testing a specific area for resistivity. About 60 percent of the book gives diagrams and specific data on how to ground almost anything.

Popovic, Branko D. *Introductory Engineering Electromagnetics.* (Illus.) Addison-Wesley, 1971. xiv+634pp. $15.95. 70-111954.

C-P   A very good textbook, useful in physics as well as in electrical engineering, and easily understood by anyone with a background in calculus, differential equations and vector analysis. A brief view of vectors is followed by a study of static electric fields. Applies energy methods and approximate solutions to realistic situations.

621.38 ELECTRONIC AND COMMUNICATION ENGINEERING

Burroughs, Lou. *Microphones: Design and Application.* (Illus.) Sagamore, 1974. 260pp. $20.00. 73-87056. ISBN 0-914130-00-5. Index.

SH-C-P   This useful guide provides explanations of microphone spatial and frequency response, construction techniques, effects of loading levels, temperature, phasing, interference, and many practical applications. For those more interested in sound than physics.

Gregor, Arthur. *Bell Laboratories: Inside the World's Largest Communications Center.* (Illus.) Scribner's, 1972. 125pp. $6.95. 70-37217. ISBN 0-684-12773-3. Index.

    JH An interesting account of the achievements of the Bell Laboratories and their predecessors. Researchers at these laboratories are shown to have achieved continuous advances in the areas of instrumentation, transmission, switching systems, etc., particularly those based on solid state physics.

Holz, Robert K., (Ed.). *The Surveillant Science: Remote Sensing of the Environment.* (Illus.) Houghton Mifflin, 1973. xii + 390pp. $7.95(p). 72-7922. ISBN 0-395-14041-2. Bibs.

    C A concise collection of key papers provides a general review. Papers describe basic characteristics of the electromagnetic spectrum, general methods of remote sensing, specific techniques, and the social implications of remote sensing. Does not cover ERTS or Skylab.

Klass, Phillip J. *Secret Sentries in Space.* (Illus.) Random House, 1971. xvi + 236pp. $7.95. 77-143994. ISBN 0-394-46972-0.

    SH A highly readable account of the early development of satellite technology and reconnaissance satellite techniques in the context of the world situation. The author argues that policy-makers in both the United States and the USSR realize that these satellites are a major stabilizing influence and reduce the possibility of war.

Kohn, Bernice. *Communications Satellites: Message Centers in Space.* Four Winds, 1975. 58pp. $5.95. 74-26872. ISBN 0-590-07356-7. Gloss.;index;CIP.

    JH-SH This excellent, information-packed introduction to satellites clearly and concisely covers the history of rocketry, principles of propulsion, early popular ideas of space travel, various communications satellites and future applications.

Martin, James. *Future Developments in Telecommunications.* (Illus.) Prentice-Hall, 1971. xiii + 412pp. $14.00. 74-156757. ISBN 0-13-345868-7. Gloss.

    SH-C A review of the state-of-the-art of telecommunications which describes 13 inventions that have the potential to affect both the technical field and society. The book is relatively jargon-free.

Paul, Günter. *The Satellite Spin-Off: The Achievements of Space Flight.* (Illus.;trans.) Luce (distr.: McKay), 1975. 272pp. $10.00. 75-11369. ISBN 0-88331-076-7. Index;CIP.

    SH-C The history, development and applications of satellites for communications and earth observation are very well covered in this well-researched and very readable book.

621.381 ELECTRONIC ENGINEERING

Alley, Charles L., and Kenneth W. Atwood. *Semiconductor Devices and Circuits.* (Illus.) Wiley, 1971. vii + 490pp. $11.95. 70-136708. ISBN 0-471-02330-2.

    C-P The operation and design considerations for many of the basic circuits in solid-state radio, hi-fi, stereo, television, radar, sonar and computers are described in this design-oriented, semiconductor electronics textbook. Suitable for technical institutes, junior colleges and self-study.

Cooper, William David. *Electronic Instrumentation and Measurement Techniques.* (Illus.) Prentice-Hall, 1970. xiii + 495pp. $12.95. 75-91197. ISBN 0-13-251686-1.

    C Thorough coverage, including AC and DC instruments, potentiometers, cathode ray oscilloscopes, waveform generators and analyzers, and good explanations of statistics involved in measurements, the binary system, and analog to digital converters.

Crowhurst, Norman H. *Basic Electronics Course.* (Illus.) Tab, 1972. 368pp. $8.95; $5.95(p). 75-178692. ISBN 0-8306-2588-7.

SH-C   The most outstanding feature is the abundance of unusual and excellent diagrams and review and examination questions. Although the approach is "basic," it covers a great deal of material.

EEE Magazine, Editors of. *Electronic Circuit Design Handbook, 4th ed.* (Illus.) Tab, 1971. 410pp. $17.95. 65-24823.

C-P   This very useful handbook contains 639 individual circuit designs from selected circuits originally published in *EEE Magazine.* They are divided by function into 19 sections which thoroughly cover amplification, oscillation, switching, and miscellaneous circuit actions.

Hallmark, Clayton. *Electronic Measurements Simplified.* (Illus.) Tab, 1974. 240pp. $7.95; $4.95(p). 73-879585. ISBN 0-8306-3702-8; 0-8306-2702-2.

SH-C   Hallmark describes electronic measurements—including electronic testing and servicing techniques—in simple terminology. This is a technical reference book which would be valuable for self-study or vocational training programs.

Kyle, James. *Electronics Unraveled: A New Commonsense Approach.* (Illus.) Tab, 1974. 229pp. $7.95; $4.95(p). 74-79583. ISBN 0-8306-4691-4; 0-8306-3691-9. Index.

JH-SH-C   A good introduction to the basic concepts of amplification and feedback and switching applications. The book concludes with six fairly simple projects using amplifiers, electronic meters and a one-transistor regenerative radio set.

Mann, Charles K., et al. *Basic Concepts in Electronic Instrumentation.* (Illus.) Harper &Row, 1974. vi + 249pp. $6.95(p). 73-16539. ISBN 0-06-042536-9. Index;CIP;bib.

C-P   The nonspecialist is provided with understandable explanations of the basic concepts underlying the design and use of electronic instruments. After a brief introduction to electrical signals and electronic systems, the authors address such topics as frequency response, filter networks and impedance matching. Problems are included.

Noll, Edward M. *Science Projects in Electronics, 2nd ed.* (Illus.) Sams, 1971. 144pp. $3.95. 74-157800. ISBN 0-672-20846-6.

JH-SH   The fundamentals of electronics technology can be learned by doing this series of laboratory experiments which starts with simple concepts of direct and alternating currents and progresses to more complicated, complete circuits involving semiconductors.

Pike, Arthur L. *Fundamentals of Electronic Circuits.* (Illus.) Prentice-Hall, 1971. xviii + 700pp. $15.50. 79-137986. ISBN 0-13-337816-0.

C   Contains numerous examples accompanied by solutions, each setting forth an application of concepts already developed. In many cases new concepts are introduced as natural by-products. The author assumes an elementary knowledge of electronics and mathematics and provides a good review in the first chapter.

Prensky, Sol D. *Electronic Instrumentation, 2nd ed.* (Illus.) Prentice-Hall, 1971. xx + 536pp. $16.95. 77-151899. ISBN 0-13-251645-4.

C   A variety of electronic instrumentation is analyzed in sufficient depth to provide the reader with an adequate understanding of the capabilities, limitations, and performance of many types of these instruments. A knowledge of basic electronics is required.

Prensky, Sol D. *Manual of Linear Integrated Circuits: Operational Amplifiers and Analog ICs.* (Illus.) Reston/Prentice-Hall, 1974. xiv + 289pp. $16.95. 73-15979. ISBN 0-87909-466-4. Index;CIP.

C-P   A technical reference manual designed for electronics technicians, engineers and hobbyists to supplement courses in basic transistor circuits. Supplies device information, tables for performance comparison, representative examples for each basic type, and specific applications.

Ristenbatt, Marlin P. *Semiconductor Circuits: Linear and Digital.* (Illus.) Prentice-Hall, 1975. xviii + 391pp. $16.95. 74-19006. ISBN 0-13-806158-0. Index;CIP.

C-P   The book is a self-contained exposition of basic electronic devices and their circuit properties. Four appendixes provide a brief review of basic electrical and mathematical concepts.

Sanborn, Paul. *Fundamentals of Transistors.* (Illus.) Doubleday, 1970. xiv + 582pp. $8.95. 67-12867. Gloss.

C-P   A programmed book which covers semiconductor theory and transistors and the basic circuits in which they are utilized. The text thoroughly discusses the physics and electrical phenomena associated with semiconducting materials such as current carriers, PN junctions, rectification, and amplification.

Sessions, Ken W. *Practical Solid-State Principles and Projects.* (Illus.) Tab, 1972. 176pp. $6.95; $3.95(p). 74-187559. ISBN 0-8306-2590-9.

SH-C   Electronics technicians, radio hams, and experimenters who wish to construct and understand practical circuits using semiconductor devices will find this book useful. A competent work also useful for the hobbyist or student.

Smith, Ralph J. *Electronics: Circuits and Devices.* (Illus.) Wiley, 1973. ix + 459pp. $13.95. 72-12833. ISBN 0-471-80181-X. Index;CIP;bib.

C-P   A fine reference book, also useful as a text for a 1-term introductory course in electronics. Strongly mathematical; freshman calculus is a prerequisite. Review questions, exercises, and problems are provided.

Tuite, Don. *Electronic Experimenter's Guidebook.* (Illus.) Tab, 1974. 182pp. $7.95; $4.95(p). 74-14322. ISBN 0-8306-4540-3; 0-8306-3540-8. Index.

SH   Practical tips on the mechanical layout and construction of electronic circuits occupy nearly half of this manual. The remainder is devoted to systematic troubleshooting of electronics equipment and guidance on component substitution in case preferred components are unavailable.

Tuite, Don. *Practical Circuit Design for the Experimenter.* (Illus.) Tab, 1974. 191pp. $8.95; $4.95(p). 74-81731. ISBN 0-8306-4726-0; 0-8306-3726-5. Index.

C   This readable handbook assists in design computations for various integrated circuits. The reader should be thoroughly proficient in high school algebra.

Ward, Brice. *Electronic Music Circuit Guidebook.* (Illus.) Tab, 1975. 223pp. $9.95. 75-20843. ISBN 0-8306-5743-6; 0-8306-4743-0(p). Index.

C   Ward has collected much information that will be helpful to a reader familiar with electronics. He clearly discusses the various characteristics of the sound from musical instruments which electronic circuits must reproduce, and explains the musical scale. Many useful circuits are completely analyzed.

Wheeler, Gershon J., and Arley L. Tripp. *Essentials of Electronics.* (Illus.) Prentice-Hall, 1971. x + 436pp. $12.95. 73-143812. ISBN 0-13-286039-2.

C   Contains a wide variety of problems and review questions. A chapter on flow graphs is included, which the authors feel are easier to interpret than mathematics. A suitable reference for libraries serving community and junior colleges.

621.38195 COMPUTERS

See also 001.64 Electronic Digital Computers.

Boyce, Jefferson C. *Digital Logic and Switching Circuits: Operation and Analysis.* (Illus.) Prentice-Hall, 1975. xii + 526pp. $14.95. 74-16763. ISBN 0-13-214478-6. Index.

C-P   Aimed at technical students with some electronic hardware background, this is a good theoretical and practical introduction. After a survey of digital logic, the author discusses simplification methods for logical circuits and troubleshooting of combinatorial logic. Applications to counters, registers, code converters and digital clocks are included.

Finkel, Jules. *Computer-Aided Experimentation: Interfacing to Minicomputers.* (Illus.) Wiley, 1975. xviii + 422pp. $21.95. 74-22060. ISBN 0-471-25884-9. Index;CIP.

SH-C-P   This textbook's emphasis is on concepts and techniques. Topics include analog to digital conversion, interface logic design, computer digital inputs, digital output, peripheral devices and minicomputer architecture, operation and programming.

Freedman, M. David. *Principles of Digital Computer Operation.* (Illus.) Wiley, 1972. x + 233pp. $8.95. 71-37953. ISBN 0-471-27731-2.

C   This introductory text is well written, contains good illustrations and effective examples. Answers the question, How do computers work? It is independent of any particular computer system and programming language and therefore rather abstract for most beginning computer science courses.

Goldstine, Herman H. *The Computer from Pascal to von Neumann.* (Illus.) Princeton Univ. Press, 1972. x + 378pp. $12.50. 70-173755. ISBN 0-691-08104-2.

SH-C-P   An excellent, comprehensive history of the people who contributed to the conception and development of digital computers during the period 1623–1957 and of analog computers through the Bush Differential Analyzer. Shows the genesis of many significant design decisions and the reasons behind them.

Halacy, Dan. *Charles Babbage: Father of the Computer.* (Illus.) Crowell-Collier, 1970. 170pp. $4.95. 79-119618.

JH-SH   Surveys the many accomplishments (difference engine, signaling devices, construction of mathematical and actuary tables) of Charles Babbage (1792–1871) and leaves the impression that Babbage would have been successful in constructing his analytical engine had he lowered his goals and compromised his ideals. A good basis for discussion of such actions and concomitant earlier availability of the computer in society.

Ilardi, Frank A. *Computer Circuit Analysis: Theory and Applications.* (Illus.) Prentice-Hall, 1976. x + 406pp. $17.50. 75-26529. ISBN 0-13-165357-1. Index;CIP.

SH-C   Ilardi lucidly introduces transistor-transistor logic (TTL) circuits to technical school students. Knowledge of basic electrical circuits is necessary, and basic algebra is helpful. Many examples.

Martin, James. *Telecommunications and the Computer, 2nd ed.* (Illus.) Prentice-Hall, 1976. xviii + 670pp. $28.00. 75-37800. ISBN 0-13-902494-8. Index;gloss.;CIP.

C-P   Begins with broad discussions of the future, organizations involved, terminals and codes, and person-computer dialogue. Contains detailed discussions of transmissions, switching and imperfections in data transmission systems. An excellent reference text.

Nagle, H. Troy, Jr., et al. *An Introduction to Computer Logic.* (Illus.) Prentice-Hall, 1975. xiv + 529pp. $17.95. 74-5419. ISBN 0-13-480012-5. Index;CIP.

C-P This text is intended for an introductory course in digital logic design. Boolean algebra, analysis of flip-flops, algorithms for design of switching networks and special topics such as NAND synthesis are covered in thorough fashion.

Pylyshyn, Zenon W., (Ed.). *Perspectives on the Computer Revolution.* Prentice-Hall, 1970. xx + 540pp. $10.50. 79-106236.

SH-C Some 35 papers by authors ranging from contemporary luminaries to Babbage in 1864 and Samuel Butler in 1872. Nonmathematical, but difficult reading. However, many papers are classics.

Spencer, Donald D. *Computers in Action: How Computers Work.* (Illus.) Hayden, 1974. 150pp. $7.50; $4.95(p). 74-4251. ISBN 0-8104-5862-4; 0-8104-5861-6. Index;CIP.

SH-C This introduction to the workings of computers also presents a historical review of data processing mechanics from the abacus to digital computers. Input/output processes of computing are detailed and programming languages are briefly described.

Turn, Rein. *Computers in the 1980s.* (Illus.) Columbia Univ. Press, 1974. x + 257pp. $12.00; $3.95(p). 74-1488. ISBN 0-231-03844-5; 0-231-03845-3. Index;CIP.

C-P Forecasts technology and techniques of computers for the next few years. Terminology is defined and the reader is given a thorough grounding in systems jargon. Excellent reading.

Ward, Brice. *Digital Electronics: Principles and Practice.* (Illus.) Tab, 1972. 288pp. $8.95; $5.95(p). 73-178689. ISBN 0-8306-2595-2.

C-P The first third of the book provides a survey of both basic concepts and subsystems of digital electronics, including numbering systems, logic circuits, counters, etc. The remaining two-thirds sets forth 70 instructive experiments and includes a list of components for assembling a home-built logic trainer. Intended for self-instruction, but suitable for a technical training course.

621.384–.389 RADIO AND RECORDING DEVICES

American Radio Relay League. *The Radio Amateur's Handbook, 52nd ed.* (Illus.) American Radio Relay League, 1975. 690pp. $8.50; $5.50(p). 41-3345. Index.

SH-C The material progresses at a graded pace, from Ohm's law and simple schematic diagrams to detailed construction specifications for complex high-frequency solid-state receivers and transmitters. Gives advice on the legal and technical licensing requirements of the Federal Communications Commission.

Everest, F. Alton. *Handbook of Multichannel Recording.* (Illus.) Tab, 1975. 332pp. $10.95; $7.95(p). 75-20842. ISBN 0-8306-5781-0; 0-8306-4781-3. Gloss.;index.

C A handbook on the engineering and management arts underlying modern sound recording. Covers management for track separation, the audio mixing console, monitoring, special effects, stereophonic and quadraphonic recordings, location, layout and construction of the studio complex and a low-budget, multitrack recording operation. Useful to school music clubs.

Hilbrink, W.R. *Who Really Invented Radio?* (Illus.) G.P. Putnam's, 1972. 108pp. $3.49. 76-170078.

SH Traces the contributions of just about every scientist or technologist whose work even touched the development of radio. Constitutes a small encyclopedia of names, dates, and contributions.

Westcott, Charles G., and Richard F. Dubbe. *Tape Recorders—How They Work, 3rd* *ed.* (Illus.;rev. by Norman H. Crowhurst.) Sams/Bobbs-Merrill, 1974. 240pp. $5.50(p). 73-90290. ISBN 0-672-20989-6. Gloss.;index.

SH-C   A dual-purpose book useful for tape recorder users, technicians, and students. Includes an explanation of tape recorder circuitry and theory and useful advice on how to obtain the best performance.

621.4 HEAT ENGINEERING AND PRIME MOVERS

See also 333.82 Fuels and Energy.

Berkowitz, David A., and Arthur M. Squires (Eds.).   *Power Generation and Environmental Change.* (Illus.) MIT Press, 1971. xxiii + 440pp. $16.95. 70-137468. ISBN 0-262-02072-6.

C-P   The subjects include nuclear power generation and radionuclide pollution, hydroelectric power, power generation with fossil fuels, and waste heat, each treated by two or more authors. Many of the topics are seldom discussed outside their authors' own sheltered areas of study.

Commoner, Barry, et al. (Eds.).   *Energy and Human Welfare—A Critical Analysis; Alternative Technologies for Power Production (Vol. 2).* (Illus.) Macmillan Information/Macmillan, 1975. xviii + 213pp. $14.95. ($40.00; 3 vol. set). 75-8987. ISBN 0-02-468430-9. Index;CIP.

C-P   Nine articles on such diverse topics as fuel cells, controlled fusion, geothermal power, fossil fuel reserves, energy from solid wastes and solar energy technologies. Useful as a reference and excellent supplemental reading for undergraduates. (For volumes 1 and 2, see *333.82 Fuels and Energy.)*

Halacy, D.S., Jr.   *The Energy Trap.* (Illus.) Four Winds, 1975. xi + 143pp. $6.95. 74-8592. ISBN 0-590-07283-8. Index;CIP.

JH-SH   Covers recent developments in energy sources, including ocean thermal power plants, wind-electric plants, electricity-producing satellites and land-based power stations, and solar energy.

Limburg, Peter R.   *Engines.* (Illus.) Watts, 1970. 88pp. $3.25. 74-77243.

JH   Chapters are devoted to the internal combustion, diesel, steam, turbine, and jet engines and to classification of engines and engines of the future. Very readable, with lucidly expressed ideas, and excellent line drawings.

Merrill, Richard, et al. (Eds.).   *Energy Primer: Solar, Water, Wind, and Biofuels.* (Illus.) Portola Institute, 1974. 200pp. $4.50(p). 74-81048. ISBN 0-914774-00-X. Index.

SH-GA-C   Written in the style of the *Whole Earth Catalog,* this book provides a powerful stimulus to learn more and do more about our energy system.

Mother Earth News.   *The Mother Earth News Handbook of Homemade Power.* (Illus.) Bantam, 1974. ix + 374pp. $1.95(p). ISBN 0-553-08535. Index.

SH-GA   A popular science approach to constructing homemade energy devices using wood, water, wind, solar, and methane power. This is an informal, educational, and useful guide for the layperson.

Prenis, John, (Ed.).   *Energy Book #1: Natural Sources and Backyard Applications.* (Illus.) Running Press, 1975. 112pp. $4.00(p). 74-84854. ISBN 0-914294-21-0. Bib.

SH-C   The articles range from elementary descriptions of primitive apparatus to highly professional papers by engineers and physicists, with about 40 percent devoted to solar energy, 25 percent to wind power and 10 percent to methane and biogas generation.

Ruedisili, Lon C., and Morris W. Firebaugh (Eds.). *Perspectives on Energy: Issues, Ideas, and Environmental Dilemmas.* (Illus.) Oxford Univ. Press, 1975. xii + 527pp. $10.95; $6.95(p). 74-22886. ISBN 0-19-501879-6.

    C   A reasonable and balanced picture of the technological uses, problems and controversies related to energy and energy development is presented. The problems of generating an integrated, long-term national energy policy are discussed. (Breeder reactors and nuclear safeguards are slighted.)

Stoker, H. Stephen, et al. *Energy: From Source to Use.* (Illus.) Scott, Foresman, 1975. 337pp. $3.95(p). 74-78225. ISBN 0-673-07947-3. Index.

    SH-C   Treats the subject of energy in its full scope, covering the history of energy development, nature of energy conversion and specific sources such as gas and coal.

Stoner, Carol Hupping, (Ed.). *Producing Your Own Power: How to Make Nature's Energy Sources Work for You.* (Illus.) Rodale, 1974. xii + 322pp. $8.95. 74-10765. ISBN 0-87857-088-8. Gloss.;index;CIP.

    SH-C   Nonscientific but practical how-to guide for producing alternate forms of energy for private dwellings. A valuable appendix gives pointers on energy conservation and references on constructing power plants utilizing energy from wind, water, wood or the sun. Also for teachers interested in practical projects.

Weissler, Paul. *Small Gas Engines: How to Repair and Maintain Them.* (Illus.) Harper &Row, 1975. 261pp. $10.95. 75-13334. ISBN 0-06-014564-1. Index

    JH-SH-C   The author presents clear and detailed instructions, accompanied by extensive photographs and diagrams, on how to repair the one-cylinder gasoline engines commonly used in chain saws, lawn mowers and snow blowers.

621.45–.47 WIND AND SOLAR ENERGY ENGINEERING

Beedell, Suzanne. *Windmills.* (Illus.) Scribner's, 1976. 143pp. $12.00. 75-15224. ISBN 0-684-14487-5. Index;gloss.

    SH-C-P   Examines the windmill on aesthetic, technological, and cultural levels, including history, the milling process, preservation and restoration of windmills, living in them, and their construction and operation. The emphasis is on English mills.

Brinkworth, B.J. *Solar Energy for Man.* (Illus.) Halsted/Wiley, 1973. xii + 251pp. $8.95. 73-549. ISBN 0-470-10425-2.

    JH-SH-C   Text, graphics and mathematics elegantly lay out the "first principles" of physics (including quantum theory) needed to evaluate solar energy schemes. Considers collection of solar energy, use for heating, conversion into electricity, photoelectricity, photochemistry and photobiology. There are practical designs throughout, and computations for design and for prediction of performance.

Knight, David C. *Harnessing the Sun: The Story of Solar Energy.* (Illus.) Morrow, 1976. 128pp. $5.95. 75-44301. ISBN 0-688-22070-3; 0-688-32070-8. Gloss;index;CIP.

    JH-SH   An unbiased compilation of the state of the art in solar energy utilization. Contains a thorough historic treatment. Presumes no special science or math background.

Merrigan, Joseph A. *Sunlight to Electricity: Prospects for Solar Energy Conversion by Photovoltaics.* (Illus.) MIT Press, 1975. xi + 163pp. $12.95. 75-6933. ISBN 0-262-13116-1. Index;CIP.

    C-P   The first part of this book is generally useful as a timely, concise statement of energy sources and expected energy usage in the United States to the year

2000. The rest of the book is of interest chiefly to specialists since it details current developments in photovoltaic conversion-difficulties, economic considerations and business opportunities.

Reynolds, John. *Windmills & Watermills.* (Illus.) Praeger, 1970. 196pp. $13.95. 71-101254.

SH-C   The evolution of these mills as mechanical devices is explained with the aid of good illustrations, and there is an exposition of the architectural features and styles that evolved in the structures that housed the mills.

621.48 NUCLEAR ENGINEERING

Foreman, Harry, (Ed.). *Nuclear Power and the Public.* Univ. of Minnesota Press, 1970. xviii + 273pp. $9.00. 78-139961.

SH-C   Taken from a symposium, the 13 papers and the discussion by audience and participants give layreaders a rare opportunity to learn of the issues involved and the conflicting interpretations of the technical experts.

Fuller, John G. *We Almost Lost Detroit.* (Illus.) Reader's Digest Press (distr.: Crowell), 1975. ix + 272pp. $8.95. 75-17870. ISBN 0-88349-070-6. Index;CIP.

SH-C-P   The book reads like a novel, focusing on the partial core meltdown of Detroit Edison's Fermi liquid metal fast breeder reactor in 1966. Flashbacks to other nuclear incidents and accidents are included. A valuable supplement to energy-and policy-related topics.

Gofman, John W., and Arthur R. Tamplin. *Poisoned Power: The Case Against Nuclear Power Plants.* Rodale, 1971. 368pp. $6.95. 70-155715.

SH-C   Suggests that faith in nuclear fission reactors as our principal source of electricity for the future is a grave mistake. The nub of the argument is that widespread use of these reactors in population centers is bound to result in the inadvertent or sabotage-induced release of radioactive materials.

Kruger, Paul, and Carel Otte (Eds.). *Geothermal Energy: Resources, Production, Stimulation.* (Illus.) Stanford Univ. Press, 1973. x + 359pp. $17.50. 72-85700. ISBN 0-8047-0822-3.

GA-C-P   One of the few comprehensive sources on the topic. Available resources, methods of exploration, characteristics, problems of development, stimulation of geothermal systems (including nuclear detonations), and the state of the art are discussed.

Mann, Martin. *Peacetime Uses of Atomic Energy, 3rd rev. ed.* (Illus.) Crowell, 1975. xi + 196pp. $8.95. 75-5621. ISBN 0-690-00118-5. Index; CIP.

JH-SH   Uses of atomic energy in earth excavation, agriculture, and health protection are covered. Mann explains the basis of nuclear fission energy and various types of nuclear fission reactors and discusses portable nuclear power supplies and nuclear fusion.

Michelsohn, David Reuben. *Atomic Energy for Human Needs.* (Illus.) Messner, 1973. 189pp. $5.29. 73-5392. ISBN 0-671-32629-5; 0-671-32630-9.

JH-SH   Concerns the accomplishments of nuclear technology and its potential for the future. Covers the use of radioactive isotopes in archeology, medicine, and the plant and animal sciences, as well as the use of nuclear techniques in various other analysis and measurement problems.

Seaborg, Glenn T., and William R. Corliss. *Man and Atom: Building a New World Through Nuclear Technology.* (Illus.) Dutton, 1971. 411pp. $10.00. 70-122795. ISBN 0-525-15099-4.

GA   A tract to promote wider public acceptance of nuclear energy technology, with emphasis on the production of electric power by fission reactors. The authors state that we must have a lot more electric power soon and they promote fission reactors as the only feasible way of providing this needed power in the near future.

Union of Concerned Scientists.   *The Nuclear Fuel Cycle: A Survey of the Public Health, Environmental, and National Security Effects of Nuclear Power, rev. ed.* (Illus.) MIT Press, 1975. xvii + 291pp. $4.95(p). 75-25759. ISBN 0-262-21005-3; 0-262-71003-X. Gloss.;index;CIP.

C-P   Covers hazards (and deaths) in uranium mining and milling and in fuel reprocessing; transportation and disposal of wastes; possible nuclear accidents in water-cooled reactors; and nuclear safeguards. All authors are resolutely against nuclear power. The facts chosen are uncontested, but bias arises from the choice of facts.

621.9 MACHINE TOOLS

Johnson, Olaf A.   *Design of Machine Tools.* (Illus.) Chilton, 1971. viii + 262pp. $8.50. 70-147253. ISBN 0-8019-5508-4.

SH-C   The subject is machine tools and how they were designed, on commission and on time for the great corporations, while exceeding the limits of technological expectation for the era. The essential elements and the simple logic of the designs are described in a most enjoyable style.

Kratfel, Edward R., and George R. Drake.   *Modern Shop Procedures.* (Illus.) Reston, 1974. xii + 356pp. $9.95; $7.94(p). 73-19644. ISBN 0-87909-506-7; 0-87909-505-9. Gloss.;index;CIP.

SH   A good introduction to shop practices for technicians, apprentices and do-it-yourselfers. Describes measuring instruments, hand-tools and small electrical tools and their use and care. Sketching preparation and interpretation of engineering drawings are covered.

627.7 UNDERWATER OPERATIONS

Dean, Anabel.   *Submerge! The Story of Divers and Their Crafts.* (Illus.) Westminster, 1976. 111pp. $6.50; $4.50(p). 75-33917. ISBN 0-664-32569-6; 0-664-34009-1. Gloss.; index;CIP.

JH-SH   Underwater vehicles from the earliest submarine to the more recent submersibles for rescue and work are described, as are underwater habitats. Stories of undersea explorers demonstrate the dangers divers face. Most valuable for suggested experiments which follow descriptions of each type of equipment and elucidate the principles by which the equipment works.

Hampton, T.A.   *The Master Diver and Underwater Sportsman.* (Illus.) Arco, 1970. 159pp. $5.95. 79-121369. ISBN 0-668-02353-8.

SH-C   A concise, no-nonsense book about diving, with an impressive number of facts in text, drawings and photographs. Briefly discussed are the physiology of diving, equipment, air supply, protective clothing, visibility, underwater cutting, welding, blasting, seamanship for divers and step-by-step procedures in aqualung and standard diving.

Limburg, Peter R., and James B. Sweeney.   *Vessels for Underwater Exploration.* (Illus.) Crown, 1973. 154pp. $5.95. 73-78878. ISBN 0-517-5034-7. Index;gloss.;bib.

JH-SH   The high points of exploration of the sea, from diving for sponges and pearls to submarines and modern oceanographic submersibles, diving suits, diving bells and scuba gear; individuals who contributed to exploration and research; and historical and current developments are described.

Piccard, Jacques. *The Sun Beneath the Sea.* (Illus.) Scribner's, 1971. xxxix + 405pp. $12.50. 76-123854. ISBN 0-684-31101-1.

SH-C The first section is a diary of the first mesoscaph, with details of its construction and operation. The second section treats in a similar fashion the mesoscaph *Ben Franklin.* The final section deals with the month-long drift of the *Ben Franklin,* submerged in the Gulf Stream in 1969. A fine documentary, interestingly written.

Sweeney, James B. *A Pictorial History of Oceanographic Submersibles.* (Illus.) Crown, 1970. 310pp. $9.95. 75-108071.

SH A well-prepared and fascinating history of submarines and other under-sea vehicles from the earliest times up to the TEKITE I undersea experiment. The many photographs contribute much. Excellent as a reference.

Woods, J.D., and J.N. Lythgoe (Eds.). *Underwater Science: An Introduction to Experiments by Divers.* (Illus.) Oxford Univ. Press, 1971. xiii + 330pp. $13.00. ISBN 0-19-217622-6.

SH-C-P A guide to observation and experimentation written by nine British scientist-divers. Chapters on apparatus, diver performance, spatial perception and vision show the problems of working underwater and suggest solutions. Among the applications are studies in fish behavior, botany, archeology, and microoceanography.

628 MUNICIPAL ENGINEERING—WATER AND AIR

Fair, Gordon M., John C. Geyer, and Daniel A. Okun. *Elements of Water Supply and Wastewater Disposal, 2nd ed.* (Illus.) Wiley, 1971. vii + 752pp. $15.95. 72-151032. ISBN 0-471-25115-1.

C-P The problems of water supply, treatment, quality, transfer, and disposal are well integrated into the applied engineering procedures necessary for managing water. A valuable reference.

Stevens, Leonard A. *Clean Water: Nature's Way to Stop Pollution.* (Illus.) Dutton, 1974. xii + 289pp. $10.00. 73-79556. ISBN 0-87690-103-8.

C Case studies of water purification systems, their operations, and problems of waste disposal cover such factors as community involvement, misguided public opinion, poor plant maintenance, and lack of trained, responsible personnel. Suitable for those interested in a broad approach to water disposal.

Warren, Charles E. *Biology and Water Pollution Control.* (Illus.) Saunders, 1971. xvi + 434pp. $11.00. 70-126463. ISBN 0-7216-9120-3.

C-P This is a detailed exposition of the kinds of water pollution that endanger aquatic life (primarily animal), and the need for establishing water quality standards based on biological research. Extensive discussions cover morphology and physiology of a variety of animals; ecology of individual organisms, populations and communities; and methods, including biological, of water purification. Historical perspectives and social implications of decisions provide insight into the problems involved.

Williamson, Samuel J. *Fundamentals of Air Pollution.* (Illus.) Addison-Wesley, 1973. xv + 472pp. $14.95. 75-186842. Index;gloss.;bibs.

C-P Topics cover the atmosphere, its energy balance, stability, wind, turbulence and diffusion; stationary and mobile pollution sources and their control;

aerosols; and the adverse effects of pollution. Appendixes review the physical and chemical principles of light scattering by small particles, blackbody radiation, molecular absorption, coriolis forces and organic compounds. Presumes a year of college-level physics and chemistry; includes problems.

629.13 AERONAUTICS

Botting, Douglas. *Shadow in the Clouds: The Story of Airships.* (Illus.) Penguin, 1975. 48pp. $1.75(p). ISBN 0-14061-061-2.

JH-SH   This timely book outlines the major events in airship history using excellent old photographs. Many interesting facts and statistics are included. Excellent collateral reading for science classes.

Collier, Basil. *A History of Air Power.* (Illus.) Macmillan, 1974. 358pp. $10.95. 73-20991. Index.

SH   Beginning with rockets and lighter-than-air balloons, Collier traces the history of the conquest of the air and its military uses. A rare depth of knowledge and research is apparent in this lucidly written book.

Dwiggins, Don. *The Sky Is Yours: You and the World of Flight.* (Illus.) Childrens, 1973. 79pp. $4.79. 73-9735. ISBN 0-516-08875-0. Gloss.

SH   An exciting picture of the diverse jobs, from aerial survey flying and flight testing to air traffic control and weather forecasting. Concludes with a provocative chapter on building your own airplane. Dynamic photographs and drawings, and an appendix of information sources on careers and aviational organizations are included.

Genett, Ann. *Contributions of Women: Aviation.* (Illus.) Dillon, 1975. 115pp. $6.95. 74-19004. ISBN 0-87518-089-2.

JH   Interesting insights into the lives of six of America's famous female aviators: Amelia Earhart, Anne Lindbergh, Jackie Cochran, Jerrie Mock, Jerrie Cobb and Emily Howell. The author writes with humor and clarity as she describes the seriousness and dedication these women needed to achieve their goals.

Gibbs-Smith, C.H.   *Flight Through the Ages: A Complete, Illustrated Chronology from the Dreams of Early History to the Age of Space Exploration.* (Illus.) Crowell, 1974. 240pp. $17.95. 74-8389. ISBN 0-690-00607-1. Index.

JH-SH-C   This chronology traces interest in flight from its mythical idealization to the technological achievements of Skylab. Handsomely illustrated, the book is encyclopedic in scope.

Olney, Ross R. *Air Traffic Control.* (Illus.) Nelson, 1972. 124pp. $4.95. 76-181672. ISBN 0-8407-6154-6. Gloss.

JH-SH   Leads the reader from the earliest attempts to control the air lanes to the present-day complex systems. The explanations of terms as they occur in the text is noteworthy. Similarly, the drawings and photographs emphasize the complexity of the system and give realistic visual impressions.

Walsh, John Evangelist. *One Day at Kitty Hawk: The Untold Story of the Wright Brothers and the Airplane.* (Illus.) Crowell, 1975. x + 305pp. $10.00. 75-12740. ISBN 0-690-00103-7. Index;CIP.

SH-C   Walsh details the events leading to the first successful powered flight at Kitty Hawk on December 17, 1903, and the first public demonstrations of the perfected flyer by the Wright Brothers in 1908. A number of photographs are included, and there are 40 pages of detailed notes and reference sources.

629.132 PRINCIPLES OF FLIGHT

Benkert, Joseph W. *Introduction to Aviation Science.* (Illus.) Prentice-Hall, 1971. xx + 604pp. $14.95. 71-135754. ISBN-0-13-478222-4.

SH   A wealth of practical information for the student pilot is brought together in one well-written book, some of which is taken from publications of the Federal Aviation Administration. A useful text or reference for practical flight training and operation, it includes airplane flight mechanics, control and stability, power plants, instruments, meteorology, navigation, and the physiology of flight.

Grey, Jerry. *The Facts of Flight.* (Illus.) Westminster, 1973. 135pp. $4.95; $2.95(p). 72-12793. ISBN 0-664-32526-2; 0-664-34004-0.

JH-SH-C   An elementary, informally written, clear exposition on the principles of flight, dealing with the social, economic and ecological aspects of air transportation; lift, drag, and thrust; ground effect, soaring, and manpowered flight; and transonic and supersonic flight. Frequent comparisons of manned flight and bird flight are made.

Halacy, D.S., Jr. *The Complete Book of Hang Gliding.* (Illus.) Hawthorn, 1975. 183pp. $6.95. 74-18691. ISBN 0-8015-3272-8; 0-8015-3273-6(p). Index.

Siposs, George. *Hang Gliding Handbook—Fly Like a Bird.* (Illus.) Tab, 1975. 210pp. $8.95; $5.95(p). 75-1699. ISBN 0-8306-5776-2; 0-8306-4776-7. Index.

SH   Siposs provides a stronger basic background in the fundamental mechanics, aerodynamics and stress analysis that would be helpful for every hang-glider pilot. Siposs is also more detailed in his discussions of the materials and techniques of construction than is Halacy. Halacy gives a better account of the history of the sport and its antecedents as well as providing better directories of organizations, publications, manufacturers, dealers and schools.

Mrazek, James E. *Sailplanes and Soaring: The Beginner's Lift into Discovering the Wonders and World of Silent Flight.* (Illus.) Stackpole, 1973. 159pp. $5.95. 73-11. ISBN 0-8117-1503-5.

JH-SH-C   Gives the history of and information sources on soaring facts about sailplanes, including nomenclature, controlling of the sailplane, flying techniques, and a list of do's and don'ts for the crew. Covers purchase of a sailplane, insurance and the private glider pilot rating.

Siberry, Mervyn. *Instruments of Flight: A Guide to the Pilot's Flight Panel of a Modern Airliner.* (Illus.) Crane, Russak, 1974. 124pp. $7.50. 73-91533. ISBN 0-8448-0301-4. Index.

JH-SH-C   An interesting and lucid description of the safety tools used by airplane cockpit crews. The book's organization simulates the instruments' usage patterns, and includes descriptions of various flight maneuvers and their importance to passenger safety.

Van Sickle, Neil D., (Ed.). *Modern Airmanship, 4th ed.* (Illus.) Van Nostrand Reinhold, 1971. xiii + 909pp. $15.95. 71-143545.

SH-C-P   This encyclopedia of knowledge is intended primarily for the pilot—it covers basic aerodynamics, stability, propulsion, meteorology, navigation and flight techniques. Recommended for courses in aeronautical science and flight, and flight-testing.

629.133 AIRCRAFT TYPES

Dwiggins, Don. *Riders of the Winds: The Story of Ballooning.* (Illus.) Hawthorn, 1973. x + 180pp. $6.95. 73-371.

JH-SH Ballooning as an expensive sport and for scientific purposes is discussed, along with aeronautics, meteorology and other necessary scientific lore. Appendixes of this well-researched volume give specific information on balloon clubs, balloon schools and balloon makers.

Gablehouse, Charles. *Helicopters and Autogiros.* (Illus.) Lippincott, 1969. xi + 254pp. $6.95. 73-85764.

SH-C Traces the fascinating history of the helicopter from ancient times to the present, its use in rescue operations, as a gunship, and for air transportation. Technical considerations include rotating wings, the rotor mechanism, autogiros and V/STOL craft. A very readable book by a knowledgeable author.

Hellman, Hal. *Helicopters and Other VTOL's.* (Illus.) Doubleday, 1970. 140pp. $3.95. 76-126383.

JH-SH The evolution of the helicopter is explained, with appropriate sketches and photographs to describe in simple terms the aerodynamics and control systems of helicopters and other VTOL's. Their uses for military, civil, rescue, and transport purposes are described.

Lowry, Peter, and Field Griffith. *Model Rocketry: Hobby of Tomorrow.* (Illus.) Doubleday, 1972. xii + 160pp. $4.50. 74-143815.

JH-SH The guide includes discussions of the history of model rocketry, rocket construction, launching systems, recovery methods, and safety precautions, as well as more advanced topics such as staging, payloads and boost gliders. Should be in all school libraries.

Robinson, Douglas H. *Giants in the Sky: A History of the Rigid Airship.* (Illus.) Univ. of Wash. Press, 1973. xxix + 376pp. $15.00. 72-11546. ISBN 0-295-95249-0. Gloss.

JH-SH-C An exciting adventure story as well as a clear, accurate, in-depth and fascinating account of the 161 airships built and flown in Germany, Britain, the United States and France. These unique crafts are traced from creation to final disposition, and an appendix tabulates the technical specification and chronological highlights of each ship.

Stine, G. Harry. *Handbook of Model Rocketry, rev. 4th ed.* (Illus.) Follett, 1976. 352pp. $9.95; $6.95(p). 75-13852. ISBN 0-695-80615-7; 0-695-80616-5. Indexes.

JH-SH-C This should be owned by every hobbyist in model rockets. Describes how to get started, tools and techniques, construction, launch techniques, performance, stability and recovery. The rocket safety code is explicitly outlined. Copiously illustrated.

629.2 LAND VEHICLES

Ayres, Robert U., and Richard P. McKenna. *Alternatives to the Internal Combustion Engine: Impacts on Environmental Quality.* (Illus.) Johns Hopkins Univ. Press, 1972. xvi + 324pp. $12.00. 74-181555. ISBN 0-8018-1369-7.

C-P A status report on current research into internal combustion engine (ICE) alternatives. Energy conservation requirements for autos and general modes of energy conservation are discussed, and the gas turbine, the Rankine-cycle engine and a hybrid form are covered in detail. Professional and well-documented.

Buel, Ronald A. *Dead End: The Automobile in Mass Transportation.* Prentice-Hall, 1972. 231pp. $6.95. 75-175808. ISBN 0-13-19680-3.

SH-C A comprehensive analysis of the role and impact of the automobile on the American environment, and on our social, economic and political institutions. The book is more impressive in displaying the full scope of the problem than in offering solutions.

Butterworth, W.E. *Wheels and Pistons: The Story of the Automobile.* (Illus.) Four Winds, 1971. 192pp. $5.95. 74-124190.

JH-SH   Traces the history of the automobile from the first steam-driven vehicle to research on the external combustion engine. A fascinating and highly readable account.

Cantonwine, Charles R. *Battery Chargers and Testers: Operation, Repair, Maintenance.* (Illus.) Chilton, 1971. viii + 349pp. $9.50. 73-153135. ISBN 0-8019-5621-8.

SH   Although primarily a service manual, this book contains enough information so that it is useful as a general reference to anyone concerned with the maintenance of batteries and battery service equipment. Appendixes on equipment sources, brand names, and technical abbreviations and symbols.

Corbett, Scott. *What About the Wankel Engine?* (Illus.) Four Winds, 1974. 80pp. $6.50. 74-8593. ISBN 0-590-07369-9. Gloss;index;CIP.

JH-SH   An easily understandable explanation of both the Wankel engine and other internal combustion engines. Also recommended for those concerned about the fuel requirements and environmental benefits of the Wankel engine.

Dark, Harris Edward. *Auto Engines of Tomorrow: Power Alternatives for Cars to Come.* (Illus.) Indiana Univ. Press, 1975. ix + 180pp $8.95. 74-6518. ISBN 0-253-10490-4. Index;CIP.

SH-C   The development of the automobile is vividly described. Many novel but unlikely schemes of automotive propulsion are also discussed. The realities of economy, environment, and resource availability are all considered.

Dark, Harris Edward. *The Wankel Rotary Engine: Introduction and Guide.* (Illus.) Indiana Univ. Press, 1974. xii + 145pp. $6.95. 73-16676. ISBN 0-253-19021-5. Index;CIP.

JH-SH-C   Readable account of the business deals and engineering innovations which brought the Wankel-powered Mazda to America. Compares the Wankel to the performance of conventional piston engines, and comments on the Mazda's unique driving characteristics. Peripheral reading in the history of technology.

Henkel, Stephen C. *Bikes: A How-To-Do-It Guide to Selection, Care, Repair, Maintenance, Decoration, Safety, and Fun on Your Bicycle.* (Illus.) Chatham (distr: Viking), 1972. 96pp. $4.95(p). 73-172453. ISBN 85699-033-7.

JH-SH   Touching on everything from the history of bicycles to easy repair jobs, it is an all-around handbook that amateur bike riders should enjoy reading and using. Many descriptive illustrations are given which are extremely helpful to those who know little about mechanics. Most instructions are written in an easy-to-follow, step-by-step manner.

Jackson, Dorothy. *What Every Woman Should Know About Her Car.* (Illus.) Chilton, 1974. 201pp. $7.95; $4.95(p). 74-17326. ISBN 0-8019-6000-2; 0-8019-6001-0. Gloss;index;CIP.

GA   Information on automobile loans and insurance is included, as are instructions on how to change a tire. The cars with which the author deals are new models. A valuable reference.

Lord Montagu of Beaulieu and Anthony Bird. *Steam Cars: 1770–1970.* (Illus.) St. Martin's, 1971. xiii + 250pp. $8.95. 70-137685. Index.

SH-C   From the steam pumping engine of Thomas Savery in 1698 and the Cugnot steam truck in 1771 to recent experimental models of steam vehicles, the authors cover in mechanical and historical detail the journey of steam vehicles. Of interest to car buffs and serious students of engine design and automobile construction; an excellent reference.

Stambler, Irwin. *Automobile Engines of Today and Tomorrow.* (Illus.) Grosset&Dunlap, 1972. 152pp. $5.95. 72-75786. ISBN 0-448-21447-4; 0-448-26216-9.

   SH This well-written, well-researched book provides a reasonably complete treatment of vehicular power plants in an interesting and informative fashion. It covers recent innovations, diesel engines, gas turbines, vapor-cycle external combustion engines, electric power plants, etc.

Woodforde, John. *The Story of the Bicycle.* (Illus.) Universe, 1971. 175pp. $4.95. 74-128402. ISBN 0-87663-135-9.

   JH-SH Traces the evolution of the bicycle from the late 18th century in an interesting fashion, and details its transformation from a hobby horse to today's 10-gear models.

629.4 ASTRONAUTICS

Ferdman, Saul, (Ed.). *The Second Fifteen Years in Space: Vol. 31, Science and Technology.* (Illus.) American Astronautical Society (distr.: Univelt), 1973. viii +201pp. $15.00. ISBN 87703-064-2.

   C-P An overview of the impact of space on our national life (from the viewpoints of industry and science) is followed by discussions of solar power, technology of rocket engines, computer control and data communications. Following this are three talks on the international outlook for European space projects after 1980, and the views of the developing countries on space.

Gallant, Roy A. *Man's Reach for the Stars.* (Illus.) Doubleday, 1971. xiii+201pp. $5.95. 74-129895.

   SH A description of the importance of getting to the moon sets the scene for a well-planned discussion of exploring space. The enthusiasm of some outstanding scientists and administrators is shown, and the problems considered will help the reader understand some of the judgments and decisions that were made.

Moore, Patrick. *The Next Fifty Years in Space.* (Illus.) Taplinger, 1976. 144pp. $12.95. 75-26326. ISBN 0-8008-5528-0. Index.

   SH-C This handsome book provides a record of space exploration up to the present and extrapolates future activities beyond the year 2000. The style is easy flowing and wide ranging. Worthwhile reading as entertainment, but not as strict science.

Riabchikov, Evgeny. *Russians in Space.* (Ed. by Nikolai P. Kamanin;trans.;illus.) Doubleday, 1971. 300pp. $10.00. 70-144291.

   JH-SH-C An eminently readable story of the Russian space program from Tsiolkovsky's early calculations in the 19th century to the manning of a space station in 1971. Not a reference book, it is primarily about Russian space people, and deserves a wide readership in this country. The 153 excellent photographs contain the technical content of the book.

Taylor, L.B., Jr. *For All Mankind: America's Space Programs of the 1970s and Beyond.* (Illus.) Dutton, 1974. xii+207pp. $8.95. 74-9109. ISBN 0-87690-115-1. Index.

   SH-C-P The many facets of the NASA space program are brought together in a concise, readable form which includes the results of space explorations to date, present uses of space, practical benefits which have resulted, and plans for future space utilization.

Von Braun, Wernher, and Frederick I. Ordway, III. *History of Rocketry and Space*

Travel, 3rd rev. ed. (Illus.) Crowell, 1975. xi + 308pp. $25.00. 74-13813. ISBN 0-690-00588-1. Index;CIP.

JH-SH-C　About half of this now classic history of developments in rocketry is devoted to space probes launched since 1958. The many illustrations will capture the reader's interest, and the book is an excellent reference.

629.43 SPACE PROBES

Bradbury, Ray, et al. *Mars and the Mind of Man.* (Illus.) Harper&Row, 1973. xiii + 143pp. $7.95. 72-9746. ISBN 0-06-010443-0.

SH-C　A record of the conversation of two scientists (Carl Sagan and Bruce Murray), two writers of science fiction and fact (Ray Bradbury and Arthur Clarke) and a science journalist (Walter Sullivan) as Mariner 9 sent over 7000 photographs of Mars; and a collection of essays about the achievement of Mariner 9 and its meaning for us, emphasizing humankind's need for exploration.

Caidin, Martin. *Destination Mars.* (Illus.) Doubleday, 1972. 295pp. $7.95. 74-186011.

JH-SH　Traces the development of our knowledge of the planet, with particular attention to the results obtained by Mariners 4, 6, 7 and 9 (U.S.) and Mars 1, 2 and 3 (U.S.S.R.). Useful collateral reading, interestingly and clearly written. Good insight into the objectives of the planetary programs.

Sharpe, Mitchell R. *Satellites and Probes: The Development of Unmanned Spaceflight.* Doubleday, 1970. 192pp. $5.95. 70-99578.

JH-SH　Offers a sound, very comprehensive introduction with emphasis on hardware and accomplishments but with glimpses of the involved personalities. The core of the book encompasses the structure, control and recovery of scientific, communications, weather, navigational and military satellites.

Smolders, Peter L. *Soviets in Space.* (Illus.;trans.) Taplinger, 1974. 286pp. $9.95. 73-16177. ISBN 0-8008-7340-8. Index.

SH-GA　Answers the post-Sputnik era question, "How did they do it?" and also describes the why of Soviet space efforts, including a review of the history of Soviet interest in space beginning in 1921. Details both feats and failures of the Soviet space program.

Strong, James. *Search the Solar System: The Role of Unmanned Interplanetary Probes.* (Illus.) Crane, Russak, 1973. 160pp. $8.50. 73-80427. ISBN 0-8448-0213-1.

SH　This is a review of what space probes have achieved, the objectives for the next half-decade, and speculation on what may be accomplished during centuries to come.

629.45 PILOTED SPACE FLIGHT

Booker, P.J., G.C. Frewer, and G.K.C. Pardoe. *Project Apollo: The Way to the Moon.* (Illus.) American Elsevier, 1970. viii + 212pp. $5.50. 72-101222. ISBN 0-444-19705-2.

SH-C　A straightforward account of how engineers and technologists carried Project Apollo through to its successful completion. Detailed descriptions of the Apollo spacecraft, the rocket vehicles, and the launch equipment are provided. Other space flights are briefly described.

Collins, Michael. *Carrying the Fire: An Astronaut's Journeys.* (Illus.) Farrar, Straus&Giroux, 1974. xvii + 478pp. $10.00. 74-7211. ISBN 0-374-11917-1. CIP.

SH-C　This ex-astronaut's vivid conversational style makes for a fascinating autobiographical account of the journey into space.

Cortright, Edgar M., (Ed.). *Apollo Expeditions to the Moon.* (Illus.) NASA (distr.: U.S. GPO), 1975. xi + 313pp. $8.90 (Stock No. 033-000-00630-6). 75-600071. Index;CIP.

    SH-C   Well-presented and illustrated memoirs of individuals associated with the Apollo program. The precepts of the program are explained, an account of engineering and management tasks is given, and the human drama is described. The 17 contributors include astronauts, administrators, science writers, scientists and one lawyer.

Gagarin, Yuri, and Vladimir Lebedev. *Survival in Space.* (Illus.;trans.) Praeger, 1969. viii + 166pp. $5.95. 74-83336.

    JH-SH   A considerable part of the book is devoted to describing human reaction to stress and how these reactions are tested to determine if an applicant would be suitable as an astronaut. The chapter on human-machine interrelationships is excellent. The limitations of machines are discussed.

Gurney, Gene. *Americans to the Moon.* (Illus.) Random House, 1970. 147pp. $3.95. 77-103405. Index.

    JH   The scope and detail of the book make it useful for reference on Apollo missions 8–12. An appendix contains capsule biographies of 18 American astronauts; dates, objectives, and participants in Apollo missions.

Lewis, Richard S. *The Voyages of Apollo: The Exploration of the Moon.* (Illus.) Quadrangle, 1974. xi + 308pp. $12.50. 74-77942. ISBN 0-8129-0477-X. Index;CIP;bibs.

    JH-SH-C   Apollo flights 12 through 17 are described as a modern epic depicting a process of intellectual development and possibly an advance in the evolution of humans.

Sharpe, Mitchell R. *"It is I, Sea Gull": Valentina Tereshkova, First Woman in Space.* (Illus.) Crowell, 1975. 214pp. $5.95. 74-14698. ISBN 0-690-00646-2. Index;CIP.

    JH-SH   Tereshkova of the Soviet Union was the first woman astronaut to go into space. This is the story of her life from her tomboy childhood on a collective farm to becoming an aerospace engineer. Although the author deals with some technical subjects, the book is easy to read.

629.7 ASTRONAUTICAL ENGINEERING

Branley, Franklyn M. *Experiments in the Principles of Space Travel, rev. ed.* (Illus.) Crowell, 1973. 113pp. $4.50. 72-7543. ISBN 0-690-27792-X.

    JH-SH   After a brief introduction covering astronomical distances, there are chapters on measurement of distances, rocket and spaceship power, pressure and gravity. A useful reference for students or teachers.

Clark, John D. *Ignition! An Informal History of Liquid Rocket Propellants.* (Illus.) Rutgers, 1972. xiv + 214pp. $10.00. 72-185390. ISBN 0-8135-0725-1.

    SH-C-P   A delightful account dealing almost exclusively with the chemistry of rocket propellants which has little to say about engine hardware, injector design, etc. Reveals the foibles and follies and heroic aspects of modern research and development.

Gatland, Kenneth. *Missiles and Rockets.* (Illus.) Macmillan, 1975. 256pp. $6.95. 75-15641. ISBN 0-02-542860-8. Index;CIP.

    SH-C-P   This pocket encyclopedia provides almost all available information on missiles and rockets, including 84 color photographs, other illustrations, and tables.

Gilfillan, Edward S., Jr. *Migration to the Stars: Never Again Enough People.* Luce (distr.: McKay), 1975. xiv + 226pp. $8.95. 74-28616. ISBN 0-088331-072-4. CIP.

   JH-SH-C   This bizarre but thought-provoking book gives an accurate, detailed description of the construction and operation of an enormous spacecraft circling earth and crafts to be used for exploratory outer space missions, developed in the context of the philosophical necessity of human colonization of the galaxy.

Hendrickson, Walter B., Jr. *Manned Spacecraft to Mars and Venus: How They Work.* (Illus.) Putnam's, 1975. 128pp. $4.29. 74-16627. SBN GB-399-60928-8; TR-399-20438-5. Index.

   JH-SH-C   Hendrickson discusses the technical and scientific planning involved in an extended manned flight, beginning with a detailed account of the most economical flight paths to Mars, Venus and Mercury. A discussion of power plants is included.

Hendrickson, Walter B., Jr. *Who Really Invented the Rocket?* (Illus.) Putnam's, 1974. 128pp. $4.29. 73-189238. SBN 399-60852-4. Index.

   SH-GA   A clear and accurate history of rocketry, beginning with the self-propelled "fire-arrows" of tenth century China and culminating in the development of the modern rocket to the moon. Well researched; a useful reference.

Mallan, Lloyd. *Suiting Up for Space: The Evolution of the Space Suit.* (Illus.) Day, 1971. ix + 262pp. $9.95. 75-89308.

   SH-C   Written in news-reporter style, this account of the 30-year development of space suits contains very little science, but it provides a great many facts and encourages readers to apply what they know about mechanics, thermodynamics, chemistry, and biology.

Schneider, William C., and Thomas E. Hanes (Eds). *The Skylab Results: Parts 1 and 2.* (Illus.) American Astronautical Society, 1975. xxviii + 1146pp. total. $30.00ea.; $60.00set. ISBN 87703-072-3.

   SH-C-P   The two compilations of papers describe almost every aspect of Skylab. The emphasis is on lessons learned in management, operations and engineering, as well as a description of the scientific experiments and their results.

630 Agriculture and Related Technologies

631 GENERAL AGRICULTURAL TECHNIQUES AND MATERIALS

Donahue, Roy L., John C. Shickluna, and Lynn S. Robertson. *Soils: An Introduction to Soils and Plant Growth, 3rd ed.* (Illus.) Prentice-Hall, 1971. xx + 587pp. $9.75. 73-159573. ISBN 0-13-821876-5. Gloss.

   SH-C   The inclusion of discussions of moon "soils" and soils of the tropics gives this text a universal scope. Theory is separated from practice for better understanding of each and for consideration of these areas separately. Appendixes provide definitions and conversion factors.

Douglas, J. Sholto. *Hydroponics: The Bengal System; with Notes on Other Methods of Soilless Cultivation, 5th ed.* (Illus.) Oxford Univ. Press, 1975. xvi + 185pp. $4.25. SBN 19-560566-7. Index.

   JH-SH-C   This is a layperson's how-to book on hydroponics. It includes chapters on apparatus needed for all sizes of units, nutrient mixtures, commercial possibilities, pests and diseases, and notes on heating, effluents, plant hormones, lighting, seed treatment, mechanization and record keeping.

Helfman, Elizabeth S.   *Our Fragile Earth.* (Illus.) Lothrop, Lee&Shepard, 1972. 160pp. $4.95. 72-1094.

> **JH**   A generalized treatise on farming practices and their effect on land. The impact of both nature and humans on soil quality is traced. Also discusses erosion, dust storms, land disfigurement, strip mining, and military destruction such as defoliation and bombing.

Hicks, Cedric Stanton.   *Man and Natural Resources: An Agricultural Perspective.* (Illus.) St. Martin's, 1975. 122pp. $16.95. 75-15274. Index.

> **C**   The author makes a convincing, eloquent case for a fundamental change in our attitude from economic to biologic emphasis. Teachers, students, and public health officials will find this book useful.

Keen, Martin L.   *The World Beneath Our Feet: The Story of Soil.* (Illus.) Messner, 1974. 96pp. $5.79. 74-7148. ISBN 0-671-32673-2; 0-671-32674-0. Gloss.;index;CIP.

> **JH-SH**   Keen outlines the chemical, biological, and mechanical factors which play important roles in the formation and characterization of various types of soil. The impact of earlier mismanagement of the land and the resulting decline of civilizations is clearly shown.

Limburg, Peter R.   *The Story of Corn.* (Illus.) Messner, 1971. 96pp. $4.29. 70-160724. ISBN 0-671-32469-1; LB 0-671-32470-5.

> **JH**   Corn and its relation to humans is intriguingly traced from an Indian myth through its discovery and use by early settlers to its modern improvements and uses. The book lucidly discusses these developments from a sound scientific basis and introduces the reader to genetics, cytology, physiology, and plant morphology.

McCoy, J.J.   *To Feed a Nation: The Story of Farming in America.* (Illus.) Nelson, 1971. 193pp. $4.95. 71-145925. ISBN 0-8407-6134-1.

> **JH-SH**   From an ecological perspective, McCoy traces the development of agricultural practices from those of the early Native Americans to present-day agribusiness, and provides some insights into the myriad problems that beset the farmer.

Mangelsdorf, Paul C.   *Corn: Its Origin, Evolution and Improvement.* (Illus.) Belknap/Harvard Univ. Press, 1974. xiv + 262pp. $20.00. 72-15454. SBN 674-17175-6. Index.

> **C-P**   This corn expert gives an exhaustive review of corn as a nutritional, cultural, and economic basis of successive American societies, focusing on the uniqueness of corn as a plant which exists only in association with humans. A valuable reference.

Mitchell, Roger L.   *Crop Growth and Culture.* (Illus.) Iowa State Univ. Press, 1970. vii + 349pp. $9.50. 72-88006. ISBN 0-8138-0377-2.

> **C**   Photosynthesis, respiration, mineral nutrition, growth regulators, and other topics in plant physiology are analyzed to give an understanding of seed production and germination, seedling development, and growth and differentiation of roots and shoots. Environmental factors are also discussed.

Raskin, Edith.   *World Food.* (Illus.) McGraw-Hill, 1971. 160pp. $4.72. 79-154837.

> **JH-SH**   The author examines local cultures and customs in the production of foodstuffs, including animal domestication. Photographs illustrate the diversity of farming methods.

Reynolds, Peter J.   *Farming in the Iron Age.* (Illus.) Cambridge Univ. Press, 1976. 48pp. $2.95(p). 75-43569. ISBN 0-521-21084-4.

JH   The author describes farming in Britain as evidenced by archeological excavations. The reader is also shown that agricultural practices during present times cannot be performed by the modern farmer without the aid of complex machines and chemicals. Lucid, accurate and readable.

Rossiter, Margaret W.   *The Emergence of Agricultural Science: Justus Liebig and the Americans, 1840–1880.* (Illus.) Yale Univ. Press, 1975. xiv + 275pp. $15.00. 74-29737. ISBN 0-300-01721-9. Index

SH-C-P   An excellent view of the development of agricultural science and the agricultural experiment stations is provided. The scientific and financial trials and tribulations of prominent agricultural scientists are brought to life.

Schlebecker, John T.   *Whereby We Thrive: A History of American Farming, 1607–1972.* (Illus.) Iowa State Univ. Press, 1975. x + 342pp. $12.95. 74-19455. ISBN 0-8138-0090-0. Index;CIP.

SH-C-P   This is an historical narration of what American farmers achieved for themselves and what others achieved for them. Aspects of land, markets, technology and science, with special reference to theoretical data, economic and political information, handling and transportation are recounted.

Scientific American.   *Plant Agriculture: Readings from* **Scientific American.** (Illus.) Freeman, 1970. 246pp. $10.00; $4.95(p). 71-99048. ISBN 0-7167-0995-3.

C   Twenty-five articles deal with the origin of agriculture; crops; plant growth and development; plant ecology; world food needs and availability; biological, geographical, physical, social, and economic factors influencing crop production.

Ucko, Peter J., and G.W. Dimbleby.   *The Domestication and Exploitation of Plants and Animals.* (Illus.) Aldine, 1969. xxvi + 581pp. $17.50. 70-87945.

C-P   Covers very thoroughly all aspects of the origin, development and consequences of animal and plant domestication, especially in the Near East and the Old World. Also covers methods of investigation, particular taxonomic groups, and an analysis of the evolution of human nutrition with respect to dietary changes resulting from exploitation of cultigens.

Winkler, A.J., et al.   *General Viticulture.* (Illus.) Univ. of Calif. Press, 1975. xxi + 710pp. $27.50. 73-87507. ISBN 0-502-02591-1. Index.

C-P   This is the bible of viticulture. There are sections on the development of the various components of the mature grapes, improving the quality of table fruit, vineyard operations, sprinkler irrigation, frost and heat control, mechanical harvesting, and wine grapes and propagation. Many illustrations and tables.

632.9 PEST CONTROL

Beatty, Rita Gray.   *The DDT Myth: Triumph of the Amateurs.* John Day, 1973. xxii + 201pp. $6.95; $3.95(p). 72-12084. ISBN 0-381-98242-4; 0-381-90007-X.

JH-SH-C   Argues that DDT is the safest, cheapest and most effective material for control of diseases which are transmitted by insects, and saves as many lives as penicillin. Typhus, yellow fever and bubonic plague have been controlled by DDT, as have many agricultural pests, and Beatty claims it is safe for people. Makes a persuasive case for not banning DDT.

Crafts, Alden S.   *Modern Weed Control.* (Illus.) Univ. of Calif. Press, 1975. viii + 440pp. $15.75. 74-76383. ISBN 0-520-02733-7. Indexes.

SH-C   Sections on the general nature of weeds, the cost of weeds to agriculture, weed control measures (chemical and biological), and a catalog of weed control materials for use in crop plantings and aquatic environments. Recommended as a general reference or text.

DeBach, Paul. *Biological Control by Natural Enemies.* (Illus.) Cambridge Univ. Press, 1974. 323pp. $14.95; $5.95(p). 73-90812. ISBN 0-521-20380-5; 0-521-09835-1. Index;bib.

   C This valuable reference presents a comprehensive study of the workings and potentiality of biological control of pests. DeBach makes a strong case for further research on biological control ecology.

Gillett, James W., (Ed.). *The Biological Impact of Pesticides in the Environment.* (Illus.) Oregon State Univ. Press, 1970. ix + 210pp. $10.00. 70-123894.

   SH-C-P These papers present research on several aspects of transport and accumulation of chlorinated hydrocarbons in the environment, and their effects on metabolism and physiology of fish, birds, and mammals. Also included are several reviews of the role and problems of pesticides in the environment. An excellent reference.

Graham, Frank, Jr. *Since Silent Spring.* Houghton Miflin, 1970. 333pp. $6.95. 77-82948.

   SH This first-class sequel to Rachel Carson's *Silent Spring* provides absorbing information about the steps that led to her book, many insights into her personality, and a factual, low-key account of subsequent chemical pollution of the environment.

Metcalf, Robert L., and William H. Luckmann (Eds). *Introduction to Insect Pest Management.* (Illus.) Wiley, 1975. xiv + 587pp. $19.95. 74-34133. ISBN 0-471-59855-0. Index;CIP.

   C-P This is a highly practical, wide-ranging approach to pest management. Such topics as plant resistance, parasitoids and predators, diseases, insecticides, attractants, repellents, genetic control, sampling and measuring are covered, and a lucid and complete statement of the conceptual foundations is provided.

Pfadt, Robert E., (Ed.). *Fundamentals of Applied Entomology, 2nd ed.* (Illus.) Macmillan, 1971. x + 693pp. $14.95. 73-133562. Gloss.

   SH-C-P This book, written for students of agriculture, is also suitable for an elementary entomology course. Important terms are defined, facts and phenomena are described in considerable depth, and problem-solving procedures are suggested. Includes a list of common names of insects.

Pringle, Laurence. *Pests and People: The Search for Sensible Pest Control.* (Illus.) Macmillan, 1972. 118pp. $4.95. 71-165104. Index;gloss.;bib.

   JH-SH The emphasis is on biological control as opposed to the use of biocides. Problems in adopting biological control are presented. Excellent collateral reading but not a reference or text.

Van Emden, Helmut F. *Pest Control and Its Ecology.* (Illus.) Arnold (distr.: Crane, Russak), 1975. 59pp. $2.75(p). ISBN 0-7131-2471-7; 0-7131-2472-5(p).

   C This book provides background information concerning the various alternatives for pest control and their ecological impact. The author puts the various methods—chemical and biological, including pest resistant crop varieties—into perspective.

Whorton, James. *Before Silent Spring: Pesticides and Public Health in Pre-DDT America.* Princeton Univ. Press, 1974. xv + 288pp. $12.50. 74-11071. ISBN 0-691-08139-5. Index;CIP.

   SH-C-P Whorton traces the history of pesticide development in the early 1960s, and describes the leadership provided by some outstanding people in the regulatory positions who helped protect the public against the use of toxic substances. The history of the Food and Drug Administration is presented.

635 HORTICULTURE

Flemer, William. *Nature's Guide to Successful Gardening and Landscaping.* (Illus.) Crowell, 1972. xi + 331pp. $8.95. 72-78264. ISBN 0-690-57455-X.

SH-C-P   Authoritative chapters on the garden as an ecosystem and on propagation of plants and the natural control of insects. The major portion of the book describes plantings suitable for various kinds of gardens. Each plant is identified by scientific and common name, many are pictured, and there are useful tips and diagrams on special culture techniques.

Franz, Maurice, (Ed.). *The Calendar of Organic Gardening: A Guidebook to Successful Gardening Through the Year.* (Illus.) Rodale, 1973. 191pp. $6.95. 73-2280. SBN 0-87857-067-5.

SH-C   The reader is told how to garden precisely, day-by-day and month-by-month, in the continental United States where seven climate zones exist for vegetable, ornamental, orchard and bush gardens and greenhouse cultivation. Two frost zone maps are attached.

Huxley, Anthony. *Plant and Planet.* (Illus.) Viking, 1975. xiv + 432pp. $14.95. 75-2403. ISBN 0-670-55886-9. Index;CIP.

SH-C   Huxley summarizes much of the basic detail of how plants function, concentrating on the more extraordinary aspects of that functioning. He successfully combines basic aspects of plant physiology, plant evolution, economic botany and current ecological concerns. Suitable for the amateur botanist and curious gardener.

Hyams, Edward. *A History of Gardens and Gardening.* (Illus.) Praeger, 1971. ix + 345pp. $25.00 70-109474.

C-P   Hyams covers recorded time and all continents in this fascinating history of the art of cultivation of plants for esthetic purposes. Hundreds of color and black-and-white illustrations. A fine coffee-table book for the horticulturist and an excellent reference.

Lees, Carlton B. *Gardens, Plants and Man.* (Illus.) Prentic-Hall, 1970. 252pp. $19.95. 67-10531.

SH-C-P   A wide scope of application for students and amateurs. It covers the history of botany and horticulture, art and ecology. The message is how people from earliest culture have appreciated the esthetic value of plants, imported and improved them, and made gardens and parks with them.

Leighton, Ann. *Early American Gardens: "For Meate or Medicine."* (Illus.) Houghton Mifflin, 1970. xviii + 441pp. $10.00. 68-26957.

SH-C   Recounts the people, times, and origins of much of early American gardening and wild plants. The forms and shapes of early gardens and the plants grown in them are discussed.

Milne, Lorus J., and Margery Milne. *Because of a Flower.* (Illus.) Atheneum, 1975. 152pp. $6.95. 74-19292. ISBN 0-689-30452-8. Index;CIP.

JH-SH   A welcome item, this book uses the flower as a center for the study of nature. The authors explore the general nature of flowers, fruits and seeds, and their relationships with animals. Written in a conversational style which is pleasant to read and attractively illustrated.

Powell, Thomas, and Betty Powell. *The Avant Gardener: A Handbook and Sourcebook of All That's New and Useful in Gardening.* (Illus.) Houghton Mifflin, 1975. vii + 263pp. $12.95. 74-34599. ISBN 0-395-20460-7. Index;CIP.

C   *The Avant Gardener* is a needed reference tool for those for whom horticulture is more than a hobby but less than a profession.

Wyman, Donald, (Ed.).   *Wyman's Gardening Encyclopedia.* (Illus.) Macmillan, 1971. xv + 1222pp. $17.50. 69-18250. Bib.

SH-C-P   A valuable reference for anyone needing information about a specific plant, type of garden, botanical term, garden product, tool, process, etc.

635.9 FLOWERS AND ORNAMENTAL PLANTS

Faust, Joan Lee.   *The New York Times Book of House Plants.* (Illus.) Quadrangle, 1973. 274pp. $9.95. 72-91701. ISBN 0-8129-0309-9.

SH-C   Helpful information on the care of houseplants, a calendar of what to do each month and thumbnail sketches of 150 popular plants. Covers gardening under lights, office plants, bottle gardens and terrariums, the forcing of bulbs, propagation, houseplant societies and where to get plants and supplies.

Forsberg, Junius L.   *Diseases of Ornamental Plants, rev. ed.* (Illus.) Univ. of Illinois Press, 1975. 220pp. $9.95; $4.00(p). 75-6053. ISBN 0-252-00560-0. Gloss.;index.

C-P   A wealth of information about the causes and control of plant diseases and use of fungicides and soil sterilization. Specific plants and their diseases are listed in alphabetical order. Useful to amateurs and professionals.

Graf, Alfred Byrd.   *Exotic Plant Manual: Fascinating Plants to Live with—Their Requirements, Propagation and Use.* (Illus.) Scribner's, 1970. 840pp. $27.50. 77-115116.

GA   This manual presents some 3600 photographs of exotic indoor plants most likely to be found in commercial cultivation. There is also an ingenious set of symbols that indicate basic cultural conditions, brief descriptions, summaries of cultural practices for indoor plants, and gardening in the tropics.

Hebb, Robert S.   *Low Maintenance Perennials.* (Illus.) Quadrangle/N.Y. Times, 1975. 220pp. $4.95(p). 75-8421. ISBN 0-8129-0587-3. Index.

SH-GA-C   This is a handbook designed to guide readers quickly and efficiently to flowering plants recommended for their ease of culture. Some 1000 species, varieties or cultivars are considered.

Hoke, John.   *Terrariums.* (Illus.) Watts, 1972. 90pp. $3.75. 70-189761. ISBN 0-531-00777-4.

JH-SH   Begins with a brief and interesting account of the development of the idea of a self-contained environment for the support of life. Then construction of several sizes and types of terrariums is outlined, supplemented with very good photographs. The types of plants, where to find them and how to collect them are discussed.

Kramer, Jack.   *The Complete Book of Terrarium Gardening.* (Illus.) Scribner's, 1974. ix + 146pp. $9.95. 73-17253. ISBN 0-684-13703-8.

JH-SH-C   Includes a section on children making their own terrariums, explanations of the principles of a terrarium, the basics of construction and practical details for successful terrariums, lists and illustrations of plants according to size and environment, plant diseases, and much more.

Lamb, Edgar, and Brian Lamb.   *The Pocket Encyclopedia of Cacti and Succulents in Color.* (Illus.) Macmillan, 1970. 217pp. $4.95. 77-93722.

JH-SH-C   Oriented to the hobbyist interested in cacti and other succulent plants. The approach is practical and tells how to grow these plants and keep them healthy and alive, as well as how to identify them.

Lavine, Sigmund A., *Wonders of the Cactus World.* (Illus.) Dodd, Mead, 1974. 78pp. $4.50. 73-6035. ISBN 0-396-06830-8. Index.

JH-SH-C   This technical work is easy to read and interesting. It covers the botany, life and growth, care and culture of cacti and their uses and light and water requirements. Scientific names are used abundantly. For students, hobbyists and even botanists.

McDonald, Elvin. *The World Book of House Plants, rev. ed.* (Illus.) Funk&Wagnalls, 1975. 320pp. $8.95. 74-23165. ISBN 0-308-10087-5. Gloss.;index;CIP.

GA   Written in a very readable style, this book gives the novice a quick run-through of the basics needed for good plant care and maintenance. Two outstanding features are ample use of charts and tables and an illustrated listing of 700 plants.

Meyer, Mary Hockenberry. *Ornamental Grasses: Decorative Plants for Home and Garden.* (Illus.) Scribner's, 1975. vii + 136pp. $9.95. 75-11720. ISBN 0-684-14300-3. Index;CIP.

C   Discusses the origin of the ornamental grasses and includes a long list of specimens, each with botanical name, common name and place of origin. Uses of grasses are explained and suggestions for proper culturing and maintenance are provided. Useful for amateurs and professionals.

Morton, Julia F. *Exotic Plants.* (Illus.) Golden, 1971. 160pp. $4.95. 79-149094. Index;bib.;gloss.

JH-SH   Describes briefly and illustrates nearly 400 species from tropical and sub-tropical climates. The emphasis is on ornamental plants, and the colored illustrations feature flowers, foliage and, in some cases, fruit and habitat sketches.

Northen, Henry T., and Rebecca T. Northen. *Greenhouse Gardening, 2nd ed.* (Illus.) Ronald, 1973. viii + 388pp. $9.50. 75-190208.

SH-C   A complete, well-written, well-illustrated "how-to" book. Covers greenhouse management, plant insect pests and diseases, and individual plants.

Padilla, Victoria. *Bromeliads: A Descriptive Listing of the Various Genera and the Species Most Often Found in Cultivation.* (Illus.) Crown, 1973. 134pp. $12.50. 72-84287. ISBN 0-517-500450. Bib.

SH-C   This is a beautifully illustrated book which will be useful for the identification of bromeliads—especially cultivated plants. It is designed to help the beginner in plant selection.

Venning, Frank D. *Cacti.* (Illus.) Golden/Western, 1974. 160pp. $1.95(p). 74-76431. Index.

SH-C   Pocket-sized, easily portable book for the amateur botanist. Contains approximately 200 descriptions, each accompanied by color illustrations, providing useful information on the history and uses of various species of cacti.

636 DOMESTIC ANIMALS

Acker, Duane. *Animal Science and Industry, 2nd ed.* (Illus.) Prentice-Hall, 1971. xii + 604pp. $10.95. 70-124513.

SH-C   Significant changes throughout the world in technology, legislation, consumers, products of animal agriculture, and animal science education are incorporated in this revision of an interestingly written, broad text.

Bowman, John C. *An Introduction to Animal Breeding.* Arnold (distr.: Crane, Russak), 1974. 76pp. $2.50(p). ISBN 0-7131-2445-8; 0-7131-2446-6.

C-P   This work is almost exclusively concerned with commercial livestock and poultry breeding and selection. The principles are fundamental to any animal population. Appropriate auxiliary reading for an introductory course in genetics.

Gay, Margaret Cooper. *How to Live with a Cat, 3rd ed.* (Illus.; rev. by J.R. Sterling) Simon&Schuster, 1969. 272pp. $5.95. ISBN 0-671-20199-9.

JH-SH-GA   An indispensable manual which tells how to get, buy, feed, raise and care for kittens and cats; advises concerning accidents, ailments, sex life, crime and punishment, "cat intelligence," and more.

McDowell, R.E., et al. *Improvement of Livestock Production in Warm Climates.* (Illus.) Freeman, 1972. x + 711pp. $19.50. 74-170395. ISBN 0-7167-0825-6. Bib.

C-P   Discusses problems associated with tropical and subtropical livestock production. Covers physical environment and animal physiology, nutrition, management and breeding. Primary emphasis is on the cow, but there are chapters on water-buffalo and sheep, and a discussion of handling meat and milk in the tropics.

Manolson, Frank, and Robert C. Williams. *My Cat's in Love, Or How to Survive Your Feline's Sex Life, Pregnancy, and Kittening.* (Illus.) St. Martin's, 1970. viii + 204pp. $6.95. 72-121859.

SH-C   This excellent book by two veterinary surgeons tells cat lovers and breeders what they ought to know about cats, their raising, care, feeding, training, illnesses, treatments, etc.

Marlin, Herb, and Sam Savitt. *How to Take Care of Your Horse Until the Vet Comes: A Horse Health and First Aid Guide.* (Illus.) Dodd, Mead, 1975. 96pp. $4.95 75-11727. ISBN 0-396-07145-7. Index;CIP.

JH-SH-GA-C   A quick first aid reference. While the style of presentation is simple enough for a 10- or 12-year old, the technical information is accurate and well presented.

Ricciuti, Edward R. *Shelf Pets: How to Take Care of Small Wild Animals.* (Illus.) Harper&Row, 1971. xi + 132pp. $4.50. 76-135782. ISBN 0-06-024993-5.

JH   A useful and much-needed contribution on the care of creatures that youngsters are apt to bring home and which can be looked after reasonably easily. The largest animals mentioned are hamsters, guinea pigs, and turtles. There are excellent instructions for making simple terraria and aquaria.

Rogers, Cyril. *Seed Eating Birds as Pets.* (Illus.) Scribner's 1975. 105pp. $7.95. 74-4611. ISBN 0-684-13897-2. Index.

JH-SH-C   This is a general reference book on bird genetics and cage and aviary birds. Information on feeding, accommodation, care, suggestions for collections of foreign birds, and sickness and disease.

Smythe, R.H. *The Dog: Structure and Movement.* (Illus.) Arco, 1970. 144pp. $5.95. 70-113948. ISBN 0-668-02319-8.

SH-C   Any dog owner or fancier could profit from this book, which deals primarily with the anatomy of the dog as it is related to observed movements.

638.1 BEEKEEPING

Aebi, Ormond, and Harry Aebi. *The Art and Adventure of Beekeeping.* (Illus.) Unity, 1975. v + 184pp. $9.95; $4.95(p). 74-14661. ISBN 0-913300-39-X; 0-913300-38-1. Index;CIP.

JH-SH-C   Bee's life stages, reproductive habits, seasonal needs, and how to work with honeybees for the most pleasure and profit are described.

Morse, Roger A. *Bees and Beekeeping.* (Illus.) Comstock/Cornell Univ. Press, 1975. 295pp. $13.50. 74-27440. ISBN 0-8014-0884-9. Index.

SH   An excellent general overview of the art of beekeeping. Covers elementary
     biology, communication and social control and colony and apiary manage-
ment, and even gives a good overall picture of agriculture in the northern United
States. For amateurs and professionals.

639.2–.3 COMMERCIAL FISHING AND AQUACULTURE

Bardach, John E., John H. Ryther, and William O. McLarney. *Aquaculture: The Farming and Husbandry of Freshwater and Marine Organisms.* (Illus.) Wiley, 1972. xii + 868pp. $37.50. 72-2516. ISBN 0-471-04825.

SH-C-P   Rather than being a complete text of aquaculture, this is a highly read-
     able, interesting, and instructive book that will provoke discussion and
could serve as a standard text for students of agriculture.

Brown, Joseph E. *The Sea's Harvest: The Story of Aquaculture.* (Illus.) Dodd, Mead, 1975. 96pp. $4.95. 75-9646. ISBN 0-396-07153-8. Index;CIP.

JH-SH   A readable summary of the field of mariculture, this work provides much
     of the basic information needed for understanding the goals and limita-
tions of this potentially significant industry.

Euller, John. *Whaling World.* (Illus.) Doubleday, 1970. 119pp. $3.95. 79-121578.

JH-SH   A brief, factual, sometimes thrilling history of the whaling industry from
     prehistoric times to the present. Discusses the different kinds of whales,
their natural history, and some of their unique biology. The book is largely an indict-
ment of the destruction of these magnificent mammals.

Hanson, Joe A., (Ed.). *Open Sea Mariculture: Perspectives, Problems, and Prospects.* (Illus.) Dowden, Hutchinson&Ross, 1974. xi + 410pp. $24.00. 74-13103. ISBN 0-87933-130-5. Index;CIP.

C   This is a comprehensive report on the opportunities and problems of the con-
     trolled growth of marine organisms away from the shoreline. The volume con-
sists of perspectives; descriptions of the oceanic environment; biological, engineering and technological considerations; and research recommendations.

Jensen, Albert C. *The Cod.* (Illus.) Crowell, 1972. 182pp. $7.95. 70-187933. ISBN 0-690-19688-1.

SH-C   A remarkably complete and authoritative account of the cod. Its natural
     history, geographical distribution and economic importance, the develop-
ment of fishing methods and techniques and the greatly increased exploitation by
factory ships all have a role in the story.

639.34 AQUARIUMS

Cox, Graham F. *Tropical Marine Aquaria.* (Illus.) Grosset&Dunlap, 1972. 159pp. $3.99. 73-175348. ISBN 0-448-00814-9; 0-448-04181-2.

SH   A basic handbook giving details for the design of tanks, aeration, temperature
     control, lighting, and filtration, and for both a natural system and a synthetic
system under continuous filtration. Thumbnail descriptions of tropical invertebrates
and fishes, full-color illustrations, diseases, and "do's and don't's" for beginners.

Cust, George, and Peter Bird. *Tropical Freshwater Aquaria.* (Illus.) Grosset&Dun-
lap, 1971. 159pp. $3.99. 70-145742. ISBN 0-448-00869-6.

SH   Information that a beginning aquarist can understand, but also an easy refer-
     ence for experienced fanciers, with 345 colored illustrations of likely fish for

home aquaria; plus plants, parasites, equipment, procedural care, and the basic biology of aquaria.

Dutta, Reginald. *Fell's Beginner's Guide to Tropical Fish (Fish Tanks, Coldwater Aquarium Fish, Pond Fish, Ponds, and Marines).* (Illus.) Fell, 1975 (c.1971). 179pp. $5.95. 75-4357. ISBN 0-8119-0254-4. Index.

JH-SH-C   This basic manual covers important aspects of the aquarium hobby, including the setting up of the tank, conditioning of the water, lighting, water temperature, feeding, breeding, etc.

Halstead, Bruce W., and Bonnie L. Landa. *Tropical Fish: A Guide for Setting Up and Maintaining an Aquarium for Tropical Fish and Other Animals.* (Illus.) Golden/ Western, 1975. 160pp. $6.60; $1.95(p). 74-80977. ISBN 0-307-24361-3. Index.

JH-SH-C-P   With lots of illustrations and clear writing, this book provides details of setting up and maintaining an aquarium, how to treat diseases and care and feeding of many species. A good choice for both beginning and experienced aquarists.

Hoedeman, J.J. *Naturalists' Guide to Fresh-Water Aquarium Fish.* (Illus.;trans.) Sterling, 1974. 1152pp. $30.00. 72-95209. ISBN 0-8069-3722-X; 0-8069-3723-8(p). Gloss.;index.

SH-C-P   Two interrelated books bound under one cover give both a basic intro-duction to ichthyology and a complete illustrated catalog of virtually all freshwater fish found in home aquaria.

Neugebauer, Wilbert. *Marine Aquarium Fish Identifier.* (Illus.;trans.) Sterling, 1975 (c. 1973). 256pp. $4.95. 74-82341. ISBN 0-8069-3724-6; 0-8069-3725-4. Indexes.

JH-SH   Fish which can be kept in salt water aquariums are described, classified into families by shape and other factors, and identified by color plates. Useful to both the amateur and professional.

Paysan, Klaus. *Aquarium Fish from Around the World.* (Illus.) Lerner, 1970. 106pp. $6.95. 73-102892. ISBN 0-8225-0561-4.

JH   The introduction explains how to set up the aquarium with fish groupings as well as plantings. Fifty fish are discussed and beautifully illustrated. Informa-tion is provided on mating habits, dietary requirements, size, scientific names, water temperature, interesting vignettes, etc.

Schiotz, Arne. *A Guide to Aquarium Fishes and Plants.* (Illus.;trans.) Lippincott, 1972. 223pp. $6.95. 75-38541. ISBN 0-397-00866-X. Bib.

SH-GA-C   Over 350 fish and 50 plant species are arranged by family, including freshwater and marine. The original paintings are of excellent quality. Includes scientific and common names, and a map showing the world distribution of each family.

Simon, Hilda. *Strange Breeding Habits of Aquarium Fish.* (Illus.) Dodd, Mead, 1975. 160pp. $5.95. 74-11798. ISBN 0-396-07025-6. Gloss.;index.

JH-SH-C   The intermediate-to-advanced aquarist will be intrigued by the strange breeding habits of eight groups of marine, estuarine and freshwater fishes. Diverse topics such as nest building, mouth brooding and postnatal care are clearly presented. Not a how-to-raise fish book, but a good guide to observing.

Spotte, Stephen. *Marine Aquarium Keeping: The Science, Animals, and Art.* (Illus.) Wiley, 1973. xv + 171pp. $9.95. 73-4425. ISBN 0-471-81759-7.

JH-SH   Spotte discusses sea water, filtration, aeration, feeding and general main-tenance, with especially valuable chapters on which organisms can be maintained successfully and the diseases frequently encountered.

639.9 CONSERVATION OF BIOLOGICAL RESOURCES

See also 301.3 Ecology and Community; 333.7 Conservation of Resources; and 591.042 Rare and Extinct Animals.

Frome, Michael. *Battle for the Wilderness.* Praeger, 1974. 246pp. $8.95. 73-15175. ISBN 0-275-19110-0. Index;CIP.

SH-C Frome addresses the problems concerning the preservation or conservation of all our important wilderness areas. First he offers a definition and discussion of wilderness in many different lights, and then he provides suggestions on how to save the wilderness.

Gunter, Pete. *The Big Thicket: A Challenge for Conservation.* (Illus.) Chatham (distr.: Viking), 1971. xvii + 172pp. $12.50. 73-184310. ISBN 85699-044-2.

SH-C An ecological monograph on the Big Thicket region of east Texas. Catalogs the continuing destruction of its many diverse ecosystems, but shows that a National Park is potentially viable in this region. Considers the biology and the economics of the area and the social impact on it.

Jenkins, Alan C. *Wild Life in Danger.* (Illus.) St. Martin's, 1974. 168pp. $5.95. 73-82074.

JH-SH A remarkably well illustrated review of human predation on other vertebrates. Not a scientific study of these animals, but some discussion of methods of studying populations, predation and migration is included. Metric units have been added to this edition. There is a strong argument for strict conservation measures, but no emotionalism.

Laycock, George. *Wild Animals: Safe Places.* (Illus.) Four Winds, 1973. 168pp. $5.95. 72-87082.

JH-SH Covers three national parks (Yellowstone, Mt. McKinley and the Everglades) and the Hawaiian Island National Wildlife Refuge. Describes the animals found in these wilderness areas.

Line, Les, (Ed.). *What We Save Now: An Audubon Primer of Defense.* (Illus.) Houghton Mifflin, 1973. xix + 438pp. $10.00. 74-132785. ISBN 0-395-16613-6.

JH-SH-C Twenty-five examples of human-induced ecological disasters—from strip mining to gypsy moth plagues—illustrate the thesis that the world is changing for the worse. Interesting and provocative; suitable for reference.

Margolin, Malcolm. *The Earth Manual.* (Illus.) Houghton Mifflin, 1975. ix + 190pp. $10.00; $5.95(p). 75-1324. ISBN 0-395-20425-9; 0-395-20541-7. Index;CIP.

SH The author outlines in a straightforward, interesting way how to take care of wood lots, parks, small farms and big backyards to render them more livable by encouraging wilderness. Teachers looking for field trip ideas will like this book.

Mitchell, John G. *Losing Ground.* Sierra Club, 1975. 227pp. $8.95. 75-1073. ISBN 0-87156-128-X. CIP.

JH-SH-GA A collection of essays describing recent visits to locales in which the author had previously lived, fished, or fought against the destruction of the environment. The book reads easily and would be excellent supplementary reading in ecology.

Myers, Norman. *The Long African Day.* (Illus.) Macmillan, 1972. 404pp. $25.00. 71-182022.

SH-C The successes and shortcomings of the National Parks system in preserving the natural ecosystems in East Africa are critically discussed. Examined are the implications of the activities and the impact of various species, including humans, on each other.

Nash, Roderick. *Wilderness and the American Mind, rev. ed.* Yale Univ. Press, 1973. xvi + 300pp. $10.00; $2.95(p). 72-91303. ISBN 0-300-01648-4; 0-300-01649-2. Bib.

C-P   Nash traces wilderness attitudes to ancient roots and gives sharp definition to the conflicts and contradictions in the wilderness movement today. An essential reference and a delight to read.

Olsen, Jack. *Slaughter the Animals, Poison the Earth.* (Illus.) Simon&Schuster, 1971. 287pp. $6.95. 70-156160. ISBN 0-671-20996-5.

JH-SH-C-P   Targets are the sheepmen, cattle ranchers and government agencies that have joined forces to exterminate wildlife. Ostensibly their mass poisoning is directed against coyotes and other predators, but the author demonstrates that the methods used are unselective. The case has been argued before, but seldom so persuasively.

Philip, H.R.H., Duke of Edinburgh, and James Fisher. *Wildlife Crisis.* (Illus.) Cowles, 1970. 256pp. $14.95. 76-116129. ISBN 0-402-12511-8.

SH-C   One of the most pictorially magnificent books available on the importance of conserving wildlife and on the practical, esthetic and academic significance of preventing its extinction. The necessity of creating an administrative system capable of formulating and implementing a comprehensive conservation policy is emphasized. A primary reference.

Regenstein, Lewis. *The Politics of Extinction: The Shocking Story of the World's Endangered Wildlife.* (Illus.) Macmillan, 1975. xxiv + 280pp. $9.95. 74-30233.

SH-C   Regenstein supplies details about the people, businesses, government agencies, conservation and lobbying groups involved with political, commercial and institutional interests which endanger many animal groups. A valuable and important book.

Scheffer, Victor B. *A Voice for Wildlife.* (Illus.) Scribner's, 1974. x + 245pp. $8.95. 73-19290. ISBN 0-684-13714-3. Index;CIP.

SH-C   Scheffer's contention is that our relationships with wildlife should be determined on the basis of ecological harmony and ethical recognition of wildlife's rights and not on the basis of their commercial or other consumptive usefulness.

Shuttlesworth, Dorothy E. *The Wildlife of South America.* (Illus.) Hastings House, 1974. 120pp. $5.95. 74-817. ISBN 0-8038-8069-3. Index;CIP.

JH-SH   Intended to stir the reader's interest in the conservation of South American wildlife, this book also gives solid background information on various species and their habitats.

Sutton, Ann, and Myron Sutton. *Yellowstone: A Century of the Wilderness Idea.* (Illus.) Macmillan, 1972. 219pp. $25.00. 72-183207.

SH   The color photographs are outstanding; the text is concise, delightful and accurate. Covers history, natural history, the problems of establishing and maintaining national parks worldwide, and the difficulties encountered in species preservation.

Zimmerman, David R. *To Save a Bird in Peril.* (Illus.) Coward, McCann&Geoghegan, 1975. 286pp. $9.95. 74-30604. SBN 698-10671-7. Gloss.;index;bib.

JH-SH-C   This book will be enjoyed by professional as well as amateur ornithologists and conservationists. The writing is fresh, conversational, filled with interesting facts and accounts of bird conservation, appropriately footnoted and varied.

640 Domestic Arts and Sciences

640.73 CONSUMER EDUCATION

Berger, Melvin. *Consumer Protection Labs.* (Illus.) John Day, 1975. 127pp. $6.95. 75-6686. ISBN 0-381-99622-0 RB. Index;CIP.

    JH-SH   Berger describes nine types of scientific laboratories that affect consumer products. Examples are government food, food processing, safety testing, and agricultural labs. An excellent resource for high school libraries and guidance collections.

641.1 APPLIED NUTRITION

See also 613 General and Personal Hygiene and 614 Public Health.

American Medical Association. *Nutrients in Processed Foods: Fats, Carbohydrates.* (Illus.) Publishing Sciences Group, 1975. xx + 206pp. $16.00. 73-85400. ISBN 0-88416-013-0.

    C-P   Deals with fat and fatty acid substitution, sugars and complex carbohydrates and functional and nutritional considerations. An excellent reference.

American Medical Association. *Nutrients in Processed Foods: Vitamins and Minerals.* (Illus.) Publishing Sciences Group, 1974. xvii + 193pp. $16.00. 73-85400. ISBN 0-88416-006-8.

    SH-C-P   Evaluates the health costs and benefits of fortification of processed foods. Recommended for dieticians, nutritionists, food scientists, and consumers.

Ashley, Richard, and Heidi Duggal. *Dictionary of Nutrition.* St. Martin's, 1975. 236pp. $8.95. 73-87397.

    SH-C   This is a concise, practical interest book which provides basic nutritional information. The authors detail the compositions and nutrient values of over 400 foods and discuss related factors which affect nutrition.

Fenten, Barbara, and D.X. Fenten. *Natural Foods.* (Illus.) Watts, 1974. 66pp. $3.95. 73-10381. ISBN 0-531-02675-2; 0-531-02409-1(p). Index;CIP.

    JH-SH   Sophisticated coverage of natural foods from an advocate's viewpoint. Beginning with a general introduction, the Fentens discuss food additives, give pointers for shopping for natural foods, and teach the reader how to select a good natural foods store.

Guthrie, Helen Andrews. *Introductory Nutrition, 3rd ed.* (Illus.) Mosby, 1975. viii + 576pp. $10.95. 74-20986. ISBN 0-8016-2000-7. Gloss.;index;CIP.

    C   Components of food are discussed and a brief historical sketch of each is provided. The applied nutrition section covers selection of an adequate diet, dietary standards, nutrition in pregnancy and lactation, and alternative food patterns as well as chemical and physiological aspects of food utilization.

Hyde, Margaret O., and Elizabeth Held Forsyth. *What Have You Been Eating? Do You Really Know?* McGraw-Hill, 1975. 146pp. $5.72. 74-31376. ISBN 0-07-031564-7. Index;CIP.

    JH-SH   A reasonably unbiased treatment of some of the current concerns about nutrition, human needs for food, and the food supply, along with some guidelines for thoughtful consumer choices. Topics include food faddism, additives, diets, hunger, food in different cultures, etc.

Lamb, Lawrence E.   *What You Need to Know About Food and Cooking for Health.*
Viking, 1973. xiii + 412pp. $10.00. 74-186736. SBN 670-75923-6.

JH-SH-C-P   A treatise on the art and science of food preparation and eating, with
a detailed listing of the calories contributed by the various food components. Tells ''how to eat most of the things you like and at the same time have a
healthier, more natural diet.''

Lamb, Mina W., and Margarette L. Harden.   *The Meaning of Human Nutrition.*
(Illus.) Pergamon, 1973. vii + 284pp. $11.50; $6.50(p). 72-11636. ISBN 0-08-017078-1;
0-08-017079-X.

JH-SH-C   This textbook covers the basis for identification of reliable authors,
speakers and publications; the involvement of the consumer in providing adequate nutrition through food purchase; the gross assessment of nutritional
status; dietary supplements; the impact of culture on dietary practices; the relationship of nutrition in growth and development; and a discussion of energy, proteins,
mineral supplements and vitamins.

Margolius, Sidney.   *Health Foods: Facts and Fakes.* Walker, 1973. 293pp. $6.95. 75-
186187. ISBN 0-8027-0375-5. Index;gloss.;bib.

SH-C   Causes of growth of the health food and cosmetic movement are explored,
and detrimental effects of health faddism, both economic and physical, are
noted. Margolius evaluates evidence from various sources and substantiates his conclusions by extensive observations and documentation. Appendix covers vitamins
and minerals, food additives, recommended daily allowances for nutrients.

Mayer, Jean, (Ed.).   *U.S. Nutrition Policies in the Seventies.* Freeman, 1973.
xi + 256pp. $7.95; $3.95(p). 72-6548. ISBN 0-7167-0599-0; 0-7167-0596-6.

SH-C   This publication, an outgrowth of a White House Conference, summarizes
the nutritional policies of this country. Topics range from obesity to effects
of lack of food among the poor, nutritionally related diseases, new types of foods,
and consumer concerns. The authors convincingly point out that malnutrition exists
in the United States and that 20 to 30 million people cannot afford a sound diet.

Stare, Fredrick J., and Margaret McWilliams.   *Living Nutrition.* (Illus.) Wiley, 1973.
vii + 467pp. $10.95. 73-4554. ISBN 0-471-82075-X.

SH-C   A textbook for students and practitioners covering sociopsychological aspects of nutrition as well as nutrition from the physiological viewpoint.

Taylor, Ronald L.   *Butterflies in My Stomach; Or, Insects in Human Nutrition.* (Illus.)
Woodbridge, 1975. 224pp. $8.95. 75-13593. ISBN 0-912800-08-8. CIP.

C   Taylor's book is a laudable attempt to educate people about the untapped
source of protein which insects represent. Elementary and high school
teachers also will find valuable information.

Wellford, Harrison.   *Sowing the Wind: A Report from Ralph Nader's Center for Study of
Responsive Law on Food Safety and the Chemical Harvest.* Grossman, 1972.
xxiii + 384pp. $7.95. 76-112513. SBN 670-65945-2.

SH-C-P   Reveals the hidden dangers in food which may be tainted by inadequate
inspection of the addition of many potentially harmful substances in
order to increase its durability, appearance, and weight. A valuable reference.

White, Philip L., and Nancy Selvey (Eds.).   *Let's Talk About Food: Answers to Your
Questions About Foods and Nutrition.* Publishing Sciences Group, 1974. 282pp. $6.95.
73-85401. ISBN 0-88416-008-4. Index.

SH   This common-sense approach to nutrition and diet is based on material
adapted from *Today's Health.* A question-and-answer format provides the
lay reader with medical and dietary information.

641.3 FOODS

See also 338.1 Agricultural Production.

Brothwell, Don, and Patricia Brothwell. *Food in Antiquity.* (Illus.) Praeger, 1969. 248pp. $8.50. 69-19524.

    SH-C A survey of foods through prehistory and the early historic period, with emphasis on the Old World. Attention is paid to both domesticated and wild food resources, and there is a discussion of diet, diseases, and techniques for recovering and analyzing food remains.

Codd, L.W., et al. (Eds.). *Chemical Technology: An Encyclopedic Treatment: Edible Oils and Fats, Animal Food Products, Material Resources, Vol. 8.* (Illus.) Barnes&Noble, 1975. xxvii + 600pp. $45.00. 68-31037. ISBN 06-491109-8. Index.

    SH-C-P Includes historical, geographical and analytical aspects of the subject as well as the technological and economical. Clear and concise.

Codd, L.W., et al. (Eds.). *Chemical Technology: An Encyclopedic Treatment: Vegetable Food Products and Luxuries, Vol. 7.* (Illus.) Barnes&Noble, 1975. xxi + 905pp. $45.00. 68-31037. ISBN 06-491108-X. Index.

    SH-C-P Information on sources, manufacture, processing and uses of food products of vegetable origin is presented. Topics range from growing conditions through structure and production to methods of preservation.

Crane, Eva, (Ed.). *Honey: A Comprehensive Survey.* (Illus.) Crane, Russak, 1975. xvi + 608pp. $42.50. 72-83309. ISBN 0-8448-0062-7. Indexes.

    C-P Discusses not only the historical aspects of honey, but also technical and practical aspects, including its use in foods, wines, pharmaceutical items, etc. Maps, tables and illustrations are included.

Esterer, Arnulf K., and Louise A. Esterer. *Food.* (Illus.) Messner, 1969. 190pp. $3.64. 78-83152. ISBN 671-32186-2.

    JH This excellent fact-filled treatise provides a history of the earliest forms of food used by humans and the shift from hunting to cultivation. Topics such as the diversity of things used as food, processing and preservation, modern dietetics, and our constant striving to improve food quality and to increase its quantity are also covered.

Riedman, Sarah R. *Food for People, 2nd rev. ed.* (Illus.) Abelard-Schuman, 1976. 228pp. $6.95. 75-33193. ISBN 0-200-00161-2. Index;CIP.

    JH-SH A cornucopia of facts about food and how it affects people. Some of the information is presented in a detective-story mode. Discusses food eaten by people in different parts of the world, simple biochemistry and biology (stressing the need for balanced diets and the dangers of fad diets), world hunger, and possible new sources of protein-rich food. Introduces some rather difficult concepts in an easy-to-understand manner.

Scientific American. *Food: Readings from* Scientific American. (Illus.) Freeman, 1973. 268pp. $11.00; $5.50(p). 73-3138. ISBN 0-7167-0876-0; 0-7167-0875-2. Bibs.

    SH-C This collection covers foods essential to good health, immediate sources of food, and future resources. A good, quick source work, particularly for science teachers.

643.6 APPLIANCE REPAIR

Darr, Jack. *The Home Appliance Clinic: Controls, Cycle Timers, Wiring and Repair.* (Illus.) Tab, 1974. 195pp. $7.95; $4.95(p). 74-14324. ISBN 0-8306-4745-7; 0-8306-3745-1. Index.

SH-C-P   Descriptions of circuits and common repair problems make this easy-to-read book helpful to electronic repair specialists. Though not complete, it includes controls, heaters, timers, driers and refrigerators. Ideal as a text for training personnel and as a ready reference.

Heiserman, David L. *Handbook of Small Appliance Troubleshooting and Repair.* (Illus.) Prentice-Hall, 1974. xii + 340pp. $17.00. 73-14989. ISBN 0-13-381749-0. Index;CIP.

SH-C   Well-organized manual for the handyperson. Identifies potential problems in electrical systems and keys possible subassembly malfunctions to overall appliance repair. Recommended also for technical education courses to demonstrate organized approaches to appliance repair.

660 Industrial Technologies

Abisch, Roz, and Boche Kaplan. *Textiles.* Watts, 1975. 64pp. $3.90. 74-13430. ISBN 0-531-00824-X. Gloss.;index;CIP.

JH-SH   A fascinating narrative on the development of textiles from the handmade variety of the Stone Age through the modern mills that use synthetic fibers. Uses archeological evidence and helpful illustrations.

Allen, P.W. *Natural Rubber and the Synthetics.* (Illus.) Wiley, 1972. 255pp. $15.75. 72-5094. ISBN 0-470-02329-5.

C-P   An excellent evaluation, from a worldwide viewpoint, of the scientific, technical, economic and social aspects of both the natural and synthetic rubber industries.

Bockris, J.O'M., and Z. Nagy. *Electrochemistry for Ecologists.* (Illus.) Plenum, 1974. xvi + 204pp. $12.50. 73-84003. ISBN 0-306-30749-9. Index;CIP.

SH-C-P   Firmly establishes the notion that electrochemical technology will someday play a major role in manufacturing, recycling, and transportation, without the ecological hazards of today's energy sources. A clear presentation of technical data makes this book useful to ecologists as well as chemists.

Cottrell, Alan. *An Introduction to Metallurgy, 2nd ed.: SI Units.* (Illus.) Crane, Russak, 1975. xii + 548pp. $29.50; $14.00(p). 75-21731. ISBN 0-8448-0767-2. Index.

C-P   A complete survey of the field; outlines the various relevant scientific disciplines and relates these to metallurgy as an applied science. A good reference book.

Hahn, James, and Lynn Hahn. *Plastics.* (Illus.) Watts, 1974. 65pp. $3.45. 73-21944. ISBN 0-531-02702-3. Gloss.;index;CIP.

JH-SH   Requires a basic knowledge of chemistry. Covers the history, chemical make-up, and uses of plastics, and explains the physical behavior of plastic materials. Also discusses important aspects of plastics disposal.

Lyttle, Richard B. *Paints, Inks and Dyes.* (Illus.) Holiday House, 1974. 178pp. $6.50. 73-16876. ISBN 0-8234-0240-1. Gloss.;index;CIP.

JH-SH   A comprehensive introduction, including background and current technology. Presents safe experimental projects.

Milby, Robert V. *Plastics Technology.* (Illus.) McGraw-Hill, 1973. x + 581pp. $12.95. 72-7408. ISBN 07-041918-3.

SH-C   The various procedures for molding, extrusion, laminating, fabricating and casting of thermoset and thermoplastic materials are described and pictured in detail. Brief descriptions of the various types of plastics and conventional test procedures. Review questions are given.

Perry, Robert H., and Cecil H. Chilton (Eds.). *Chemical Engineers' Handbook, 5th ed.* (Illus.) McGraw-Hill, 1973. xx + 2434pp. $35.00. 73-7866. ISBN 0-07-0490478-9.

   C-P   This is a definitive basic reference work for the chemical engineering profession. It contains the most recent and best information on the design and performance of equipment and systems involving the unit processes of chemical engineering.

Schuler, Frederic W., and Lilli Schuler. *Glassforming: Glassmaking for the Craftsman.* (Illus.) Chilton, 1970. xv + 151pp. $12.50. 71-135056. ISBN 0-8019-5558-0.

   SH-C-P   Anyone interested in glassmaking as an art form should have this volume, with its colorplates, outstanding illustrations, and easy-to-follow instructions.

Van Thoor, T.J.W., (Ed.). *Chemical Technology: Vol. 5: Natural Organic Materials and Related Synthetic Products.* (Illus.) Barnes&Noble, 1972. xxxii + 898pp. $42.50 ($37.50 in 8 vol. set). 68-31037. ISBN 06-491106-3.

   SH-C-P   A significant source of immediate information for technical personnel, students of technology, economists, and the general reader. The most frequently asked questions about technology are answered in a brief but lucid manner. Gives important perspectives on technology outside the United States.

711 AREA PLANNING

Carpenter, Philip L., et al. *Plants in the Landscape.* (Illus.) Freeman, 1975. viii- + 481pp. $16.00. 74-32292. ISBN 0-7167-0778-0. Indexes;CIP.

   C   Three sections provide a historical perspective and an introduction to landscape plants and design. Four sections cover contracting and preparation of plans; landscape construction; maintenance; and applications in urban, suburban, and rural environments.

Faludi, Andreas, (Ed.) *A Reader in Planning Theory.* Pergamon, 1973. xii + 399pp. $15.75; $10.00(p). 72-11536. ISBN 0-08-017066-8; 0-08-017067-6.

   C-P   These "theories of planning" include explicit statements of values as well as the use of the usual data gathering activities to help urban decision-makers fulfill their responsibilities.

Farris, Martin T., and Forrest E. Harding. *Passenger Transportation.* (Illus.) Prentice-Hall, 1976. xiv + 290pp. $12.95. 75-35709. ISBN 0-13-652750-7. Index;CIP.

   C-P   A qualitative approach that considers engineering, geography, planning, economic and pricing constraints, regulatory systems, marketing problems, social benefits and costs, and political considerations. Nontechnical; appropriate for engineering courses in transportation and city and regional planning courses.

Goodman, Robert. *After the Planners.* (Illus.) Simon&Schuster, 1971. 231pp. $6.95. 74-154100. ISBN 0-671-20981-7.

   SH-C-P   The author denounces the architects, roadbuilders, urban renewal experts, and others who he claims have forced the poor and underprivileged into physical environments in which there is no opportunity for change. His suggested solutions include helping those affected to take a part in the planning.

Heller, Alfred, (Ed.). *The California Tomorrow Plan.* (Illus.) William Kaufmann, 1972. 120pp. $7.95; $2.50(p). 72-85217. ISBN 0-913232-01-7; 0-913232-02-5.

   SH-C-P   *California TWO,* an outline of the most feasible and satisfying process for getting a better state that a group of experienced citizens could envisage, is the focus of this well-organized study. A must for college students and elected officials concerned with the future of the environment under federalism, but it is written so simply that its use in high schools can be recommended.

Herbert, David. *Urban Geography: A Social Perspective.* (Illus.) Praeger, 1973. 320pp. $10.00; $3.95(p). 72-83562.

   C   Analyzes the internal structure of the modern city from the viewpoint of social
        geography, with only a limited treatment of economic factors. A wide range of examples is utilized, mainly Western world cities. Straightforward and easily understood; a valuable reference.

Johnson, James H. *Urban Geography: An Introductory Analysis, 2nd ed.* (Illus.) Pergamon, 1973. xv + 203pp. $6.95; $3.95(p). 67-21274. ISBN 0-08-016927-9; 08-016928-7.

   GA-C   One of the best. The emphasis is on spatial patterns of land use in cities:
        urban growth, urban form, demographic and occupational characteristics, location and size, the city center and residential suburbs, manufacturing areas, and theories of urban structure. Recreation and environmental aspects are not discussed. A rather segmented approach, not systems-oriented.

Munzer, Martha E., and John Vogel, Jr. *New Town: Building Cities from Scratch.* (Illus.) Knopf, 1974. 150pp. $6.95. 74-158. ISBN 0-394-92673-0. Index;CIP.

   JH-SH   Outlines the development of "new town" philosophy and cites specific
        examples in Columbia, Md.; Reston, Va.; and Jonathan, Minn. Valuable for discussions of government practices, economics and social policy.

Perin, Constance. *With Man in Mind: An Interdisciplinary Prospectus for Environmental Design.* MIT Press, 1970. 185pp. $7.50. 76-123251. ISBN 0-262-16042-0.

   C-P   This interesting theory of environmental design, centering about people and
        their priorities, is rather academic and stuffy but full of interesting analyses of differing approaches to behavioral and social research which have a bearing upon human environment, particularly in the inner city.

Reilly, William K., (Ed.). *The Use of Land: A Citizens' Policy Guide to Urban Growth.* (Illus.) Crowell, 1973. 318pp. $10.00; $3.95(p). 73-8215. ISBN 0-690-00267-X.

   SH-C-P   Many facets of the urban development problem are explored here, and
        solutions are recommended. Gives an in-depth analysis of the complex relationships among growth factors, property rights, laws, traditions and public welfare.

Saarinen, Thomas F. *Environmental Planning: Perception and Behavior.* (Illus.) Houghton Mifflin, 1976. xii + 262pp. $6.50(p). 75-19533. ISBN 0-395-20618-9. Indexes.

   C-P   Begins with studies of human-environment relationships in individual or per-
        sonal spaces, and the same studies continue on to buildings and groups, then neighborhoods, cities, regions, and countries and finally the entire world. The thoughts and writings of over 300 experts in fields such as environmental psychology, perception, behavior, physiology, design and science, ecology, biomedicine, systems, etc. are cited.

Yanev, Peter. *Peace of Mind in Earthquake Country: How to Save Your Home and Life.* (Illus.) Chronicle, 1974. 304pp. $9.95. 74-7406. ISBN 0-87701-050-1. Index;CIP.

   SH-C   Yanev considers location and structural features of buildings as well as
        risks within the building and human reactions. The aim is to help individual property owners survive serious loss, but public administrators would also be interested.

721 ARCHITECTURAL DESIGN AND CONSTRUCTION

Birch, Jack W., and B. Kenneth Johnstone. *Designing Schools and Schooling for the Handicapped: A Guide to the Dynamic Interaction of Space, Instructional Materials, Facilities, Educational Objectives and Teaching Methods.* Charles C. Thomas, 1975. 227pp. $14.50. 74-23213. ISBN 0-398-03362-5. Index;CIP.

    C-P   This is a timely guide to intelligent planning of the educational and architectural needs of all types of children. An essential text for educational administrators, architects, school board members and laypersons.

Fitch, James Marston. *American Building 2: Environmental Forces that Shape It, 2nd rev. ed.* (Illus.) Houghton Mifflin, 1972. xi + 349pp. $15.00. 65-10689. ISBN 0-395-12667-3.

    C-P   Strikes at the fundamental weaknesses of the traditional approach taken by architects in the analysis and solution of a broad range of architectural problems. Presents a well-documented case detailing the shortcomings of modern buildings throughout the world. A sound and penetrating analysis.

Gardiner, Stephen. *Evolution of the House: An Introduction.* (Illus.) Macmillan, 1974. xiii + 298pp. $10.95. 73-10784. ISBN 0-02-542500-5. Index;CIP.

    C   Presents a philosophical orientation for the developments in home construction and design from prehistoric shelters to the sophisticated models of today. Casual, enjoyable reading.

Givoni, B. *Man, Climate and Architecture.* (Illus.) American Elsevier, 1969. xiii + 364pp. $14.00. 69-15822. ISBN 444-20039-8.

    C-P   An extremely well-organized reference or text on the design principles to be used in different climates. The sections on physiological and sensory responses and biophysical effects of environmental factors are most timely.

Jencks, Charles. *Architecture 2000: Predictions and Methods.* (Illus.) Praeger, 1971. 128pp. $7.50; $3.95(p). 76-12859.

    C-P   This profusely illustrated book covers, in a somewhat technical manner, the evolution of architecture in the 20th century, and projects to the last quarter of the century. Begins with a general consideration of methods of forecasting; the remainder of the book describes actual predictions and relates them to biology, sociology, other technologies, and psychology.

Leckie, Jim, et al. *Other Homes and Garbage: Designs for Self-Sufficient Living.* (Illus.) Sierra Club, 1975. 302pp. $9.95(p). 75-8913. ISBN 0-87156-141-7. CIP.

    SH-GA-C   Leckie covers a lot of ground, including alternative architecture, small-scale generation of electricity, solar heating, waste handling, water supply, agriculture and aquaculture.

Myller, Rolf. *From Idea Into House.* (Illus.) Atheneum, 1974. 64pp. $6.95.

    JH-SH-C   This popular-science primer examines the details of home design and construction points to remember when buying land, and what a family should look for. Budgeting, agreements with contractors, and other legal matters are also explained.

Oliver, Paul. *Shelter in Africa.* (Illus.) Praeger, 1971. 240pp. $15.00. 77-151805.

    C   A series of case studies which explain the meaning of building in Africa. Each study is devoted to a particular culture and geographic location and explains the relationships between shelter and religion, life style, available building materials, climate and other factors. Constitutes a comprehensive view of African cultures and their historical conflict with European influences.

Paine, Roberta M. *Looking at Architecture.* (Illus.) Lothrop, Lee&Shepard, 1974. 127pp. $6.95. 73-17718. ISBN 0-688-41553-9; 0-688-51553-3. Gloss.;index;CIP.

   JH-SH   Surveys the world's architectural practice, including examples from Africa, China, India and Mexico. Paine comments on the respective social background of the builders and includes a glossary of building materials.

Pawley, Martin. *Garbage Housing.* (Illus.) Wiley, 1975. 120pp. $17.50. 75-19239. ISBN 0-470-67278-1.

   SH-C   Pawley discusses building with garbage and devotes most of the book to a description of the most thorough experiment on the use of garbage for building, the Heineken Brewery WOBO project. Export beer bottles were redesigned as building blocks for self-help housing but the project was not fully implemented.

Setzekorn, William David. *Architecture.* (Illus.) Dillon, 1974. 91pp. $4.95. 74-4725. ISBN 0-87518-072-8. Index;CIP.

   SH   Setzekorn reviews the scope of architecture and shows the diversity of duties in commercial offices. He stresses the expanded capabilities of teamwork in the design process.

Steadman, Philip. *Energy, Environment and Building.* (Illus.) Cambridge Univ. Press, 1975. ix + 287pp. $14.95; $5.95(p). 74-29352. ISBN 0-521-20694-4; 0-521-09926-9.

   SH-C-P   This is a nontechnical and well-documented compendium on energy and the environment in the design, construction and utilization of buildings. Alternative sources of energy and conservation measures in fossil fuel consumption are emphasized.

Vale, Brenda, and Robert Vale. *The Autonomous House: Design and Planning for Self-Sufficiency.* (Illus.) Universe Books, 1975. 224pp. $10.00. 75-15354. ISBN 0-87663-254-1. Index.

   C-P   The authors review the state of the art for energy self-sufficiency for western-style homes. Solar and wind power are examined objectively and specific problems are discussed.

771 PHOTOGRAPHIC EQUIPMENT AND PROCESSES

Clulow, Frederick W. *Color: Its Principles and Their Application.* (Illus.) Morgan&Morgan, 1972. 236pp. $14.00.

   SH-C   This textbook, with color photographs depicting fundamental color principles and processes, discusses the properties of colored materials, color vision, color mixing processes, color measurement, and color reproduction processes in a practical and readily understood manner.

Craven, George M. *Object and Image: An Introduction to Photography.* (Illus.) Prentice-Hall, 1975. viii + 250pp. $10.95. 74-9739. ISBN 0-13-628925-8. Index;CIP.

   C   Craven discusses the history of photography. The functional aspects of photography are discussed, and Craven has provided an excellent introduction to the mechanics of black-and-white photography. Contemporary styles are examined.

Feininger, Andreas. *Basic Color Photography.* (Illus.) Amphoto, 1972. 128pp. $6.95. 72-159329. ISBN 0-8174-0542-9.

   SH-C   The basic approach to choosing cameras, films, lenses and other equipment used in color photography is discussed in terms of the photographer's objective. Feininger also deals with techniques, equipment usage, and the "art" of photography.

Fineman, Mark B.   *The Home Darkroom, 2nd ed.* (Illus.) AMPHOTO, 1976. 96pp. $3.95(p). 72-79608. ISBN 0-8174-0555-0. Index.

GA   Using a simple, profusely illustrated format, the author provides step-by-step instructions for the entire black-and-white developing and printing process, including how to convert a bathroom into a dark room and how to select, handle and store the chemicals, papers and equipment.

Horder, Alan, (Ed.).   *The Manual of Photography.* (Illus.) Chilton, 1971. 596pp. $15.95. 70-155805. ISBN 0-8019-5655-2.

C-P   A reference for student, amateur and professional. It outlines the photographic process, characteristics of light, physics and geometry of lenses, production of images, camera construction, operation, exposure, and chemistry of developing and printing for both black-and-white and color photography. Circuit diagrams, charts, graphs, and negatives illustrate a wide range of exposure and processing errors. Assumes some knowledge of chemistry and physics.

Jenkins, Reese V.   *Images and Enterprise: Technology and the American Photographic Industry, 1839 to 1925.* (Illus.) Johns Hopkins Univ. Press, 1975. xviii + 371pp. $20.00. 75-11348. ISBN 0-8018-1588-6. Index;CIP.

SH-C-P   Brings out the intrigue and drama that took place as entrepreneurs battled for leadership in an expanding photographic services market. Centers around the basic innovations in photographic technology that gave rise to forces of change within the industry. Of interest primarily to photographers and historians of technology.

Jonas, Paul.   *Manual of Darkroom Procedures and Techniques.* (Illus.) Chilton, 1971. 160pp. $7.95. 74-164637. ISBN 0-8019-5628-5.

SH   A short, simple handbook of basic procedures, with explanations of the theory behind each. The appendixes list equipment and chemicals available from the major manufacturers.

Noren, Catherine.   *Photography: How to Improve Your Technique.* (Illus.) Watts, 1973. 63pp. 73-5687. ISBN 0-531-02640-X. Gloss.

JH   Covers very basic photographic technique in an uncomplicated presentation of choice of equipment, use of cameras and film and simple darkroom work. Numerous comparison photographs are well chosen to illustrate various effects discussed in the text.

Rothschild, Norman, and Wright Kennedy.   *Filter Guide for Color and Black & White, 3rd ed.* (Illus.) Chilton, 1971. 126pp. $3.50. 70-131961. ISBN 0-8019-5611-0.

GA   Combines detailed elementary instructions on when and how to use all kinds of filters with tables summarizing all the information that even the most advanced photographer would want to know about a very wide range of filters. Numerous illustrations of the effects under discussion.

Sullivan, George.   *Understanding Photography.* (Illus.) Warne, 1972. 128pp. $4.50. 72-83127. Index;gloss.;bib.

SH   Nearly all facets of photography, including history, camera styles, developing and printing, exposure, depth of field, composition and lighting, and career opportunities, are discussed.

Time-Life Books, Editors of.   *Life Library of Photography.* (Illus.) Time-Life Books (distr.: Morgan&Morgan), 1970. $9.95ea. Book 1: *The Camera.* 236pp. 70-93302. Book 2: *Light and Film.* 227pp. 77-116582. Book 3: *The Print.* 235pp. 70-124382. Book 4: *Color.* 235pp. 78-130209. *Photographer's Handbook.* 64pp. $1.50. 73-114406.

SH-C   Planned as a self-study and reference set, these books are concerned with elementary matters, the latest developments in the art, and its equipment, accessories, facilities and techniques. These splendid books will appeal to amateur and professional photographers, and they will be useful additions to reference collections.

778 PHOTOGRAPHIC TECHNIQUES AND APPLICATIONS

Angel, Heather. *Nature Photography: Its Art and Techniques.* (Illus.) Fountain/ Scribner's, 1974 (c. 1972). 222pp. $14.95. ISBN 0-852-42015-2. Indexes.

SH-C   A practical approach with numerous illustrations to introduce the beginner to nature photography. Emphasizes British resources and organizations, but offers much for the American audience.

Bauer, Erwin A. *Hunting with a Camera: A World Guide to Wildlife Photography.* (Illus.) Winchester, 1974. viii + 324pp. $12.95. 74-78698. ISBN 0-87691-143-2. Index.

SH-C-P   The author begins with some basic information on photographing wildlife and buying equipment. Specific places for photographing animals are described in considerable detail. Could be used for excursion planning.

Elam, Jane. *Photography: Simple and Creative With and Without a Camera.* (Illus.) Van Nostrand Reinhold, 1975. 96pp. $9.95; $5.95(p). 75-12164. ISBN 0-442-22280-7; 0-442-22281-5. Gloss.;CIP.

JH-SH-C   The book gets away from photography as a limited technical operation and shows how it can be used as an exciting, experimental and creative medium. A fine introduction to the artistic side of photography.

Happé, L. Bernard. *Basic Motion Picture Technology.* (Illus.) Hastings House, 1971. x + 362pp. $10.00. ISBN 0-8038-0729-5.

C   Traces why and how film evolved to the current sizes and styles. Advanced discussions of recording, processing, production, optics, chemistry, and electromechanical reproduction. A reference classic.

Horvath, Joan. *Filmmaking for Beginners.* (Illus.) Nelson, 1974. 162pp. 74-701. ISBN 0-8407-6375-1. Index;CIP.

JH-SH   Much basic information in a simple readable format. Gives current data on recent 8mm equipment.

Jacobs, Lou, Jr. *You and Your Camera.* (Illus.) Lothrop, Lee&Shepard, 1971. 159pp. $4.95. 70-133627.

JH-SH   Jacobs' major concern is with the use of the cartridge film, 35mm, and reflex cameras. Composition and lighting are discussed thoroughly. There is a chapter on amateur motion picture-making, and notes on developing, printing, and enlarging in the home darkroom.

Marchington, John, and Anthony Clay. *An Introduction to Bird and Wildlife Photography: In Still and Movie.* (Illus.) Faber&Faber, 1974. 149pp. £2.95. ISBN 0-571-10171-2. Index.

SH-C   A problem-solving approach for those attempting wildlife field work on a limited budget and with limited expertise.

Papert, Jean. *Photomacrography: Art and Techniques.* Chilton, 1971. x + 117pp. $7.95. 78-134240. ISBN 0-8019-5626-9.

SH-C   This is an excellent book for beginners in photography and for others of greater sophistication who want to branch out into photomacrography. Brief but adequate discussions of equipment and techniques.

Parsons, Christopher. *Making Wildlife Movies: A Beginner's Guide.* (Illus.) Stackpole, 1971. 224pp. $7.95. 75-144109. ISBN 0-8117-0966-3.

    JH-SH-C This is a thorough examination of nature filming—equipment, films, field techniques, script-writing, and editing—for use by the beginner. It is also a presentation of attitudes toward both problem solving and recording living events.

Scheffer, Victor B. *The Seeing Eye.* (Illus.) Scribner's, 1971. 48pp. $5.95. 70-140773. ISBN 0-684-92311-4.

    JH Conveys by beautiful photographs and accompanying text an appreciation of the fact that nature is orderly in its three elements of beauty—form, texture and color. A good book for all who are interested in art, photography and nature.

Zakia, Richard D. *Perception and Photography.* (Illus.) Prentice-Hall, 1974. 160pp. $10.50. 74-3402. ISBN 0-13-656934-X. Index;CIP.

    SH-C-P Utilizes Gestalt psychology as a springboard to imagery in photography. Excellent photographs are used to illustrate parallels between Gestalt and information and visual theory. The synthesis of these factors is used to clarify the means toward visual perceptions achieved by photography.

796.5 OUTDOOR LIFE

Angier, Bradford. *Survival with Style.* (Illus.) Stackpole, 1972. 320pp. $6.95. 72-7369. ISBN 0-8117-1717-8.

    JH-SH-C A compendium of information on survival in a variety of wilderness situations. Topics treated are food, making shelters, finding and storing water, staying warm, getting and preparing game, weather reading, fire building, making clothes, direction finding, signaling, and first aid.

Kirk, Ruth. *Exploring Crater Lake Country,* (Illus.) Univ. of Wash. Press, 1975. 74pp. $4.95(p). 75-9506. ISBN 0-295-95393-4; 0-295-95397-7(p). Index;CIP.

    GA A fine introduction to a wild and diverse region of Oregon, this work includes historical and present geology, aboriginal and immigrant human history, plants, wildlife and advice on camping and hiking. An outstanding guide to outdoor enjoyment.

Kulish, John W., with Aino Kulish. *Bobcats Before Breakfast.* (Illus.) Stackpole, 1969. x + 189pp. $5.95. 72-88179. ISBN 0-8117-0250-2.

    JH-SH The author relates what he learned as a 20th century frontiersman living "with and off nature." Readers will find excitement and entertainment in the word pictures of beavers, otters, bobcats, and other animals, and of woodcraft, canoeing, hunting, fishing, and trapping.

McClurg, David R. *The Amateur's Guide to Caves & Caving: Skill-Building Ways to Finding and Exploring the Underground Wilderness.* (Illus.) Stackpole, 1973. 191pp. $5.95; $2.95(p). 72-14152. ISBN 0-8117-0094-1.

    SH This volume assumes no knowledge of caving and outlines in a simple and at times whimsical manner all the steps necessary for successful cave exploring. One chapter deals with the scientific aspects of caving.

Norton, Boyd. *The Grand Tetons.* (Illus.) Viking, 1974. 128pp. $17.95. 74-7507. ISBN 670-34777-9.

    SH-C Norton's sensitive treatment of the magnificent Tetons is a masterpiece of writing and photographic art. In a brief text, the geologic formation, exploration, and recreational environment of the Tetons are vividly described. A book well worth its cost.

Geography and History

904.5 EVENTS OF NATURAL ORIGIN

Briggs, Peter. *Rampage: The Story of Disastrous Floods, Broken Dams and Human Fallibility.* (Illus.) McKay, 1973. x + 211pp. $6.95. 73-76557.

JH-SH This is a comprehensive world history of floods ranging from the Great Flood of biblical times to the devastating Buffalo Creek flood in Virginia in 1972. A good case against strip mining is offered. Shows that the United States has been very slow and ineffective in flood control.

Brown, Walter R., and Billye W. Cutchen. *Historical Catastrophes: Floods.* (Illus.) Addison-Wesley, 1975. 175pp. $5.75. 74-23684. ISBN 0-201-00762-2. Index;CIP.

JH-SH A description of eight historic floods is provided; floods in myths, legends and religious writings are reviewed, and some methods of flood control are discussed.

910.02 PHYSICAL GEOGRAPHY

Amedeo, Douglas, and Reginald G. Golledge. *An Introduction to Scientific Reasoning in Geography.* (Illus.) Wiley, 1975. xvi + 431pp. $17.95. 75-1411. ISBN 0-471-02537-2. Indexes;CIP.

C-P A potpourri of logic and scientific method, philosophy of science, elementary statistics and case studies.

Fellows, Donald K. *The Environment of Mankind: An Introduction to Physical Geography.* (Illus.) Hamilton, 1975. xi + 484pp. $12.95. 74-32158. ISBN 0-471-25718-4. Index;CIP.

SH-C The earth's major natural subsystems—climate, water, surface, etc.—are presented in this text that gives an overview of the interrelations of these phenomena. The author has successfully simplified extremely complex processes and made them meaningful to students.

Flawn, Peter T. *Environmental Geology.* (Illus.) Harper&Row, 1970. xix + 313pp. $13.95. 75-103915. Gloss.

SH-C An excellent book on applied ecology with primary attention to our relationship to our geological habitat, developed from the standpoint of both economics and engineering. Useful as a primary text in environmental or urban geology; a collateral text in engineering and economics; and a reference for professionals.

Fried, John J. *Life Along the San Andreas Fault.* (Illus.) Saturday Review, 1973. x + 269pp. $7.95. 72-88649. ISBN 0-8415-0233-1.

SH-C Relates various aspects of urban development to potential earthquake hazards and gives a good summary of the theories of earthquake causes and of research on methods for prediction and prevention of earthquakes. A well-documented piece of reporting.

McKenzie, Garry D., and Russel O. Utgard (Eds.). *Man and His Physical Environment: Readings in Environmental Geology, 2nd ed.* Burgess, 1975. x + 388pp. $6.95(p). 74-20042. ISBN 8087-1348-5. Gloss.

C-P An excellent collection of brief modern classics on humans and their environment. The articles are nontechnical and include many of the most important writings on this broad topic. Recommended for everyone from high school teachers to members of Congress.

Ordway, Frederick I., III. *Pictorial Guide to Planet Earth.* (Illus.) Crowell, 1975. x + 191pp. $12.95. 74-34291. ISBN 0-690-621193-0. Index;CIP.

   SH-C   A thorough review of the applications of remote sensing to earth sciences and resources. Beginning with a historical survey of earth observation from space, the author covers uses of the data in hydrology, agriculture, weather, mapping and other areas.

Strahler, Arthur N. *Introduction to Physical Geography, 3rd ed.* (Illus.) John Wiley, 1973. viii + 468pp. $11.95. 72-8958. ISBN 0-471-83172-7. Index;bib.

   C   Covers geographic grids and time, oceans and meteorology, hydrology, climates, soils, natural vegetation, earth's crust (including plate tectonics), and geomorphology. New topics include global balances of radiation, heat, and water; flow and storage of energy; and environmental impact. Contains review questions.

Strahler, Arthur N. *Physical Geography, 4th ed.* (Illus.) Wiley, 1975. ix + 699pp. $15.95. 74-9994. ISBN 0-471-83160-3. Index;CIP.

   GA-C-P   This new edition now includes the seventh approximation soils classification system, the science of remote sensing, and human modifications to the physical environment. An excellent text.

Strahler, Arthur N., and Alan H. Strahler. *Environmental Geoscience: Interaction Between Natural Systems and Man.* (Illus.) Hamilton, 1973. ix + 571pp. $12.95. 72-10325. ISBN 0-471-83163-8. Bib.

   JH-SH-C   Deals with the fundamental principles of earth science necessary for understanding and analyzing environmental problems. The major sections describe important atmospheric, hydrospheric and lithospheric processes, plus two chapters on the biosphere. An analysis of humans' effect on the overall energy systems is provided at the end of each section.

910.03 HUMAN GEOGRAPHY

Chisholm, Michael. *Human Geography: Evolution or Revolution?* (Illus.) Penguin, 1975. 207pp. $4.95(p). ISBN 0-1402-1883-1. Indexes.

   C-P   Chisholm deals with the origins and growth of modern geography; the major focus is on what sort of science geography is or should be. Gives literature surveys of static and dynamic patterns and theories of social structure and process.

Cox, Kevin R. *Man, Location, and Behavior: An Introduction to Human Geography.* (Illus.) Wiley, 1972. 399pp. $10.95. 72-4790. ISBN 0-471-18150-1.

   SH-C   Rich in localization theory and presented almost exclusively in a literary, nonmathematical way. Definitions are clear and the range of subject matter is covered accurately and thoroughly. Not intended to be a view of Earth's physical geographic reality.

de Blij, Harm J. *Geography: Regions & Concepts.* (Illus.) Wiley, 1971. 642pp. $12.50. 73-127662. ISBN 0-471-20060-3.

   C   The author has skillfully woven geographic concepts into a regional framework, using the thread of historical development. All areas of the world are covered, but the Western world is given better treatment than the rest.

Gould, Peter, and Rodney White. *Mental Maps.* (Illus.) Penguin, 1974. 204pp. $2.95(p). ISBN 0-14-02-1688-X. Index.

   SH-C-P   Broad-ranging study of the geography of human perception. Uses numerous examples to underline the importance of human perceptions based on personal "mental maps" and their effects on human behavior. Particularly interesting for cartographers and geographers.

Harper, Robert A., and Theodore H. Schmudde. *Between Two Worlds: A New Introduction to Geography.* (Illus.) Houghton Mifflin, 1973. xiii + 586pp. $13.95. 72-6890. ISBN 0-395-12075-6. Bib.

    C   Geography is viewed from a systems perspective—the traditional closed human system and the modern, worldwide, interconnected system. Different parts of the world are analyzed within a framework of case studies of several industrialized and developing countries. Suitable for an introductory geography text.

Hessel, Milton. *Man's Journey Through Time: The Important Events in Each Area of the Earth in Each Period of History.* (Illus.) Simon&Schuster, 1974. 224pp. $9.95. 73-8852. SBN 671-65209-5. Index.

    JH-SH-C   This is a comprehensive outline of history, complete with drawings, maps and diagrams covering the history of the earth in the solar system as well as human history.

Hoyt, Joseph Bixby. *Man and the Earth, 3rd ed.* (Illus.) Prentice-Hall, 1973. x + 496pp. $12.95. 71-171962. ISBN 0-13-550947-5.

    C   Hoyt has combined the physical and human aspects of geography with both systemic and regional concepts. Will be useful for high school teachers.

Orme, A.R. *Ireland.* (Illus.) Aldine, 1970. xviii + 276pp. $5.95; $2.95(p). 70-91729. ISBN 0-202-10030; 0-202-10035.

    JH-SH-C-P   An unusual and well-written study of the effects of humans upon the land. Orme describes how successive waves of immigration have shaped the land, and how the growth of centers of habitation have affected the natural landscape, the flow of rivers and watersheds, and the fertility of the soil. The book transfers the history of humanity into the history of the ecosystem.

Prescott, J.R.V. *Political Geography.* (Illus.) Methuen (distr.: St. Martin's), 1975 (c.1972). 124pp. $10.95. SBN 416-07000-0; 416-07010-8(p). Index.

    C-P   In this brief, highly readable introduction to political geography, Prescott outlines the possible relations between the two fields and describes methods used in verifying or quantifying them. Many examples are used as well as the results of substantive studies. Highly recommended for the general reader, geographer and political scientist.

Weiner, J.S. *The Natural History of Man.* (Illus.) Universe, 1971. xii + 254pp. $12.50. 72-103105. ISBN 0-87663-147-2.

    SH-C   An excellent review. A long section on cultural development shows how humans became a force in the environment—first with agriculture and then with industrialization and mass societies. The present age is one of "microclimatological bioengineering" with new stresses on humans. Useful as a general statement on the subject.

910.4 Accounts of Travel

Doolittle, Jerome, et al. *Canyons and Mesas.* (Illus.) Time-Life, 1974. 184pp. $7.95. 74-77772. Index.

    JH-SH-C   The authors examine the spectacular country of the Colorado Plateau, presenting the geology, meteorology, botany and zoology of the area in exquisite prose.

Heyerdahl, Thor. *Fatu-Hiva: Back to Nature.* (Illus.) Doubleday, 1975. vii + 276pp. $10.00. 74-33646. ISBN 0-385-08921-X. Index;CIP.

    JH-SH-C   This is an account of the Heyerdahls' travels around the island of Fatu-Hiva, their constant battle with the environment, ventures into taboo grounds and unfriendly encounters with natives. Sixty-one black-and-white photographs.

Laycock, George. *Death Valley.* (Illus.) Four Winds/Scholastic, 1976. 113pp. $6.95. 75-42276. ISBN 0-590-07399-0. Index;CIP.

JH   This is an attractive guidebook and introduction to Death Valley—its exploration, early inhabitants, mineral resources, flora and fauna. Also shows how early travelers coped with the Mojave Desert and describes the development of the borax industry. Includes a detailed map.

Moser, Don, et al. *The Snake River Country.* (Illus.) Time-Life, 1974. 184pp. $7.95. 74-80283. Index.

JH-SH-C   This publication is primarily concerned with the southern two-thirds of Idaho, with parts of the neighboring western and eastern states included in the geomorphic region. An excellent physiographic and cultural map is provided.

Ogburn, Charlton. *The Southern Appalachians: A Wilderness Quest.* (Illus.) Morrow, 1975. 245pp. $14.95. 74-34293. ISBN 0-688-00341-9. Index;CIP.

C   Ogburn writes of his wanderings through the Appalachian wilderness, mixing geology, botany, ecology and anthropology. He describes his feelings about the area and his philosophy of life. For conservationists and classes in ecology, geology or geography.

White, Jon Manchip. *A World Elsewhere: One Man's Fascination with the American Southwest.* (Illus.) Crowell, 1975. 320pp. $8.95. 74-32338. ISBN 0-690-00720-5. Index;CIP.

JH-SH-C   In the form of letters to a blind friend, White presents his views of the American Southwest. After an overview of the geography and history, he introduces the wonders of this land, including the unique animals and plants and their unusual survival techniques.

910.9 EXPLORATION

Ashe, Geoffrey, Thor Heyerdahl, et al. *The Quest for America.* (Illus.) Praeger, 1971. 298pp. $15.00. 78-151832.

SH-C   A wide variety of topics concerning contact between the Old and New Worlds, exploration, and speculation. Thought-provoking and alternative hypotheses are cited, including the myth of St. Brendan, the Norse situation, the Japan and Ecuador situation, and the role of the Portuguese in the discovery of America. History, myth, archeology, art and architecture, and revelation are all used in this excellent book.

Bowman, John S. *The Quest for Atlantis.* (Illus.) Doubleday, 1971. 182pp $4.50. 72-139007.

SH-C   The discussion is initiated with definitions of Atlantis—one literal, another allegorical. The story begins with Plato; his possible sources are examined, as are those of subsequent writers from each of the major periods. This work is both a history of Western thought and a chronicle of the Atlantis problem.

Dodge, Ernest S. *Beyond the Capes: Pacific Exploration from Captain Cook to the Challenger (1776-1877).* (Illus.) Little, Brown, 1971. xv + 429pp. $12.50. 70-149464.

SH-GA-C   An excellent book which tells the story of the exploration of the South Seas for the 100 years after Captain Cook's death. Of great interest to the zoogeographer, oceanographer, historian of science, South Seas buff, or anyone interested in early ocean exploration.

Gordon, Bernard L. *Man and Sea: Classic Accounts of Marine Explorations.* Natural History Press, 1970. xxiv + 498pp. $9.95. 76-116208.

SH-C   In this book are drawn together the ancient tales and the recent scientific reports of men who love the sea. Chapters are specific and short, each by an authority. Relaxing yet informative reading in such diverse areas as nutrition, climatology, housing, engineering, geology, and agriculture.

Heyerdahl, Thor. *The* **Ra** *Expeditions.* (Illus.; trans.) Doubleday, 1971. xi + 341pp. $10.00. 72-139031.

JH-SH-C   Adventure, anthropology, and archeology are blended somewhat unevenly in this account of an attempt to test the theory that elements of Mediterranean civilization could have crossed to the Americas by reed boats. The story is absorbing and stimulating, even if the theories and conclusions are open to dispute.

Linklater, Eric. *The Voyage of the* **Challenger.** (Illus.) Doubleday, 1972. 288pp. $15.00. 70-186381.

SH-C   This account is based principally on the descriptions of Sub-Lieutenant Lord George Campbell and of Henry Mosely, naturalist on the *Challenger.* The illustrations, many by staff members or the official artist, are outstanding. Fascinating reading, not just because it gives insight into oceanography and exploration, but also because it shows the workings of an inquiring mind.

Parker, John. *Discovery: Developing Views of the Earth from Ancient Times to the Voyages of Captain Cook.* (Illus.) Scribner's, 1972. viii + 216pp. $6.50. 74-37190. ISBN 684-12725-3.

JH-SH-GA   A fascinating account of attempts to explore and understand the physical nature of the world. Describes our historical efforts to locate ourselves in relation to neighbors—nearby and across the seas. An excellent job of turning history into exciting stories.

Savoy, Gene. *On the Trail of the Feathered Serpent.* (Illus.) Bobbs-Merrill, 1974. xi + 216pp. $10.00. 72-89709. ISBN 0-672-51668-3.

SH   Stimulating adventure story in which the author attempts to sail from Peru to Mexico via reed boat in order to demonstrate that the ancient Inca, Aztec and Mayan civilizations could have maintained maritime communication. Excellent photographs.

912 ATLASES

Espenshade, Edward, Jr. and Joel Morrison (Eds.) *Goode's World Atlas, 14th ed.* Rand McNally, 1976. 372pp. $12.95; $7.95(p). ISBN 0-528-83020; 0-528-63470-4.

JH-SH-C   Presents the culmination of content updating and expansion to meet the needs of a wide range of users. Features include a pronouncing index to some 36,000 place names and approximately 100 special subject (thematic) maps. A very fine atlas.

Fitzgerald, Ken. *The Space-Age Photographic Atlas.* (Illus.) Crown, 1970. x + 246pp. $7.95. 70-93399.

JH-SH-C   This atlas of planet Earth was compiled from photographs taken from Gemini and Apollo spacecraft and from aircraft. Key maps serve as visual aids to interpretation of the pictures representing political areas, terrain, cities and landforms. A brief summary of the earth sciences and an appendix on the uses of aerial photography enrich this modern geographic reference book.

Hammond. *Ambassador World Atlas.* (Illus.) Hammond, 1971. xvi + 480pp. $14.95. 72-654261.

Citation World Atlas. (Illus.) Hammond, 1971. vii + 352pp. $10.95. 75-654314.

Medallion World Atlas. (Illus.) Hammond, 1971. xvi + 656pp. $24.95. 71-654313.

World Atlas—Hallmark Edition. (2 vols.;illus.) Hammond, 1971. xvi + 656pp. $39.95. 79-654260.

JH-SH-C   Each work contains 320 pages of maps with current place names. They include continental, area, country, state and provincial maps with keyed indexes to places, and topographic, land use, and resource maps, all in color. Other common features are photographs, flag reproductions, information of language, U.S. population figures and Zip Codes, glossaries, word indexes, tables of world statistics, geographical terms, and an explanation of the various types of map projections. The *Hallmark* and *Ambassador* include a section on environment and life, an illustrated historical atlas of the Bible lands and of the world, and a somewhat more detailed historical atlas of the United States.

Rand McNally. *Cosmopolitan World Atlas, Planet Earth Edition.* (Illus.) Rand McNally, 1971. 408pp. $19.95. 72-654253.

JII-SH-C   The revised and enlarged edition of this well-known and widely used atlas has outstanding photographs obtained during various NASA space missions which explain the place of the earth in the solar system, the phases of the earth, its atmosphere, and panoramas from far out in space that provide true photographs of the size of traditional conventional maps. There are brief illustrated discussions of earth geology, human geography and the moon landing.

915 ASIA

Fairservis, Walter A., Jr. *Before Buddha Came.* (Illus.) Scribner's, 1972. 133pp. $6.95. 75-37213. ISBN 0-684-12779-2. Index;bib.

SH   An account, based on archeological research, of how the peoples lived during the period in which the ancient civilizations of the Far East were evolving. Four succinct chapters cover China, Korea, Japan and Central Asia. The book includes a tabular chronology of the various dynasties, kingdoms and cultures in the four areas.

Golann, Cecil Paige. *Our World: The Taming of Israel's Negev.* (Illus.) Messner, 1970. 128pp. $4.50. 72-123557. ISBN 0-671-32353-9.

JH-SH   The book deals with the physical geography of the Negev, the rapid growth in the area's population after its annexation by Israel in 1948, and the irrigation systems that were put into place to make the desert agriculturally productive.

Karmon, Yehuda. *Israel: A Regional Geography.* (Illus.) Wiley-Interscience, 1971. xi + 345pp. $20.75. 70-116162. ISBN 0-471-45870-8.

C   The book covers Israel's physical base and history; examines the present population, including origin, social stratification, and economic activities and regional development of some 35 regions; and analyzes those areas which have been under the control of Jordan and Egypt from 1948 to 1967. Well documented.

Parker, W.H. *The Soviet Union.* (Illus.) Aldine, 1970. xi + 188pp. 79-91731. ISBN 0-202-10019; 0-202-10033. Index;bib.

SH-C   This brief geography of the Soviet Union combines mature understanding of the country with successful communication in nontechnical language. The book is well equipped with maps, photographs, and footnotes. Organization, although firmly rooted in historical geography, is primarily regional, based on the great latitudinal belts of vegetation.

Spencer, J.E., and William L. Thomas. *Asia, East by South: A Cultural Geography,* *2nd ed.* (Illus.) Wiley, 1971. xv + 669pp. $16.50 74-138920. ISBN 0-471-81545-4.

C-P   A discussion of the physical and cultural environments of Asia from India-Pakistan through Southeast Asia, China and Japan. The cultural environment is handled better than is the physical environment. The emphasis on history and prehistory in certain sections sets the stage for the appreciation of current environmental patterns.

Trewartha, G.T. *The Less Developed Realm.* (Illus.) Wiley, 1972. xi + 499pp. $12.95; $7.95(p). 76-173680. ISBN 0-471-88794-3; 0-471-88795-1.

C-P   The second in a planned trilogy, this volume focuses on the regional characteristics of Asia (excluding Japan and Asiatic Russia), Africa and most of Latin America. The first volume considered planetary geography, and the third volume the highly industrialized countries of the world. An excellent series by one of the foremost geographers.

Tuan, Yi-Fu. *China.* (Illus.) Aldine, 1970. xi + 225pp. $5.95. 71-86829. ISBN 0-202-10022; 0-202-10034(p).

C-P   A brief scholarly historical geography of China. Following a review of China's physical geography, more than half of the study is devoted to the historical evolution of China's cultural geography. The remaining quarter deals with the period 1850 to the present with only cursory treatment of Mao's China.

916 AFRICA

Bradford, Ernle. *Mediterranean: Portrait of a Sea.* (Illus.) Harcourt Brace Jovanovich, 1971. 574pp. $10.00. 70-153682. ISBN 0-15-158584-9.

C   The sea, climate, geology, flora and fauna, and other aspects of the geography of the region are neatly and interestingly summarized. The remainder of the book condenses more than 4,000 years of the history and the culture of civilizations that have flowered and declined along the shores of the Mediterranean.

Carrington, Richard. *The Mediterranean: Cradle of Western Civilization.* (Illus.) Viking, 1971. 287pp. $16.95. 75-156754. ISBN 0-670-46559-3.

SH-C   This richly illustrated book examines the geographical and cultural evolution of the main civilizations of this region: Egypt, Phoenicia, Crete, Etruria, Greece, Rome, and Byzantium. Effects of Christian and Moslem religions are noted, and the natural history of the area is described. Recommended for travelers and for the history and geography student.

Sweeney, Charles. *Naturalist in the Sudan.* (Illus.) Taplinger, 1974. 240pp. $8.50. 73-1778. ISBN 0-80008-5466-7.

SH-C   An entertaining blend of adventure, discovery, keen scientific observation and subjective discovery. Offers much new and useful information in the areas of natural history, anthropology, ecology, and ethnology. Sidelights on several diverse peoples of the Sudan also appear.

Watson, Jane Werner. *The Niger: Africa's River of Mystery.* (Illus.) Garrard, 1971. 96pp. $2.59. 76-135581. ISBN 0-8116-6374-4.

JH-SH   The geography, exploration, and recent political history of the Niger area are presented in exciting, descriptive language. Excellent maps, numerous photographs.

917–918 NORTH AND SOUTH AMERICA

Ault, Phil. *These Are the Great Lakes.* (Illus.) Dodd, Mead, 1972. 175pp. $4.50. 72-1533. ISBN 0-396-06607-0. Index;bib.

JH-SH   A fascinating journey from the formation of the fresh-water lakes in glacial times through geological history, Indian history and legend, trials of the explorers, to current problems of pollution resulting from human exploitation.

Carter, Hodding. *Man and the River: The Mississippi.* (Illus.) Rand McNally, 1970. 174pp. $14.95. 72-122391.

JH-SH-C   This story of the Mississippi River is followed by an outstanding color pictorial review of the river from its origin. The author has skillfully blended many historical, geological, geographical, biological, and economic facts into an appealing story.

Kelley, Don Greame. *Edge of a Contintent: The Pacific Coast from Alaska to the Baja.* (Illus.) American West, 1972. 288pp. $17.50. 78-119004. ISBN 0-910118-19-1.

JH-SH   Kelley briefly describes the land, sea, climate, plants, animals, and exploratory history of the entire coast as a whole; the islands, bays and gulfs; the northern forest and southern desert; and the geology, glaciers, rivers and human inhabitants. Projects the spirit of exploration, relates sensitively and sensibly the interaction of humans and their environment and covers reasonably well the greatly different geographic provinces.

Malkus, Alida. *The Amazon: River of Promise.* (Illus.) McGraw-Hill, 1970. 128pp. $5.95. 77-107447.

JH-SH   This book is packed with information about the Amazon Basin and Brazil in particular. The historic and modern economic development is the primary focus; summary presentation is made of the geography; tropical forest; examples of the flora, fauna, primitive tribes; and colonization by the Spanish and the Portuguese.

Paterson, J.H. *North America: A Geography of Canada and the United States, 5th ed.* (Illus.) Oxford Univ. Press, 1975. xiii + 368pp. $9.50. 74-21826. Index;bib.

C-P   Recommended as a basic text for advanced college students, nonspecialist adults and laypersons who wish to gain general insights into North America as a region. Seven chapters are devoted to systematic studies of North America, and 14 to regional studies.

Wood, Frances, and Dorothy Wood. *America: Land of Wonders.* (Illus.) Dodd, Mead, 1973. viii + 214pp. $4.95. 72-3155. ISBN 0-396-06529-5.

JH-SH   This delightful book, complete with photographs that stimulate the imagination, takes the reader across the American continent, defining and describing "natural wonders," such as waterways, the very scenic Appalachian Trail, deserts, mountains and rivers. Includes notes on the history of each area, and ecological facts and problems.

919 OCEANIA

Cumberland, Kenneth B., and James S. Whitelaw. *New Zealand.* (Illus.) Aldine, 1970. xiii + 194pp. $5.95; $2.95(p). 73-110624. ISBN 0-202-10039; 0-202-10040.

C   The impact of the various settlers on the geography of New Zealand is presented in this study of the ecological impact of each successive group of settlers. Students and travelers will find a wealth of information presented in a readable, lively and lucid style.

Dos Passos, John. *Easter Island: Island of Enigmas.* (Illus.) Doubleday, 1971. xi + 150pp. $6.95. 78-111160.

> **SH-C**   Raises the questions and reviews the observations of major European and American travelers to the Pacific island and discusses the civil war of the 17th century, the decline of the old culture and the demoralization of the natives at the hands of Europeans and the Peruvian slave raiders. A fine chapter on the Pascuenses today closes an excellent book.

Holton, George, (Photographs) **and Kenneth E. Read** (Text). *The Human Aviary: A Pictorial Discovery of New Guinea.* (Illus.) Scribner's, 1971. 64pp. $6.95. 78-143954. ISBN 0-684-12385-1.

> **SH-C**   Although brief, the text and photographs provide realistic glimpses of New Guinea wildlife, topography, and culture. Domestic arrangements, customs, activities, celebrations, and the like are beautifully described.

Perry, Roger. *The Galápagos Islands.* (Illus.) Dodd, Mead, 1972. 92pp. $4.00. 72-724. ISBN 0-396-06576-7. Index.

> **JH-SH**   An excellent introduction to the biology, geography and geology of the Galápagos, and presents their human and geological history. Contains a good map and many well-selected black-and-white photographs portraying the unusual plants and animals. Offers a clear picture of problems in conserving the islands' uniqueness.

Robinson, William Albert. *Return to the Sea.* (Illus.) John de Graff, 1972. 232pp. $8.95. 75-185573. ISBN 8286-0060-0.

> **JH-SH**   Provides a personal glimpse of Polynesia by an author who is adventurer, disease fighter, and amateur anthropologist. Includes an analysis of future problems in Polynesia and an excellent description of the ecology, underscoring the coming deadly effect of industrial civilization.

919.98 POLAR REGIONS

Brent, Peter. *Captain Scott and the Antarctic Tragedy.* (Illus.) Saturday Review Press, 1974. 223pp. $12.50. 73-75732. ISBN 0-8715-0258-7. Index.

> **SH-C**   A well-researched treatment of Scott the hero which also strips away the mass of publicity and exposes Scott the man. Drawings and water colors by Scott's companion leave the reader with an appropriate feeling of solitude and desolation which matches the heroic tragedy of the book.

Cameron, Ian. *Antarctica: The Last Continent.* (Illus.) Little, Brown, 1974. 256pp. $14.95. 74-7626. Index.

> **JH-SH-C**   All phases of the Antarctic experience, from notions of *Terra Australis* to the Antarctic Treaty, are discussed in a lively, interesting fashion. The major expeditions and explorers as well as lesser known ones are detailed. Especially noteworthy are the histories of the sealing and whaling industries. The illustrations are well chosen, and this book stands out in its field.

Halle, Louis J. *The Sea and the Ice: A Naturalist in Antarctica.* (Illus.) Houghton Mifflin, 1973. xv + 286pp. $8.95. 72-6813. ISBN 0-395-15470-7.

> **JH-SH-C**   Geographic, physiographic and historical facts are interspersed in this account of life in the world's harshest environment. Special attention is paid to the Cape pigeon, wandering albatross, royal albatross, mollyhawks, sooty shearwater, prions, antarctic fulmar, Wilson's storm petrel, emperor penguin, Adélie penguin, skuas, whale, krill and seals.

Lewis, Richard S., and Philip M. Smith (Eds.). *Frozen Future: A Prophetic Report from Antarctica.* (Illus.) Quadrangle, 1973. xiv + 455pp. $12.50. 70-164424.

C   Assesses the American stake in Antarctica since the IGY; political coopera-
    tion, scientific research projects, management and development under these
special conditions, scientific and operational activities, including first-hand accounts
by the scientists involved and the future of the area.

Lyttle, Richard B. *Polar Frontiers: A Background Book on the Arctic, the Antarctic, and Mankind.* Parents', 1972. viii + 264pp. $4.95. 72-741. ISBN 0-8193-0600-2; 0-8193-0601-0. Index;bib.

JH-SH-C   Lyttle describes in episodic fashion the natural history of the polar
          areas, showing their wide ranges of variation, their discovery and early
exploitation, and the current state of knowledge about them, including some ongoing
research. Useful reading for a world geography curriculum.

Maxwell, A.E., and Ivar Ruud. *The Year-Long Day: One Man's Arctic.* (Illus.) Lippincott, 1976. 240pp. $8.95. 75-40412. ISBN 0-397-01131-8. CIP.

JH-GA   The exciting story of Ruud's 6 years, including long, dark winters, on a
        totally uninhabited island in the Arctic. The unbelievable difficulty of
simply surviving the arctic winter by oneself is realistically depicted.

Montfield, David. *A History of Polar Exploration.* (Illus.) Dial, 1974. 208pp. $17.50.
73-17943. ISBN 0-8037-3738-6. Index.

SH-GA-C   Narrative accounts are given of the explorers, their motives, their
          vehicles and equipment, the conditions of the journeys, and the en-
vironmental factors. Many black-and-white and color plates are included.

Neider, Charles, (Ed.). *Antarctica: Authentic Accounts of Life and Exploration of the World's Highest, Driest, Windiest, Coldest and Most Remote Continent.* (Illus.) Random House, 1972. x + 463pp. $10.00. 73-37072. ISBN 0-394-46831-7.

SH-C   Amundsen, Scott, Bellinghausen, Shackleton, Wilkes and other famous
       explorers tell their stories of Antarctic exploration in their own words. The
author's introductory account of his trip to Antarctica is masterful description. On
balance this is an excellent accounting of Antarctic exploration.

Ponting, Herbert. *Scott's Last Voyage: Through the Antarctic Camera of Herbert Ponting.* (Illus.; ed. by Ann Savours) Praeger, 1975. 160pp. $12.50. 74-13507. ISBN
0-275-52670-4. CIP.

JH-SH-C   Despite the technological difficulties, Ponting's photographs of Scott's
          last expedition of the Antarctic in 1910 capture every aspect of that
heroic, tragic adventure. His motion pictures and color and black-and-white photographs are accompanied by a detailed, sympathetic and interesting account of the
journey. For those interested in the history of photography, and the history and
geography of Antarctica.

Scott, Jack Denton. *Journey into Silence.* Reader's Digest Press (distr.: Crowell),
1976. 200pp. $7.95. 75-33837. ISBN 0-88349-083-8. Index;CIP.

JH-SH-C   The story of Scott's sea voyage from the northern shore of Norway to
          the "Chinese Wall Front Glacier," where he discovered pure silence.
Full of adventure, with patches of history, geography, geology, oceanography, biology, ecology, natural history, sociology and technology. Beautifully told.

930.1 ARCHEOLOGY

Bass, George F. *Archaeology Beneath the Sea.* (Illus.) Walker, 1975. 238pp. $12.95.
74-24795. ISBN 0-8027-0477-8. Index.

SH-C A first-person account of the discovery of a Bronze Age merchant ship, some 3200 years old (still the oldest shipwreck ever found), and the subsequent discovery of a Byzantine merchantman and a Roman trading ship. Also provides an excellent summary of other projects in the Aegean Sea. Bass shares both the excitement of discovery and the frustrations of working with makeshift equipment and limited funds.

Berger, Rainer. *Scientific Methods in Medieval Archaeology.* (Illus.) Center for Medieval and Renaissance Studies, 1970. xvii + 459pp. $20.00. 75-99771. ISBN 0-520-01626-2.

C-P Illustrates techniques in archeological and related research as applied to Europe and the Near East from 300 to 1600 A.D. Methods described are radiocarbon dating, dendrochronology, thermoluminescence, tephrochronology, archemagnetic dating, chemical studies of ceramics and glass, trace element analysis, rapid x-ray fluorescence analysis, and magnetic prospecting.

Binford, Lewis R. *An Archaeological Perspective.* Seminar, 1972. xii + 463pp. $11.95. 76-182629.

C These papers cover a variety of topics—from historical site materials through the Acheulian era of over 400,000 years ago. Special attention is devoted to problems of understanding the origins of agriculture and reconstructing prehistoric social organization. Useful as a means of becoming acquainted with a major trend in today's archeology.

Brennan, Louis A. *Beginner's Guide to Archaeology: The Modern Digger's Step-by-Step Introduction to the Expert Ways of Unearthing the Past.* (Illus.) Stackpole, 1973. 318pp. $9.95. 73-4193. ISBN 0-8117-0418-1. Bibs.

C Covers the needs for archeological exploration, provides basic techniques and precautions, and outlines the archeological history of the Americas. Contains a section on techniques for writing good reports.

Brodrick, A. Houghton, (Ed.). *Animals in Archaeology.* (Illus.) Praeger, 1972. 180pp. $15.00. 75-172995.

C An attempt to gain a developmental perspective on the interrelationships between the animals who hunted man, the animals hunted by man, and those animals woven into the period of time denoted by archaeological record. Recommended for those who enjoy archaeology—the illustrations are excellent, the organization is simple, and the text is lucid and relatively nontechnical.

Coles, John. *Archaeology By Experiment.* (Illus.) Scribner's, 1974. 182pp. $3.50(p). 74-3668. ISBN 0-684-13818-2; 0-684-14078-0. Index.

SH-C-P Summarizes experiments on archeological artifacts and suggests alternative rationales for those attempting to interpret the functions of archeological remains. Reveals some errors in past interpretations, makes extinct ways of life more vivid, and increases the reader's appreciation of the knowledge and ingenuity of original populations.

Fairservis, Walter A., Jr. *The Threshold of Civilization: An Experiment in Prehistory.* (Illus.) Scribner's, 1975. xii + 256pp. $12.50; $4.95(p). 74-14892. ISBN 0-684-12775-X. Index;CIP.

SH-C-P Through a combination of philosophical, anthropological and archeological fare, Fairservis offers an enlightening concept of the development of civilization. He defines two concepts of man—technological versus sociocultural— and examines the relationship of toolmaking, speech and culture to the development of civilization.

Hawkes, Jacquetta, (Ed.). *Atlas of Ancient Archaeology.* (Illus.) McGraw-Hill, 1974. 272pp. $19.50. 73-22453. ISBN 0-07-027293-X. Index;CIP.

JH-SH-C-P Both students of archeology and travelers in Eurasia and Africa will find this atlas interesting. A short summary of world prehistory is excellent, and the sections on Old World sites thorough and accurate. New World sites, however, are thinly represented and scholars should approach the data with some skepticism.

Johnstone, Paul. *The Archaeology of Ships.* (Illus.) Walck, 1974. 135pp. $8.95. 73-19474. ISBN 0-8098-3532-0. Index;CIP.

SH-C In 10 chapters, arranged chronologically in order of the ages of the ships involved, the author sketches the techniques and procedures that have been used to investigate and reconstruct a series of wooden ships whose remains have been preserved through time.

Magnusson, Magnus. *Introducing Archaeology.* (Illus.) Walck, 1973. 127pp. $4.95(p). 72-6908. ISBN 0-8093-3109-0. Index;bib.

JH-SH-C This particularly exciting book by an outstanding archeologist describes major discoveries which illustrate the development of the field from the "treasure hunting" of the early 19th century to today's scientific discipline. Includes a discussion of forgeries in the art market and the useful role which amateurs can play in archeology.

Marshack, Alexander. *The Roots of Civilization: The Cognitive Beginnings of Man's First Art, Symbol and Notation.* (Illus.) McGraw-Hill, 1972. 413pp. $17.50. 70-140958. ISBN 07-040535-2.

C-P The author entered the field with no professional biases and began to study patterns of lines and dots that had been ignored by most other investigators. The building up of the evidence is presented as a detective story, which contributes significantly to the readability and intelligibility of a highly complicated argument. Opens a new line of thinking that indicates use of a much more sophisticated system of time keeping than has previously been postulated for this early period.

Mazonowicz, Douglas. *Voices from the Stone Age: A Search for Cave and Canyon Art.* (Illus.) Crowell, 1974. viii + 211pp. $12.95. 74-8590. ISBN 0-690-00574-1. Index;CIP.

JH-SH An anecdotal, personal account of an archeologist's years of exploration and interpretation of prehistoric art. In addition to his thoughts and feelings, the author gives brief but complete explanations of such sites as Altamira. Enjoyable for readers of all backgrounds.

Robbins, Maurice, with Mary B. Irving. *The Amateur Archaeologist's Handbook, 2nd ed.* (Illus.) Crowell, 1973. xiv + 288pp. $7.95. 73-245. ISBN 0-690-05569-2. Gloss.;bib.

SH-C-P This book is good for educating interested diggers in the "how-to" of planning, excavating and preserving North American sites. The technical background material is well researched and up-to-date. Five appendixes: a list of sites open to the public, archeological societies, museums, relevant college courses, and state antiquities laws.

Watson, Patty Jo, Steven A. Leblanc, and Charles L. Redman. *Explanation in Archeology: An Explicitly Scientific Approach.* (Illus.) Columbia Univ., 1971. xviii + 191pp. $6.00. 73-158340. ISBN 0-231-3544-6.

C-P The authors summarize and elaborate on recent archeological debates. Basically, they attempt to explain the approach of the so-called "new archeologists" who are seeking to apply scientific explanation and procedure to their studies. Will undoubtedly be a standard textbook for teaching archeological theory.

Woodbury, Richard B.   *Alfred V. Kidder.* (Illus.) Columbia Univ. Press, 1973. 200pp. $8.00; $2.95(p). 72-10082. ISBN 0-231-03484-9; 0-231-03485-7.

SH-C-P   This excellent, accurate and clearly written biography provides a concise description of the life of one of the most renowned of American archeologists. Half of the book outlines Kidder's life in informative detail. The other half reprints portions of his major contributions to the professional literature.

930.13 ARCHEOLOGY OF THE ANCIENT WORLD

Hadingham, Evan.   *Circles and Standing Stones: An Illustrated Exploration of Megalith Mysteries of Early Britain.* (Illus.) Walker, 1975. vii + 240pp. $12.50. 74-82169. ISBN 0-8027-0463-8. Index;bib.

C   The author presents clear and concise descriptions of the monuments and major finds in both recent excavations and classic sites. He pays vigorous attention to a history of their various interpretations. A useful tour guide and introduction to the early antiquities.

Hawkins, Gerald S.   *Beyond Stonehenge.* (Illus.) Harper&Row, 1973. xiii + 319pp. $10.00. 72-79671. SBN 06-011786-9.

SH   A description of the author's personal experiences and knowledge of architectural sites such as Stonehenge, describing the evolution of historical thought about these sites and his own conceptions. Can be used as a supplement in astronomy or ancient history courses.

Masson, V.M., and V.I. Sarianidi.   *Central Asia: Turkmenia Before the Achaemenids.* (Illus.;trans. & ed. by Ruth Tringham) Praeger, 1972. 219pp. $12.50. 70-131350.

SH-C   A general summary of the history of humans in Central Asia, from the Caspian Sea to the Pamir plateau. The area is within the shadow of the Middle East but also contributed to the continuing development of civilization in that area and beyond. A useful reference on a little known region of the world.

Trippett, Frank, et al.   *The First Horsemen.* (Illus.) Time-Life (dist.: Little, Brown), 1975. 166pp. $7.95. 74-80650. Index.

SH-C   Combining the techniques of both archeology and ethnology, the story of the role of the horse in Eurasia from 1000 B.C. to the present is vividly described in both text and pictures. Excellent collateral reading for any discussion on the history of man.

930.132 ANCIENT EGYPT

Edwards, I.E.S.   *The Pyramids of Egypt, rev. ed.* (Illus.) Viking, 1972. 240pp. $14.00. 73-186741. Bib.

JH-SH-C   Summarizes thoroughly, accurately and interestingly the best established opinion on the history of the development of pyramids. The first two sections deal with the social and religious structure of ancient Egypt and the final section with the technology of pyramid building. Excellent illustrations.

Harris, James E., and Kent R. Weeks.   *X-Raying the Pharaohs.* (Illus.) Scribner's, 1973. 195pp. $10.00 72-1180. SBN 684-13016-5.

SH   A popular account of the x-ray investigation of the mummies of pharoahs, other royalty and priests; part of the project of the University of Michigan School of Dentistry to study medical aspects of large collections of skeletal materials recovered in recent years in Nubia. An excellent series of color and black-and-white plates are included.

James, T.G.H.   *The Archaeology of Ancient Egypt.* (Illus.) Walck, 1973. 144pp. $4.95 (p). 72-6999. ISBN 0-8098-3110-4.

JH-SH-C   Makes history and the study of history understandable to young readers. Discusses the Egyptian civilization, the early discoveries of Petrie, the disastrous enthusiasm of Belzoni, the Pyramids and their explorers, Egyptian customs, language, etc. The book is interesting, informative, well written and exceedingly well produced. The illustrations and style are outstanding.

Lauer, Jean-Philippe.   *Saqqara: The Royal Cemetery of Memphis: Excavations and Discoveries Since 1850.* (Illus.) Scribner's, 1976. 248pp. $25.00 75-33508. ISBN 0-684-14551-0. Index.

C-P   A summary of the discoveries made at Saqquara, an immense, ancient necropolis in Egypt dating principally to the late fourth and third millennia B.C. The author is an architect and archeologist. For students of Egyptian art and nonspecialists with some knowledge of Egyptian history and art.

Macaulay, David.   *Pyramid.* (Illus.) Houghton Mifflin, 1975. 80pp. $7.95. 75-9964. ISBN 0-395-21407-6. Gloss.;CIP.

JH   This book briefly covers all aspects of construction of both the pyramid and the related structures of the burial complex. The author tells the story of the tomb building activities of an imaginary king of the 5th dynasty, including such commonly neglected details as site selection, leveling methods and techniques of stone quarrying and transport.

Mendelssohn, Kurt.   *The Riddle of the Pyramids.* (Illus.) Praeger, 1974. 224pp. $12.95. 73-18485. Index;CIP:bib.

GA   Suggests that the construction of the pyramids was a giant works-project-administration effort through which the rulers of Egypt attained power and welded together many small tribes into one great nation. Excellent factual information and illustrations; suitable for advanced young readers, adults and even professionals in the field.

Michalowski, K.   *Karnak.* (Illus.) Praeger, 1970. 119pp. $7.50. Bib.

SH-C   An excellent compact history of the famous ancient temple center and a synopsis of Egyptian history as it relates to the architectural development of the site.

Pace, Mildred Mastin.   *Wrapped for Eternity: The Story of the Egyptian Mummy.* (Illus.) McGraw-Hill, 1974. 192pp. $6.95. 73-14539. ISBN 0-07-048053-2.

JH   Pace writes about the technique of embalming, the religion of the Egyptian civilization, the history of the archeologists' search for burial places and the skill of unwrapping a mummy. The black-and-white drawings are fascinating. Very easy to read.

Wilson, John A.   *Thousands of Years: An Archaeologist's Search for Ancient Egypt.* (Illus.) Scribner's, 1972. 218pp. $9.95. 73-179442. ISBN 0-684-12728-8.

SH-C-P   Makes a strong plea for de-emphasizing the material aspects of archeology and for emphasizing the importance of human behavior. There are numerous glimpses into the interactions of many of the great scholars of Egyptology. An argument is made which directly and clearly indicates the relevance of ancient history and prehistory to our understanding of the present human condition. For all who are or will be involved in the study of humans.

930.133 ANCIENT ISRAEL

Avi-Yonah, Michael, (Ed.)   *Encyclopedia of Archaeological Excavations in the Holy*

Land: Vol. 1. (Illus.) Prentice-Hall, 1975. 341pp. $25.00ea. 73-14997. ISBN 0-13-275115-1. CIP. *Vol. 2.* 1976. 305pp. ISBN 0-13-275123-2.

C-P This encyclopedia covers archeological sites, geographical areas and special topics such as churches. Entries are written by the excavators and specialists and provide considerable detail about sites. Very useful reference for beginning to advanced students and specialists in Eastern Mediterranean archeology.

Eisenberg, Azriel, and Dov Peretz Elkins. *Treasures from the Dust.* (Illus.) Abelard-Schuman, 1972. ix + 149pp. $6.95. 78-15683. ISBN 0-200-71827-4.

JH-SH-C Gives insight into the technical difficulties, the hard labor, the frustrations and the ultimate rewards of archeology. Describes 15 archeological finds in the Near East which have allowed better understanding and dating of events described in early writings of Judaism and Christianity. Presumes a wide knowledge of terms, ancient cultures and the Bible.

Harker, Ronald. *Digging Up the Bible Lands.* (Illus.) Walck, 1973. 127pp. $4.95(p). 72-6954. ISBN 0-8098-3111-2.

JH-SH-C Sketches the history of ancient Israel and describes eight sites which have yielded important data for our understanding of the Bible. The emphasis is on dramatic finds rather than broad-scale archeological and cultural information.

Kenyon, Kathleen M. *Archaeology in the Holy Land, 3rd ed.* (Illus.) Praeger, 1970. 360pp. $8.50. 70-88898.

C-P The time covered is from earliest Neolithic settlements to pre-Hellenistic Palestine. A satisfactory compromise between highly technical reporting and popularizing of the material; uses written documents and archeology to piece together the story of conflict of cultures and the rise and fall of contending peoples and civilizations.

Noble, Iris. *Treasure of the Caves: The Story of the Dead Sea Scrolls.* (Illus.) Macmillan, 1971. 214pp. $5.95. 69-11303.

GA Shows archeology as an exciting, scholarly, tedious, satisfying profession involving considerably more academic investigation than actual excavation, or even intrigues with treasure seekers. The story of the discovery and interpretation of the Dead Sea Scrolls is set within the establishment of modern Israel.

Pearlman, Moshe. *The First Days of Israel: In the Footsteps of Moses.* (Illus.) World, 1973. 230pp. $19.95. 72-88113. ISBN 0-529-04843.

GA Describes ancient Israel from Moses to Gideon (from the Exodus through most of the period of the Judges). Using the most recent and thoroughgoing scholarship of historians, archeologists and biblical specialists, Pearlman shows what must have happened, given the historical, environmental, sociological and psychological circumstances. Utterly fascinating, with beautiful illustrations.

Yadin, Yigael. *Hazor: The Rediscovery of a Great Citadel of the Bible.* (Illus.) Random House, 1975. 280pp. $20.00. 74-5406. ISBN 0-394-49454-7. Index;CIP.

SH-C-P *Hazor* is a personal account, written for the lay reader. It tells the reader what was found in the excavations and how it was found. The reader joins with the author in moments of anxiety and excitement.

930.135 MESOPOTAMIA

Adams, Robert McC., and Hans J. Nissen. *The Uruk Countryside: The Natural Setting of Urban Societies.* (Illus.) Univ. of Chicago Press, 1972. x + 241pp. $17.50. 78-179489. ISBN 0-226-00500-3.

C-P   Concerned with conditions accompanying the first known proliferation of urban life and with Uruk, a prime example of early urbanization, the authors relate changes in regional ecology over time to the associated major changes in social structure during a 7000 year period. A technical but lucid presentation of the methodologies of economic geography applied to archeology.

Glubok, Shirley, (Ed.).   *Digging in Assyria.* (Abridged and adapted from *Nineveh and Its Remains,* by Austen Henry Layard) (Illus.) Macmillan, 1970. 124pp. $7.95. 79-103679.

JH-SH-C   The present volume contains many of Layard's original drawings, numerous fine photographs of Assyrian sculptures from the British Museum and the Metropolitan Museum of Art, a map of the area in which Layard worked, and the ground plans of his excavations at Numrud (Calah) and Kuyunjik (Nineveh). Describes the importance of archeology to history.

Hamblin, Dora Jane, et al.   *The First Cities.* (Illus.) Time-Life, 1973. 160pp. $7.95. 73-83187.

JH-SH-C   The book illuminates the beginnings of urban living in prehistoric times, concentrating on the cities of Jericho, Catal Huyuk, Uruk and Moenjo-Daro. The photographs and paintings are superbly reproduced here.

Hicks, Jim, et al.   *The Empire Builders.* (Illus.) Time-Life (distr.: Little, Brown), 1974. 160pp. $7.95. 74-75832. Index.

JH-SH   A popular archeological account which concentrates on the imperialistic drives of the Hittites, a poorly understood early population of the Middle East. Excellent illustrations highlight this concise summary of Hittite military and political history.

Lansing, Elizabeth.   *The Sumerians: Inventors and Builders.* (Illus.) McGraw-Hill, 1971. 176pp $7.71. 78-154231. ISBN 0-07-036537-9.

JH-SH   Sumeria was a hearth area and had a long list of first achievements, including writing. The story of the discovery of the wedge-shaped marks on the clay tablets and the final reading of these marks, a major achievement for archeology, is told.

930.137 ROMAN EMPIRE

Ashbee, Paul.   *Ancient Scilly: From the First Farmers to the Early Christians.* (Illus.) David&Charles, 1975. 352pp. $18.50. ISBN 0-7153-6588-1. Indexes;bib.

C-P   This is a complete compilation of the archeology, history and physiography of the Scilly Islands.

Barfield, Lawrence.   *Northern Italy Before Rome.* (Illus.) Praeger, 1971. 208pp. $12.50. 75-163100. Index,bib.

SH-C   The archeology of Northern Italy from the earliest hunters of the Pleistocene glacial ages until Roman times is presented. The stone tools and pottery of the early farmers and the copper, bronze and iron tools of the metal ages are described in some detail. Attention is given to temporal and geographical subdivisions of the area and contact with neighboring regions. Numerous chronological charts, maps, drawings and photographs.

Hamblin, Dora Jane.   *Pots and Robbers.* (Illus.) Simon&Schuster, 1970. 258pp. $4.95. 74-86949. ISBN 671-65087-4.

JH-SH-C   Brings Italian archeology to life. Presents revealing insights into the illegal search for antiquities, museum thefts, fakes and forgeries and police attempts to curb illegal archeology, and recounts the how, when and where of professional archeology's recovery and preservation of important finds. Included is the archeology of cities and ships both underground and underwater.

Hamblin, Dora Jane, et al. *The Etruscans.* (Illus.) Time-Life, 1975. 160pp. $5.95. 74-25453. Index;bib.

   SH-C The author describes the environment, economic and political history, religion and daily life of the Etruscans.

930.138 GREECE

Nichols, Marianne. *Man, Myth, and Monument.* (Illus.) Morrow, 1975. xii + 340pp. $12.95. 75-11971. ISBN 0-688-02943-4. Index;CIP.

   C-P Nichols presents a descriptive integration of archeology and myth in Greece from 2200 to 1100 B.C. The correspondences between myth and archeology form a necessarily discontinuous story. This stimulating book will serve well those already somewhat acquainted with Bronze Age Greece.

Roebuck, Carl, (Ed.). *The Muses at Work: Arts, Crafts and Professions in Ancient Greece and Rome.* (Illus.) MIT Press, 1969. 294pp. $12.50. 72-87305. ISBN 0-262-18034-0. Index.

   SH-C-P Sculpture, architecture, pottery making, farming, sailing, trade, music and acting in ancient Greece and (briefly) Rome; literary and artistic evidence for the arts, crafts and professions of ancient Greece; and selections from the vast archeological record bring the reader closer to an understanding of life in the ancient world.

930.1391 ANCIENT AEGEAN SEA ISLANDS

Branigan, Keith. *The Foundations of Palatial Crete: A Survey of Crete in the Early Bronze Age.* (Illus.) Praeger, 1970. xvi + 232pp. $9.50. 75-102201.

   C-P Synthesizes excavation reports and the major discoveries of the previous 20 years. Well-documented data demonstrate the evolution of Minoan civilization from its indigenous, rather than intrusive, Bronze Age beginnings. Emphasis is on the local evolution of cultural traits and on the social reconstruction of life during the Bronze Age based on these traits. Some knowledge of Minoan (Cretan) prehistory is needed.

Chadwick, John. *The Mycenaean World.* (Illus.) Cambridge Univ. Press, 1976. xvii + 201pp. $17.95; $6.95(p). 75-36021. ISBN 0-521-21077-1; 0-521-29037-6. Index;CIP.

   SH-C Chadwick discusses the Mycenaean culture, using information gleaned from his interpretation of the Mycenaean tablets plus archeological and literary resources. Indicates the character of the people, their social structure, administrative system, religion, agriculture, weaponry and military establishment.

Edey, Maitland A., et al. *Lost World of the Aegean.* (Illus.) Time-Life, 1975. 160pp. $5.95. 74-21774. Index.

   SH-C There is a brief discussion of the geography, history, archeology and of the Aegean archeologists. The major portion of the book is a description of the three cultures flowering during the three millenia in the Aegean: Cycladic, Minoan and Mycenaean. Recommended for students and upper elementary and junior high school teachers.

Herberger, Charles F. *The Thread of Ariadne: The Labyrinth of the Calendar of Minos.* (Illus.) Philosophical Library, 1972. xi + 158pp. $12.00. 72-78167. ISBN 8022-2089-1.

   C-P Herberger convincingly shows that the border of the 15th century B.C. "Toreador Fresco" in the Palace of Knossos in Crete is the key to understanding the ancient Minoan ritual calendar. Using mathematical computations to decipher fresco elements, analyses of "mythopoetic" traditions, and comparative artifactual evidence, Herberger shows that establishing cyclic rituals and forseeing eclipses was well within the scientific capabilities of the Minoans.

Hood, Sinclair. *The Minoans: The Story of Bronze Age Crete.* (Illus.) Praeger, 1971. 239pp. $9.50. 77-121075. Index;bib.

SH-C   Students in Greek experience and language courses and students of mythology will benefit from this careful, scholarly review. The annotated bibliography and the artifact descriptions will be instrumental in raising questions about the kind of cultural and social organization that must have existed to produce them.

Platon, Nicholas. *Zakros: The Discovery of a Lost Palace of Ancient Crete.* (Illus.) Scribner's, 1971. 345pp. $19.95. 70-123855. ISBN 0-684-31103-8. Index;gloss.;bib.

SH-C   The author describes his personal discovery and excavation of a 3580-year-old Minoan palace at Zakros, a harbor town on the rugged eastern coast of Crete. A chronological description of the events leading up to the discovery of the palace is followed by a room-by-room description of the archeological findings. Also includes numerous pertinent references to previous Minoan and Mycenaean finds.

930.14 EUROPE

Alexander, John. *Yugoslavia: Before the Roman Conquest.* (Illus.) Praeger, 1972. 175pp. $12.50. 79-135512.

SH-C   The archeology of Yugoslavia is presented from the earliest fossil men of the Lower Paleolithic period (before 70,000 B.C.) to the Roman Iron Age just before the time of Christ. A brief climatic and geographic background for the cultural development is presented, and tools, jewelry and art objects are profusely illustrated in drawings and black-and-white photographs.

Froncek, Thomas, et al. *The Northmen.* (Illus.) Time-Life (distr.:Little, Brown), 1974. 160pp. $7.95. 74-77815. Index.

JH-SH-C-P   This magnificently illustrated volume describes the culture and artifacts of early Scandinavian peoples. Extensive captions to the numerous photographs provide the reader with an excellent background in Norse history.

Gimbutas, Marija. *The Slavs.* (Illus.) Praeger, 1971. 240pp. $10.00. 73-121074.

C-P   A detailed synopsis of the origins and development of the Slavic peoples during the first 10 centuries A.D. The main exposition is based on the archeological periods and a reconstruction of social structure and religion; the ways of life of each period during Slavic development are related to the wider perspective of European prehistory and history. An excellent English source on later East European prehistory, of use to both general readers and students studying European archeology.

MacKendrick, Paul. *The Dacian Stones Speak.* (Illus.) Univ. of North Carolina Press, 1975. xxi + 248pp. $12.95. 73-16210. ISBN 0-8078-1226-9. Index;bib.;CIP.

C-P   The cultural history of Romania as revealed by archeology is discussed. The book is also an important contribution to Roman period archeology and a useful synthesis of Romanian archeology. A cultural and historical chronology is included.

Todd, Malcolm. *Everyday Life of the Barbarians: Goths, Franks and Vandals.* (Illus.) Putnam's, 1972. viii + 184pp. $5.00. ISBN 7134-1689-0.

SH-C-P The author concentrates on the Germanic peoples of northwestern Europe who provided an interface culture between classical Rome and the nomadic tribes of the eastern steppes. The discoveries at excavated German town-sites prove they had a rich and complex culture. A precise and clearly written text and a useful reference in history, anthropology, or archeology.

930.16 AFRICA

Buxton, David. *The Abyssinians.* (Illus.) Praeger, 1970. 259pp. $8.50. 79-112633.

JH-SH-C The major focus of this book is on the religion, religious architecture, literature, painting, metalwork, and minor arts of the Axum kingdom of the Semitic-speaking people of northern Ethiopia. A useful book on a little known area of the world, with many text figures and 128 photographs.

Diop, Chekh Anta. *The African Origin of Civilization: Myth or Reality.* (Illus.;trans.) Lawrence Hill, 1974. xvii + 316pp. $12.95; $5.95(p). ISBN 0-88208-0210; 0-88208-022-9. Index;CIP.

C Excerpts from two works by this celebrated Senegalese author on the black African origins of Egyptian civilization. Evidence is based on early historians; descriptions of physical types (dark skin, woolly hair, etc.). This theory complements Greenberg's hypothesis concerning commonalities between ancient Egyptian and African languages.

Keating, Rex. *Nubian Rescue.* (Illus.) Hawthorn, 1975. xviii + 269pp. $12.50. ISBN 0-8015-5469-1. Index;bib.

SH-C-P The author deals chronologically and in depth with the history, culture and archeology of Nubia. An account of the concentrated effort to save the ancient temples and monuments in areas which were flooded is included. Useful as a text or for reading in archeology and history and a valuable reference for teachers and advanced students.

930.17 NORTH AMERICA

Anderson, Duane. *Western Iowa Prehistory.* (Illus.) Iowa State Univ. Press, 1975. xiii + 85pp. $3.95. 74-22166. ISBN 0-8138-1765-X. CIP.

JH-SH Well-written studies of regional prehistory aimed at the general public, such as this one is, are very rare. Using the archeological evidence, this work traces peoples in the area from the early Ice Age up to historic contacts. The importance of properly gathering archeological evidence is stressed, and this book should appeal to a wide audience.

Brunhouse, Robert L. *Pursuit of the Ancient Maya: Some Archaeologists of Yesterday.* (Illus.) Univ. of New Mexico Press, 1975. viii + 252pp. $8.95. 74-27443. ISBN 0-8263-0363-3. Index.

SH-C Brunhouse devotes a chapter each to Teobert Maler, Alfred P. Maudslay, Sylvanus G. Morley, Frederick A. Mitchell-Hedges, Herbert J. Spinden, William E. Gates, and two chapters to Frans Blom. The effects of their personalities on the individual approaches to archeology are discussed.

Ceram, C.W. *The First American: A Story of North American Archaeology.* (Illus.) Harcourt Brace Jovanovich, 1971. xxi + 357pp. $9.95. 73-139460. ISBN 0-15-131250-8.

SH-C-P An admirable history of North American archeology and archeologists. The pioneers in the field receive justified recognition, and modern trends, techniques, and theories add to the perspective. Excellent supplementary reading, providing a suitable introduction to the subject.

Claiborne, Robert, et al. *The First Americans.* (Illus.) Time-Life (distr.:Little, Brown), 1973. 160pp. $7.95. 73-75294.

JH-SH-C The author begins with the settlement of the late Pleistocene Siberian pioneers in the New World. The rest of the text follows a roughly chronological-ecotype approach, with chapters on big game hunters, foragers, Eskimos, Southwestern farmers and mound-building Eastern woodlanders.

Fitting, James E., (Ed.). *The Development of North American Archaeology: Essays in the History of Regional Traditions.* (Illus.) Anchor/Doubleday, 1973. viii + 309pp. $2.50(p). 72-76998. ISBN 0-385-06047-5.

C-P The introductory chapter could well become a "module" outlining the general history of North American archeology. Eight other contributors (including Frison and Sprague) discuss the history of the Arctic, Canada, Northeastern and Southeastern U.S., Plains, Southwest and Intermontane West, California and Pacific Northwest in remarkable depth. (Some chapters are somewhat dated.)

Gorenstein, Shirley. *Not Forever on Earth: Prehistory of Mexico.* (Illus.) Scribner's, 1975. xvi + 153pp. $7.95. 73-1366. ISBN 0-684-13837-9. Gloss.;index;CIP.

SH This brief survey of the prehistory of Mexico begins with the earliest cultural evidence known and continues through the Aztecs until the Spanish conquest. A brief history of archeological work in Mexico is presented and all the major civilizations are discussed. A good introduction to Mexican prehistory.

Ingstad, Helge. *Westward to Vinland.* (Illus.) St. Martin's, 1969. 250pp. $6.95. 67-10089. Index.

SH-C A popular account of Ingstad's archeological research at L'Anse aux Meadow at the north end of Newfoundland. This site (c. A.D. 1000) may be the first Norse settlement in the New World; the arguments presented are persuasive. A wealth of information on the area, the Indian and Eskimo populations, and historical data on the Norse settlements and explorations of the period around A.D. 1000 is presented.

Pike, Donald G., and David Muench. *Anasazi: Ancient People of the Rock.* (Illus.) American West, 1974. 191pp. $18.50. 73-90795. ISBN 0-910118-49-0.

JH-SH-C The text is chronological and primarily descriptive and is accompanied by excellent color illustrations of the sites and environment. The authors provide a good introduction to Southwestern prehistory.

Rohn, Arthur H. *Wetherill Mesa Excavations: Mug House.* (Illus.) U.S. Dept. of Interior, 1971. (Supt. of Documents, Washington, DC 20402.) xix + 280pp. $5.00 70-600351.

C-P Mug House is a prehistoric community of nearly 70 rooms, eight kivas, and several open courtyards in Mesa Verde National Park, occupied during the thirteenth century by Anasazi pueblo-dwellers. Rohn's monograph on the archeology at Mug House is a detailed, useful, and well-produced volume aimed at professional and semiprofessional readers and covering both artifacts and social organization.

Scientific American. *Early Man in America: Readings from Scientific American.* (Illus.) Freeman, 1973. 93pp. $5.95; $2.95(p). 72-12251. ISBN 0-7167-0864-7; 0-7167-0863-9.

SH-C These ten articles, first published from 1951 through 1971, represent important illustrations of archeological methodology and reflect the variety of theoretical perspectives that still characterize this frontier area of New World prehistoric research. Topics range from a discussion of the Bering Strait land bridge, details on the continental route of migration and the time of entry of man into the New World, to descriptive accounts of early finds in both North and South America.

Silverberg, Robert. *The Mound Builders.* (Illus.) New York Graphic Society, 1970. 276pp. $5.95. 76-109177. ISBN 0-8212-0342-8.

SH-C An abridged version of *Mound Builders of Ancient America: The Archaeology of a Myth* (1968). Describes many misconceptions surrounding the mounds and archeological information on the groups that built the mounds. A skillful, serious presentation of numerous details which are not easily available elsewhere.

Weaver, Muriel Porter. *The Aztecs, Maya, and Their Predecessors: Archaeology of Mesoamerica.* (Illus.) Seminar, 1972. xvi + 347pp. $11.95. 70-183477. Index;gloss.

C-P An important contribution to the semi-popular literature on the origins and growth of the native civilizations of Mexico and Central America. Not light reading, but a reliable source for details. There are many handsome and appropriate illustrations but too few maps.

Willey, Gordon R., and Jeremy A. Sabloff. *A History of American Archaeology.* (Illus.) Freeman, 1974. 252pp. $9.95; $4.95(p). 73-17493. ISBN 0-7167-0267-3; 0-7167-0266-5. Index;CIP:bib.

C-P There is a brief section of definitions and basic relationships, followed by a history of American archeology, traced in chronological order as a sequence of evolutionary stages, based on the predominant intellectual emphasis of the period. The volume is well illustrated with photographs, drawings, graphs and charts. An excellent text or reference work for all archeology students.

930.18 SOUTH AMERICA

Hemming, John. *The Conquest of the Incas.* (Illus.) Harcourt Brace Jovanovich, 1970. 641pp. $12.50. 74-117573. ISBN 0-15-122560-5. Index;gloss.;bib.

SH-C-P A chronicle of events occurring during the conquest through the final Inca defeat at Vilcabamba in 1572 and the fate of the Inca descendants who survived. A chronology of events, a kinship chart of the Incas and Pizarro's descendants, a table of Inca measurements, and much more.

Karen, Ruth. *Kingdom of the Sun: The Inca: Empire Builders of the Americas.* (Illus.) Four Winds/Scholastic, 1975. 255pp. $9.95. 75-9886. ISBN 0-590-17288-3. Gloss.;index;CIP.

SH Explanations of the Inca hierarchy, social organization and agricultural systems are presented here. A composite life story of a man and a woman in Inca times is related and a short guide to Inca ruins and museums, and a summary of the available written records are included.

Katz, Friedrich. *The Ancient American Civilizations.* (Illus.;trans.) Praeger, 1972. xvi + 386pp. $15.00. 71-165845.

C Through a review of archeological literature, Katz sets forth a masterful summary of classical (ca. 200 A.D.–ca. 900 A.D.) and postclassical (Aztec and Inca) civilizations in Mesoamerica and the Andes. Reconstructs the evolution of the Aztec and Inca civilizations in relation to the earlier cultures from which they emerged and evaluates the various theories concerning the origins of American Indians.

Kendall, Ann. *Everyday Life of the Incas.* (Illus.) Putnam's, 1973. 216pp. $5.00. ISBN 0-7134-1690-4.

JH-SH-C Kendall provides information about Inca land and history, modes of living, rituals, work and craftmanship, religious life and the final civil war and Spanish conquest.

Lumbreras, Luis G.   *The Peoples and Cultures of Ancient Peru.* (Illus.;trans.) Smithsonian Institution Press, 1974. vii + 248pp. $15.00. 74-2104. ISBN 0-87474-146-7. Index:CIP.

SH-C-P   The tone of this book is set by the fact that 42 percent of the 232 figures are of pottery, only a minor part of which are utilitarian or kitchen ware. Monumental architecture and specialized crafts are the subjects of the remaining plates. The best available summary of Peruvian prehistory.

Reichel-Dolmatoff, Gerardo.   *San Agustin: A Culture of Colombia.* (Illus.) Praeger, 1972. 163pp. $12.50. 70-143979.

SH-C-P   Well suited to a general audience in style, form and content. The setting of San Agustin is well explained, in both geographical and historical terms, with much of the focus on the stone statues, tombs and barrows. The black-and-white illustrations are excellent.

Willey, Gordon R.   *An Introduction to American Archaeology. Vol. 2: South America.* (Illus.) Prentice-Hall, 1971. xiv + 559pp. $18.00. 66-10096. ISBN 0-13-477851-0.

C-P   A fundamental contribution to American archeology, this volume has an extraordinary number of excellent illustrations, charts and maps. Willey divides South America into a number of geographic areas and reviews the archeology of each in terms of the chronological framework the author has used throughout his studies of American archeology. For any serious student of archeology or of pre-Columbian art.

Indexes

Author Index

Title and Subject Index

(Subjects are italicized.)